HANDBOOK OF APPLIED ALGORITHMS

HANDBOOK OF APPLIED ALGORITHMS

SOLVING SCIENTIFIC, ENGINEERING AND PRACTICAL PROBLEMS

Edited by

Amiya Nayak

SITE, University of Ottawa
Ottawa, Ontario, Canada

Ivan Stojmenović

EECE, University of Birmingham, UK

A JOHN WILEY & SONS, INC., PUBLICATION

Published by John Wiley & Sons, Inc., Hoboken, New Jersey
Published simultaneously in Canada

Library of Congress Cataloging-in-Publication Data:

Handbook of applied algorithms: solving scientific, engineering, and practical problem / edited by Amiya Nayak & Ivan Stojmenovic.
p. cm.
ISBN 978-0-470-04492-6
1. Computer algorithms. I. Nayak, Amiya. II. Stojmenovic, Ivan.
QA76.9.A43H353 2007
005.1–dc22

2007010253

Printed in the United States of America

10 9 8 7 6 5 4 3 2 1

CONTENTS

PREFACE

Although vast activity exists, especially recent, the editors did not find any book that treats applied algorithms in a comprehensive manner. The editors discovered a number of graduate courses in computer science programs with titles such as "Design and Analysis of Algorithms, "Combinatorial Algorithms" "Evolutionary Algorithms" and "Discrete Mathematics." However, when glancing through the course contents, it appears that they were detached from the real-world applications. On the contrary, recently some graduate courses such as "Algorithms in Bioinformatics" emerged, which treat one specific application area for algorithms. Other graduate courses heavily use algorithms but do not mention them anywhere explicitly. Examples are courses on computer vision, wireless networks, sensor networks, data mining, swarm intelligence, and so on.

Generally, it is recognized that software verification is a necessary step in the design of large commercial software packages. However, solving the problem itself in an optimal manner precedes software verification. Was the problem solution (algorithm) verified? One can verify software based on good and bad solutions. Why not start with the design of efficient solutions in terms of their time complexities, storage, and even simplicity? One needs a strong background in design and analysis of algorithms to come up with good solutions.

This book is designed to bridge the gap between algorithmic theory and its applications. It should be the basis for a graduate course that will contain both basic algorithmic, combinatorial and graph theoretical subjects, and their applications in other disciplines and in practice. This direction will attract more graduate students into such courses. The students themselves are currently divided. Those with weak math backgrounds currently avoid graduate courses with a theoretical orientation, and vice versa. It is expected that this book will provide a much-needed textbook for graduate courses in algorithms with an orientation toward their applications.

This book will also make an attempt to bring together researchers in design and analysis of algorithms and researchers that are solving practical problems. These communities are currently mostly isolated. Practitioners, or even theoretical researchers from other disciplines, normally believe that they can solve problems themselves with some brute force techniques. Those that do enter into different areas looking for "applications" normally end up with theoretical assumptions, suitable for proving theorems and designing new algorithms, not having much relevance for the claimed application area. On the contrary, the algorithmic community is mostly engaged in their own problems and remains detached from reality and applications. They can rarely answer simple questions about the applications of their research. This is valid even for the experimental algorithms community. This book should attract both sides and encourage collaboration. The collaboration should lead toward modeling problems with sufficient realism for design of practical solutions, also allowing a sufficient level of tractability.

The book is intended for researchers and graduate students in computer science and researchers from other disciplines looking for help from the algorithmic community. The book is directed to both people in the area of algorithms, who are interested in some applied and complementary aspects of their activity, and people that want to approach and get a general view of this area. Applied algorithms are gaining popularity, and a textbook is needed as a reference source for the use by students and researchers.

This book is an appropriate and timely forum, where researchers from academics (both with and without a strong background in algorithms) and emerging industry in new application areas for algorithms (e.g., sensor networks and bioinformatics) learn more about the current trends and become aware of the possible new applications of existing and new algorithms. It is often not the matter of designing new algorithms, but simply the recognition that certain problems have been already solved efficiently. What is needed is a starting reference point for such resources, which this book could provide.

Handbook is based on a number of stand-alone chapters that together cover the subject matter in a comprehensive manner. The book seeks to provide an opportunity for researchers, graduate students, and practitioners to explore the application of algorithms and discrete mathematics for solving scientific, engineering, and practical problems. The main direction of the book is to review various applied algorithms and their currently "hot" application areas such as computational biology, computational chemistry, wireless networks, and computer vision. It also covers data mining, evolutionary algorithms, game theory, and basic combinatorial algorithms and their applications. Contributions are made by researchers from United States, Canada, United Kingdom, Italy, Greece, Cyprus, France, Denmark, Spain, and India.

Recently, a number of application areas for algorithms have been emerging into their own disciplines and communities. Examples are computational biology, computational chemistry, computational physics, sensor networks, computer vision, and others. Sensor networks and computational biology are currently among the top research priorities in the world. These fields have their own annual conferences and books published. The algorithmic community also has its own set of annual meetings, and journals devoted to algorithms. Apparently, it is hard to find a mixture of the two communities. There are no conferences, journals, or even books with mixed content, providing forum for establishing collaboration and providing directions.

BRIEF OUTLINE CONTENT

This handbook consists of 18 self-contained chapters. Their content will be described briefly here.

Many practical problems require an exhaustive search through the solution space, which are represented as combinatorial structures such as permutations, combinations, set partitions, integer partitions, and trees. All combinatorial objects of a certain kind need to be generated to test all possible solutions. In some other problems, a randomly generated object is needed, or an object with an approximately correct ranking among all objects, without using large integers. Chapter 1 describes fast algorithms for generating all objects, random object, or object with approximate ranking, for basic types of combinatorial objects.

Chapter 2 presents applications of combinatorial algorithms and graph theory to problems in chemistry. Most of the techniques used are quite general, applicable to other problems from various fields. The problem of cell growth is one of the classical problems in combinatorics. Cells are of the same shape and are in the same plane, without any overlap. The central problem in this chapter is the study of hexagonal systems, which represent polyhexes or benzenoid hydrocarbons in chemistry. An important issue for enumeration and exhaustive generation is the notion of *isomorphic* or *equivalent* objects. Usually, we are interested in enumerating or generating only one copy of equivalent objects, that is, only one representative from each isomorphism class. Polygonal systems are considered different if they have different shapes; their orientation and location in the plane are not important. The main theme in this chapter is isomorph-free exhaustive generation of polygonal systems, especially polyhexes. In general, the main algorithmic framework employed for exhaustive generation is backtracking, and several techniques have been developed for handling isomorphism issues within this framework. This chapter presents several of these techniques and their application to exhaustive generation of hexagonal systems.

Chapter 3 describes some graph-theoretic models in chemistry and molecular biology. RNA, proteins, and other structures are described as graphs. The chapter defines and illustrates a number of important molecular descriptors and related concepts. Algorithms for predicting biological activity of given molecule and its structure are discussed. The ability to predict a molecule's biological activity by computational means has become more important as an ever-increasing amount of biological information is being made available by new technologies. Annotated protein and nucleic databases and vast amounts of chemical data from automated chemical synthesis and high throughput screening require increasingly more sophisticated efforts. Finally, this chapter describes popular machine learning techniques such as neural networks and support vector machines.

A major paradigm shift in molecular biology occurred recently with the introduction of gene-expression microarrays that measure the expression levels of thousands of genes at once. These comprehensive snapshots of gene activity can be used to investigate metabolic pathways, identify drug targets, and improve disease diagnosis. However, the sheer amount of data obtained using the high throughput microarray experiments and the complexity of the existing relevant biological knowledge is beyond the scope of manual analysis. Chapter 4 discusses the bioinformatics algorithms that help analyze such data and are a very valuable tool for biomedical science.

Activities of contemporary society generate enormous amounts of data that are used in decision-support processes. Many databases have current volumes in the hundreds of terabytes. The difficulty of analyzing this kind of data volumes by human operators is clearly insurmountable. This lead to a rather new area of computer science, data mining, whose aim is to develop automatic means of data analysis for discovering new and useful patterns embedded in data. Data mining builds on several disciplines: statistics, artificial intelligence, databases, visualization techniques, and others and has crystallized as a distinct discipline in the last decade of the past century. The range of subjects in data mining is very broad. Among the main directions of this branch of computer science, one should mention identification of associations between data items, clustering, classification, summarization, outlier detection, and so on. Chapters 6 and 7 concentrate on two classes of data mining algorithms: clustering algorithms and identification of association rules.

Data stream processing has recently gained increasing popularity as an effective paradigm for processing massive data sets. A wide range of applications in computational sciences generate huge and rapidly changing data streams that need to be continuously monitored in order to support exploratory analyses and to detect correlations, rare events, fraud, intrusion, unusual, or anomalous activities. Relevant examples include monitoring network traffic, online auctions, transaction logs, telephone call records, automated bank machine operations, and atmospheric and astronomical events. Due to the high sequential access rates of modern disks, streaming algorithms can also be effectively deployed for processing massive files on secondary storage, providing new insights into the solution of several computational problems in external memory. Streaming models constrain algorithms to access the input data in one or few sequential passes, using only a small amount of working memory and processing each input item quickly. Solving computational problems under these restrictions poses several algorithmic challenges. Chapter 8 is intended as an overview and survey of the main models and techniques for processing data streams and of their applications.

Frequency assignment is a well-known problem in operations research for which different mathematical models exist depending on the application-specific conditions. However, most of these models are far from considering actual technologies currently deployed in GSM networks, such as frequency hopping. In these networks, interferences provoked by channel reuse due to the limited available radio spectrum result in a major impact of the quality of service (QoS) for subscribers. In Chapter 9, the authors focus on optimizing the frequency planning of a realistic-sized, real-world GSM network by using evolutionary algorithms.

Methods from game theory and mechanism design have been proven to be a powerful mathematical tool in order to understand, control and efficiently design dynamic, complex networks, such as the Internet. Game theory provides a good starting point for computer scientists in order to understand selfish rational behavior of complex networks with many agents. Such a scenario is readily modeled using game theory techniques, in which players with potentially different goals participate under a common setting with well-prescribed interactions. Nash equilibrium stands out as the predominant concept of rationality in noncooperative settings. Thus, game theory and its notions of equilibria provide a rich framework for modeling the behavior of selfish agents in these kinds of distributed and networked environments and offering mechanisms to achieve efficient and desirable global outcomes in spite of the selfish behavior. In Chapter 10, we review some of the most important algorithmic solutions and advances achieved through game theory.

Real-time face detection in images received growing attention recently. Recognition of other objects, such as cars, is also important. Applications are in similar and content-based real-time image retrieval. The task is currently achieved by designing and applying automatic or semisupervised machine learning algorithms. Chapter 11 will review some algorithmic solutions to these problems. Existing real-time object detection systems appear to be based primarily on the AdaBoost framework, and this chapter will concentrate on it. Emphasis is given on approaches that build fast and reliable object recognizers in images based on small training sets. This is important in cases where the training set needs to be built manually, as in the case of detecting back of cars, studied as a particular example.

Existing computer vision applications that demonstrated their validity are mostly based on shape analysis. A number of shapes, such as linear or elliptic ones, are well studied. More complex classification and recognition tasks require new shape descriptors. Chapter 12 reviews some algorithmic tools for measuring and detecting shapes. Since shape descriptors are expected to be applied not only to a single object but also to a multiobject or dynamic scene, time complexity of the proposed algorithms is an issue, in addition to accuracy.

Cryptographic algorithms are extremely important for secure communication over an insecure channel and have gained significant importance in modern day technology. Chapter 13 introduces the basic concepts of cryptography, and then presents general principles, algorithms, and designs for block and stream ciphers, public key cryptography, and key agreement. The algorithms largely use mathematical tools from algebra, number theory, and algebraic geometry and have been explained as and when required.

Chapter 14 studies the issues related to secure communication among sensor nodes. The sensor nodes are usually of limited computational ability having low CPU power, small amount of memory, and constrained power availability. Thus, the standard cryptographic algorithms suitable for state of the art computers may not be efficiently implemented in sensor nodes. This chapter describes strategies that can work in constrained environment. It first presents basic introduction to the security issues in distributed wireless sensor networks. As implementation of public key infrastructure may not be recommendable in low end hardware platforms, chapter describes key predistribution issues in detail. Further it investigates some specific stream ciphers for encrypted communication that are suitable for implementation in low end hardware.

In Chapter 15, the authors consider localized algorithms, as opposed to centralized algorithms, which can be used in topology control for wireless ad hoc or sensor networks. The aim of topology control can be to minimize energy consumption, or to reduce interferences by organizing/structuring the network. This chapter focuses on neighbor elimination schemes, which remove edges from the initial connection graph in order to generate energy efficient, sparse, planar but still connected network in localized manner.

Low Earth Orbit (LEO) satellite networks are deployed as an enhancement to terrestrial wireless networks in order to provide broadband services to users regardless of their location. LEO satellites are expected to support multimedia traffic and to provide their users with some form of QoS guarantees. However, the limited bandwidth of the satellite channel, satellite rotation around the Earth, and mobility of end users makes QoS provisioning and mobility management a challenging task. One important mobility problem is the intrasatellite handoff management. Chapter 16 proposes RADAR—refined admission detecting absence region, a novel call admission control and handoff management scheme for LEO satellite networks. A key ingredient in the scheme is a companion predictive bandwidth allocation strategy that exploits the topology of the network and contributes to maintaining high bandwidth utilization.

After a brief review of conventional approaches to shortest path routing, Chapter 17 introduces an alternative algorithm that abstracts a network graph into a logical tree. The algorithm is based on the decomposition of a graph into its minimum cycle basis (a basis of the cycle vector space of a graph having least overall weight or length). A procedure that abstracts the cycles and their adjacencies into logical nodes and links correspondingly is introduced. These logical nodes and links form the next level logical graph. The procedure is repeated recursively, until a loop-free logical graph is derived. This iterative abstraction is called a logical network abstraction procedure and can be used to analyze network graphs for resiliency, as well as become the basis of a new routing methodology. Both these aspects of the logical network abstraction procedure are discussed in some detail.

With the tremendous growth of bandwidth-intensive networking applications, the demand for bandwidth over data networks is increasing rapidly. Wavelength division multiplexing (WDM) optical networks provide promising infrastructures to meet the information networking demands and have been widely used as the backbone networks in the Internet, metropolitan area networks, and high capacity local area networks. Efficient routing on WDM networks is challenging and involves hard optimization problems. Chapter 18 introduces efficient algorithms with guaranteed performance for fundamental routing problems on WDM networks.

ACKNOWLEDGMENTS

The editors are grateful to all the authors for their contribution to the quality of this handbook. The assistance of reviewers for all chapters is also greatly appreciated. The University of Ottawa (with the help of NSERC) provided an ideal working environment for the preparation of this handbook. This includes computer facilities for efficient Internet search, communication by electronic mail, and writing our own contributions.

The editors are thankful to Paul Petralia and Whitney A. Lesch from Wiley for their timely and professional cooperation, and for their decisive support of this project. We thank Milos Stojmenovic for proposing and designing cover page for this book.

Finally, we thank our families for their encouragement, making this effort worthwhile, and for their patience during the numerous hours at home that we spent in front of the computer.

We hope that the readers will find this handbook informative and worth reading. Comments received by readers will be greatly appreciated.

Amiya Nayak
SITE, University of Ottawa, Ottawa, Ontario, Canada

Ivan Stojmenović
EECE, University of Birmingham, UK

November 2007

ABSTRACTS

1 GENERATING ALL AND RANDOM INSTANCES OF A COMBINATORIAL OBJECT

Many practical problems require an exhaustive search through the solution space, which are represented as combinatorial structures, such as, permutations, combinations, set partitions, integer partitions, and trees. All combinatorial objects of a certain kind need to be generated to test all possible solutions. In some other problems, a randomly generated object is needed, or an object with an approximately correct ranking among all objects, without using large integers. Fast algorithms for generating all objects, random object, or object with approximate ranking for basic types of combinatorial objects are described.

2 BACKTRACKING AND ISOMORPH-FREE GENERATION OF POLYHEXES

General combinatorial algorithms and their application to enumerating molecules in chemistry are presented and classical and new algorithms for the generation of complete lists of combinatorial objects that contain only inequivalent objects (isomorph-free exhaustive generation) are discussed. We introduce polygonal systems, and how polyhexes and hexagonal systems relate to benzenoid hydrocarbons. The central theme is the exhaustive generation of nonequivalent hexagonal systems, which is used to walk the reader through several algorithmic techniques of general applicability. The main algorithmic framework is backtracking, which is coupled with sophisticated methods for dealing with isomorphism or symmetries. Triangular and square systems, as well as the problem of matchings in hexagonal systems and their relationship to Kékule structures in chemistry are also presented.

3 GRAPH THEORETIC MODELS IN CHEMISTRY AND MOLECULAR BIOLOGY

The field of chemical graph theory utilizes simple graphs as models of molecules. These models are called molecular graphs, and quantifiers of molecular graphs are known as molecular descriptors or topological indices. Today's chemists use molecular descriptors to develop algorithms for computer aided drug designs, and computer based searching algorithms of chemical databases and the field is now more commonly known as combinatorial or computational chemistry. With the completion of the human genome project, related fields are emerging such as chemical genomics and pharmacogenomics. Recent advances in molecular biology are driving new methodologies and reshaping existing techniques, which in turn produce novel approaches to nucleic acid modeling and protein structure prediction. The origins of chemical graph theory are revisited and new directions in combinatorial chemistry with a special emphasis on biochemistry are explored. Of particular importance is the extension of the set of molecular descriptors to include graphical invariants. We also describe the use of artificial neural networks (ANNs) in predicting biological functional relationships based on molecular descriptor values. Specifically, a brief discussion of the fundamentals of ANNs together with an example of a graph theoretic model of RNA to illustrate the potential for ANN coupled with graphical invariants to predict function and structure of biomolecules is included.

4 ALGORITHMIC METHODS FOR THE ANALYSIS OF GENE EXPRESSION DATA

The traditional approach to molecular biology consists of studying a small number of genes or proteins that are related to a single biochemical process or pathway. A major paradigm shift recently occurred with the introduction of gene-expression microarrays that measure the expression levels of thousands of genes at once. These comprehensive snapshots of gene activity can be used to investigate metabolic pathways, identify drug targets, and improve disease diagnosis. However, the sheer amount of data obtained using high throughput microarray experiments and the complexity of the existing relevant biological knowledge is beyond the scope of manual analysis. Thus, the bioinformatics algorithms that help analyze such data are a very valuable tool for biomedical science. First, a brief overview of the microarray technology and concepts that are important for understanding the remaining sections are described. Second, microarray data preprocessing, an important topic that has drawn as much attention from the research community as the data analysis itself is discussed. Finally, some of the more important methods for microarray data analysis are described and illustrated with examples and case studies.

5 ALGORITHMS OF REACTION–DIFFUSION COMPUTING

A case study introduction to the novel paradigm of wave-based computing in chemical systems is presented in Chapter 5. Selected problems and tasks of computational geometry, robotics and logics can be solved by encoding data in configuration of chemical medium's disturbances and programming wave dynamics and interaction.

6 DATA MINING ALGORITHMS I: CLUSTERING

Clustering is the process of grouping together objects that are similar. The similarity between objects is evaluated by using a several types of dissimilarities (particularly, metrics and ultrametrics). After discussing partitions and dissimilarities, two basic mathematical concepts important for clustering, we focus on ultrametric spaces that play a vital role in hierarchical clustering. Several types of agglomerative hierarchical clustering are examined with special attention to the single-link and complete link clusterings. Among the nonhierarchical algorithms we present the k-means and the PAM algorithm. The well-known impossibility theorem of Kleinberg is included in order to illustrate the limitations of clustering algorithms. Finally, modalities of evaluating clustering quality are examined.

7 DATA MINING ALGORITHMS II: FREQUENT ITEM SETS

The identification of frequent item sets and of association rules have received a lot of attention in data mining due to their many applications in marketing, advertising, inventory control, and many other areas. First the notion of frequent item set is introduced and we study in detail the most popular algorithm for item set identification: the Apriori algorithm. Next we present the role of frequent item sets in the identification of association rules and examine the levelwise algorithms, an important generalization of the Apriori algorithm.

8 ALGORITHMS FOR DATA STREAMS

Data stream processing has recently gained increasing popularity as an effective paradigm for processing massive data sets. A wide range of applications in computational sciences generate huge and rapidly changing data streams that need to be continuously monitored in order to support exploratory analyses and to detect correlations, rare events, fraud, intrusion, and unusual or anomalous activities. Relevant examples include monitoring network traffic, online auctions, transaction logs, telephone call records, automated bank machine operations, and atmospheric and astronomical events. Due to the high sequential access rates of modern disks, streaming algorithms can also be effectively deployed for processing massive files on secondary storage, providing new insights into the solution of several computational problems in external memory. Streaming models constrain algorithms to access the input data in one or few sequential passes, using only a small amount of working memory and processing each input item quickly. Solving computational problems under these restrictions poses several algorithmic challenges.

9 APPLYING EVOLUTIONARY ALGORITHMS TO SOLVE THE AUTOMATIC FREQUENCY PLANNING PROBLEM

Frequency assignment is a well-known problem in operations research for which different mathematical models exist depending on the application-specific conditions. However, most of these models are far from considering actual technologies currently deployed in GSM networks, such as frequency hopping. In these networks, interferences provoked by channel reuse due to the limited available radio spectrum result in a major impact of the quality of service (QoS) for subscribers. Therefore, frequency planning is of great importance for GSM operators. We here focus on optimizing the frequency planning of a realistic-sized, real-world GSM network by using evolutionary algorithms (EAs). Results show that a (1+10) EA developed by the chapter authors for which different seeding methods and perturbation operators have been analyzed is able to compute accurate and efficient frequency plans for real-world instances.

10 ALGORITHMIC GAME THEORY AND APPLICATIONS

Methods from game theory and mechanism design have been proven to be a powerful mathematical tool in order to understand, control, and efficiently design dynamic, complex networks, such as the Internet. Game theory provides a good starting point for computer scientists to understand selfish rational behavior of complex networks with many agents. Such a scenario is readily modeled using game theory techniques, in which players with potentially different goals participate under a common setting with well prescribed interactions. The Nash equilibrium stands out as the predominant concept of rationality in noncooperative settings. Thus, game theory and its notions of equilibria provide a rich framework for modeling the behavior of selfish agents in these kinds of distributed and networked environments and offering mechanisms to achieve efficient and desirable global outcomes despite selfish behavior. The most important algorithmic solutions and advances achieved through game theory are reviewed.

11 ALGORITHMS FOR REAL-TIME OBJECT DETECTION IN IMAGES

Real time face detection images has received growing attention recently. Recognition of other objects, such as cars, is also important. Applications are similar and content based real time image retrieval. Real time object detection in images is currently achieved by designing and applying automatic or semi-supervised machine learning algorithms. Some algorithmic solutions to these problems are reviewed. Existing real time object detection systems are based primarily on the AdaBoost framework, and the chapter will concentrate on it. Emphasis is given to approaches that build fast and reliable object recognizers in images based on small training sets. This is important in cases where the training set needs to be built manually, as in the case of detecting the back of cars, studied here as a particular example.

12 2D SHAPE MEASURES FOR COMPUTER VISION

Shape is a critical element of computer vision systems, and can be used in many ways and for many applications. Examples include classification, partitioning, grouping, registration, data mining, and content based image retrieval. A variety of schemes that compute global shape measures, which can be categorized as techniques based on minimum bounding rectangles, other bounding primitives, fitted shape models, geometric moments, and Fourier descriptors are described.

13 CYPTOGRAPHIC ALGORITHMS

Cryptographic algorithms are extremely important for secure communication over an insecure channel and have gained significant importance in modern day technology. First the basic concepts of cryptography are introduced. Then general principles, algorithms and designs for block ciphers, stream ciphers, public key cryptography, and protocol for key-agreement are presented in details. The algorithms largely use mathematical tools from algebra, number theory, and algebraic geometry and have been explained as and when required.

14 SECURE COMMUNICATION IN DISTRIBUTED SENSOR NETWORKS (DSN)

The motivation of this chapter is to study the issues related to secure communication among sensor nodes. Sensor nodes are usually of limited computational ability having low CPU power, a small amount of memory, and constrained power availability. Thus the standard cryptographic algorithms suitable for state of the art computers may not be efficiently implemented in sensor nodes. In this regard we study the strategies that can work in constrained environments. First we present a basic introduction to the security issues in distributed wireless sensor networks. As implementation of public key infrastructure may not be recommendable in low end hardware platforms, we describe key predistribution issues in detail. Further we study some specific stream ciphers for encrypted communication that are suitable for implementation in low end hardware.

15 LOCALIZED TOPOLOGY CONTROL ALGORITHMS FOR AD HOC AND SENSOR NETWORKS

Localized algorithms, in opposition to centralized algorithms, which can be used in topology control for wireless ad hoc or sensor networks are considered. The aim of topology control is to minimize energy consumption, or to reduce interferences by organizing/structuring the network. Neighbor elimination schemes, which consist of removing edges from the initial connection graph are focused on.

16 A NOVEL ADMISSION FOR CONTROL OF MULTIMEDIA LEO SATELLITE NETWORKS

Low Earth Orbit (LEO) satellite networks are deployed as an enhancement to terrestrial wireless networks in order to provide broadband services to users regardless of their location. In addition to global coverage, these satellite systems support communications with hand-held devices and offer low cost-per-minute access cost, making them promising platforms for personal communication services (PCS). LEO satellites are expected to support multimedia traffic and to provide their users with some form of quality of service (QoS) guarantees. However, the limited bandwidth of the satellite channel, satellite rotation around the Earth and mobility of end-users makes QoS provisioning and mobility management a challenging task. One important mobility problem is the intra-satellite handoff management. While global positioning systems (GPS)-enabled devices will become ubiquitous in the future and can help solve a major portion of the problem, at present the use of GPS for low-cost cellular networks is unsuitable. RADAR—refined admission detecting absence region—a novel call admission control and handoff management scheme for LEO satellite networks is proposed in this chapter. A key ingredient in this scheme is a companion predictive bandwidth allocation strategy that exploits the topology of the network and contributes to maintaining high bandwidth utilization. Our bandwidth allocation scheme is specifically tailored to meet the QoS needs of multimedia connections. The performance of RADAR is compared to that of three recent schemes proposed in the literature. Simulation results show that our scheme offers low call dropping probability, providing for reliable handoff of on-going calls, and good call blocking probability for new call requests, while ensuring high bandwidth utilization.

17 RESILIENT RECURSIVE ROUTING IN COMMUNICATION NETWORKS

After a brief review of conventional approaches to shortest path routing an alternative algorithm that abstracts a network graph into a logical tree is introduced. The algorithm is based on the decomposition of a graph into its minimum cycle basis (a basis of the cycle vector space of a graph having least overall weight or length). A procedure that abstracts the cycles and their adjacencies into logical nodes and links correspondingly is introduced. These logical nodes and links form the next level logical graph. The procedure is repeated recursively, until a loop-free logical graph is derived. This iterative abstraction is called a logical network abstraction procedure and can be used to analyze network graphs for resiliency, as well as become the basis of a new routing methodology. Both these aspects of the logical network abstraction procedure are discussed in some detail.

18 ROUTING ALGORITHMS ON WDM OPTICAL NETWORKS

With the tremendous growth of bandwidth-intensive networking applications, the demand for bandwidth over data networks is increasing rapidly. Wavelength division multiplexing (WDM) optical networks provide promising infrastructures to meet the information networking demands and have been widely used as the backbone networks in the Internet, metropolitan area networks, and high-capacity local area networks. Efficient routing on WDM networks is challenging and involves hard optimization problems. This chapter introduces efficient algorithms with guaranteed performance for fundamental routing problems on WDM networks.

CONTRIBUTORS

Editors

Amiya Nayak, received his B.Math. degree in Computer Science and Combinatorics and Optimization from University of Waterloo in 1981, and Ph.D. in Systems and Computer Engineering from Carleton University in 1991. He has over 17 years of industrial experience, working at CMC Electronics (formerly known as Canadian Marconi Company), Defence Research Establishment Ottawa (DREO), EER Systems and Nortel Networks, in software engineering, avionics, and navigation systems, simulation and system level performance analysis. He has been an Adjunct Research Professor in the School of Computer Science at Carleton University since 1994. He had been the Book Review and Canadian Editor of VLSI Design from 1996 till 2002. He is in the Editorial Board of International Journal of Parallel, Emergent and Distributed Systems, and the Associate Editor of International Journal of Computing and Information Science. Currently, he is a Full Professor at the School of Information Technology and Engineering (SITE) at the University of Ottawa. His research interests are in the area of fault tolerance, distributed systems/algorithms, and mobile ad hoc networks with over 100 publications in refereed journals and conference proceedings.

Ivan Stojmenovic, received his Ph.D. degree in mathematics in 1985. He earned a third degree prize at the International Mathematics Olympiad for high school students in 1976. He held positions in Serbia, Japan, United States, Canada, France, and Mexico. He is currently a Chair Professor in Applied Computing at EECE, the University of Birmingham, UK. He published over 200 different papers, and edited three books on wireless, ad hoc, and sensor networks with Wiley/IEEE. He is currently editor of over ten journals, and founder and editor-in-chief of three journals. Stojmenovic was cited >3400 times and is in the top 0.56% most cited authors in Computer Science (Citeseer 2006). One of his articles was recognized as the Fast Breaking Paper, for October 2003 (as the only one for all of computer science), by Thomson ISI Essential Science Indicators. He coauthored over 30 book chapters, mostly very recent. He collaborated with over 100 coauthors with Ph.D. and a number of their graduate students from 22 different countries. He (co)supervised over 40 Ph.D. and master theses, and published over 120 joint articles with supervised students. His current research interests are mainly in wireless ad hoc, sensor, and cellular networks. His research interests also include parallel computing, multiple-valued logic, evolutionary computing, neural networks, combinatorial algorithms, computational geometry, graph theory, computational chemistry, image processing, programming languages, and computer science education. More details can be seen at www.site.uottawa.ca/~ivan.

Authors

Andrew Adamatzky, Faculty of Computing, Engineering and Mathematical Science University of the West of England, Bristol, BS16 1QY, UK [andrew.adamatzky@uwe.ac.uk]

Enrique Alba, Dpto. de Lenguajes y Ciencias de la Computación, E.T.S. Ing. Informática, Campus de Teatinos, 29071 Málaga, Spain [eat@lcc.uma.es www.lcc.uma.es/~eat.]

Theodoros N. Arvanitis, Electronics, Electrical, and Computer Engineering, University of Birmingham, Edgbaston, Birmingham B15 2TT, UK [T.Arvanitis@bham.ac.uk]

Kevin J. Baughan, Electronics, Electrical, and Computer Engineering, University of Birmingham, Edgbaston, Birmingham B15 2TT, UK

Costas C. Constantinou, Electronics, Electrical, and Computer Engineering, University of Birmingham, and Prolego Technologies Ltd., Edgbaston, Birmingham B15 2TT, UK [C.Constantinou@bham.ac.uk]

Camil Demetrescu, Department of Computer and Systems Science, University of Rome "La Sapienza", Via Salaria 113, 00198 Rome, Italy [demetres @dis.uniroma1.it]

Irene Finocchi, Department of Computer and Systems Science, University of Rome "La Sapienza", Via Salaria 113, 00198 Rome, Italy

Hannes Frey, Department of Mathematics and Computer Science, University of Southern Denmark, Campusvej 55, DK-5230 Odense M, Denmark [frey@imada.sdu.dk]

Qianping Gu, Department of Computing Science, Simon Fraser University, Burnaby, BC V5A 1S6, Canada [qgu@cs.sfu.ca]

Debra Knisley, Department of Mathematics, East Tennessee State University, Johnson City, TN 37614-0663, USA [knisleyd@mail.etsu.edu]

Jeff Knisley, Department of Mathematics, East Tennessee State University, Johnson City, TN 37614-0663, USA [knisleyj@etsu.edu]

Bin Liu, Electronics, Electrical, and Computer Engineering, University of Birmingham, Edgbaston, Birmingham B15 2TT, UK

Francisco Luna, Universidad de Málaga, ETS. Ing. Informática, Campus de Teatinos, 29071 Málaga, Spain [flv@lcc.uma.es]

Subhamoy Maitra, Applied Statistical Unit, Indian Statistical Institute, 203 B.T. Road, Koltkata, India [subho@isical.ac.in]

Patrick Mauroy, Universidad de Málaga, ETS. Ing. Informática, Campus de Teatinos, 29071 Málaga, Spain [Patrick.Mauroy@optimi.com]

Marios Mavronicolas, Department of Computer Science, University of Cyprus, Nicosia CY-1678, Cyprus [mavronic@cs.ucy.ac.cy]

Uros Midic, Center for Information Science and Technology, Temple University, 300 Wachman Hall, 1805 N. Broad St., Philadelphia, PA 19122, USA

Lucia Moura, School of Information Technology and Engineering, University of Ottawa, Ottawa, ON K1N 6N5, Canada [lucia@site.uottawa.ca]

Amiya Nayak, SITE, University of Ottawa, 800 King Edward Ave., Ottawa, ON K1N 6N5, Canada [anayak@site.uottawa.ca]

Antonio J. Nebro, Universidad de Málaga, ETS. Ing. Informática, Campus de Teatinos, 29071 Málaga, Spain [antonio@lcc.uma.es]

Zoran Obradovic, Center for Information Science and Technology, Temple University, 300 Wachman Hall, 1805 N. Broad St., Philadelphia, PA 19122, USA [zoran@ist.temple.edu]

Stephan Olariu, Department of Computer Science, Old Dominion University, Norfolk, Virginia, 23529, USA [olariu@cs.odu.edu]

Vicky Papadopoulou, Department of Computer Science, University of Cyprus, Nicosia CY-1678, Cyprus [viki@cs.ucy.ac.cy]

Salvador Pedraza, Universidad de Málaga, ETS. Ing. Informática, Campus de Teatinos, 29071 Málaga, Spain [Salvador.Pedraza@optimi.com]

Mona E. Rizvi, Department of Computer Science, Norfolk State University, 700 Park Avenue, Norfolk, VA 23504, USA [mrizvi@nsu.edu]

Syed R. Rizvi, Department of Computer Science, Old Dominion University, Norfolk, VA 23529, USA

Paul L. Rosin, School of Computer Science, Cardiff University, Cardiff CF24 3AA, Wales, UK [Paul.Rosin@cs.cf.ac.uk]

Bimal Roy, Applied Statistical Unit, Indian Statistical Institute, 203 B.T. Road, Kolkata, India [bimal@isical.ac.in]

Dan A. Simovici, Department of Mathematics and Computer Science, University of Massachusetts at Boston, Boston, MA 02125, USA [dsim@cs.umb.edu]

David Simplot-Ryl, IRCICA/LIFL, Univ. Lille 1, CNRS UMR 8022, INRIA Futurs, POPS Research Group, Bât. M3, Citá Scientifique, 59655 Villeneuve d'Ascq Cedex, France [David.Simplot@lifl.fr]

Paul Spirakis, University of Patras, School of Engineering, GR 265 00, Patras, Greece [spirakis@cti.gr]

Alexander S. Stepanenko, Electronics, Electrical, and Computer Engineering, University of Birmingham, Edgbaston, Birmingham B15 2TT, UK [ass@th.ph.bham.ac.uk]

Ivan Stojmenovic, SITE, University of Ottawa, Ottawa, ON K1N 6N5, Canada [ivan@site.uottawa.ca]

Milos Stojmenovic, School of Information Technology and Engineering, University of Ottawa, Ottawa, ON K1N 6N5, Canada [mstoj075@site.uottawa.ca]

Slobodan Vucetic, Center for Information Science and Technology, Temple University, 300 Wachman Hall, 1805 N. Broad St., Philadelphia, PA 19122, USA [vucetic@ist.temple.edu]

Hongbo Xie, Center for Information Science and Technology, Temple University, 300 Wachman Hall, 1805 N. Broad St., Philadelphia, PA 19122, USA

Joviša Žunić, Department of Computer Science, University of Exeter, Harrison Building North Park Road, Exeter EX4 4QF, UK [j.zunic@exeter.ac.uk]

CHAPTER 1

Generating All and Random Instances of a Combinatorial Object

IVAN STOJMENOVIC

1.1 LISTING ALL INSTANCES OF A COMBINATORIAL OBJECT

The design of algorithms to generate combinatorial objects has long fascinated mathematicians and computer scientists. Some of the earliest papers on the interplay between mathematics and computer science are devoted to combinatorial algorithms. Because of its many applications in science and engineering, the subject continues to receive much attention. In general, a list of all combinatorial objects of a given type might be used to search for a counterexample to some conjecture, or to test and analyze an algorithm for its correctness or computational complexity.

This branch of computer science can be defined as follows: Given a combinatorial object, design an efficient algorithm for generating all instances of that object. For example, an algorithm may be sought to generate all n-permutations. Other combinatorial objects include combinations, derangements, partitions, variations, trees, and so on.

When analyzing the efficiency of an algorithm, we distinguish between the cost of *generating* and cost of *listing* all instances of a combinatorial object. By generating we mean producing all instances of a combinatorial object, without actually outputting them. Some properties of objects can be tested dynamically, without the need to check each element of a new instance. In case of listing, the output of each object is required. The lower bound for producing all instances of a combinatorial object depends on whether generating or listing is required. In the case of generating, the time required to "create" the instances of an object, without actually producing the elements of each instance as output, is counted. Thus, for example, an optimal sequential algorithm in this sense would generate all n-permutations in $\theta(n!)$ time, that is, time linear in the number of instances. In the case of listing, the time to actually "output" each instance in full is counted. For instance, an optimal sequential algorithm generates all n-permutations in $\theta(nn!)$ time, since it takes $\theta(n)$ time to produce a string.

Handbook of Applied Algorithms: Solving Scientific, Engineering and Practical Problems
Edited by Amiya Nayak and Ivan Stojmenović

Let P be the number of all instances of a combinatorial object, and N be the average size of an instance. The *delay* when generating these instances is the time needed to produce the next instance from the current one. We list some desirable properties of generating or listing all instances of a combinatorial object.

Property 1. *The algorithm* lists *a*ll instances in asymptotically optimal time, *that is, in time O(NP).*

Property 2. *The algorithm* generates *all instances with* constant average delay. *In other words, the algorithm takes O(P) time to generate all instances. We say that a generating algorithm has constant average delay if the time to generate all instances is O(P); that is, the ratio T/P of the time T needed to generate all instances and the number of generated instances P is bounded by a constant.*

Property 3. *The algorithm generates all instances with* constant (worst case) delay. *That is, the time to generate the next instance from the current one is bounded by a constant. Constant delay algorithms are also called* loopless *algorithms, as the code for updating given instance contains no (repeat, while, or for) loops.*

Obviously, an algorithm satisfying Property 3 also satisfies Property 2. However, in some cases, an algorithm having constant delay property is considerably more sophisticated than the one satisfying merely constant average delay property. Moreover, sometimes an algorithm having constant delay property may need more time to generate all instances of the same object than an algorithm having only constant average delay property. Therefore, it makes sense to consider Property 3 independently of Property 2.

Property 4. *The algorithm* does not use large integers *in generating all instances of an object. In some papers, the time needed to "deal" with large integers is not properly counted in.*

Property 5. *The algorithm is the* fastest *known algorithm for generating all instances of given combinatorial object. Several papers deal with comparing actual (not asymptotic) times needed to generate all instances of given combinatorial object, in order to pronounce a "winner," that is, to extract the one that needs the least time. Here, the fastest algorithm may depend on the choice of computer. Some computers support fast recursion giving the recursive algorithm advantage over iterative one. Therefore, the ratio of the time needed for particular instructions over other instructions may affect the choice of the fastest algorithm.*

We introduce the *lexicographic order* among sequences. Let $a = a_1, a_2, \ldots, a_p$ and $b = b_1, b_2, \ldots, b_q$ be two sequences. Then a precedes $b (a < b)$ in lexicographic order if and only if there exists i such that $a_j = b_j$ for $j<i$ and either $p = i + 1 < q$ or $a_i < b_i$. The lexicographic order corresponds to dictionary order. For example, $112 < 221$ (where $i = 1$ from the definition).

For example, the lexicographic order of subsets of $\{1, 2, 3\}$ in the set representation is $\emptyset$, $\{1\}$, $\{1, 2\}$, $\{1, 2, 3\}$, $\{1, 3\}$, $\{2\}$, $\{2, 3\}$, $\{3\}$. In binary notation, the order of subsets is somewhat different: 000, 001, 010, 011, 100, 101, 110, 111, which correspond to subsets $\emptyset$, $\{3\}$, $\{2\}$, $\{2, 3\}$, $\{1\}$, $\{1, 3\}$, $\{1, 2\}$, $\{1, 2, 3\}$, respectively. Clearly the lexicographic order of instances depends on their representation. Different notations may lead to different listing order of same instances.

Algorithms can be classified into *recursive* or *iterative,* depending on whether or not they use recursion. The iterative algorithms usually have advantage of giving easy control over generating the next instance from the current one, which is often a desirable characteristic. Also some programming languages do not support recursion. In this chapter we consider only iterative algorithms, believing in their advantage over recursive ones.

Almost all sequential generation algorithms rely on one of the following three ideas:

1. *Unranking*, which defines a bijective function from consecutive integers to instances of combinatorial objects. Most algorithms in this group do not satisfy Property 4.
2. *Lexicographic updating,* which finds the rightmost element of an instance that needs "updating" or moving to a new position.
3. *Minimal change,* which generates instances of a combinatorial object by making as little as possible changes between two consecutive objects. This method can be further specified as follows:
   - Gray code generation, where changes made are theoretically minimal possible.
   - Transpositions, where instances are generated by exchanging pairs of (not necessarily adjacent) elements.
   - Adjacent interchange, where instances are generated by exchanging pairs of adjacent elements.

The algorithms for generating combinatorial objects can thus be classified into those following lexicographic order and those following a minimal change order. Both orders have advantages, and the choice depends on the application. Unranking algorithms usually follow lexicographic order but they can follow minimal change one (normally with more complex ranking and unranking functions).

Many problems require an exhaustive search to be solved. For example, finding all possible placements of queens on chessboard so that they do not attack each other, finding a path in a maze, choosing packages to fill a knapsack with given capacity optimally, satisfy a logic formula, and so on. There exist a number of such problems for which polynomial time (or quick) solutions are not known, leaving only a kind of exhaustive search as the method to solve them.

Since the number of candidates for a solution is often exponential to input size, systematic search strategies should be used to enhance the efficiency of exhaustive search. One such strategy is the *backtrack*. Backtrack, in general, works on partial solutions to a problem. The solution is extended to a larger partial solution if there is a hope to reach a complete solution. This is called an extend phase. If an extension of the current solution is not possible, or a complete solution is reached and another one is sought, it backtracks to a shorter partial solution and tries again. This is called a reduce phase. Backtrack strategy is normally related to the lexicographic order of instances of a combinatorial object. A very general form of backtrack method is as follows:

```
initialize;
repeat
    if current partial solution is extendable then extend else reduce;
    if current solution is acceptable then report it;
until search is over
```

This form may not cover all the ways by which the strategy is applied, and, in the sequel, some modifications may appear. In all cases, the central place in the method is finding an efficient test as to whether current solution is extendable. The backtrack method will be applied in this chapter to generate all subsets, combinations, and other combinatorial objects in lexicographic order.

Various algorithms for generating all instances of a combinatorial object can be found in the journal *Communications of ACM* (between 1960 and 1975) and later in *ACM Transactions of Mathematical Software* and *Collected Algorithms from ACM*, in addition to hundreds of other journal publications. The generation of ranking and unranking combinatorial objects has been surveyed in several books [6,14,21,25,30,35,40].

1.2 LISTING SUBSETS AND INTEGER COMPOSITIONS

Without loss of generality, the combinatorial objects are assumed to be taken from the set $\{1, 2, \ldots, n\}$, which is also called n-set. We consider here the problem of generating subsets in their set representation. Every subset [or (n,n)-subset] is represented in the set notation by a sequence $x_1, x_2, \ldots, x_r$, $1 \le r \le n$, $1 \le x_1 < x_2 < \ldots < x_r \le n$. An (m,n)-subset is a subset with exactly m elements.

Ehrlich [11] described a loopless procedure for generating subsets of an n-set. An algorithm for generating all (m,n)-subsets in the lexicographic order is given in the work by Nijenhius and Wilf [25]. Semba [33] improved the efficiency of the algorithm; the algorithm is modified in the work by Stojmenović and Miyakawa [37] and presented in Pascal-like notation without goto statements. We present here the algorithm from the work by Stojmenović and Miyakawa [37]. The generation goes in the following manner (e.g., let $n = 5$):

1	12	123	1234	12345
			1235	
		124	1245	
		125		
	13	134	1345	
		135		
	14	145		
	15			
2	23	234	2345	
		235		
	24	245		
	25			
3	34	345		
	35			
4	45			
5.				

The algorithm is in *extend* phase when it goes from left to right staying in the same row. If the last element of a subset is n, the algorithm shifts to the next row. We call this the *reduce* phase.

read(n); $r \leftarrow 0$; $x_r \leftarrow 0$;
repeat
    if $x_r < n$ **then** *extend* **else** *reduce;*
    print out $x_1, x_2, \ldots, x_r$
until $x_1 = n$
$extend \equiv \{x_{r+1} \leftarrow x_r + 1; r \leftarrow r + 1\}$
$reduce \equiv \{r \leftarrow r - 1; x_r \leftarrow x_r + 1\}$.

The algorithm is loopless, that is, has constant delay. To generate (m,n)-subsets, the **if** instruction in the algorithm should be changed to

if $x_r < n$ and $r < m$ **then** $\{x_{r+1} \leftarrow x_r + 1; r \leftarrow r + 1\}$ (* extend *)
    else if $x_r < n$ **then** $x_r \leftarrow x_r + 1$ (*cut *)
        else $\{r \leftarrow r - 1; x_r \leftarrow x_r + 1\}$ (* reduce *).

The new *cut* phase will be used when the algorithm goes from one subset to a subset in a lower row, skipping several subsets (having more than m elements). For example, for $m = 3$ and $n = 5$, the first three columns of the last table of subsets are

(3,5)-subsets. This illustrates the backtrack process applied on all subsets to extract (m,n)-subsets.

We now present the algorithm for generating variations. A (m,n)-variation out of $\{p_1, p_2, \ldots, p_n\}$ can be represented as a sequence $c_1c_2\ldots c_m$, where $p_1 \leq c_i \leq p_n$. Let $z_1z_2\ldots z_m$ be the corresponding array of indices, that is, $c_i = p_{z_i}$, $1 \leq i \leq m$. The next variation can be determined by a backtrack search that finds an element c_t with the greatest possible index t such that $z_t < n$, therefore increasable (the index t is called the turning point). The value of z_t is increased by 1 while the new value of z_i for $i \geq t$ is 1. The algorithm is as follows.

```
for i ← 0 to m do z_i ← 1 ;
repeat
        print out p_{z_i}, 1 ≤ i ≤ m ;
        t ← m ;
        while z_t = n do t ← t − 1 ;
        z_t ← z_t + 1;
        for i ← t + 1 to m do z_i ← 1
until t = 0.
```

We now prove that the algorithm has constant average delay property. Every step will be assigned to the current value of t; in this way the time complexity T is subdivided into m portions $T_1, T_2, \ldots, T_m$. In the process of a backtrack search and the update of elements, every portion T_i for $t \leq i \leq m$ increases by a constant amount. After the update, ith element does not change (moreover, the backtrack search does not reach it) during the next n^{m-i} variations (i.e., T_i does not increase). Therefore, on average, T_i increases by $O(1/n^{m-i})$. It follows that the average delay is, up to a constant,

$$\sum_{i=1}^{m} \frac{1}{n^{m-1}} = \frac{1}{n^m}\frac{n^{m+1}-1}{n-1} = O(1).$$

Subsets may be also represented in binary notation, where each "1" corresponds to the element from the subset. For example, subset $\{1,3,4\}$ for $n = 5$ is represented as 11010. Thus, subsets correspond to integers written in the binary number system (i.e., counters) and to bitstrings, giving all possible information contents in a computer memory. A simple recursive algorithm for generating bitstrings is given in the work by Parberry [28]. A call to bitstring (n) produces all bitstrings of length n as follows:

```
procedure bitstring( m);
        if m = 0 then print out c_i;
        else c_m ← 0; bitstring(m − 1 );
           c_m ← 1; bitstring(m − 1 ) .
```

Given an integer n, it is possible to represent it as the sum of one or more positive integers (called parts) a_i that is, $n = x_1 + x_2 + \cdots + x_m$. This representation is called an *integer partition* if the order of parts is of no consequence. Thus, two partitions of an integer n are distinct if they differ with respect to the x_i they contain. For example, there are seven distinct partitions of the integer $5: 5, 4+1, 3+2, 3+1+1, 2+2+1, 2+1+1+1, 1+1+1+1+1$. If the order of parts is important then the representation of n as a sum of some positive integers is called *integer composition*. For example, integer compositions of 5 are the following:

$$5,\ 4+1,\ 1+4,\ 3+2,\ 2+3,\ 3+1+1,\ 1+3+1,\ 1+1+3,\ 2+2+1,$$
$$2+1+2,\ 1+2+2,\ 2+1+1+1,\ 1+2+1+1,\ 1+1+2+1,$$
$$1+1+1+2,\ 1+1+1+1+1.$$

Compositions of an integer n into m parts are representations of n in the form of the sum of exactly m positive integers. These compositions can be written in the form $x_1 + \cdots + x_m = n$, where $x_1 \geq 0, \ldots, x_m \geq 0$. We will establish the correspondence between integer compositions and either combinations or subsets, depending on whether or not the number of parts is fixed.

Consider a composition of $n = x_1 + \cdots + x_m$, where m is fixed or not fixed. Let $y_1, \ldots, y_m$ be the following sequence: $y_i = x_1 + \cdots + x_i$, $1 \leq i \leq m$. Clearly, $y_m = n$. The sequence $y_1, y_2, \ldots, y_{m-1}$ is a subset of $\{1, 2, \ldots, n-1\}$. If the number of parts m is not fixed then compositions of n into any number of parts correspond to subsets of $\{1, 2, \ldots, n-1\}$. The number of such compositions is in this case $CM(n) = 2^{n-1}$. If the number of parts m is fixed then the sequence $y_1, \ldots, y_{m-1}$ is a combinations of $m-1$ out of $n-1$ elements from $\{1, \ldots, n-1\}$, and the number of compositions in question is $CO(m, n) = C(m-1, n-1)$. Each sequence $x_1 \ldots x_m$ can easily be obtained from $y_1, \ldots, y_m$ since $x_i = y_i - y_{i-1}$ (with $y_0 = 0$).

To design a loopless algorithm for generating integer compositions of n, one can use this relation between compositions of n and subsets of $\{1, 2, \ldots, n-1\}$, and the subset generation algorithm above.

1.3 LISTING COMBINATIONS

A (m,n)-combination out of $\{p_1, p_2, \ldots, p_n\}$ can be represented as a sequence $c_1, c_2, \ldots, c_m$, where $p_1 \leq c_1 < c_2 < \cdots < c_m \leq p_n$. Let $z_1, z_2, \ldots, z_m$ be the corresponding array of indices, that is, $c_i = p_{z_i}$, $1 \leq i \leq m$. Then $1 \leq z_1 < z_2 < \cdots < z_m \leq n$, and therefore $z_i \leq n - m + i$ for $1 \leq i \leq m$. The number of (m,n)-combinations is binomial coefficient $C(m, n) = n!/(m!(n-m)!)$. In this section, we investigate generating the $C(m,n)$ (m,n)-combinations, in lexicographically ascending order. Various sequential algorithms have been given for this problem.

Comparisons of combination generation techniques are given in the works by Akl [1] and Payne and Ives [29]. Akl [1] reports algorithm by Misfud [23] to be the fastest while Semba [34] improved the speed of algorithm [23].

The sequential algorithm [23] for generating (m,n)-combinations determines the next combination by a backtrack search that finds an element c_t with the greatest possible index t such that $z_t < n - m + t$, therefore increasable (the index t is called the turning point). The new value of z_i for $i \geq t$ is $z_t + i - t + 1$.

The average delay of the algorithm is $O(n/(n-m))$ [34]. The delay is constant whenever $m = o(n)$. On the contrary, the average delay may be nonconstant in some cases (e.g., when $n - m = O(\sqrt{n})$). Semba [34] modified the algorithm by noting that there is no need to search for the turning point as it can be updated directly from one combination to another, and that there is no need to update the elements with indices between t and m if they do not change from one combination to another. If $z_t < n - m + t - 1$ then all elements in the next combination will be less that their appropriate maximal values and the turning point of the next combination will be index m. In this case, a total of $d = m - t + 1$ elements change their value in the next combination. Otherwise, that is, when $z_t = n - m + t - 1$, the new value for the turning point element becomes its maximal possible value $n - m + t$, elements between t and m remain unchanged (with their maximal possible values), and the turning point for the next combination is the element with index $t - 1$. Only one element is checked in this case. The following table gives values of t and d for (4,6)-combinations.

1234	1235	1236	1245	1246	1256	1345	1346	1356	1456	2345	2346	2356	2456	3456
$t = 4$	4	3	4	3	2	4	3	2	1	4	3	2	1	0
$d = 1$	1	2	1	1	3	1	1	1	4	1	1	1	1	

The algorithm [34] is coded in FORTRAN language using goto statements. Here we code it in PASCAL-like style.

```
z0 ← 1; t ← m;
for i ← 1 to m do zi ← i;
repeat
        print out p_zi, 1 ≤ i ≤ m;
        zt ← zt + 1;
        if zt = n − m + t    then t ← t − 1
                      else for i = t + 1 to m do zi ← zt + i − t; t ← m
until t = 0.
```

The algorithm always does one examination to determine the turning point. We now determine the average number d of changed elements. For a fixed t, the number of (m,n)-combinations that have t as the turning point with $z_t < n - m + t - 1$ is $C(t, n - m + t - 2)$. This follows because $z_i = n - m + i$ when $i > t$ for each of these combinations while $z_1, z_2, \ldots, z_t$ can be any $(t, n - m + t - 2)$ -combination. The turning point element is always updated. In addition, $m - t$ elements whenever $z_t < n - m + t - 1$, which happens $C(t, n - m + t - 2)$ times. Therefore, the total number of updated elements (in addition to the turning point) to generate all combinations is

$$\sum_{t=1}^{m}(m-t)C(t,n-m+t-2)=\sum_{j=0}^{m-1}jC(n-j-2,n-m-2)$$
$$=\frac{m}{n-m}C(n-m-1,n-1)-m$$
$$=\frac{m}{n}C(m,n)-m.$$

Thus, the algorithms updates, on the average, less than $m/n+1<2$ elements and therefore the average delay is constant for any m and $n(m \leq n)$.

1.4 LISTING PERMUTATIONS

A sequence $p_1, p_2, \ldots, p_n$ of mutually distinct elements is a *permutation* of $S=\{s_1, s_2, \ldots, s_n\}$ if and only if $\{p_1, p_2, \ldots, p_n\}=\{s_1, s_2, \ldots, s_n\}=S$. In other words, an n-permutation is an ordering, or arrangement, of n given elements. For example, there are six permutations of the set $\{A, B, C\}$. These are ABC, ACB, BAC, BCA, CAB, and CBA.

Many algorithms have been published for generating permutations. Surveys and bibliographies on the generation of permutations can be found in the Ord-Smith [27] and Sedgewick [31] [27,31]. Lexicographic generation presented below is credited to L.L. Fisher and K.C. Krause in 1812 by Reingold et al. [30].

Following the backtrack method, permutations can be generated in lexicographic order as follows. The next permutation of $x_1x_2\ldots x_n$ is determined by scanning from right to left, looking for the rightmost place where $x_i<x_{i+1}$ (called again the turning point). By another scan, the smallest element x_j that is still greater than x_i is found and interchanged with x_i. Finally, the elements $x_{i+1}, \ldots, x_n$ (which are in decreasing order) are reversed. For example, for permutation 3, 9, 4, 8, 7, 6, 5, 2, 1, the turning point $x_3=4$ is interchanged with $x_7=5$ and 8, 7, 6, 4, 2, 1 is reversed to give the new permutation 3, 9, 5, 1, 2, 4, 6, 7, 8. The following algorithm is the implementation of the method for generating permutations of $\{p_1, p_2, \ldots, p_n\}$. The algorithm updates the indices z_i (such that $x_i=p_{z_i}$,), $1 \leq i \leq n$.

```
for i ← 0 to n do z_i ← i ;
i ← 1;
while i ≠ 0 do {
        print out p_{z_i}, 1 ≤ i ≤ n;
        i ← n − 1;
        while z_i ≥ z_{i+1} do i ← i − 1;
        j ← n;
        while z_i ≥ z_j do j ← j − 1;
```

```
ch ← z_i; z_i ← z_j; z_j ← ch;
v ← n; u ← i + 1;
while v>u do {ch ← z_v; z_v ← z_u; z_u ← ch; v ← v − 1;
      u ← u + 1}}.
```

We prove that the algorithm has constant average delay property. The time complexity of the algorithm is clearly proportional to the number of tests $z_i \geq z_{i+1}$ in the first *while* inside loop. If ith element is the turning point, the array $z_{i+1}, \ldots, z_n$ is decreasing and it takes $(n-1)$ tests to reach z_i. The array $z_1 z_2 \ldots z_i$ is a (*m*,*n*)-permutation. It can be uniquely completed to n-permutation $z_1 z_2 \ldots z_n$ such that $z_{i+1} > \cdots > z_n$. Although only these permutations for which $z_i < z_{i+1}$ are valid for z_i to be the turning point, we relax the condition and artificially increase the number of tests in order to simplify the proof. Therefore for each i, $1 \leq i \leq n-1$ there are at most $P(i,n) = n(n-1)\cdots(n-i+1)$ arrays such that z_i is the turning point of n-permutation $z_1 z_2 \ldots z_n$. Since each of them requires $n-i$ tests, the total number of tests is at most $\sum_{i=1}^{n-1} P(i,n)(n-i) = \sum_{i=1}^{n-1} (n(n-1)\cdots(n-i+1)(n-i)) = \sum_{i=1}^{n-1} n!/(n-i-1)! = n! \sum_{j=0}^{n-2} 1/j!$. Since $j! = 2 \cdot 3 \cdots j > 2 \times 2 \cdots \times 2 = 2^{j-1}$, the average number of tests is $< 2 + \sum_{j=2}^{n-2} 1/(2^{j-1}) = 2 + 1/2 + 1/4 + \ldots < 3$. Therefore the algorithm has constant delay property. It is proved [27] that the algorithm performs about $1.5n!$ interchanges.

The algorithm can be used to generate the permutations with repetitions. Let $n_1, n_2, \ldots, n_k$ be the multiplicities of elements $p_1, p_2, \ldots, p_k$, respectively, such that the total number of elements is $n_1 + n_2 + \cdots + n_k = n$. The above algorithm uses no arithmetic with indices z_i and we can observe that the same algorithm generates permutations with repetitions if the initialization step (the first instruction, i.e., **for** loop) is replaced by the following instructions that find the first permutation with repetitions.

```
n ← 0; z_0 ← 0;
for i ← 1 to k do
     for j ← 1 to n_i do {n ← n + 1; z_n ← j};
```

Permutations of combinations (or (*m*,*n*)-permutations) can be found by generating all (*m*,*n*)-combinations and finding all (*m*,*m*)-permutations for each (*m*,*n*)-combination. The algorithm is then obtained by combining combination and permutation generating algorithms. In the standard representation of (*m*,*n*)-permutations as an array $x_1 x_2 \ldots x_m$, the order of instances is not lexicographic. Let $c_1 c_2 \ldots c_m$ be the corresponding combination for permutation $x_1 x_2, \ldots, x_m$, that is, $c_1 < c_2 < \cdots < c_m$ and $\{c_1, c_2, \ldots, c_m\} = \{x_1, x_2, \ldots, x_m\}$. Then we can observe that the obtained order of generating (*m*,*n*)-permutations is lexicographic if they are represented as an array of $2m$ elements $c_1 c_2 \ldots c_m\, x_1 x_2 \ldots x_m$, composed of corresponding (*m*,*n*)-combination followed by the (*m*,*n*)-permutation. In other words, the order is lexicographic if corresponding combinations are compared before comparing permutations.

1.5 LISTING EQUIVALENCE RELATIONS OR SET PARTITIONS

An equivalence relation of the set $Z = \{p_1, \ldots, p_n\}$ consists of classes $\pi_1, \pi_2, \ldots, \pi_k$ such that the intersection of every two classes is empty and their union is equal to Z. Equivalence relations are often referred to as set partitions. For example, let $Z = \{A, B, C\}$. Then there are four equivalence relations of Z : $\{\{A, B, C\}\}$, $\{\{A, B\}\{C\}\}$, $\{\{A, C\}\{B\}\}$, $\{\{A\}, \{B, C\}\}$, and $\{\{A\}, \{B\}, \{C\}\}$.

Equivalence relations of Z can be conveniently represented by codewords $c_1c_2 \ldots c_n$ such that $c_i = j$ if and only if element p_i is in class π_j. Because equivalence classes may be numbered in various ways ($k!$ ways), such codeword representation is not unique. For example, set partition $\{\{A, B\}\{C\}\}$ is represented with codeword 112 while the same partition $\{\{C\}\{A, B\}\}$ is coded as 221.

In order to obtain a unique codeword representation for given equivalence relation, we choose lexicographically minimal one among all possible codewords. Clearly $c_1 = 1$ since we can choose π_1 to be the class containing p_1. All elements that are in π_1 are also coded with 1. The class containing element that is not in π_1 and has the minimal possible index is π_2 and so on. For example, let $\{\{C, D, E\}, \{B\}, \{A, F\}\}$ be a set partition of $\{A, B, C, D, E, F\}$. The first equivalence class is $\{A, F\}$, the second is $\{B\}$, and the third is $\{C, D, E\}$. The corresponding codeword is 123331.

A codeword $c_1 \ldots c_n$ represents an equivalence relation of the set Z if and only if $c_1 = 1$ and $1 \leq c_r \leq g_{r-1} + 1$ for $2 \leq r \leq n$, where $c_i = j$ if i is in π_j, and $g_r = \max(c_1, \ldots, c_r)$ for $1 \leq r \leq n$. This follows from the definition of lexicographically minimal codeword. Element p_t is either one of the equivalence classes with some other element $p_i(i<t)$ in which case c_t receives one of existing codes assigned to elements $p_1, p_2, \ldots, p_{t-1}$ or in none of previous classes, in which case it starts a new class with index one higher than previously maximal index.

Sequential algorithms [9,12,25,32] generate set partitions represented by codewords in lexicographic order. The next equivalence relation is found from the current one by a backtracking or recursive procedure in all known sequential generating techniques that maintain the lexicographic order of elements; in both cases an increasable element (one for which $x_j \leq g_j - 1$ is satisfied) with the largest possible index t is found ($t \leq n - 2$); we call this element the *turning point.* For example, the turning point of the equivalence relation 1123 is the second element ($t = 2$).

A list of codewords and corresponding partitions for $n = 4$ and $Z = \{A, B, C, D\}$ is, in lexicographic order, as follows:

1111 = {{A, B, C, D}}, 1112 = {{A, B, C}, {D}}, 1121 = {{A, B, D}, {C}},
1122 = {{A, B}, {C, D}}, 1123 = {{A, B}, {C}, {D}},
1211 = {{A, C, D}, {B}}, 1212 = {{A, C}, {B, D}},
1213 = {{A, C}, {B}, {D}}, 1221 = {{A, D}, {B, C}},
1222 = {{A}, {B, C, D}}, 1223 = {{A}, {B, C}, {D}}, 1231 = {{A, D}, {B}, {C}},
1232 = {{A}, {B, D}, {C}}, 1233 = {{A}, {B}, {C, D}}, 1234 = {{A}, {B}, {C}, {D}}.

We present an iterative algorithm from the work by Djokić et al. [9] for generating all set partitions in the codeword representation. The algorithm follows backtrack method for finding the largest r having an increasable c_r, that is, $c_r < g_{r-1} + 1$.

program setpart(n);
$r \leftarrow 1$; $c_1 \leftarrow 1$; $j \leftarrow 0$; $b_0 \leftarrow 1$; $n1 \leftarrow n-1$;
repeat
  while $r < n1$ **do** $\{r \leftarrow r+1;\ c_r \leftarrow 1;\ j \leftarrow j+1;\ b_j \leftarrow r\}$;
  for $i \leftarrow 1$ **to** $n-j$ **do** $\{c_n \leftarrow i$; print out $c_1, c_2, \ldots, c_n\}$;
  $r \leftarrow b_j$; $c_r \leftarrow c_r + 1$;
  if $c_r > r - j$ **then** $j \leftarrow j - 1$
until $r = 1$

In the presented iterative algorithm b_j is the position where current position r should backtrack after generating all codewords beginning with $c_1, c_2, \ldots, c_{n-1}$. Thus the backtrack is applied on $n-1$ elements of codeword while direct generation of the last element in its range speeds the algorithm up significantly (in most set partitions the last element in the codeword is increasable). An element of b is defined whenever $g_r = g_{r-1}$, which is recognized by either $c_r = 1$ or $c_r > r - j$ in the algorithm. It is easy to see that the relation $r = g_{r-1} + j$ holds whenever j is defined. For example, for the codeword $c = 111211342$ we have $g = 111222344$ and $b = 23569$. Array b has $n - g_n = 9 - 4 = 5$ elements.

In the algorithm, backtrack is done on array b and finds the increasable element in constant time; however, updating array b for future backtrack calls is not a constant time operation (**while** loop in the program). The number of backtrack calls is B_{n-1} (recall that B_n is the number of set partitions over n elements).

The algorithm has been compared with other algorithms that perform the same generation and it was shown to be the fastest known iterative algorithm. A recursive algorithm is proposed in the work by Er [12]. The iterative algorithm is faster than recursive one on some architectures and slower on other [9].

The constant average time property of the algorithm can be shown as in the work by Semba [32]. The backtrack step returns to position r exactly $B_r - B_{r-1}$ times, and each time it takes $n - r + 1$ for update (**while** loop), for $2 \leq r \leq n-1$. Therefore, up to a constant, the backtrack steps require $(B_2 - B_1)(n-1) + (B_3 - B_2)(n-2) + \cdots + (B_{n-1} - B_{n-2})2 < B_2 + B_3 + \cdots + B_{n-2} + 2B_{n-1}$. The update of nth element is performed $B_n - B_{n-1}$ times. Since $B_{i+1} > 2B_i$, the average delay, up to a constant, is bounded by

$$\frac{B_n + B_{n-1} + \cdots + B_2}{B_n} < 1 + \frac{1}{2} + \frac{1}{2^2} + \cdots + \frac{1}{2^{n-2}} < 2.$$

1.6 GENERATING INTEGER COMPOSITIONS AND PARTITIONS

Given an integer n, it is possible to represent it as the sum of one or more positive integers (called *parts*) x_i, that is, $n = x_1 + x_2 + \cdots + x_m$. This representation is called an *integer partition* if the order of parts is of no consequence. Thus, two partitions of an integer n are distinct if they differ with respect to the x_i they contain. For example, there are seven distinct partitions of the integer 5:

$$5,\ 4+1,\ 3+2,\ 3+1+1,\ 2+2+1,\ 2+1+1+1,\ 1+1+1+1+1.$$

In the *standard* representation, a partition of n is given by a sequence $x_1, \ldots, x_m$, where $x_1 \geq x_2 \geq \cdots \geq x_m$, and $x_1 + x_2 + \cdots + x_m = n$. In the sequel x will denote an arbitrary partition and m will denote the number of parts of x (m is not fixed). It is sometimes more convenient to use a *multiplicity* representation for partitions in terms of a list of the distinct parts of the partition and their respective multiplicities. Let $y_1 > \cdots > y_d$ be all distinct parts in a partitions, and $c_1, \ldots, c_d$ their respective (positive) multiplicities. Clearly $c_1 y_1 + \cdots + c_d y_d = n$.

We first describe an algorithm for generating integer compositions of n into any number of parts and in lexicographic order. For example, compositions of 4 in lexicographic order are the following: $1+1+1+1,\ 1+1+2,\ 1+2+1,\ 1+3,\ 2+1+1,\ 2+2,\ 3+1,\ 4$. Let $x_1 \ldots x_m$, where $x_1 + x_2 + \cdots + x_m = n$ be a composition. The next composition, following lexicographic order, is $x_1, \ldots, x_{m-1}+1, 1, \ldots, 1 (x_m - 1$ 1s). In other words, the next to last part is increased by one and the $x_m - 1$, 1s are added to complete the next composition. This can be coded as follows:

```
program composition( n);
m ← 1; x_1 ← n;
repeat
    for j ← 1 to m do print out x_1, x_2, ..., x_m;
    m ← m − 1; x_m ← x_m + 1;
    for j ← 1 to x_{m+1} − 1 do {m ← m + 1; x_m ← 1}
until m = n.
```

In antilexicographic order, a partition is derived from the previous one by subtracting 1 from the rightmost part greater than 1, and distributing the remainder as quickly as possible. For example, the partitions following $9+7+6+1+1+1+1+1+1$ is $9+7+5+5+2$. In standard representation and antilexicographic order, the next partition is determined from current one $x_1 x_2 \ldots x_m$ in the following way. Let h be the number of parts of x greater than 1, that is, $x_i > 1$ for $1 \leq i \leq h$, and $x_i = 1$ for $h < i \leq m$. If $x_m > 1$ (or $h = m$) then the next partition is $x_1, x_2, \ldots, x_{m-1}, x_m - 1, 1$. Otherwise (i.e., $h < m$), the next partition is obtained by replacing $x_h, x_{h+1} = 1, \ldots, x_m = 1$ with $(x_h - 1), (x_h - 1), \ldots, (x_h - 1), d$, containing c elements, where $0 < d \leq x_h - 1$ and $(x_h - 1)(c - 1) + d = x_h + m - h$.

We describe two algorithms from the work by Zoghbi and Stojmenovic [43] for generating integer partitions in standard representation and prove that they have constant average delay property. The first algorithm, named ZS1, generates partitions in antilexicographic order while the second, named ZS2, uses lexicographic order.

Recall that h is the index of the last part of partition, which is greater than 1 while m is the number of parts. The major idea in algorithm ZS1 is coming from the observation on the distribution of x_h. An empirical and theoretical study shows that $x_h = 2$ has growing frequency; it appears in 66 percent of cases for $n = 30$ and in 78 percent of partitions for $n = 90$ and appears to be increasing with n. Each partition of n containing a part of size 2 becomes, after deleting the part, a partition of $n - 2$ (and *vice versa*). Therefore the number of partitions of n containing at least one part of size 2 is $P(n - 2)$. The ratio $P(n - 2)/P(n)$ approaches 1 with increasing n. Thus, almost all partitions contain at least one part of size 2. This special case is treated separately, and we will prove that it suffices to argue the constant average delay of algorithm ZS1. Moreover, since more than 15 instructions in known algorithms that were used for all cases are replaced by 4 instructions in cases of at least one part of size 2 (which happens almost always), the speed up of about four times is expected even before experimental measurements. The case $x_h > 2$ is coded in a similar manner as earlier algorithm, except that assignments of parts that are supposed to receive value 1 is avoided by an initialization step that assigns 1 to each part and observation that inactive parts (these with index $> m$) are always left at value 1. The new algorithm is obtained when the above observation is applied to known algorithms and can be coded as follows.

```
Algorithm ZS1
for i ← 1 to n do x_i ← 1;
x_1 ← n; m ← 1; h ← 1; output x_1;
while x_1 ≠ 1 do {
    if x_h = 2 then {m ← m + 1; x_h ← 1; h ← h − 1}
    else {r ← x_h − 1; t ← m − h + 1; x_h ← r;
          while t ≥ r do {h ← h + 1; x_h ← r; t ← t − r}
          if t = 0 then m ← h
          else m ← h + 1
               if t>1 then {h ← h + 1; x_h ← t}}
  output x_1, x_2, ..., x_m}}.
```

We now describe the method for generating partitions in lexicographic order and standard representation of partitions. Each partition of n containing two parts of size 1 (i.e., $m - h > 1$) becomes, after deleting these parts, a partition of $n - 2$ (and *vice versa*). Therefore the number of integer partitions containing at least two parts of size 1 is $P(n - 2)$, as in the case of previous algorithm. The coding in this case is made simpler, in fact with constant delay, by replacing first two parts of size 1 by one part of size 2. The position h of last part >1 is always maintained. Otherwise, to find the next partition in the lexicographic order, an algorithm will do a backward search to find the first part that can be increased. The last part x_m cannot be increased. The next to last part x_{m-1} can be increased only if $x_{m-2} > x_{m-1}$. The element that will be increased is x_j where $x_{j-1} > x_j$ and $x_j = x_{j+1} = \ldots = x_{m-1}$. The jth part becomes $x_j + 1$, h receives value j, and appropriate number of parts equal to 1 is added to complete the sum to n. For example, in the partition $5 + 5 + 5 + 4 + 4 + 4 + 1$ the leftmost 4 is increased, and the next partition is $5 + 5 + 5 + 5 + 1 + 1 + 1 + 1 + 1 + 1 + 1 + 1$. The following is a code of appropriate algorithm ZS2:

Algorithm ZS2
for $i \leftarrow 1$ **to** n **do** $x_i \leftarrow 1$; output x_i, $i = 1, 2, \ldots, n$;
$x_0 \leftarrow 1$; $x_1 \leftarrow 2$; $h \leftarrow 1$; $m \leftarrow n - 1$; output x_i, $i = 1, 2, \ldots, m$;
while $x_1 \neq n$ **do** {
  if $m - h > 1$ **then** $\{h \leftarrow h + 1; x_h \leftarrow 2; m \leftarrow m - 1\}$
    else $\{j \leftarrow m - 2;$
      while $x_j = x_{m-1}$ **do** $\{x_j \leftarrow 1; j \leftarrow j - 1\}$;
      $h \leftarrow j + 1$; $x_h \leftarrow x_{m-1} + 1$;
      $r \leftarrow x_m + x_{m-1}(m - h - 1)$; $x_m \leftarrow 1$;
      if $m - h > 1$ **then** $x_{m-1} \leftarrow 1$;
      $m \leftarrow h + r - 1$;
  output $x_1, x_2, \ldots, x_m\}$.

We now prove the constant average delay property of algorithms ZS1 and ZS2.

Theorem 1 *Algorithms ZS1 and ZS2 generate unrestricted integer partitions in standard representation with constant average delay, exclusive of the output.*

Proof. Consider part $x_i \geq 3$ in the current partition. It received its value after a backtracking search (starting from last part) was performed to find an index $j \leq i$, called the turning point, that should change its value by 1 (increase/decrease for lexicographic/antilexicographic order) and to update values x_i for $j \leq i$. The time to perform both backtracking searches is $O(r_j)$, where $r_j = n - x_1 - x_2 - \cdots - x_j$ is the remainder to distribute after first j parts are fixed. We decide to charge the cost of the backtrack search evenly to all "swept" parts, such that each of them receives constant $O(1)$ time. Part x_i will be changed only after a similar backtracking step "swept" over ith part or recognized ith part as the turning point (note that ith part is the turning point in at least one of the two backtracking steps). There are $\mathrm{RP}(r_i, x_i)$ such partitions that keep all x_j intact. For $x_i \geq 3$ the number of such partitions, is $\geq r_i^2/12$. Therefore the average number of operations that are performed by such part i during the "run" of $\mathrm{RP}(r_i, x_i)$, including the change of its value, is $O(1)/RP(r_i, x_i) \leq O(1)/r_i^2 = O(1/r_i^2) < q_i/r_i^2$, where q_i is a constant. Thus the average number of operations for all parts of size ≥ 3 is $\leq q_1/r_1^2 + q_2/r_2^2 + \cdots + q_s/r_s^2 \leq q(1/r_1^2 + \cdots + 1/r_s^2) < q(1/n^2 + 1/(n-1)^2 + \cdots + 1/1^2) < 2q$ (the last inequality can be obtained easily by applying integral operation on the last sum), which is a constant. The case that was not counted in is when $x_i \leq 2$. However, in this case both algorithms ZS1 and ZS2 perform constant number of steps altogether on all such parts. Therefore the algorithm has overall constant time average delay. ■

The performance evaluation of known integer partition generation methods is performed in the work by Zoghbi and Stojmenovic [43]. The results show clearly that both algorithms ZS1 and ZS2 are superior to all other known algorithms that generate partitions in the standard representation. Moreover, both algorithms SZ1 and ZS2 were even faster than any algorithm for generating integer partitions in the multiplicity representation.

1.7 LISTING t-ARY TREES

The t-ary trees are data structures consisting of a finite set of n nodes, which either is empty ($n = 0$) or consists of a root and t disjoint children. Each child is a t-ary subtree, recursively defined. A node is the parent of another node if the latter is a child of the former. For $t = 2$, one gets the special case of rooted binary trees, where each node has a left and a right child, where each child is either empty or is a binary tree. A computer representation of t-ary trees with n nodes is achieved by an array of n records, each record consisting of several data fields, t pointers to children and a pointer to the parent. All pointers to empty trees are nil. The number of t-ary trees with n nodes is $B(n, t) = (tn)!/(n!(tn - n)!)/((t - 1)n + 1)$ (cf. [19,42]).

If the data fields are disregarded, the combinatorial problem of generating binary and, in general, t-ary trees is concerned with generating all different shapes of t-ary trees with n nodes in some order. The lexicographic order of trees refers to the lexicographic order of the corresponding tree sequences. There are over 30 ingenious generating algorithms for generating binary and t-ary trees. In most references, tree sequences are generated in lexicographic order. Each of these generation algorithms causes trees to be generated in a particular order. Almost all known sequential algorithms generate tree sequences, and the inclusion of parent–child relations requires adding a decoding procedure, usually at a cost of greatly complicating the algorithm and/or invalidating the run time analysis. Exceptions are the works by Akl et al. [4] and Lucas et al. [22].

Parent array notation [4] provides a simple sequential algorithm that extends trivially to add parent–children relations. Consider a left-to-right breadth first search (BFS) labeling of a given tree. All nodes are labeled by consecutive integers $1, 2, \ldots, n$ such that nodes on a lower level are labeled before those on a higher level, while nodes on the same level are labeled from left to right. Children are ordered as $L = 1, \ldots, t$. Parent array $p_1, \ldots, p_n$ can be defined as follows: $p_1 = 1$, $p_i = t(j - 1) + L + 1$ if i is the Lth child of node j, $2 \le i \le n$, and it has property $p_{i-1} < p_i \le ti - t + 1$ for $2 \le i \le n$. For example, the binary tree on Figure 1.1 has parent array 1, 3, 4, 5, 7, 8; the 3-ary tree on Figure 1.1 has parent array 1, 2, 3, 4, 8, 10, 18.

The algorithm [4] for generating all parent arrays is extended from the work by Zaks [42] to include parent–children relations (the same sequence in the works by Zaks [42] and Akl et al. [4] refers to different trees). The Lth children of node i is denoted by $child_{i,L}$ (it is 0 if no such child exist) while $parent_i$ denotes the parent

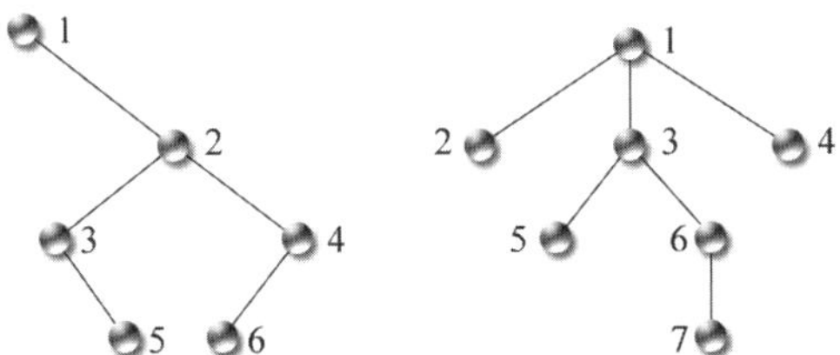

FIGURE 1.1 Binary tree 1, 3, 4, 5, 7, 8 and ternary tree 1, 2, 3, 4, 8, 10, 18.

node of i. Integer division is used throughout the algorithm. The algorithm generates tree sequences in lexicographic order.

```
for i ← 1 to n do
    for L ← 1 to t do child_{i,L} ← 0;
for i ← 1 to n do {p_i ← i; parent_i ← (i − 2)/t + 1;
            L ← p_i − 1 − t(parent_i − 1); child_{(i−2)/t+1,L} ← i}
repeat
    report t-ary tree;
    j ← n;
    while p_j < 2j − 1 and j>1 do {i ← parent_j;
                 L ← p_i − 1 − t(i − 1); child_{i,L} ← 0; j ← j − 1}
    p_j ← p_j + 1;
    for i ← j + 1 to n do p_i ← p_{i−1} + 1;
    for i ← j to n do {k ← (p_i − 2)/t + 1; parent_i ← k;
            L ← p_i − 1 − t(k − 1); child_{k,L} ← i}
until p_1 = 2 .
```

Consider now generating t-ary trees in the children array notation. A tree is represented using a children array $c_1 c_2, c_3, \ldots, c_{tn}$ as follows:

- The jth children of node i is stored in $c_{(i-1)t+j+1}$ for $1 \leq i \leq n-1$ and $1 \leq j \leq t$; missing children are denoted by 0. The array is, for convenience, completed with $c_1 = 1$ and $c_{(n-1)t+2} = \cdots = c_{nt} = 0$ (node n has no children).

For example, the children array notations for trees in FIGURE 1.1 are 102340560000 and 12340005060000007000. Here we give a simple algorithm to generate children array tree sequences, for the case of t-ary trees (generalized from the work by Akl et al. [4] that gives corresponding generation of binary trees).

The rightmost element of array c that can be occupied by an integer $j>0$, representing node j, is obtained when j is tth child of node $j-1$, that is, it is $c_{(j-1)t+1}$. We say that an integer j is mobile if it is not in $c_{(j-1)t+1}$ and all (nonzero) integers to its right occupy their rightmost positions. A simple sequential algorithm that uses this notation to generate all t-ary trees with n nodes is given below. If numerical order $0 < 1 < \cdots < n$ is assumed, the algorithm generates children array sequences in antilexicographic order. Alternatively, the order may be interpreted as lexicographic if $0, 1, \cdots, n$ are treated as symbols, ordered as "1" < "2" < ... < "n" < "0". Numeric lexicographic order may be obtained if 0 is replaced by a number larger than n (the algorithm should always report that number instead of 0).

```
for i ← 1 to n do c_i ← i;
for i ← n + 1 to tn do c_i ← 0;
repeat
    print out c_1, ..., c_tn;
    i ← (n − 1)t;
```

while $\left(c_i = 0 \text{ or } c_i = \frac{k-1}{t} + 1\right)$ and $(i > 1)$ **do** $i \leftarrow i - 1$;
$c_{i+1} \leftarrow c_i$;
$c_i \leftarrow 0$;
for $k \leftarrow 1$ **to** $n - c_{i+1}$ **do** $c_{i+k+1} \leftarrow c_{i+k} + 1$;
for $k \leftarrow i + n - c_{i+1} + 2$ **to** $(n-1)t + 1$ **do** $c_k \leftarrow 0$
until $i = 1$.

We leave as an exercise to design an algorithm to generate well-formed parenthesis sequences. This can be done by using the relation between well-formed parenthesis sequences and binary trees in the children representation, and applying the algorithm given in this section.

An algorithm for generating B-trees is described in the work by Gupta et al. [16]. It is based on backtrack search, and produces B-trees with worst case delay proportional to the output size. The order of generating B-trees becomes lexicographic if B-trees are coded as a B-tree sequence, defined in [5]. The algorithm [16] has constant expected delay in producing next B-tree, exclusive of the output, which is proven in the work by Belbaraka and Stojmenovic [5]. Using a decoding procedure, an algorithm that generates the B-tree data structure (meaning that the parent–children links are established) from given B-tree sequence can be designed, with constant average delay.

1.8 LISTING SUBSETS AND BITSTRINGS IN A GRAY CODE ORDER

It is sometimes desirable to generate all instances of a combinatorial object in such a way that successive instances differ as little as possible. An order of all instances that minimizes the difference between any two neighboring instances is called *minimal change order*. Often the generation of objects in minimal change order requires complicated and/or computationally expensive procedures. When new instances are generated with the least possible changes (by a single insertion of an element, single deletion or single replacement of one element by another, interchange of two elements, updating two elements only, etc.), corresponding sequences of all instances of a combinatorial objects are refereed to as *Gray codes*. In addition, the same property must be preserved when going from the last to first sequence. In most cases, there is no difference between minimal change and Gray code orders. They may differ when for a given combinatorial object there is no known algorithm to list all instances in Gray code order. The best existing algorithm (e.g., one in which two instances differ at two positions whereas instances may differ in one position only) then is referred to achieving minimal change order but not in Gray code order.

We describe a procedure for generating subsets in binary notation, which is equivalent to generating all bitstrings of given length. It is based on a backtrack method and sequence comparison rule. Let $e_1 = 0$ and $e_i = x_1 + x_2 + \cdots + x_{i-1}$ for $1 < i \leq n$. Then the sequence that follows $x_1x_2\ldots x_n$ is $x_1x_2\ldots x_{i-1}x_i'x_{i+1}\ldots x_n$, where i is the largest index such that $e_i + x_i$ is even and $'$ is complement function

($0' = 1,\ 1' = 0$; also $x' = x + 1 \bmod 2$).

```
read( n);
for i ← 0 to n do {x_i ← 0; e_i ← 0};
repeat
        print out x_1, x_2, ..., x_n;
        i ← n;
        while x_i + e_i is odd do i ← i − 1;
        x_i ← x'_i;
        for j ← i + 1 to n do e_j ← e'_j
until i = 0 .
```

The procedure has $O(n)$ worst case delay and uses no large integers. We will prove that it generates Gray code sequences with constant average delay. The element x_i changes 2^{i-1} times in the algorithm, and each time it makes $n - i + 1$ steps back and forth to update x_i. Since the time for each step is bounded by a constant, the time to generate all Gray code sequences is $\sum_{i=1}^{n} c2^{i-1}(n - i + 1)$. The average delay is obtained when the last number is divided by the number of generated sequences 2^n, and is therefore

$$c\sum_{i=1}^{n} 2^{-n+i-1}(n - i + 1) = c\sum_{i=1}^{n} 2^{-i}i = c\left(2 - \frac{n}{2^n} - \frac{1}{2^{n-1}}\right) < 2c.$$

An algorithm for generating subsets in the binary notation in the binary reflected Gray code that has constant delay in the worst case is described in the work by Reingold et al. [30]. Efficient loopless algorithms for generating k-ary trees are described in the Xiang et al. [41].

1.9 GENERATING PERMUTATIONS IN A MINIMAL CHANGE ORDER

In this section we consider generating the permutations of $\{p_1, p_2, \ldots, p_n\}$ $(p_1 < \cdots < p_n)$ in a minimum change order. We present one that is based on the idea of adjacent transpositions, and is independently proposed by Johnson [18] and Trotter [39]. It is then simplified by Even [14]. In the work by Even [14], a method by Ehrlich is presented, which has constant delay. The algorithm presented here is a further modification of the technique, also having constant delay, and suitable as a basis for a parallel algorithm [36].

The algorithm is based on the idea of generating the permutations of $\{p_1, p_2, \ldots, p_n\}$ from the permutations of $\{p_1, p_2, \ldots, p_{n-1}\}$ by taking each such permutation and inserting p_n in all n possible positions of it. For example, taking the permutation $p_1 p_2 \ldots p_{n-1}$ of $\{p_1, p_2, \ldots, p_{n-1}\}$ we get n permutations of $\{p_1, p_2, \ldots, p_n\}$ as follows:

$$
\begin{array}{l}
p_1\ p_2\ \cdots\ p_{n-2}\ p_{n-1}\ p_n \\
p_1\ p_2\ \cdots\ p_{n-2}\ p_n\quad p_{n-1} \\
p_1\ p_2\ \cdots\ p_n\quad p_{n-2}\ p_{n-1} \\
\quad\quad\quad\cdot \\
\quad\quad\quad\cdot \\
\quad\quad\quad\cdot \\
p_n\ p_1\ \cdots\ p_{n-3}\ p_{n-2}\ p_{n-1}.
\end{array}
$$

The nth element sweeps from one end of the $(n-1)$ -permutation to the other by a sequence of adjacent swaps, producing a new n-permutation each time. Each time the nth element arrives at one end, a new $(n-1)$ -permutation is needed. The $(n-1)$-permutations are produced by placing the $(n-1)$th element at each possible position within an $(n-2)$ -permutation. That is, by applying the algorithm recursively to the $(n-1)$ elements.

The first permutation of the set $\{p_1, p_2, \ldots, p_n\}$ is $p_1, p_2, \ldots, p_n$. Assign a *direction* to every element, denoted by an arrow above the element. Initially all arrows point to the left. Thus if the permutations of $\{p_1, p_2, p_3, p_4\}$ are to be generated, we would have

$$\overleftarrow{p_1}\ \overleftarrow{p_2}\ \overleftarrow{p_3}\ \overleftarrow{p_4}.$$

Now an element is said to be *mobile* if its direction points to a smaller adjacent neighbor. In the above example, p_2, p_3 and p_4 are mobile, while in

$$\vec{p}_3\ \overleftarrow{p_2}\ \overleftarrow{p_1}\ \vec{p}_4$$

only p_3 is mobile. The algorithm is as follows:

While there are mobile elements **do**
(i) find the largest mobile element; call it p_m
(ii) reverse the direction of all elements larger than p_m
(iii) switch p_m with the adjacent neighbor to which its direction points
endwhile.

The straightforward implementation of the algorithm leads to an algorithm that exhibits a linear time delay. The algorithm is modified to achieve a constant delay. After initial permutation, the following steps are then repeated until termination:

1. Move element p_n to the left, by repeatedly exchanging it with its left neighbor, and do (i) and (ii) in the process.
2. Generate the next permutation of $\{p_1, p_2, \ldots, p_{n-1}\}$ (i.e., do step (iii)).
3. Move element p_n to the right, by repeatedly exchanging it with its right neighbor, and do (i) and (ii) in the process.

4. Generate the next permutation of $\{p_1, p_2, \ldots, p_{n-1}\}$ (i.e., do step (iii)).

For example, permutations of {1, 2, 3, 4} are generated in the following order:

1234, 1243, 1423, 4123	move element 4 to the left
4132	132 is the next permutation of 123, with 3 moving to the left
1432, 1342, 1324	move 4 to the right
3124	312 is the next permutation following 132, with 3 moving to the left
3142, 3412, 4312	4 moves to the left
4321	321 is the next permutation following 312; 2 in 12 moves to the left
3421, 3241, 3214	4 moves to the right
2314	231 follows 321, where 3 moves to the right
2341, 2431, 4231	4 moves to the left
4213	213 follows 231, 3 moved to the right
2413, 2143, 2134	4 moves to the right.

The constant delay is achieved by observing that the mobility of p_n has a regular pattern (moves $n-1$ times and then some other element moves once). It takes $n-1$ steps to move p_n to the left or right while (i), (ii), and (iii) together take $O(n)$ time. Therefore, if steps (i), (ii), and (iii) are performed after p_n has already finished moving in a given direction, the algorithm will have constant average delay. If the work in steps (i) and (ii) [step (iii) requires constant time] is evenly distributed between consecutive permutations, the algorithm will achieve constant worst case delay. More precisely, finding largest mobile element takes $n-1$ steps, updating directions takes also $n-1$ steps. Thus it suffices to perform two such steps per move of element p_n to achieve constant delay per permutation.

The current permutation is denoted $d_1, d_2, \ldots, d_n$. The direction is stored in a variable a, where $a_i = -1$ for left and $a_i = 1$ for right direction. When two elements are interchanged, their directions are also interchanged implicitly. The algorithm terminates when no mobile element is found. For algorithm conciseness, we assume that two more elements p_0 and p_{n+1} are added such that $p_0 < p_1 < \ldots < p_n < p_{n+1}$. Variable i is used to move p_n from right to left ($i = n, n-1, \ldots, 2$) or from left to right ($i = 1, 2, \ldots, n-1$). The work in steps (i) and (ii) is done by two "sweeping" variables l (from left to right) and r (from right to left). They update the largest mobile elements dlm and drm, respectively, and their indices lm and rm, respectively, that they detect in the sweep. When they "meet" ($l = r$ or $l = r - 1$) the largest mobile element dlm and its index lm is decided, and the information is broadcast (when $l > r$) to all other elements who use it to update their directions. Obviously the sweep of variable i coincides with either the sweep of l or sweep of r. For clarity, the code below considers these three sweeps separately. The algorithm works correctly for $n>2$.

```
procedure output;
        { for s ← 1 to n do write( d[s]); writeln}

procedure exchange ( c, b: integer);
        { ch ← d[c + b]; d[c + b] ← d[c]; d[c] ← ch; ch ← a[c + b];
a[c + b] ← a[c]; a[c] ← ch };

procedure updatelm; {
        l ← l + 1; if (d[l] = p_n) or (d[l + dir] = p_n) then l ← l + 1;
        if l > r then {
                if d[l − 1] ≠ p_n then l1 ← l − 1 else l1 ← l − 2;
                if d[l + 1] ≠ p_n then l2 ← l + 1 else l2 ← l + 2;
                if (((a[l] = −1) and (d[l1] < d[l])) or ((a[l] = 1) and
(d[l2] < d[l]))) and (d[l]>dlm)
                        then {lm ← l; dlm ← d[l]};};
        if ((l = r) or (l = r − 1)) and (drm>dlm) then {lm ← rm;
dlm ← drm};
        if (l>r) and (d[r]>dlm) then a[r] ← −a[r];
        r ← r − 1; if (d[r] = p_n) or (d[r + dir] = p_n) then r ← r − 1;
        if l < r then {
                if d[r − 1] ≠ p_n then l1 ← r − 1 else l1 ← r − 2;
                if d[r − 1] ≠ p_n then l2 ← r + 1 else l2 ← r + 2;
                if (((a[r] = −1) and (d[l1] < d[r])) or
((a[r] = 1) and (d[l2] < d[r]))) and (d[r]>drm)
                        then { rm ← r; drm ← d[r] }; };
                if ((l = r) or (l = r − 1)) and (drm>dlm) then
{ lm ← rm; dlm ← drm };
                if (lεr) and (d[r]>dlm) then a[r] − a[r];
                exchange( i, dir);
                if i + dir = lm then lm ← i;
                if i + dir = rm then rm ← i;
                output; };
read( n); for i ← 0 to n + 1 do read p_i;
d[0] ← p_{n+1}; d[n + 1] ← p_{n+1}; d[n + 2] ← p_0;
for i ← 1 to n do { d[i] ← p_i; a[i] ← −1};
repeat
                output;
                l ← 1; r ← n + 1; lm ← n + 2; dlm ← p_0; rm ← n + 2;
drm ← p_0; dir ← −1;
                for i ← n downto 2 do updatelm;
                exchange (lm, a[lm]);
```

```
                output;
                l ← 1; r ← n + 1; lm ← n + 2; dlm ← p0;
   drm ← p0; rm ← n + 2; dir ← 1;
                for i ← 1 to n − 1 do updatelm;
                exchange (lm, a[lm]);
   until lm = n + 2.
```

1.10 RANKING AND UNRANKING OF COMBINATORIAL OBJECTS

Once the objects are ordered, it is possible to establish the relations between integers $1, 2, \ldots, N$ and all instances of a combinatorial object, where N is the total number of instances under consideration. The mapping of all instances of a combinatorial object into integers is called *ranking*. For example, let *f(X)* be ranking procedure for subsets of the set $\{1, 2, 3\}$. Then, in lexicographic order, $f() = 1$, $f(\{1\}) = 2$, $f(\{1, 2\}) = 3$, $f(\{1, 2, 3\}) = 4$, $f(\{1, 3\}) = 5$, $f(\{2\}) = 6$, $f(\{2, 3\}) = 7$ and $f(\{3\}) = 8$. The inverse of ranking, called *unranking*, is mapping of integers $1, 2, \ldots, N$ to corresponding instances. For instance, $f^{-1}(4) = \{1, 2, 3\}$ in the last example.

The objects can be enumerated in a systematic manner, for some combinatorial classes, so that one can easily construct the sth element in the enumeration. In such cases, an unbiased generator could be obtained by generating a random number s in the appropriate range *(1,N)* and constructing the sth object. In practice, random number procedures generate a number r in interval [0,1); then $s = \lceil rN \rceil$ is required integer.

Ranking and unranking functions exist for almost every kind of combinatorial objects, which has been studied in literature. They also exist for some objects listed in minimal change order. The minimal change order has more use when all instances are to be generated since in this case either the time needed to generate is less or the minimal change order of generating is important characteristics of some applications. In case of generating an instance at random, the unranking functions for minimal change order is usually more sophisticated than the corresponding one following lexicographic order. We use only lexicographic order in ranking and unranking functions presented in this chapter.

In most cases combinatorial objects of given kind are represented as integer sequences. Let $a_1a_2 \ldots a_m$ be such a sequence. Typically each element a_i has its range that depends on the choice of elements $a_1, a_2, \ldots, a_{i-1}$. For example, if $a_1a_2 \ldots a_m$ represents a *(m,n)*-combination out of $\{1, 2, \ldots, n\}$ then $1 \le a_1 \le n - m + 1$, $a_1 < a_2 \le n - m + 2, \ldots,\ a_{m-1} < a_m \le n$. Therefore element a_i has $n - m + 1 - a_{i-1}$ different choices.

Let $N(a_1, a_2, \ldots, a_i)$ be the number of combinatorial objects of given kind whose representation starts with $a_1a_2 \ldots a_i$. For instance, in the set of (4,6)-combinations we have $N(2, 3) = 3$ since 23 can be completed to (4,6)-combination in three ways: 2345, 2346, and 2356.

To find the rank of an object $a_1 a_2 \ldots a_m$, one should find the number of objects preceding it. It can be found by the following function:

```
function rank(a_1, a_2, ..., a_m )
  rank ← 1 ;
    for i ← 1 to m do
        for each x < a_i
            rank ← rank + N(a_1, a_2, ..., a_{i-1}, x).
```

Obviously in the last **for** loop only such values x for which $a_1 a_2 \ldots a_{i-1} x$ can be completed to represent an instance of a combinatorial object should be considered (otherwise adding 0 to the rank does not change its value). We now consider a general procedure for unranking. It is the inverse of ranking function and can be calculated as follows.

```
procedure unrank ( rank, n, a_1, a_2, ..., a_m)
    i ← 0 ;
    repeat
            i ← i + 1;
            x ← first possible value;
            while N(a_1, a_2, ..., a_{i-1}, x) ≤ rank do
                    {rank ← rank − N(a_1, a_2, ..., a_{i-1}, x);
                    x ← next possible value};
            a_i ← x
    until rank = 0;
a_1 a_2 ... a_m ← lexicographically first object starting by a_1 a_2 ... a_i.
```

We now present ranking and unranking functions for several combinatorial objects. In case of ranking combinations out of $\{1, 2, \ldots, n\}$, x is ranged between $a_{i-1} + 1$ and $a_i - 1$. Any (m, n)-combination that starts with $a_1 a_2 \ldots a_{i-1} x$ is in fact a $(m - i, n - x)$- combination. The number of such combinations is $C(m - i, n - x)$. Thus the ranking algorithm for combinations out of $\{1, 2, \ldots, n\}$ can be written as follows ($a_0 = 0$ in the algorithm):

```
function rankcomb (a_1, a_2, ..., a_m)
    rank ← 1 ;
    for i ← 1 to m do
        for x ← a_{i-1} + 1 to a_i − 1 do
            rank ← rank + C(m − i, n − x).
```

In lexicographic order, $C(4, 6) = 15$ (4,6)-combinations are listed as 1234, 1235, 1236, 1245, 1246, 1256, 1345, 1346, 1356, 1456, 2345, 2346, 2356, 2456, 3456. The rank of 2346 is determined as $1 + C(4 - 1, 6 - 1) + C(4 - 4, 6 - 5) = 1 + 10 + 1 = 12$ where last two summands correspond to combinations that start with 1 and 2345, respectively. Let us consider a larger example. The rank of 3578 in

(4,9)-combinations is $1 + C(4-1, 9-1) + C(4-1, 9-2) + C(4-2, 9-4) + C(4-3, 9-6) = 104$ where four summands correspond to combinations starting with 1, 2, 34, and 356, respectively.

A simpler formula is given in the work by Lehmer [21]: the rank of combination $a_1a_2 \ldots a_m$ is $C(m, n) - \sum_{j=1}^{m} C(j, n-1-a_{m-j+1})$. It comes from the count of the number of combinations that follow $a_1a_2 \ldots a_m$ in lexicographic order. These are all combinations of j out of elements $\{a_{m-j+1}+1,\ a_{m-j+1}+2, \ldots, a_n\}$, for all $j,\ 1 \leq j \leq m$. In the last example, combinations that follow 3578 are all combinations of 4 out of $\{4, 5, 6, 7, 8, 9\}$, combinations with first element 3 and three others taken from $\{6, 7, 8, 9\}$, combinations which start with 35 and having two more elements out of set $\{8, 9\}$ and combination 3579.

The function calculates the rank in two nested **for** loops while the formula would require one **for** loop. Therefore general solutions are not necessarily best in the particular case. The following unranking procedure for combinations follows from general method.

```
procedure unrankcomb (rank, n, a1, a2, ..., am)
    i ← 0; a0 ← 0;
    repeat
        i ← i + 1;
        x ← a(i−1) + 1;
        while C(m − i, n − x) ≤ rank do
                {rank ← rank − C(m − i, n − x); x ← x + 1};
        ai ← x
    until rank = 0;
    for j = i + 1 to m do aj ← n − m + j.
```

What is 104th (4,9)-combination? There are $C(3, 8) = 56$ (4,9)-combinations starting with a 1 followed by $C(3, 7) = 35$ starting with 2 and $C(3, 6) = 20$ starting with 3. Since $56 + 35 \leq 104$ but $56 + 35 + 20 > 104$ the requested combination begins with a 3, and the problem is reduced to finding $104 - 56 - 35 = 13$th (3,6)-combination. There are $C(2, 5) = 10$ combinations starting with 34 and $C(2, 4) = 6$ starting with a 5. Since $13 > 10$ but $13 < 10 + 6$ the second element in combination is 5, and we need to find $13 - 10 = 3$rd (2,4)-combination out of $\{6, 7, 8, 9\}$, which is 78, resulting in combination 3578 as the 104th (4,9)-combination.

We also consider the ranking of subsets. The subsets in the set and in the binary representation are listed in different lexicographic orders. In binary representation, the ranking corresponds to finding decimal equivalent of an integer in binary system. Therefore the rank of a subset $b_1, b_2, \ldots, b_n$ is $b_n + 2b_{n-1} + 4b_{n-2} + \cdots + 2^{n-1}b_1$. For example, the rank of 100101 is $1 + 4 + 32 = 37$. The ranks are here between *0* and $2^n - 1$ since in many applications empty subset (here with rank 0) is not taken into consideration. The ranking functions can be generalized to variations out of $\{0, 1, \ldots, m-1\}$ by simply replacing all "2" by "m" in the rank expression. It corresponds to decimal equivalent of a corresponding number in number system with base m.

Similarly, the unranking of subsets in binary representation is equivalent to converting a decimal number to binary one, and can be achieved by the following procedure that uses the mod or remainder function. The value rank mod 2 is 0 or 1, depending whether rank is even or odd, respectively. It can be generalized for m-variations if all "2" are replaced by "m".

function unranksetb($n,\ a_1a_2\ldots a_m$)
  $rank \leftarrow m;\ a_0 \leftarrow 0;$
  for $i \leftarrow m$ **downto** 1 **do**
    $\{b_i \leftarrow rank \bmod 2;\ rank \leftarrow rank - b_i 2^{n-i}\}.$

In the set representation, the rank of n-subset $a_1a_2\ldots a_m$ is found by the following function from the work by Djokić et al. [10].

function rankset($n,\ a_1a_2\ldots a_m$)
  $rank \leftarrow m;\ a_0 \leftarrow 0;$
  for $i \leftarrow 1$ **to** $m - 1$ **do**
    for $j \leftarrow a_i + 1$ **to** $a_{i+1} - 1$ **do**
      $rank \leftarrow rank + 2^{n-j}.$

The unranking function [10] gives n-subset with given rank in both representations but the resulting binary string $b_1b_2\ldots b_n$ is assigned its rank in the lexicographic order of the set representation of subsets.

function unranksets($rank,\ n,\ a_1a_2\ldots a_m$)
  $m \leftarrow 0;\ k \leftarrow 1;$ **for** $i \leftarrow 1$ **to** n **do** $b_i \leftarrow 0$;
  repeat
    if $rank \le 2^{n-k}$ **then** $\{b_k \leftarrow 1;\ m \leftarrow m+1;\ a_m \leftarrow k\};$
    $rank \leftarrow rank - (1 - b_k)2^{n-k} - b_k;$
    $k \leftarrow k + 1$
  until $k > n$ or $rank = 0$.

As noted in the work by Djokić [10], the rank of a subset $a_1a_2\ldots a_m$ among all (m, n)-subsets is given by

$$\text{ranks}(a_1a_2\ldots a_m) = \text{rankcomb}(a_1a_2\ldots a_m) + \text{rankcomb}(a_1a_2\ldots a_{m-1}) + \cdots + \text{rankcomb}(a_1a_2) + \text{rankcomb}(a_1).$$

Let $L(m,\ n) = C(1, n) + C(2,\ n) + \cdots + C(n,\ m)$ be the number of $(m,\ n)$-subsets. The following unranking algorithm [10] returns the subset $a_1a_2\ldots a_m$ with given rank.

function unranklim ($rank,\ n,\ m,\ a_1a_2\ldots a_r$)
  $r \leftarrow 0;\ i \leftarrow 1;$
  repeat
    $s \leftarrow t - 1 - L(m - r - 1,\ n - i);$

```
        if s>0 then t ← s else {r ← r + 1; a_r ← i; rank ← rank − 1};
        i ← i + 1
    until i = n + 1 or rank = 0.
```

Note that the (*m, n*)-subsets in lexicographic order also coincide with a minimal change order of them. This is a rare case. Usually it is trivial to show that lexicographic order of instances of an object is not a minimal change order.

Ranking and unranking functions for integer compositions can be described by using the relation between compositions and either subsets or combinations (discussed above).

A ranking algorithm for n-permutations is as follows [21]:

```
function rankperm(a_1 a_2 ... a_n )
    rank ← 1 ;
    for i ← 1 to n do
            rank ← rank + k(n − i)! where k = |{1, 2, ..., a_i − 1}\
{a_1, a_2, ..., a_{i−1}}|.
```

For example, the rank of permutation 35142 is $1 + 2 \times 4! + 3 \times 3! + 1 \times 1! = 68$ where permutations starting with 1, 2, 31, 32, 34, and 3512 should be taken into account. The unranking algorithm for permutations is as follows [21]. Integer division is used (i.e., $13/5 = 2$).

```
procedure unrankperm(rank, n, a_1 a_2 ... a_n)
    for i ← 1 to n do {
            ⌊k ← (rank − 1)/(n − i)!⌋;
            a_i ← kth element of {1, 2, ..., n}\{a_1, a_2, ..., a_{i−1}};
            rank ← rank − (k − 1)(n − i)!}.
```

The number of instances of a combinatorial object is usually exponential in size of objects. The ranks, being large integers, may need $O(n)$ or similar number of memory location to be stored and also $O(n)$ time for the manipulation with them. Avoiding large integers is a desirable property in random generation in some cases. The following two sections offer two such approaches.

1.11 RANKING AND UNRANKING OF SUBSETS AND VARIATIONS IN GRAY CODES

In a Gray code (or minimal change) order, instances of a combinatorial object are listed such that successive instances differ as little as possible. In this section we study Gray codes of subsets in binary representation. Gray code order of subsets is an ordered cyclic sequence of 2^n n-bit strings (or codewords) such that successive codewords differ by the complementation of a single bit. If the codewords are considered to be vertices of an n-dimensional binary cube, it is easy to conclude that Gray code order of subsets corresponds to a Hamiltonian path in the binary cube. We will occasionally refer in the sequel to nodes of binary cubes instead of subsets. Although a binary cube may have various Hamiltonian paths, we will define only one such path, called the binary-reflected Gray code [17] that has a number of advantages, for example, easy generation and traversing a subcube in full before going to other subcube. The (binary reflected) Gray code order of nodes of n-dimensional binary cube can be defined in the following way:

- For $n = 1$ the nodes are numbered $g(0) = 0$ and $g(1) = 1$, in this order,
- If $g(0), g(1), \ldots, g(2^n - 1)$ is the Gray code order of nodes of an n-dimensional binary cube, then $g(0) = 0g(0)$, $g(1) = 0g(1), \ldots, g(2^n - 1) = 0g(2^n - 1)$, $g(2^n) = 1g(2^n - 1)$, $g(2^n + 1) = 1g(2^n - 2), \ldots, g(2^{n+1} - 2) = 1g(1)$, $g(2^{n+1} - 1) = 1g(0)$ is a Gray code order of nodes of a $(n + 1)$-dimensional binary cube.

As an example, for $n = 3$ the order is $g(0) = 000$, $g(1) = 001$, $g(2) = 011$, $g(3) = 010$, $g(4) = 110$, $g(5) = 111$, $g(6) = 101$, $g(7) = 100$. First, let us see how two nodes u and v can be compared in Gray code order. We assume that a node x is represented by a bitstring $x_1 \geq x_2 \ldots x_n$. This corresponds to decimal node *address* $x = 2^{n-1}x_1 + 2^{n-2}x_2 + \cdots + 2x_{n-1} + x_n$ where $0 \leq x \leq 2^n - 1$. Let i be the most significant (or leftmost) bit where u and v differ, that is, $u[l] = v[l]$ for $l < i$ and $u[i] \neq v[i]$. Then $u < v$ if and only if $u[1] + u[2] + \cdots + u[i]$ is an even number. For instance, $11100 < 10100 < 10110$.

The above comparison method gives a way to find Gray code address t of a node u (satisfying $g(t) = u$), using the following simple procedure; it ranks the Gray code sequences.

```
procedure rank_GC(n, u, t);
    sum ← 0; t ← 0;
    for l ← 1 to n do {
        sum ← sum + u[l];
        if sum is odd then t ← t + 2^(n−l) }.
```

The inverse operation, finding the binary address u of node having Gray code address t $(0 \leq t \leq 2^n - 1)$, can be performed by the following procedure; it unranks the Gray code sequences.

```
procedure unrank_GC( n,u,t);
    sum ← 0; q ← t; size ← 2^n;
    for l ← 1 to n do {
        size ← size/2;
        if q ≥ size then {q ← q − size; s ← 1} else s ← 0 ;
        if sum + s is even then u[l] ← 0 else u[l] ← 1;
        sum ← sum + u[l]}.
```

The important property of the Gray code order is that corresponding nodes of a binary cube define an edge of the binary cube whenever they are neighbors in the Gray code order (this property is not valid for the lexicographic order $0, 1, 2, \ldots, 2^n - 1$ of binary addresses).

The reflected Gray code order for subsets has been generalized for variations [7,15]. Gray codes of variations have application in analog to digital conversion of data.

We establish a n-ary reflected Gray code order of variations as follows. Let $x = x_1 \geq x_2 \ldots x_m$ and $y = y_1 y_2 \ldots y_m$ be two variations. Then $x < y$ iff there exist i, $0 \leq i \leq m$, such that $x_j = y_j$ for $j < i$ and either $x_1 + x_2 + \ldots + x_{i-1}$ is even and $x_i < y_i$ or $x_1 + x_2 + \cdots + x_{i-1}$ is odd and $x_i > y_i$. We now prove that the order is a minimal change order. Let x and y be two consecutive variations in given order, $x < y$, and let $x_j = y_j$ for $j < i$ and $x_i \neq y_i$. There are two cases. If $x_i < y_i$ then $X_i = x_1 + x_2 + \cdots + x_{i-1}$ is even and $y_i = x_i + 1$. Thus X_{i+1} and Y_{i+1} have different parity, since $Y_{i+1} = X_{i+1} + 1$. It means that either $x_{i+1} = y_{i+1} = 0$ or $x_{i+1} = y_{i+1} = n - 1$ (the $(i + 1)$th element in x is the maximum at that position while the $(i + 1)$ –the element in y is the minimum at given position, and they are the same because of different parity checks). Similarly we conclude $Y_j = X_j + 1$ and $x_j = y_j$ for all $j > i + 1$. The case $x_i > y_i$ can be analyzed in analogous way, leading to the same conclusion.

As an example, 3-ary reflected Gray code order of variations out of $\{0, 1, 2\}$ is as follows (the variations are ordered columnwise):

000 122 200
001 121 201
002 120 202
012 110 212
011 111 211
010 112 210
020 102 220
021 101 221
022 100 222.

It is easy to check that, at position $i (1 \leq i \leq m)$, each element repeats n^{m-i} times. The repetition goes as follows, in a cyclic manner: 0 repeats n^{m-i} times, 1 repeats n^{m-i} times, ..., $n - 1$ repeats n^{m-i} times, and then these repetitions occur in reverse order, that is $n - 1$ repeats n^{m-i} times, ..., 0 repeats n^{m-i} times.

Ranking and unranking procedures for variations in the n-ary reflected Gray code are described in the work by Flores [15].

1.12 GENERATING COMBINATORIAL OBJECTS AT RANDOM

In many cases (e.g., in probabilistic algorithms), it is useful to have means of generating elements from a class of combinatorial objects uniformly at random (an unbiased generator). Instead of testing new hypothesis on all objects of given kind, which may be time consuming, several objects chosen at random can be used for testing, and likelihood of hypothesis can be established with some certainty. There are several ways of choosing a random object of given kind. All known ways are based on the correspondence between integer or real number(s) and combinatorial objects. This means that objects should be ordered in a certain fashion. We already described two general ways for choosing a combinatorial object at random. We now describe one more way, by using *random number series*. This method uses a series of random numbers in order to avoid large integers in generating a random instance of an object. Most known techniques in fact generate a series of random numbers. This section will present methods for generating random permutations and integer partitions. A random subset can easily be generated by flipping coin for each of its elements.

1.12.1 Random Permutation and Combination

There exist a very simple idea of generating a random permutation of $A = \{a_1, \ldots, a_n\}$. One can generate an array $x_1, x_2, \ldots, x_n$ of random numbers, sort them, and obtain the destination indices for each element of A in a random permutation. The first m elements of the array can be used to determine a random (m, n)-combination (the problem of generating combinations at random is sometimes called random sampling). Although very simple, the algorithm has $O(n \log n)$ time complexity [if random number generation is allowed at most $O(\log n)$ time]. We therefore describe an alternative solution that leads to a linear time performance. Such techniques for generating permutations of $A = \{a_1, \ldots, a_n\}$ at random first appeared in the works by the Durstenfeld [8] and Hoses [24], and repeated in the works by Nijeshius [25] and Reingold [30]. The algorithm uses a function random (x) that generates a random number x from interval (0,1), and is as follows.

for $i \leftarrow 1$ **to** $n - 1$ **do** {
  random(x_i);
  $c_i \lfloor x_i(n - i + 1) \rfloor + 1$;
  $j \leftarrow i - 1 + c_i$;
  exchange a_i with a_j }.

As an example, we consider generating a permutation of $\{a, b, c, d, e, f\}$ at random. Random number $x_1 = 0.7$ will choose $\lfloor 6 \times 0.7 \rfloor + 1 = 5$th element e as the first element in a random permutation, and decides the other elements considering the set $\{b, c, d, a, f\}$ (e exchanged with a). The process is repeated: another random number, say $x_2 = 0.45$, chooses $\lfloor 5 \times 0.45 \rfloor + 1 = 3$rd element d from $\{b, c, d, a, f\}$ to be the second element in a random permutation, and b and d are exchanged. Thus, random permutation begins with e, d, and the other elements are decided by continuing same process on the set $\{c, b, a, f\}$.

Assuming that random number generator takes constant time, the algorithm runs in linear time. The same algorithm can be used to generate combinations at random. The first m iterations of the **for** loop determine (after sorting, if such output is preferable) a combination of m out of n elements.

Uniformly distributed permutations cannot be generated by sampling a finite portion of a random sequence and the standard method [8] does not preserve randomness of the x-values due to computer truncations. Truncation problems appear with other methods as well.

1.12.2 Random Integer Partition

We now present an algorithm from the work by Nijenhius and Wilf [26] that generates a random integer partition. It uses the distribution of the number of partitions *RP(n,m)* of n into parts not greater than m.

First, we determine the first part. An example of generating random partition of 12 will be easier to follow than to show formulas. Suppose a random number generator gives us $r_1 = 0.58$. There are 77 partitions of 12. In lexicographic order, the random number should point to $0.58 \times 77 = 44.66$th integer partition. We want to avoid rounding and unranking here. Thus, we merely determine the largest part such. Looking at the distribution *RP(12,m)* of partitions of 12 (Section 1.2), we see that all integer partitions with ranks between 35 and 47 have the largest part equal to 5. What else we need in a random partition of 12? We need a random partition of $12 - 5 = 7$ such that its largest part is 5 (the second part cannot be larger than the first part). There are $\mathrm{RP}(7, 5) = 13$ such partitions. Let the second random number be $r_2 = 0.78$. The corresponding partition of 7 has the rank $0.78 \times 13 = 10.14$. Partitions of 7 ranked between 9 and 11 have the largest part equal to 4. It remains to find a random partition of $7 - 4 = 3$ with largest part 4 (which in this case is not a real restriction). There are $\mathrm{RP}(3, 3) = 3$ partitions as candidates let $r_3 = 0.20$. Then $0.20 \times 3 = 0.6$ points to the third (and remaining) parts of size 1. However, since the random number is taken from open interval (0,1), in our scheme the partition $n = n$ will never be chosen unless some modification to our scheme is made. Among few possibilities, we choose that the value < 1 as the rank actually points to the available partition with the maximal rank. Thus, we decide to choose partition $3 = 3$, and the random partition of 12 that we obtained is $12 = 5 + 4 + 3$.

An algorithm for generating random rooted trees with prescribed degrees (where the number of nodes of each down degree is specified in advance) is described in the work by Atkinson [3]. A linear time algorithm to generate binary trees uniformly at random, without dealing with large integers is given in the work by Korsch [20]. An algorithm for generating valid parenthesis strings (each open parenthesis has its matching closed one and vice versa) uniformly at random is described in the work by Arnold and Sleep [2]. It can be modified to generate binary trees in the bitstring notation at random.

1.13 UNRANKING WITHOUT LARGE INTEGERS

Following the work by Stojmenovic [38], this section describes functions mapping the interval $[0 \ldots 1)$ into the set of combinatorial objects of certain kind, for example, permutations, combinations, binary and t-ary trees, subsets, variations, combinations with repetitions, permutations of combinations, and compositions of integers. These mappings can be used for generating these objects at random, with equal probability of each object to be chosen. The novelty of the technique is that it avoids the use of very large integers and applies the random number generator only once. The advantage of the method is that it can be applied for both random object generation and dividing all objects into desirable sized groups.

We restrict ourselves to generating only one random number to obtain a random instance of a combinatorial object but request no manipulation with large integers. Once a random number g in [0,1) is taken, it is mapped into the set of instances of given combinatorial object by a function $f(g)$ in the following way. Let N be the number of all instances of a combinatorial object. The algorithm finds the instance x such that the ratio of the number of instances that precede x and the total number of instances is $\leq g$. In other words, it finds the instance $f(g)$ with the ordinal number $\lfloor gN \rfloor + 1$. In all cases that will be considered in this section, each instance of given combinatorial object may be represented as a sequence $x_1 \ldots x_m$, where x_i may have integer values between 0 and n (m and n are two fixed numbers), subject to constraints that depend on particular case.

Suppose that the first $k-1$ elements in given instance are fixed, that is, $x_i = a_i$, $1 \leq i < k$. We call them $(k-1)$-fixed instances. Let $a'_1 < \cdots < a'_h$ be all possible values of x_k of a given $(k-1)$-fixed instance. By $S(k, u)$, $S(k, \leq u)$, and $S(k, \geq u)$, we denote the ratio of the number of $(k-1)$-fixed instances for which $x_k = a'_u(x_k \leq a'_u$, and $x_k \geq a'_u$ respectively) and the number of $(k-1)$-fixed instances. In other words, these are the probabilities (under uniform distribution) that an instance for which $x_i = a_i$, $1 \leq i < k$, has the value in variable x_k which is $= a'_u$, $\leq a'_u$, and $\geq a'_u$, respectively.

Clearly, $S(k, u) = S(k, \leq u) - S(k, \leq u-1)$ and $S(k, \geq u) = 1 - S(k, \leq u-1)$. Thus

$$\frac{S(k, u)}{S(k, \geq u)} = \frac{S(k, \leq u) - S(k, \leq u-1)}{1 - S(k, \leq u-1)}.$$

Therefore

$$S(k, \leq u) = S(k, \leq u-1) + (1 - S(k, \leq u-1))\frac{S(k, u)}{S(k, \geq u)}.$$

Our method is based on the last equation. The large numbers can be avoided in cases when $S(k, u)/S(k, \geq u)$ is explicitly found and is not a very large integer. This condition is satisfied for combinations, permutations, t-ary trees, variations, subsets, and other combinatorial objects.

Given g from $[0, \ldots, 1)$, let l be chosen such that $S(1, \leq u-1) < g \leq S(1, \leq u)$. Then $x_1 = a'_u$ and the first element of combinatorial object ranked g is decided. To decide the second element, the interval $[S(1, \leq u-1) \ldots S(1, \leq u))$ containing g can be linearly mapped to interval $[0 \ldots 1)$ to give the new value of g as follows:

$$g \leftarrow \frac{g - S(1, \leq u-1)}{S(1, \leq u) - S(1, \leq u-1)}.$$

The search for the second element proceeds with the new value of g. Similarly the third, ..., mth elements are found. The algorithm can be written formally as follows, where $p\prime$ and p stand for $S(k, \leq u-1)$ and $S(k, \leq u)$, respectively.

```
procedure object( m, n, g);
  p′ ← 0;
  for k ← 1 to m do {
      u ← 1;
      p ← S(k, 1);
      while p ≤ g do {
          p′ ← p;
          u ← u + 1;
          p ← p′ + (1 − p′) S(k, u) / S(k, ≥ u) }
      x_k ← a′_u;
      g ← (g − p′) / (p − p′) }.
```

Therefore the technique does not involve large integers iff $S(k, u)/S(k, \geq u)$ is not a large integer for any k and u in the appropriate ranges (note that $S(k, \geq 1) = 1$).

The method gives theoretically correct result. However, in practice the random number g and intermediate values of p are all truncated. This may result in computational imprecision for larger values of m or n. The instance of a combinatorial object obtained by a computer implementation of above procedure may differ from the theoretically expected one. However, the same problem is present with other known methods (as noted in the previous section) and thus this method is comparable with others in that sense. Next, in applications, randomness is practically preserved despite computational errors.

1.13.1 Mapping [0 . . . 1) Into the Set of Combinations

Each (m, n)-combination is specified as an integer sequence $x_1, \ldots, x_m$ such that $1 \leq x_1 < \cdots < x_m \leq n$. The mapping $f(g)$ is based on the following lemma. Recall that $(k\text{-}1)$-fixed combinations are specified by $x_i = a_i$, $1 \leq i < k$. Clearly, possible values for x_k are $a'_1 = a_{k-1} + 1, a'_2 = a_{k-1} + 2, \ldots, a'_h = n$ (thus $h = n - a_{k-1}$).

Lemma 1. The ratio of the number of $(k-1)$-fixed (m,n)-combinations for which $x_k = j$ and the number of $(k-1)$-fixed combinations for which $x_k \geq j$ is $(m-k+1)/(n-j+1)$ whenever $j > a_{k-1}$.

Proof. Let $y_{k-i} = x_i - j$, $k < i \leq n$. The $(k-1)$-fixed (m,n)-combinations for which $x_k = j$ correspond to $(m-k, n-j)$ -combinations $y_1, \ldots, y_{m-k}$, and their number is $C(m-k, n-j)$. Now let $y_{k-i+1} = x_i - j + 1, k \leq i \leq n$. The $(k-1)$-fixed combinations for which $x_k \geq j$ correspond to $(m-k+1, n-j+1)$ -combinations $y_1 \ldots y_{m-k+1}$, and their number is $C(m-k+1, n-j+1)$. The ratio in question is

$$\frac{C(m-k, n-j)}{C(m-k+1, n-j+1)} = \frac{m-k+1}{n-j+1}. \blacksquare$$

Using the notation introduced in former section for any combinatorial objects, let $u = j - a_{k-1}$. Then, from Lemma 1 it follows that

$$\frac{S(k, u)}{S(k, \geq u)} = \frac{m-k+1}{n-u-a_{k-1}+1}$$

for the case of (m,n)-combinations, and we arrive at the following procedure that finds the (m,n)-combination with ordinal number $\lfloor gC(m,n) \rfloor + 1$. The procedure uses variable j instead of u, for simplicity.

```
procedure combination( m,n,g);
  j ← 0; p' ← 0;
  for k ← 1 to m do {
        j ← j + 1;
        p ← (m − k + 1)/(n − j + 1);
        while p ≤ g do {
              p' ← p;
              j ← j + 1;
              p ← p' + (1 − p')(m − k + 1)/(n − j + 1) }
     x_k ← j;
     g ← (g − p')/(p − p') }.
```

A random sample of size m out of the set of n objects, that is, a random (m,n)-combination can be found by choosing a real number g in $[0, \ldots, 1)$ and applying the map $f(g) = \text{combination}(m,n,g)$.

Each time the procedure combination (m,n,g) enters for or while loop, the index j increases by 1; since j has n as upper limit, the time complexity of the algorithm is $O(n)$, that is, linear in n. Using the correspondences established in Chapter 1, the same procedure may be applied to the case of combinations with repetitions and compositions of n into m parts.

1.13.2 Random Permutation

Using the definitions and obvious properties of permutations, we conclude that, after choosing $k-1$ beginning elements in a permutation, each of the remaining $n-k+1$ elements has equal chance to be selected next. The list of unselected elements is kept in an array *remlist*. This greatly simplifies the procedure that determines the permutation $x_1 \ldots x_n$ with index $\lfloor gP(n) \rfloor + 1$.

```
procedure permutation( n,g);
for i ← 1 to n do remlist_i ← i;
for k ← 1 to n do {
        u ← ⌊g(n − k + 1)⌋ + 1;
        x_k ← remlist_u;
        for i ← u to n − k do remlist_i ← remlist_{i+1};
        g ← g(n − k + 1) − u + 1}.
```

The procedure is based on the same choose and exchange idea as the one used in the previous section but requires one random number generator instead of a series of n generators. Because the lexicographic order of permutations and the ordering of real numbers in $[0 \ldots 1)$ coincide, the list of remaining elements is kept sorted, which causes higher time complexity $O(n^2)$ of the algorithm.

Consider an example. Let $n = 8$ and $g = 0.1818$. Then $\lfloor 0.1818 * 8! \rfloor + 1 = 7331$ and the first element of 7331st 8-permutation is $u = \lfloor 0.1818 \times 8 \rfloor + 1 = 2$; the remaining list is 1,3,4,5,6,7,8 ($7331 - 1 \times 5040 = 2291$; this step is for verification only, and is not part of the procedure). The new value of g is $g = 0.1818 \times 8 - 2 + 1 = 0.4544$, and new u is $u = \lfloor 0.4544 \times 7 \rfloor + 1 = 4$; the second element is 4th one in the remaining list, which is 5; the remaining list is 1,3,4,6,7,8. Next update is $g = 0.4544 \times 7 - 3 = 0.1808$ and $u = \lfloor 0.1808 \times 6 \rfloor + 1 = 2$; the 3rd element is the 2nd in the remaining list, that is, 3; the remaining list is 1,4,6,7,8. The new iteration is $g = 0.1808 \times 6 - 1 = 0.0848$ and $u = \lfloor 0.0848 \times 5 \rfloor + 1 = 1$; the 4th element is 1st in the remaining list, that is, 1; the remaining list is 4,6,7,8. Further, $g = 0.0848 \times 5 = 0.424$ and $u = \lfloor 0.424 \times 4 \rfloor + 1 = 2$; the 5th element is 2nd in the remaining list, that is, 6; the new remaining list is 4,7,8. The next values ot g and u are $g = 0.424 \times 4 - 1 = 0.696$ and $u = \lfloor 0.696 \times 3 \rfloor + 1 = 3$; the 6th element is 3rd in the remaining list, that is, 8; the remaining list is 4,7. Finally, $g = 0.696 \times 3 - 2 = 0.088$ and $u = \lfloor 0.088 \times 2 \rfloor + 1 = 1$; the 7th element is 1st in the remaining list, that is, 4; now 7 is left, which is the last, 8th element. Therefore, the required permutation is 2,5,3,1,6,8,4,7.

All (m,n)-permutations can be obtained by taking all combinations and listing permutations for each combination. Such an order that is not lexicographic one, and (m,n)-permutations are in this case refereed to as the permutations of combinations. Permutation of combinations with given ordinal number can be obtained by running the procedure combination first, and continuing the procedure permutation afterwards, with the new value of g that is determined at the end of the procedure combination.

1.13.3 Random *t*-Ary Tree

The method requires to determine $S(k, 1)$, $S(k, u)$, and $S(k, \geq u)$. Each element b_k has two possible values, that is, $b_k = a'_1 = 0$ or $b_k = a'_2 = 1$; thus it is sufficient to find $S(k,1)$ and $S(k, \geq 1)$. $S(k, \geq 1)$ is clearly equal to 1. Let the sequence $b_k \dots b_{tn}$ contains q ones, the number of such sequences is $D(k-1, q)$. Furthermore, $D(k,q)$ of these sequences satisfy $b_k = 0$. Then

$$S(k, 1) = \frac{D(k, q)}{D(k-1, q)} = \frac{(t(n-q) - k + 1)(tn - k - q + 2)}{(t(n-q) - k + 2)(tn - k + 1)}.$$

This leads to the following simple algorithm that finds the t-ary tree $f(g)$ with the ordinal number $\lfloor gB(t, n) \rfloor + 1$.

procedure tree(t, n, g);
  $p' \leftarrow 0$; $q \leftarrow n$;
  for $k \leftarrow 1$ to tn do {
    $b_k \leftarrow 0$;
    $p \leftarrow \frac{(t(n-q) - k + 1)(tn - k - q + 2)}{(t(n-q) - k + 2)(tn - k + 1)}$;
    if $p \leq g$ **then** {
      $p' \leftarrow p$;
      $b_k \leftarrow 1$;
      $q \leftarrow q - 1$;
      $p \leftarrow 1$ }
    $g \leftarrow \frac{g - p}{p - p'}$ }

The time complexity of the above procedure is clearly linear, that is, $O(tn)$.

1.13.4 Random Subset and Variation

There is a fairly simple mapping procedure for subsets in binary representation. Let $g = 0.\, a_1 \dots a_n a_{n+1} \dots$ be number g written in the binary numbering system. Then the subset with ordinal number $\lfloor gS(n) \rfloor + 1$ is coded as $a_1 \dots a_n$. Using a relation between subsets and compositions of n into any number of parts, described procedure can be also used to find the composition with ordinal number $\lfloor gCM(n) \rfloor + 1$.

A mapping procedure for variations is a generalization of the one used for subsets. Suppose that the variations are taken out of the set $\{0, 1, \dots, n-1\}$. Let $g{=}0.\, a_1 a_2 \dots a_m a_{m+1} \dots$ be the number g written in the number system with the base n, that is, $0 \leq a_i \leq n-1$ for $1 \leq i \leq m$. Then the variation indexed $\lfloor gV(m, n) \rfloor + 1$ is coded as $a_1 a_2 \dots a_m$.

If variations are ordered in the n-ary reflected Gray code then the variation indexed $\lfloor gV(m, n) \rfloor + 1$ is coded as $b_1 b_2 \dots b_m$, where $b_1 = a_1$, $b_i = a_i$ if $a_1 + a_2 + \dots + a_{i-1}$ is even and $b_i = n - 1 - a_i$ otherwise ($2 \leq i \leq m$).

REFERENCES

1. Akl SG. A comparison of combination generation methods. ACM Trans Math Software 1981;7(1):42–45.
2. Arnold DB. Sleep MR. Uniform random generation of balanced parenthesis strings. ACM Trans Prog Lang Syst 1980;2(1):122–128.
3. Atkinson M. Uniform generation of rooted ordered trees with prescribed degrees. Comput J 1993;36(6):593–594.
4. Akl SG, Olariu S, Stojmenovic I. A new BFS parent array encoding of t-ary trees, Comput Artif Intell 2000;19:445–455.
5. Belbaraka M, Stojmenovic I. On generating B-trees with constant average delay and in lexicographic order. Inform Process Lett 1994;49(1):27–32.
6. Brualdi RA. Introductory Combinatorics. North Holland; 1977.
7. Cohn M. Affine m-ary gray codes, Inform Control 1963;6:70–78.
8. Durstenfeld R. Random permutation (algorithm 235). Commun ACM 1964;7:420.
9. Djokić B, Miyakawa M, Sekiguchi S, Semba I, Stojmenović I. A fast iterative algorithm for generating set partitions. Comput J 1989;32(3):281–282.
10. Djokić B, Miyakawa M, Sekiguchi S, Semba I, Stojmenović I. Parallel algorithms for generating subsets and set partitions. In: Asano T, Ibaraki T, Imai H, Nishizeki T, editors. Proceedings of the SIGAL International Symposium on Algorithms; August 1990; Tokyo, Japan. Lecture Notes in Computer Science. Volume 450. p 76–85.
11. Ehrlich G. Loopless algorithms for generating permutations, combinations and other combinatorial configurations. J ACM 1973;20(3):500–513.
12. Er MC. Fast algorithm for generating set partitions. Comput J 1988;31(3):283–284.
13. Er MC. Lexicographic listing and ranking t-ary trees. Comp J 1987;30(6):569–572.
14. Even S. Algorithmic Combinatorics. New York: Macmillan; 1973.
15. Flores I. Reflected number systems. IRE Trans Electron Comput 1956;EC-5:79–82.
16. Gupta UI, Lee DT, Wong CK. Ranking and unranking of B-trees. J Algor 1983;4: 51–60.
17. Heath FG. Origins of the binary code. Sci Am 1972;227(2):76–83.
18. Johnson SM. Generation of permutations by adjacent transposition, Math Comput 1963;282–285.
19. Knuth DE. The Art of Computer Programming, Volume 1: Fundamental Algorithms. Reading, MA: Addison-Wesley; 1968.
20. Korsch JF. Counting and randomly generating binary trees. Inform Process Lett 1993;45:291–294.
21. Lehmer DH. The machine tools of combinatorics. In: Beckenbach E, editor. Applied Combinatorial Mathematics. Chapter 1. New York: Wiley; 1964. p 5–31.
22. Lucas J, Roelants van Baronaigien D, Ruskey F. On rotations and the generation of binary trees. J Algor 1993;15:343–366.
23. Misfud CJ, Combination in lexicographic order (Algorithm 154). Commun ACM 1963;6(3):103.
24. Moses LE, Oakford RV. Tables of Random Permutations. Stanford: Stanford University Press; 1963.

25. Nijenhius A, Wilf H. Combinatorial Algorithms. Academic Press; 1978.
26. Nijenhius A, Wilf HS. A method and two algorithms on the theory of partitions. J Comb Theor A 1975;18:219–222.
27. Ord-Smith RJ. Generation of permutation sequences. Comput J 1970;13:152–155 and 1971;14:136–139.
28. Parberry I. Problems on Algorithms. Prentice Hall; 1995.
29. Payne WH, Ives FM. Combination generators. ACM Transac Math Software 1979;5(2):163–172.
30. Reingold EM, Nievergelt J, Deo N. Combinatorial Algorithms. Englewood Cliffs, NJ: Prentice Hall; 1977.
31. Sedgewick R. Permutation generation methods. Comput Survey 1977;9(2):137–164.
32. Semba I. An efficient algorithm for generating all partitions of the set $\{1, \ldots, n\}$. J Inform Process 1984;7:41–42.
33. Semba I. An efficient algorithm for generating all k-subsets $(1 \leq k \leq m \leq n)$ of the set $\{1, 2, \ldots, n\}$ in lexicographic order. J Algor 1984;5:281–283.
34. Semba I. A note on enumerating combinations in lexicographic order. J Inform Process 1981;4(1):35–37.
35. Skiena S. Implementing Discrete Mathematics: Combinatorics and Graph Theory with Mathematica. Addison-Wesley; 1990.
36. Stojmenovic I. Listing combinatorial objects in parallel. Int J Parallel Emergent Distrib Syst 2006;21(2):127–146.
37. Stojmenović I, Miyakawa M. Applications of a subset generating algorithm to base enumeration, knapsack and minimal covering problems. Comput J 1988;31(1):65–70.
38. Stojmenović I. On random and adaptive parallel generation of combinatorial objects. Int J Comput Math 1992;42:125–135.
39. Trotter HF, Algorithm 115. Commun ACM 1962;5:434–435.
40. Wells MB, Elements of Combinatorial Computing. Pergamon Press; 1971.
41. Xiang L, Ushijima K, Tang C. Efficient loopless generation of Gray codes for k-ary trees. Inform Process Lett 2000;76:169–174.
42. Zaks S. Lexicographic generation of ordered trees. Theor Comput Sci 1980;10:63–82.
43. Zoghbi A, Stojmenović I. Fast algorithms for generating integer partitions. Int J Comput Math 1998;70:319–332.

CHAPTER 2

Backtracking and Isomorph-Free Generation of Polyhexes

LUCIA MOURA and IVAN STOJMENOVIC

2.1 INTRODUCTION

This chapter presents applications of combinatorial algorithms and graph theory to problems in chemistry. Most of the techniques used are quite general, applicable to other problems from various fields.

The problem of cell growth is one of the classical problems in combinatorics. Cells are of the same shape and are in the same plane, without any overlap. If h copies of the same shape are connected (two cells are connected by sharing a common edge), then they form an h-mino, polyomino, animal, or polygonal system (various names given in the literature for the same notion). Three special cases of interest are triangular, square, and hexagonal systems, which are composed of equilateral triangles, squares, and regular hexagons, respectively. Square and hexagonal systems are of genuine interest in physics and chemistry, respectively. The central problem in this chapter is the study of hexagonal systems. Figure 2.1 shows a molecule and its corresponding hexagonal system.

Enumeration and exhaustive generation of combinatorial objects are central topics in combinatorial algorithms. *Enumeration* refers to counting the number of distinct objects, while *exhaustive generation* consists of listing them. Therefore, exhaustive generation is typically more demanding than enumeration. However, in many cases, the only available methods for enumeration rely on exhaustive generation as a way of counting the objects. In the literature, sometimes "enumeration" or "constructive enumeration" are also used to refer to what we call here "exhaustive generation."

An important issue for enumeration and exhaustive generation is the notion of *isomorphic* or *equivalent* objects. Usually, we are interested in enumerating or generating only one copy of equivalent objects, that is, only one representative from each isomorphism class. Polygonal systems are considered different if they have

Handbook of Applied Algorithms: Solving Scientific, Engineering and Practical Problems
Edited by Amiya Nayak and Ivan Stojmenović

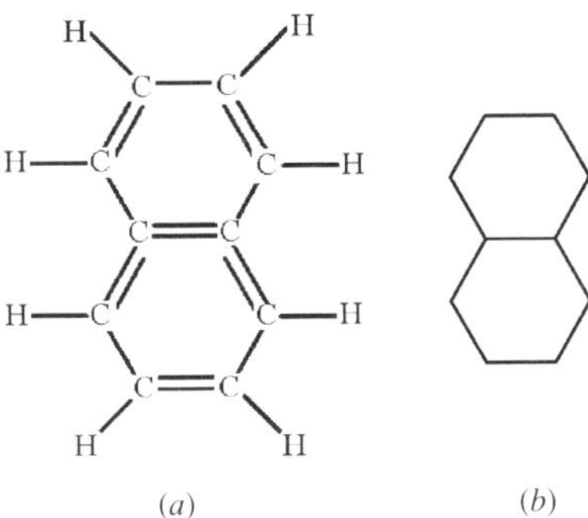

FIGURE 2.1 (*a*) A benzenoid hydrocarbon and (*b*) its skeleton graph.

different shapes; their orientation and location in the plane are not important. For example, the two hexagonal systems in Figure 2.2b are isomorphic. The main theme in this chapter is isomorph-free exhaustive generation of polygonal systems, especially polyhexes.

Isomorph-free generation provides at the same time computational challenges and opportunities. The computational challenge resides in the need to recognize or avoid isomorphs, which consumes most of the running time of these algorithms. On the contrary, the fact that equivalent objects do not need to be generated can substantially reduce the search space, if adequately exploited. In general, the main algorithmic framework employed for exhaustive generation is backtracking, and several techniques have been developed for handling isomorphism issues within this framework. In this chapter, we present several of these techniques and their application to exhaustive generation of hexagonal systems.

In Section 2.2, we present benzenoid hydrocarbons, a class of molecules in organic chemistry, and their relationship to hexagonal systems and polyhexes. We also take a close look at the parameters that define hexagonal systems, and at the topic of symmetries in hexagonal systems. In Section 2.3, we introduce general algorithms for isomorph-free exhaustive generation of combinatorial structures, which form the

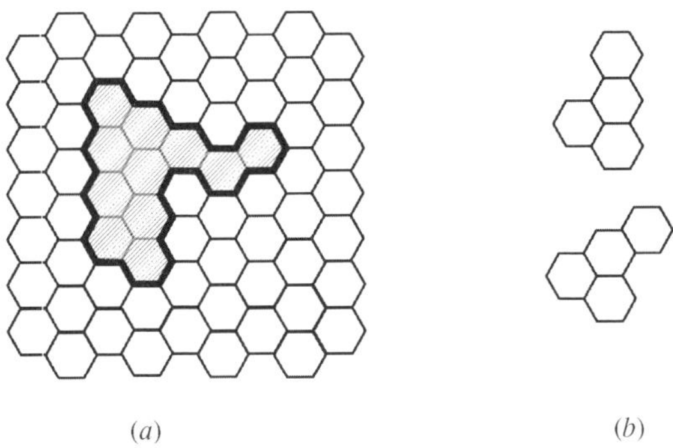

FIGURE 2.2 Hexagonal systems with (*a*) $h = 11$ and (*b*) $h = 4$ hexagons.

theoretical framework for the various algorithms presented in the sections that follow. In Section 2.4, we provide a historical overview of algorithms used for enumeration and generation of hexagonal systems. In Sections 2.5–2.7, we present some of the main algorithmic techniques used for the generation of polyhexes. We select algorithms that illustrate the use of different general techniques, and that were responsible for breakthroughs regarding the sizes of problems they were able to solve at the time they appeared. Section 2.5 presents a basic backtracking algorithm for the generation of hexagonal, square, and triangular systems. In Section 2.6, we describe a lattice-based algorithm that uses a "cage" to reduce the search space. In Section 2.7, we present two algorithms based on McKay's canonical construction path, each combined with a different way of representing a polyhex. Finally, Section 2.8 deals with a different problem involving chemistry, polygonal systems, and graph theory, namely perfect matchings in hexagonal systems and the Kekulé structure of benzenoid hydrocarbons.

2.2 POLYHEXES AND HEXAGONAL SYSTEMS

2.2.1 Benzenoid Hydrocarbons

We shall study an important class of molecules in organic chemistry, the class of benzenoid hydrocarbons. A benzenoid hydrocarbon is a molecule composed of carbon (C) and hydrogen (H) atoms. Figure 2.1a shows a benzenoid called naphthalene, with molecular formula $C_{10}H_8$ (i.e., 10 carbon atoms and 8 hydrogen atoms). In general, a class of benzenoid isomers is defined by a pair of invariants (n, s) and written as the chemical formula C_nH_s, where n and s are the numbers of carbons and hydrogens, respectively. Every carbon atom with two neighboring carbon atoms bears a hydrogen, while no hydrogen is attached to the carbon atoms with three neighboring carbon atoms. A simplified representation of the molecule as a (skeleton) graph is given in Figure 2.1b. Carbon atoms form six-membered rings, and each of them has four valences. Hydrogen atoms (each with one valence) and double valences between carbon atoms are not indicated in the corresponding graph, which has carbon atoms as vertices with edges joining two carbon atoms linked by one or two valences. In the sequel, we shall study the skeleton graphs, which will be called polyhex systems.

A polyhex (system) is a connected system of congruent regular hexagons such that any two hexagons either share exactly one edge or are disjoint. The formula C_6H_6 is represented by only one hexagon and is the simplest polyhex, called benzene. Presently, we shall be interested only in the class of geometrically planar, simply connected polyhexes. A polyhex is geometrically planar when it does not contain any overlapping edges, and it is simply connected when it has no holes. The geometrically planar, simply connected polyhexes may conveniently be defined in terms of a cycle on a hexagonal lattice; the system is found in the interior of this cycle, which represents the boundary (usually called the "perimeter") of the system. With the aim of avoiding confusion, we have adopted the term "hexagonal system" (HS) for a geometrically planar, simply connected polyhex (see Fig. 2.2a for an HS with

$h = 11$ hexagons). A plethora of names has been proposed in the literature for what we just defined (or related objects), such as benzenoid systems, benzenoid hydrocarbons, hexagonal systems, hexagonal animal, honeycomb system, fusene, polycyclic aromatic hydrocarbon, polyhex, and hexagonal polyomino, among others.

A polyhex in plane that has holes is called circulene; it has one outer cycle (perimeter) and one or a few inner cycles. The holes may have the size of one or more hexagons. Coronoids are circulenes such that all holes have the size of at least two hexagons. There are other classes of polyhexes; for instance, a helicenic system is a polyhex with overlapping edges or hexagons if drawn in a plane (or a polyhex in three-dimensional space). Fusenes are generalizations of polyhexes in which the hexagons do not need to be regular.

2.2.2 Parameters of a Hexagonal System

We shall introduce some parameters and properties of HSs in order to classify them. The leading parameter is usually the number of hexagons h in an HS (it is sometimes called the "area"). For example, HSs in Figures 2.1b, 2.2a and b have $h = 2, 11$, and 4 hexagons, respectively. The next parameter is the perimeter p, or the number of vertices (or edges) on its outer boundary. The HSs in Figures 2.1b, 2.2a and b have perimeter $p = 10, 32$, and 16, respectively. A vertex of an HS is called *internal* (*external*) if it does not (does, respectively) belong to the outer boundary. A vertex is internal if and only if it belongs to three hexagons from the given HS. The number of internal vertices i of HSs in Figures 2.1b, 2.2a and b is $i = 0, 7$ and 1, respectively. Let the total number of vertices and edges in HSs be $n = p + i$ and m, respectively. From Euler theorem, it follows that $n - m + h = 1$. There are p external and $m - p$ internal edges. Since every internal edge belongs to two hexagons, we obtain $6h = 2(m - p) + p$, that is, $m = 3h + p/2$. Therefore, $n - 2h - p/2 = 1$ and $i = 2h - p/2 + 1$ [31]. It follows that p must be even, and that i is odd if and only if p is divisible by 4.

Consider now the relation between invariants n and s of a benzenoid isomer class C_nH_s and other parameters of an HS. The number of vertices is $n = i + p = 2h + p/2 + 1 = 4h - i + 2$. We shall find the number of hydrogen atoms s, which is equal to the number of degree-2 vertices in an HS (all such vertices belong to the perimeter). Let t be the number of tertiary (degree 3) carbon atoms on the perimeter. Therefore, $p = s + t$ since each vertex on the perimeter has degree either 2 or 3. We have already derived $m = 3h + p/2$. Now, if one assigns each vertex to all its incident edges, then each edge will be "covered" twice; since each internal vertex has degree 3, it follows that $2m = 3i + 3t + 2s$. Thus, $6h + p = 3i + 3t + 2s$, that is, $3t = 6h + p - 3i - 2s$. By replacing $t = p - s$, one gets $3p - 3s = 6h + p - 3i - 2s$, which implies $s = 2p - 6h + 3i$. Next, $i = 2h - p/2 + 1$ leads to $s = p/2 + 3$. It is interesting that s is a function of p independent of h. The reverse relation reads $p = 2s - 6$, which, together with $p = s + t$, gives another direct relation $t = s - 6$. Finally, $h = (n - s)/2 + 1$ follows easily from $2h = n - p/2 - 1$ and $p = 2s - 6$. Therefore, there exists a one-to-one correspondence between pairs (h, p) and (n, s). More precisely, the number of different HSs corresponding to the same benzenoid isomer class C_nH_s is equal to the number of (nonisomorphic) HSs with area $h = (n - s)/2 + 1$ and perimeter

$p = 2s - 6$. The study of benzenoid isomers is surveyed by Brunvoll et al. [9] and Cyrin et al. [15].

We shall list all the types of chemical isomers of HSs for increasing values of $h \leq 5$; $h = 1$: C_6H_6; $h = 2$: $C_{10}H_8$; $h = 3$: $C_{13}H_9$, $C_{14}H_{10}$; $h = 4$: $C_{16}H_{10}$, $C_{17}H_{11}$, $C_{18}H_{12}$; $h = 5$: $C_{19}H_{11}$, $C_{20}H_{12}$, $C_{21}H_{13}$, $C_{22}H_{14}$.

The number of edges m of all isomers with given formula C_nH_s is $m = (3n - s)/2$. The number of edges m and number of internal vertices i are sometimes used as basic parameters; for example, $n = (4m - i + 6)/5$, $s = (2m - 3i + 18)/5$.

The *Dias parameter* is an invariant for HSs and is defined as the difference between the number of vertices and number of edges in the graph of internal edges, obtained by deleting the perimeter from a given HS, reduced by 1. In other words, it is the number of tree disconnections of internal edges. The number of vertices of the graph of internal edges is $i + t$ (only s vertices with degree 2 on the perimeter do not "participate"), and the number of internal edges is $m - p$. Thus, the Dias parameter for an HS is $d = i + t - m + p - 1 = h - i - 2 = p/2 - h - 3$. The pair of invariants (d, i) plays an important role in connection with the periodic table for benzenoid hydrocarbons [19,21]. The other parameters of an HS can be expressed in terms of d and i as follows: $n = 4d + 3i + 10$, $s = 2d + i + 8$, $h = d + i + 2$, and $p = 4d + 2i + 10$. The pair (d, i) can be obtained from pair (n, s) as follows: $d = (3s - n)/2 - 7, i = n - 2s + 6$.

There are several classifications of HSs. They are naturally classified with respect to their area and perimeter. Another classification is according to the number of internal vertices: *catacondensed* systems have no internal vertices ($i = 0$), while *pericondensed* systems have at least one internal vertex ($i > 0$). For example, HSs in Figures 2.1a, 2.3b, c and d are catacondensed, while HSs in Figures 2.2a,b and 2.3a are pericondensed. An HS is catacondensed if and only if $p = 4h + 2$. Thus, the perimeter of a catacondensed system is an even number not divisible by 4. All catacondensed systems are Hamiltonian, since the outer boundary passes through all vertices. Catacondensed HSs are further subdivided into *unbranched* (also called chains, where each hexagon, except two, has two neighbors) and *branched* (where at least one hexagon has three neighboring hexagons). Pericondensed HSs are either *basic* or *composite*, depending on whether they cannot (or can, respectively) be cut into two pieces by cutting along only one edge.

2.2.3 Symmetries of a Hexagonal System

We introduce the notion of *free* and *fixed* HSs. Free HSs are considered distinct if they have different shapes; that is they are not congruent in the sense of Euclidean geometry. Their orientation and location in the plane are of no importance. For example, the two systems shown in Figure 2.2b represent the same free HS. Different free HSs are nonisomorphic. Fixed HSs are considered distinct if they have different shapes or orientations. Thus, the two systems shown in Figure 2.2b are different fixed HSs.

The key to the difference between fixed and free HSs lies in the symmetries of the HSs. An HS is said to have a certain symmetry when it is invariant under the transformation(s) associated with that symmetry. In other words, two HSs are considered to be the same fixed HSs, if one of them can be obtained by translating the other, while two

HSs are considered the same free HSs, if one of then can be obtained by a sequence of translations and rotations that may or may not be followed by a central symmetry. A regular hexagon has 12 different transformations that map it back to itself. These are rotations for 0°, 60°, 120°, 180°, 240°, 300°, and central symmetry followed by the same six rotations. Let us denote the identity transformation (or rotation for 0°) by ε, rotation for 60° by ρ, and central symmetry by μ (alternatively, a mirror symmetry can be used). Then, these 12 transformation can be denoted as ε, ρ, ρ^2, ρ^3, ρ^4, ρ^5, μ, $\rho\mu$, $\rho^2\mu$, $\rho^3\mu$, $\rho^4\mu$, and $\rho^5\mu$, respectively. They form a group generated by ρ and μ. When these transformations are applied on a given HS, one may or may not obtain the same HS, depending on the kinds of symmetries that it has. The transformations of an HS that produce the same fixed HS form a subgroup of the transformation group $G = \{\varepsilon, \rho, \rho^2, \rho^3, \rho^4, \rho^5, \mu, \rho\mu, \rho^2\mu, \rho^3\mu, \rho^4\mu, \rho^5\mu\}$. Every free HS corresponds to $1, 2, 3, 4, 6$, or 12 fixed HSs, depending on its symmetry properties. Thus, the HSs are classified into symmetry groups of which there are eight possibilities, which are defined here as subgroups of G: $D_{6h} = G$, $C_{6h} = \{\varepsilon, \rho, \rho^2, \rho^3, \rho^4, \rho^5\}$, $D_{3h} = \{\varepsilon, \rho^2, \rho^4, \mu, \rho^2\mu, \rho^4\mu\}$, $C_{3h} = \{\varepsilon, \rho^2, \rho^4\}$, $D_{2h} = \{\varepsilon, \rho^3, \mu, \rho^3\mu\}$, $C_{2h} = \{\varepsilon, \rho^3\}$, $C_{2v} = \{\varepsilon, \mu\}$, and $C_s = \{\varepsilon\}$. The number of fixed HSs for each free HS under these symmetry groups are specifically (in the same order): 1, 2, 2, 4, 3, 6, 6, and 12. Note that the number of elements in the subgroup multiplied by the number of fixed HSs for each free HS is 12 for each symmetry group. For example, HS in Figure 2.1b has symmetry group D_{2h}, while HSs in Figure 2.2a and b are associated with C_s (have no symmetries). Examples of HSs with other symmetry groups are given in Figure 2.3.

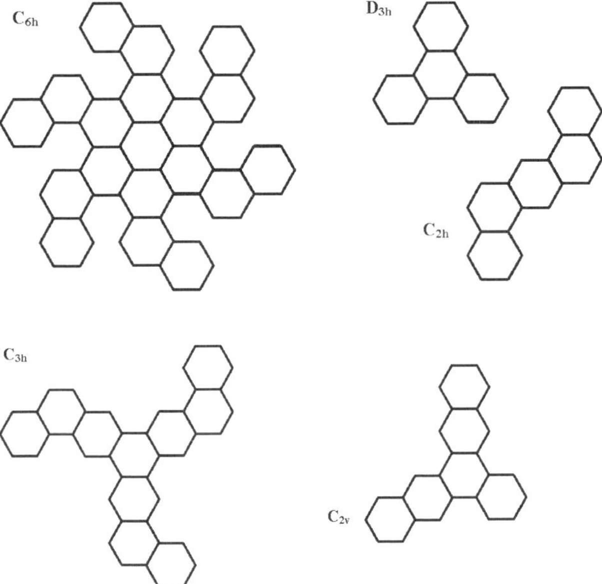

FIGURE 2.3 Hexagonal systems and their symmetry groups.

Let $H(h)$ and $N(h)$ denote the number of fixed and free (nonisomorphic) HSs with h hexagons, respectively. Furthermore, $N(h)$ can be split into the numbers for the different symmetries, say $N(G, h)$, where G indicates the symmetry group. Then $H(h) = N(D_{6h}, h) + 2N(C_{6h}, h) + 2N(D_{3h}, h) + 4N(C_{3h}, h) + 3N(D_{2h}, h) + 6N(C_{2h}, h) + 6N(C_{2v}, h) + 12N(C_s, h)$. For the free HSs, $N(h) = N(D_{6h}, h) + N(C_{6h}, h) + N(D_{3h}, h) + N(C_{3h}, h) + N(D_{2h}, h) + N(C_{2h}, h) + N(C_{2v}, h) + N(C_s, h)$. Eliminating $N(C_s, h)$, we get

$$N(h) = \frac{1}{12}\left[11N(D_{6h}, h) + 10N(C_{6h}, h) + 10N(D_{3h}, h) + 8N(C_{3h}, h) \right.$$
$$\left. +9N(D_{2h}, h) + 6N(C_{2h}, h) + 6N(C_{2v}, h) + H(h)\right]. \tag{2.1}$$

As we will see later, some algorithms use the above formula in order to compute $N(h)$ via computing the quantities on the right-hand side and avoiding the often costly computation of $N(C_s, h)$.

2.2.4 Exercises

1. Let $n = p + i$ be the number of vertices and m be the number of edges of an HS. Show that $m = 5h + 1 - i$.
2. Prove that the maximal number of internal vertices of a HS, for fixed area h, is $2h + 1 - \sqrt{12h - 3}$ [30,37]. Also, show that the perimeter of an HS satisfies $2\sqrt{12h - 3} \leq p \leq 4h + 2$.
3. Prove that $0 \leq \Delta \leq \lfloor h/3 \rfloor$ and $1/2(1 - (-1)^i) \leq \Delta \leq i$ [9].
4. Prove the following upper and lower bounds for the Dias parameter [9]: $\sqrt{12h - 3} - h - 3 \leq d \leq h - 2$.
5. Prove that $2h + 1 + \sqrt{12h - 3} \leq n \leq 4h + 2$ [37].
6. Prove that $3 + \sqrt{12h - 3} \leq s \leq 2h + 4$ [33].
7. Prove that $3h + \left\lceil \sqrt{12h - 3} \right\rceil \leq m \leq 5h + 1$ [30,37].
8. Prove that the possible values of s are within the range [30,37] $2\left\lceil 1/2(n + \sqrt{6n}) \right\rceil - n \leq s \leq n + 2 - 2\lceil (n - 2)/4 \rceil$.
9. Prove that $n - 1 + \lceil (n - 2)/4 \rceil \leq m \leq 2n - \left\lceil (n + \sqrt{6n})/2 \right\rceil$ [37].
10. Show that $s + 3\lceil s/2 \rceil - 9 \leq m \leq s + \left\lfloor (s^2 - 6s)/12 \right\rfloor - 2$ [15].
11. Prove that $\lceil (m - 1)/5 \rceil \leq h \leq m - \left\lceil (2m - 2 + \sqrt{4m + 1})/3 \right\rceil$ [37].
12. Prove that $1 + \left\lceil (2m - 2 + \sqrt{4m + 1})/3 \right\rceil \leq n \leq m + 1 - \lceil (m - 1)/5 \rceil$ [37].
13. Show that $3 - 2m + 3\left\lceil (2m - 2 + \sqrt{4m + 1})/3 \right\rceil \leq s \leq m + 3 - 3\lceil (m - 1)/5 \rceil$ [9].
14. Let $d(r, s)$ be the distance between the vertices r and s in an HS (which is the length of the shortest path between them) [32]. The Wiener index W is the sum of all distances (between all pairs of vertices) in a given HS. Show that if B_1 and B_2 are catacondensed HSs with an equal number of hexagons, then $W(B_1) = W(B_2) \pmod 8$.

2.3 GENERAL ALGORITHMS FOR ISOMORPH-FREE EXHAUSTIVE GENERATION

In this section, we present general algorithms for generating exactly one representative of each isomorphism class of any kind of combinatorial objects. The reader is referred to the works by Brinkmann [6] and McKay [46] for more information on this type of methods and to the survey by Faulon et al. [26] for a treatment of these methods in the context of enumerating molecules.

The algorithms in this section generate combinatorial objects of size $n+1$ from objects of size n via backtracking, using a recursive procedure that should be first called with parameters of an empty object, namely $X = [\ ]$ and $n = 0$. They are presented in a very general form that can be tailored to the problem at hand. In particular, procedures `IsComplete(`X`)` and `IsExtendible(`X`)` can be set to ensure that all objects of size up to n or exactly n are generated, depending on the application. In addition, properties of the particular problem can be used in order to employ further prunings, which cannot be specified in such a general framework but which are of crucial importance.

The basic algorithms we consider here (Algorithms BasicGenA and BasicGenB) exhaustively generate all objects using backtracking and only keep one representative from each isomorphism class. They both require a method for checking whether the current object generated is the one to be kept in its isomorphism class. In Algorithm BasicGenA, this is done by remembering previously generated objects, which are always checked for isomorphism against the current object.

```
Algorithm BasicGenA (X = [x_1, x_2, ..., x_n], n)
    redundancyFound = false
    if (IsComplete(X)) then
        if (for all Y ∈ GenList: ¬ AreIsomorphic(X, Y)) then
            GenList = GenList ∪ {X}
            process X
        else redundancyFound = true
    if ((¬redundancyFound) and (IsExtendible(X))) then
        for all extensions of X: X' = [x_1, x_2, ..., x_n, x']
            if (IsFeasible(X')) then
                BasicGenA(X', n + 1)
```

The third line of Algorithm BasicGenA is quite expensive in terms of time, since an isomorphism test `AreIsomorphic(`X, Y`)` between X and each element Y in $GenList$ must be computed; see the works by Kocay [43] and McKay [44] for more information on isomorphism testing and by McKay [45] for an efficient software package for graph isomorphism. In addition, memory requirements for this algorithm become a serious issue as all the previously generated objects must be kept.

In Algorithm BasicGenB, deciding whether the current object is kept is done by a rule specifying who is the *canonical representative* of each isomorphism class. Based on this rule, the current object is only kept if it is canonical within its isomorphism class. A commonly used rule is that the canonical object be the lexicographically smallest one in its isomorphism class. In this case, a simple method for canonicity testing (a possible implementation of procedure `IsCanonical(`X`)` below) is one that generates all objects isomorph to the current object X by applying all possible symmetries, and rejecting X if it finds a lexicographically smaller isomorph.

```
Algorithm BasicGenB (X = [x_1, x_2, ..., x_n], n)
    redundancyFound = false
    if (IsComplete(X)) then
        if (IsCanonical(X)) then process X
                              else redundancyFound = true
    if ((¬redundancyFound) and (IsExtendible(X))) then
        for all extensions of X: X' = [x_1, x_2, ..., x_n, x']
            if (IsFeasible(X')) then
                BasicGenB(X', n + 1)
```

In Algorithm BasicGenB, the pruning given by the use of flag *redundancyFound* assumes that the canonicity rule guarantees that a complete canonical object that has a complete ancestor must have a canonical complete ancestor. This is a reasonable assumption, which is clearly satisfied when using the "lexicographically smallest" rule.

The next two algorithms substantially reduce the size of the backtracking tree by making sure it contains only one copy of each nonisomorphic partial object. That is, instead of testing isomorphism only for complete objects, isomorphism is tested at each tree level. Faradzev [24] and Read [50] independently propose an *orderly generation* algorithm. This algorithm also generates objects of size n by extending objects of size $n-1$ via backtracking. Like in Algorithm BasicGenB, it uses the idea that there is a canonical representative of every isomorphism class that is the object that needs to be generated (say, the lexicographically smallest). When a subobject of certain size is generated, canonicity testing is performed, and if the subobject is not canonical, the algorithm backtracks. Note that the canonical labeling and the extensions of an object must be defined so that each canonically labeled object is the extension of exactly one canonical object. In this way, canonical objects of size n are guaranteed to be the extension of exactly one previously generated canonical object of size $n-1$.

```
Algorithm OrderlyGeneration (X = [x_1, x_2, ..., x_n], n)
    if (IsComplete(X)) then process X.
    if (IsExtendible(X)) then
        for all extensions of X: X' = [x_1, x_2, ..., x_n, x']
            if (IsFeasible(X')) then
                if (IsCanonical(X')) then
                    OrderlyGeneration(X', n + 1)
```

McKay [46] proposes a related but distinct general approach, where generation is done via a *canonical construction path*, instead of a canonical representation. In this method, objects of size n are generated from objects of size $n-1$, where only *canonical augmentations* are accepted. So, in this method the canonicity testing is substituted by testing whether the augmentation from the smaller object is a canonical one; the canonicity of the augmentation is verified by the test `IsParent(X,X')` in the next algorithm. The canonical labeling does not need to be fixed as in the orderly generation algorithm. Indeed, the relabeling of an object of size $n-1$ must not affect the production of an object of size n via a canonical augmentation.

```
Algorithm McKayGeneration1 (X = [x1, x2, ..., xn], n)
    if (IsComplete(X)) then process X.
    if (IsExtendible(X)) then
        for all inequivalent extensions of X: X' = [x1, x2, ..., xn, x']
            if (IsFeasible(X')) then
                if (IsParent(X,X')) then   /* if augmentation is canonical */
                    McKayGeneration1(X', n + 1)
```

The previous algorithm may appear simpler than it is, because a lot of its key features are hidden in the test `(IsParent(X,X'))`. This test involves several concepts and computations related to isomorphism. We delay discussing more of these details until they are needed in the second application of this method in Section 2.7.2. The important and nontrivial fact established by McKay regarding this algorithm is that if X has two extensions X'_1 and X'_2 for which X is the parent, then it is enough that these objects be inequivalent extensions to guarantee that they are inequivalent. In other words, Algorithm McKayGeneration1 produces the same generation as Algorithm McKayGeneration2 below:

```
Algorithm McKayGeneration2 (X = [x1, x2, ..., xn], n)
    if (IsComplete(X)) then process X.
    if (IsExtendible(X)) then
        S = ∅
        for all extensions of X: X' = [x1, x2, ..., xn, x']
            if (IsFeasible(X')) then
                if (IsParent(X,X') then   /* if augmentation is canonical */
                    S = S ∪ {X'}
        Remove isomorph copies from S
        for all X' ∈ S do
            McKayGeneration2(X', n + 1)
```

Indeed, McKay establishes that in Algorithm McKayGeneration2 the isomorph copies removed from set S must come from symmetrical extensions with respect to the parent object X, provided that the function `IsParent(X,X')` is defined as prescribed in his article [46]. Algorithm McKayGeneration1 is the stronger, more efficient version of this method, but for some applications it may be more convenient to use the simpler form of Algorithm McKayGeneration2. McKay's method is related to the *reverse search* method of Avis and Fukuda [1]. Both are based on the idea of having a rule for deciding parenthood for objects, which could otherwise be generated as extensions of several smaller objects. However, they differ in that Avis and Fukuda's method is not concerned with eliminating isomorphs, but simply repeated objects.

Note that all the given algorithms allow for generation from scratch when called with parameters $X = [\ ]$ and $n = 0$, as well as from the complete isomorph-free list of objects at level n by calling the algorithm once for each object. In the latter case, for Algorithms BasicGenB and OrderlyGeneration, the list of objects at level n must be canonical representatives, while for Algorithms BasicGenA and McKayGeneration, any representative of each isomorphism class can be used.

2.4 HISTORICAL OVERVIEW OF HEXAGONAL SYSTEM ENUMERATION

In this section, we concentrate on the main developments in the enumeration and generation of hexagonal systems, which are geometrically planar and simply connected polyhexes, as defined earlier. A similar treatment can be found in the article by Brinkmann et al. [8]. For more information on the enumeration and generation of hexagonal systems and other types of polyhexes, the reader is referred to the books by Dias [19,20], Gutman and Cyvin [17,33,34], Gutman et al. [36], and Trinajstic [59]. For a recent treatment on generating and enumerating molecules, see the survey by Faulon et al. [26].

The enumeration of HSs is initiated by Klarner [40] who lists all HSs for $1 \leq h \leq 5$ and is followed by a race for counting HSs for larger values of h. The presence of faster computers and development of better algorithms enabled the expansion of known generation and enumeration results.

The first class of algorithms is based on the **boundary code**. Knop et al. [42] used this method for counting and even drawing HSs for $h \leq 10$. Using the same approach, HSs were exhaustively generated for $h = 11$ [53] and $h = 12$ [38]. The boundary code is explained in Section 2.5, where we give a basic backtracking algorithm (following the framework of Algorithm BasicGenB) for the generation of triangular, square, and hexagonal systems.

The next generation of algorithms uses the **dualistic angle-restricted spanning tree (DAST) code** [49], which is based on the dualistic approach associated with a general polyhex [3]. This approach was used for generating all HSs with $h = 13$ [47], $h = 14$ [48], $h = 15$ [49], and $h = 16$ [41]. This method uses a graph embedded on the regular hexagonal lattice containing the HS. Each vertex is associated with the center of a hexagon, and two vertices are connected if they share an edge. This graph is rigid; that is, angles between adjacent edges are fixed. Therefore, any spanning tree of this graph completely determines the HS. DAST algorithms exhaustively generate canonical representatives of dualist spanning trees using again a basic backtracking algorithm.

The next progress was made by Tosic et al. [56], who propose a lattice-based method that uses a "cage," which led to the enumeration of HSs for $h = 17$. This is a completely different method from the previous ones. The lattice-based approach focuses on counting the number of HSs on the right-hand side of equation (2.1) in order to compute $N(h)$. This algorithm accomplishes this by generating nonisomorphic HSs with nontrivial symmetry group based on a method of Redelmeier [51], and by generating all fixed HSs by enclosing them on a triangular region of the hexagonal lattice, which they call a cage. The **cage** algorithm is described in Section 2.6.

The **boundary edge code** algorithm by Caporossi and Hansen [12] enabled the generation of all HSs for $h = 18$ to $h = 21$. The **labeled inner dual algorithm** by Brinkmann et al. [7] holds the current record for the *exhaustive generation* of polyhexes, having generated all polyhexes for $h = 22$ to $h = 24$. Each of these two algorithms use a different representation for the HSs, but both use the generation by canonical path introduced by McKay [46] given by the framework of Algorithms McKayGeneration1 and McKayGeneration2 from Section 2.3. Both algorithms are described in Section 2.7.

TABLE 2.1 Results on the Enumeration and Exhaustive Generation of HSs

h	$N(h)$	Algorithm	Type	Year	Reference
1	1	–	–		
2	1	–	–		
3	3	–	–		
4	7	–	–		
5	22	–	–		
6	81	–	–		
7	331	–	–		
8	1453	–	–		
9	6505	–	–	1965	[40]
10	30086	BC	G	1983	[42]
11	141229	BC	G	1986	[53]
12	669584	BC	G	1988	[38]
13	3198256	DAST	G	1989	[47]
14	15367577	DAST	G	1990	[48]
15	74207910	DAST	G	1990	[49]
16	359863778	DAST	G	1990	[41]
17	1751594643	CAGE	E	1995	[56]
18	8553649747	BEC	G		
19	41892642772	BEC	G		
20	205714411986	BEC	G		
21	1012565172403	BEC	G	1998	[12]
22	4994807695197	LID	G		
23	24687124900540	LID	G		
24	122238208783203	LID	G	2002	[7]
25	606269126076178	FLM	E		
26	3011552839015720	FLM	E		
27	14980723113884739	FLM	E		
28	74618806326026588	FLM	E		
29	372132473810066270	FLM	E		
30	1857997219686165624	FLM	E		
31	9286641168851598974	FLM	E		
32	46463218416521777176	FLM	E		
33	232686119925419595108	FLM	E		
34	1166321030843201656301	FLM	E		
35	5851000265625801806530	FLM	E	2002	[60]

Finally, Vöge et al. [60] give an algorithm that enables a breakthrough on the enumeration of HSs, allowing for the counting of all HSs with $h = 25$ to $h = 35$. Like the cage algorithm, they use a lattice-based approach, but instead of brute force generation of all fixed HSs, they employ transfer matrices and the **finite lattice method** by Enting [23] to compute $H(h)$. Their algorithm is based on counting using generating functions, so they enumerate rather than exhaustively generate HSs.

Table 2.1 provides a summary of the results obtained by enumeration and exhaustive generation algorithms. For each h, it shows in order: the number $N(h)$ of free HSs with h hexagons, the first algorithmic approach that computed it, whether the algorithm's type was exhaustive generation (G) or enumeration (E), publication year, and reference. When the year and reference are omitted, it is to be understood that it can be found in the next row for which these entries are filled.

2.5 BACKTRACKING FOR HEXAGONAL, SQUARE, AND TRIANGULAR SYSTEMS

In this section, we presents a basic backtracking algorithm, based on the boundary code, for listing all nonisomorphic polygonal systems. This algorithm is applicable for hexagonal [53], triangular [22], and square [54] systems. First, each of these "animals" is decoded as a word over an appropriate alphabet. A square system can be drawn such that each edge is either vertical or horizontal. If a counterclockwise direction along the perimeter of a square system is followed, each edge can be coded with one of four characters, say from the alphabet {0, 1, 2, 3}, where 0, 1, 2, and 3 correspond to four different edge orientations (see Fig. 2.4b). For example, the square system in Figure 2.4a can be coded, starting from the bottom-left corner, as the word 00100122100010122101223221233230333. The representation of a square system is obviously not unique, since it depends on the starting point.

Similarly, each hexagonal or triangular system can be coded using words from the alphabet {0, 1, 2, 3, 4, 5}, where each character corresponds to one of six possible edge orientations, as indicated in Figure 2.4d. Figure 2.4c shows a triangular system that can be coded, starting from bottommost vertex and following counterclockwise order, as 11013242345405; the hexagonal system in Figure 2.4e can be coded, starting from the bottom-left vertex and following counterclockwise direction, as 01210123434505.

Let $l_i(u)$ denote the number of appearances of the letter i in the word u. For example, $l_4(01210123434505) = 2$, since exactly two characters in the word are equal to 4.

Lemma 1 [54] *A word u corresponds to a square system if and only if the following conditions are satisfied:*

1. $l_0(u) = l_2(u)$ *and* $l_1(u) = l_3(u)$, *and*
2. *for any nonempty proper subword* w *of* u, $l_0(w) \neq l_2(w)$ *or* $l_1(w) \neq l_3(w)$.

Proof. A given closed path along the perimeter can be projected onto Cartesian coordinate axes such that 0 and 2 correspond to edges in the opposite directions (and,

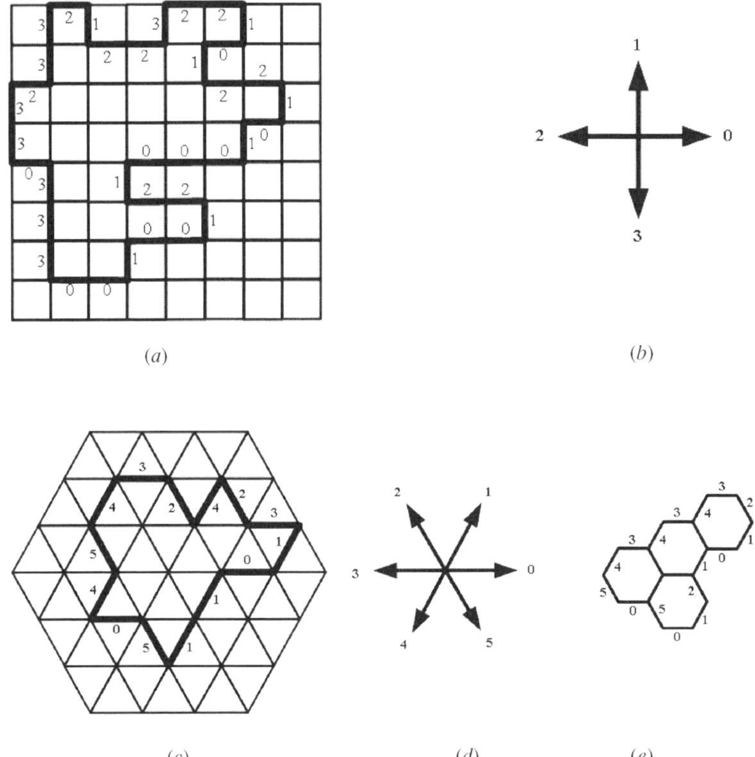

FIGURE 2.4 Boundary codes for polygonal systems.

similarly, edges 1 and 3), as indicated in Figure 2.4b. Since the number of projected "unit" edges in direction 0 must be equal to the number of projected unit edges in direction 2, it follows that $l_0(u) = l_2(u)$. Similarly, $l_1(u) = l_3(u)$. To avoid self-intersections along the perimeter, both equalities shall not be met simultaneously for any proper subword of u. ■

Lemma 2 [53] *A word $u = u_1u_2...u_p$ corresponds to a hexagonal system if and only if the following conditions are satisfied:*

1. $l_0(u) = l_3(u)$, $l_1(u) = l_4(u)$, *and* $l_2(u) = l_5(u)$,
2. *for any nonempty proper subword w of u, $l_0(w) \neq l_3(w)$ or $l_1(w) \neq l_4(w)$, or $l_2(w) \neq l_5(w)$, and*
3. $u_{i+1} = u_i \pm 1 \pmod 6$, $i = 1, 2, ..., p-1$.

Proof. Condition 3 follows easily from the hexagonal grid properties. To verify condition 1, consider, for example, a vertical line passing through the middle of each horizontal edge (denoted by 0 or 3). Each such vertical line intersects only edges marked by 0 or 3, and no other edge. Therefore, in order to return to the starting point of the perimeter, each path along the boundary must make equal number of moves to the right and to the left; thus, the number of 0s and 3s in a hexagonal system is equal. The other two equalities in 1 follow similarly. Condition 2 assures that no self-intersection of the boundary occurs. ■

Lemma 3 [22] *A word u corresponds to a triangular system if and only if the following conditions are satisfied:*

1. $l_0(u) - l_3(u) = l_4(u) - l_1(u) = l_2(u) - l_5(u)$, *and*
2. *no proper subword of u satisfies condition 1.*

Proof. Project all edges of a closed path onto a line normal to directions 2 and 5. All edges corresponding to characters 2 and 5 have zero projections while the length of projections of edges 0, 1, 3, and 4 are equal; edges 0 and 1 have equal sign, which is opposite to the sign of projections of edges 3 and 4. The sum of all projections for a closed path is 0 and therefore $l_0(u) + l_1(u) = l_3(u) + l_4(u)$. Analogously, $l_1(u) + l_2(u) = l_4(u) + l_5(u)$. ■

The same polygonal system can be represented by different words. Since the perimeter can be traversed starting from any vertex, there are p words in the clockwise and p words in the counterclockwise direction for the same fixed polygonal system $u_1u_2...u_p$. In addition, central symmetry and rotations can produce additional isomorphic polygonal systems. In the case of hexagonal and triangular systems, each free polygonal system corresponds to at most 12 fixed ones, as discussed above (the symmetry groups for hexagonal and triangular systems coincide). Thus, each HS or TS (triangular system) may have up to $24p$ isomorphic words (words that define the same free system). They can be generated by repeated application and combination of the following transformations: $\alpha(u_1u_2...u_p) = u_2u_3...u_pu_1$, $\beta(u_1u_2...u_p) = u_pu_{p-1}...u_2u_1$ and $\sigma(u_1u_2...u_p) = \sigma(u_1)\sigma(u_2)...\sigma(u_p)$, where σ is an arbitrary element of the transformation group G described above. G is generated by permutations $\mu = 123450$ ($\mu(t) = t + 1 \pmod 6$) and $\rho = 345012$ ($\rho(t) = 3 + t \pmod 6$))

In the case of square systems, each word has similarly up to $2p$ words obtained by starting from an arbitrary vertex and following (counter) clockwise direction, and up to eight isomorphic systems corresponding to the symmetry group of a square. The group is generated by a rotation of $\pi/4$ and a central symmetry, which correspond to permutations $\mu = 1230$ ($\mu(t) = t + l \pmod 4$) and $\rho = 2301$ ($\rho(t) = 2 + t \pmod 4$), respectively. The transformation group contains eight elements $\{\varepsilon, \mu, \mu^2, \mu^3, \rho, \mu\rho, \mu^2\rho, \mu^3\rho\}$.

In summary, each polygonal system can be coded by up to $24p$ words and only one of them shall be selected to represent it. We need a procedure to determine whether or not a word that corresponds to a polygonal system is the representative among all words that correspond to the same polygonal system. As discussed in Section 2.3, Algorithm BasicGenA is time and space inefficient when used for large computations, where there are millions of representatives. Instead, we employ

Algorithm BasicGenB. We may select, say, the lexicographically first word among all isomorphic words as the canonical representative.

We shall now determine the *area* of a polygonal system, that is the number of polygons in its interior. Given a closed curve, it is well known that the curvature integration gives the area of the interior of the curve. Let (x_i, y_i) be the Cartesian coordinates of the vertex where the ith edge (corresponding to the element u_i in the word u) starts. Then, the area obtained by curvature integration along the perimeter of a given polygonal system that is represented by a word $u = u_1u_2 \dots u_n$ is $P = 1/2 \sum_{i=1}^{p} (x_{i+1} - x_i)(y_{i+1} - y_i) = 1/2 \sum_{i=1}^{p} (x_i y_{i+1} - x_{i+1} y_i)$. The number of polygons h in the interior of a polygonal system is then obtained when P is divided by the area of one polygon, namely $\sqrt{3}/4$, $3\sqrt{3}/2$, and 1 for triangular, hexagonal, and square systems, respectively, where each edge is assumed to be of length 1. It remains to compute the coordinates (x_i, y_i) of vertices along the perimeter. They can be easily obtained by projecting each of the unit vectors corresponding to directions 0, 1, 2, 3, 4, and 5 of triangular/hexagonal and 0, 1, 2, and 3 of square system onto the Cartesian coordinates.

Let $u = u_1u_2...u_j$ be a given word over the appropriate alphabet. If it represents a polygonal system, then conditions 1 and 2 are satisfied from the appropriate lemma (Lemma 1, 2, or 3). Condition 1 means that the corresponding curve is closed and condition 2 that it has no self-intersections. Suppose that condition 2 is satisfied but not condition 1; that is, the corresponding curve has no self-intersections and is not closed. We call such a word *addable*. It is clear that u can be completed to a word $u' = u_1u_2...u_p$, for some $p > j$, representing a polygonal system if and only if u is addable. If u is addable, then it can be extended to a word $u_1u_2...u_ju_{j+1}$, where u_{j+1} has the following possible values: $u_j - 1, u_j + 1 \pmod 6$ for hexagonal, $u_j + 4$, $u_j + 5$, u_j, $u_j + 1$, and $u_j + 2 \pmod 6$ for triangular (note that obviously $u_{j+1} \neq u_j + 3 \pmod 6$), and $u_j - 1$, u_j, and $u_j + 1 \pmod 4$ for square (note that $u_{j+1} \neq u_j + 2 \pmod 4$) systems.

```
Algorithm BacktrackS_{j,h}(p)
  Procedure GenPolygonalSystem(U = [u_1, ..., u_j], j, p) {
   if (U = [u_1, ..., u_j] represents a polygonal system) then
     if (U = [u_1, ..., u_j] is a canonical representative) then {
       find its area h;
       S_{j,h} <- S_{j,h} + 1;
       print u_1, ..., u_j
       }
   else
     if (U = (u_1, ..., u_j is addable) and (j < p) then
       for all feasible values of u_{j+1} with respect to U do
         GenPolygonalSystem([u_1, ..., u_j, u_{j+1}], j + 1, p)
  }
  begin main
    u_1 <- 0;
    GenPolygonalSystem([u_1], 1, p)
  end main
```

TABLE 2.2 Number of Square and Triangle Systems with h Polygons

	h												
	1	2	3	4	5	6	7	8	9	10	11	12	13
S	1	1	2	5	12	25	107	363	1248	4460			
T	1	1	1	3	4	12	24	66	159	444	1161	3226	8785

Algorithm Backtrack$S_{j,h}(p)$ determines the numbers $S_{j,h}$ of polygonal systems with perimeter j and area h, for $j \leq p$ (i.e., for all perimeters $\leq p$ simultaneously). Due to symmetry and lexicographical ordering for the choice of a canonical representative, one can fix $u_1 = 0$. This algorithm follows the framework given by Algorithm BasicGenB in Section 2.3.

This algorithm was used to produce the numbers $S_{p,h}$ and the results were obtained for the following ranges: $p \leq 15$ for triangular [22], $p \leq 22$ for square [54], and $p \leq 46$ for hexagonal [53] systems. Using the relation $p \leq 4h + 2$ for hexagonal, $p \leq h + 2$ for triangular, and $p \leq 2h + 2$ for square systems, the numbers of polygonal systems with exactly h polygons are obtained for the following ranges of h: $h \leq 13$ (triangular), $h \leq 10$ (square), and $h \leq 11$ (hexagonal systems). These numbers are given for square and triangular systems in Table 2.2. The data for hexagonal systems can be found in the corresponding entries in Table 2.1. Table 2.3 gives some enumeration results [53] for the number of nonisomorphic HSs with area h and perimeter p.

TABLE 2.3 Hexagonal Systems with Area h and Perimeter p

	h								
	1	2	3	4	5	6	7	8	9
$p = 6$	1	—	—	—	—	—	—	—	—
$p = 8$	—	—	—	—	—	—	—	—	—
$p = 10$	—	1	—	—	—	—	—	—	—
$p = 12$	—	—	1	—	—	—	—	—	—
$p = 14$	—	—	2	1	—	—	—	—	—
$p = 16$	—	—	—	1	1	—	—	—	—
$p = 18$	—	—	—	5	3	3	1	—	—
$p = 20$	—	—	—	—	6	4	3	1	—
$p = 22$	—	—	—	—	12	14	10	9	4
$p = 24$	—	—	—	—	—	24	25	21	15
$p = 26$	—	—	—	—	—	36	68	67	55
$p = 28$	—	—	—	—	—	—	106	144	154
$p = 30$	—	—	—	—	—	—	118	329	396
$p = 32$	—	—	—	—	—	—	—	453	825
$p = 34$	—	—	—	—	—	—	—	411	1601
$p = 36$	—	—	—	—	—	—	—	—	1966
$p = 38$	—	—	—	—	—	—	—	—	1489
Σ	1	1	3	7	22	81	331	1435	6505

2.5.1 Exercises

1. Prove that $p \leq h + 2$ for triangular systems.
2. Prove that $p \leq 2h + 2$ for square systems.
3. Find the projections of each unit vector corresponding to directions 0, 1, 2, 3, 4, and 5 of triangular/hexagonal and 0, 1, 2, and 3 of square system onto the x and y coordinate axes.
4. An unbranched catacondensed HS can be coded as a word $u = u_1u_2...u_p$ over the alphabet $\{0, 1, 2, 3, 4, 5\}$, where u_i corresponds to the vector joining ith and $(i + l)$th hexagon in the HS (the vector notation being as defined in Fig. 2.4). Prove that a word u is the path code of an unbranched catacondensed HS if and only if for every subword y of u, $|l_0(y) + l_5(y) - l_3(y) - l_2(y)| + |l_1(y) + l_2(y) - l_4(y) - l_5(y)| > 1$. Show that there always exist a representative of an equivalence class beginning with 0 and having 1 as the first letter different from 0 [55].
5. Describe an algorithm for generating and counting unbranched catacondensed HSs [55].
6. The test for self-intersection given as condition 2 in Lemmas 1–3 requires $O(n)$ time (it suffices to apply it only for subwords that have different beginning but the same ending as the tested word). Show that one can use an alternative testing that will require constant time, by using a matrix corresponding to the appropriate grid that stores 1 for every grid vertex occupied by a polygon and 0 otherwise.
7. Design an algorithm for generating and counting branched catacondensed HSs [11].
8. Design an algorithm for generating and enumerating coronoid hydrocarbons, which are HSs with one hole (they have outer and inner perimeters) [10].
9. Let $u_1u_2 \ldots u_p$ be a boundary code of an HS as defined above. Suppose that an HS is traced along the perimeter in the counterclockwise direction. A new boundary code $x = x_1x_2...x_p$ is defined over the alphabet $\{0, 1\}$ such that $x_i = 0$ if $u_i = u_{i-1} + l \pmod 6$ and $x_i = 1$ if $u_i = u_{i-1} - 1 \pmod 6$ (where $y_0 = y_p$). Show that the number of 1s is t while the number of 0s is s, where s and t are defined in Section 2.2.2. Design an algorithm for generating and counting HSs based on the new code.
10. Design an algorithm for generating HSs with area h which would be based on adding a new hexagon to each HS of area $h - 1$.
11. Let h, p, i, m, n, and d be defined for square (triangular, respectively) systems analogously to their definitions for HSs. Find the corresponding relations between them.

2.5.2 Open Problems

Find a closed formula or a polynomial time algorithm to compute the number of nonisomorphic hexagonal (triangular, square) systems with area h.

2.6 GENERATION OF HEXAGONAL SYSTEMS BY A CAGE ALGORITHM

This section describes an algorithm by Tosic et al. [56] that enumerates nonisomorphic hexagonal systems and classifies them according to their perimeter length. This algorithm therefore performs the same counting as the one in the previous section but is considerably faster (according to the experimental measurements), and was the first to enumerate all HSs with $h \leq 17$.

The algorithm is a lattice-based method that uses the results of the enumeration and classification of polyhex hydrocarbons according to their various kinds of symmetry and equation (2.1). These enumerations are performed by separate programs, which are not discussed here. Known results on the enumeration and classification of HSs according to symmetries are surveyed by Cyrin et al. [14]. In the present computation, the symmetry of the HSs is exploited by adopting the method of Redelmeier [51]. This method is improved in some aspects by using a boundary code (see the previous section) for the HSs. The exploitation of symmetry involves separate enumeration of the fixed HSs on one hand ($H(h)$) and free HSs of specific (nontrivial) symmetries on the other (other values on the right-hand side of equation (2.1)).

The easiest way to handle a beast (HS) is to put it in a cage. A *cage* is a rather regular region of the hexagonal grid in which we try to catch all relevant hexagonal systems. This algorithm uses a triangular cage. Let Cage(n) denote a triangular cage with n hexagons along each side. Figure 2.5 shows Cage(9) and exemplifies how a coordinate system can be introduced in Cage(n).

It is almost obvious that each hexagonal system that fits inside a cage can be placed in the cage in such a way that at least one of its hexagons is on the x-axis of the cage, and at least one of its hexagons is on the y-axis of the cage. We say that such HSs are *properly placed* in the cage. Thus, we generate and enumerate all HSs that are properly placed in the cage.

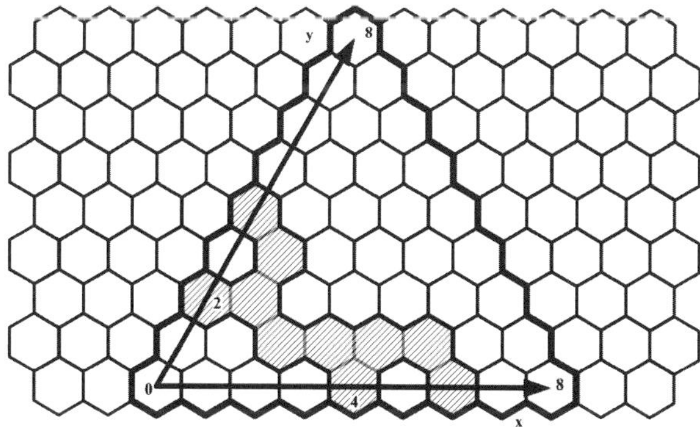

FIGURE 2.5 A hexagonal system properly placed in a cage.

Let B be a free HS with h hexagons and let G_B be its symmetry group. It can be easily shown that B can be properly placed in Cage(h) in exactly $|G_B|$ ways. Therefore, we can use equation (2.1) in order to determine $N(h)$. This requires the knowledge of $N(D_{6h}, h)$, $N(C_{6h}, h)$, $N(D_{3h}, h)$, $N(C_{3h}, h)$, $N(D_{2h}, h)$, $N(C_{2h}, h)$, and $N(C_{2v}, h)$), which are found by separate generation algorithms not discussed here, as well as of $H(h)$, the total number of fixed hexagons, which is determined by the algorithm discussed in this section. By using this approach, we completely avoid isomorphism tests, which are considered to be the most time-consuming parts of similar algorithms. Note that this is sufficient for enumeration, but if we need exhaustive generation, isomorphism tests would be required.

One needs Cage(h) to be able to catch all properly placed HSs with up to h hexagons. However, it turns out that the beasts are not that wild. Almost all hexagonal systems with h hexagons appear in Cage($h-1$). This allows a significant speedup due to the reduction in the search space. Those HSs that cannot be properly placed in Cage($h-1$) can easily be enumerated (see Exercise 3). Therefore, we can restrict our attention to Cage($h-1$), when dealing with hexagonal systems with h hexagons.

Let p and q be the smallest x- and y-coordinates (respectively) of all (centers of) hexagons of an HS that is properly placed in Cage($h-1$). Hexagons with coordinates $(p, 0)$ and $(0, q)$ (with respect to the coordinate system of the cage) are named key hexagons. Let $H(p, q)$ denote the set of all HSs with $\leq h$ hexagons that are properly placed in Cage($h-1$) and their key hexagons on x- and y-axes have coordinates $(p, 0)$ and $(0, q)$, respectively. Figure 2.5 shows one element of $H(4, 2)$.

The family $\{H(p, q) : 0 \leq p \leq h-2, 0 \leq q \leq h-2\}$ is a partition of the set of all hexagonal systems that are well placed in Cage($h-1$). Because of symmetry, it can be verified that $|H(p, q)| = |H(q, p)|$, for all $p, q \in \{0, 1, \ldots, h-2\}$. Thus, the job of enumeration of all properly placed hexagons is reduced to determining $|H(p, q)|$ for all $p \geq q$.

Given the numbers $0 \leq p \leq q \leq h-2$ and Cage($h-1$), determining $|H(p, q)|$ reduces to generating all hexagons systems from $H(p, q)$. We do that by generating their boundary line. A quick glance at Figure 2.5 reveals that the boundary line of a hexagonal system can be divided into two parts: the *left part* of the boundary (from the readers point of view), which starts on the y-axis below the key hexagon and finishes at the first junction with x-axis, and the rest of the boundary, which we call the *right part* of the boundary.

We recursively generate the left part of the boundary line. As soon as it reaches the x-axis, we start generating the right part. We maintain the length of the boundary line as well as the area of the hexagonal system. The trick that gives the area of the hexagonal system is simple: hexagons are counted in each row separately, starting from y-axis, such that their number is determined by their x-coordinate. Each time the boundary goes up (down), we add (subtract, respectively) the corresponding x-coordinate. When following the contour of HS in counterclockwise direction (i.e., in the direction of generating HS, see Fig. 2.5), there remain some hexagons out of HS to the left of the vertical contour line that goes down while hexagons to the left of the vertical line that goes up belong to the HS. The "zigzag" movements do not interfere with the area. Once the generation is over, the area of the HS gives the number of hexagons circumscribed in this manner. The area count is used to eliminate HSs with more than h hexagons, which appear during the generation of systems with h hexagons that belong to $H(p, q)$.

However, it would be a waste of time (and computing power) to insist on generating elements of $H(p, q)$ strictly. This would require additional tests to decide whether the left part of the boundary has reached x-axis precisely at hexagon p or not. In addition, once we find out we have reached the x-axis at hexagon, say, $p + 2$, why should we ignore it for the calculation of $H(p + 2, q)$? We shall therefore introduce another partition of the set of all properly placed HSs.

Given h and Cage($h - 1$), let $H^*(q) = \bigcup_{j=0}^{h-2} H(j, q)$, for all $q = 0, 1, \ldots, h - 2$. It is obvious that $\{H^*(q) : 0 \leq q \leq h - 2\}$ is a partition of the set of all HSs with h hexagons that are properly placed in Cage($h - 1$). Instead of having two separate phases (generating $H(p, q)$ and adding appropriate number to total), we now have one phase in which generating and counting are put together. We should prevent appearances of hexagonal systems from $H(p, q)$ with $p < q$. This requires no computational overhead because it can be achieved by forbidding some left and some down turns in the matrix representing the cage. On the contrary, avoiding the forbidden turns accelerates the process of generating the boundary line.

The algorithm is a school example of backtracking, thus facing all classical problems of the technique: Even for small values of h the search tree misbehaves, so it is essential to cut it as much as possible. One idea that cuts some edges of the tree is based on the fact that for larger values of q there are some parts of the cage that cannot be reached by hexagonal systems with $\leq h$ hexagons, but can easily be reached by useless HSs that emerge as a side effect. That is why we can, knowing q, forbid some regions of the cage.

The other idea that reduces the search tree is counting the boundary hexagons. A *boundary hexagon* is a hexagon that has at least one side in common with the boundary line and that is in the interior of the hexagonal system we are generating. It is obvious that boundary hexagons shall be part of the HS, so we keep track of their number. We use that number as a very good criterion for cutting off useless edges in the search tree. The idea is simple: further expansion of the left/right part of the boundary line is possible if and only if there are less than h boundary hexagons the boundary line has passed by.

The next idea that speeds up the algorithm is *living on credit*. When we start generating the left part of the boundary, we do not know where exactly is it going to finish on the x-axis, but we know that it is going to finish on the x-axis. In other words, knowing that there is one hexagon on the x-axis that is going to become a part of the HS, we can count it as a boundary hexagon in advance. It represents a credit of the hexagonal bank, which is very eagerly exploited. Thus, many useless HSs are discarded before the left part of the boundary lands on the x-axis.

All these ideas together represent the core of the algorithm, which can be outlined as follows.

```
Algorithm CageAlgorithm(h)
  procedure ExpandRightPart(ActualPos,BdrHexgns) {
    if (EndOfRightPart) then {
      n ←NoOfHexagons()
      if (n ≤ h) then {
        determine p;
        if (p = q) then total[n] ← total[n] + 1
        else total[n] ← total[n] + 2
      }
    }
    else {
        FindPossible(ActualPos,FuturePos)
        while (RightPartCanBeExpanded(ActuallPos, FuturePos))
            and (BdrHexgns≤ h) do {
          ExpandRightPart(FuturePos,update(BdrHexgns))
          CalcNewFuturePos(ActualPos,FuturePos)
        }
    }
  }
  procedure ExpandLeftPart(ActualPos,BdrHexgns) {
    if (EndOfLeftPart) then
      ExpandRightPart (RightlnitPos(q), updCredit(BdrHexgns))
    else {
      FindPossible(ActualPos,FuturePos)
          while (LeftPartCanBeExpanded(ActualPos, FuturePos)) and
          (BdrHexgns ≤h) do {
        ExpandLeftPart(FuturePos,update(BdrHexgns))
        CalcNewFuturePos(ActualPos,FuturePos)
      }
    }
  }
  begin main
    initialize Cage(h-1);
    total[1..h] ← 0
    for q ← 0 to h − 2 do {
      initialize y-axis key hexagon(q)
      ExpandLeftPart(LeftInitPos(q),InitBdrHexgns(q))
    }
  end main
```

2.6.1 Exercises

1. Design algorithms for counting square and triangular systems, using analogous ideas as these presented in this section for HSs.

2. Design algorithms for generating all HSs with area h and perimeter p, which belong to a given kind of symmetry of HSs (separate algorithms for each of these symmetry classes).
3. Prove that the number of HSs with h hexagons that cannot be placed properly in Cage($h-1$) is $(h^2 - h + 4)2^{h-3}$. Show that, among them, there are $(h^2 - 3h + 2)2^{h-4}$ pericondensed (with exactly one inner vertex) and $(h^2 + h + 6)2^{h-4}$ catacondensed HSs [56].

2.7 TWO ALGORITHMS FOR THE GENERATION OF HSs USING MCKAY'S METHOD

2.7.1 Generation of Hexagonal Systems Using the Boundary Edge Code

Caporossi and Hansen [12] give an algorithm, based on Algorithm McKayGeneration2 seen in Section 2.3, for isomorph-free generation of hexagonal systems represented by their boundary edge code (BEC). Their algorithm was the first to generate all the HSs with $h = 18$ to $h = 21$ hexagons.

We first describe the BEC representation of an HS, exemplified in Figure 2.6. Select an arbitrary external vertex of degree 3, and follow the boundary of the HS recording the number of boundary edges of each hexagon it traverses. Then, apply circular shifts and/or a reversal, in order to obtain a lexicographically maximum code. Note that each hexagon can appear one, two or three times as digits in the BEC code. Caporossi and Hansen [12] prove that an HS always start with a digit greater than or equal to 3.

Now, two aspects of the algorithm need specification: How to determine which sub-HS (of order $h-1$) of an HS of order h will be selected to be its parent in the generation tree, and how hexagons are added to existing HSs to create larger HSs.

In Figure 2.7, we show the generation tree explored by this algorithm for $h = 4$. Note that, for example, from the HS with code 5351 we can produce six nonisomorphic HSs, but only three of them are kept as its legitimate children. The rule for determining the parent of an HS is to remove the hexagon corresponding to the first digit of its BEC code. In other words, the parent of an HS is the one obtained by

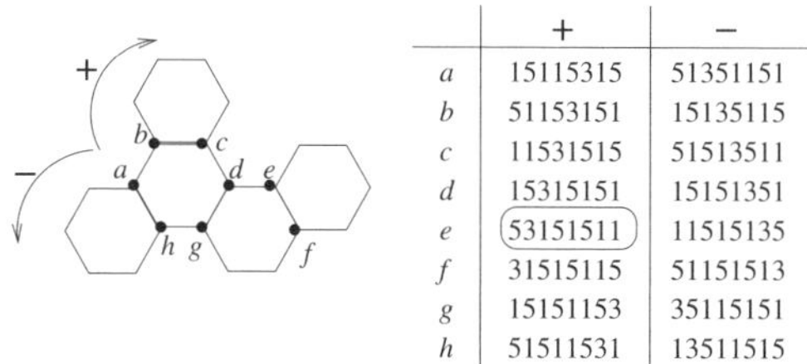

	+	−
a	15115315	51351151
b	51153151	15135115
c	11531515	51513511
d	15315151	15151351
e	53151511	11515135
f	31515115	51151513
g	15151153	35115151
h	51511531	13511515

FIGURE 2.6 Boundary edge code for a hexagonal system.

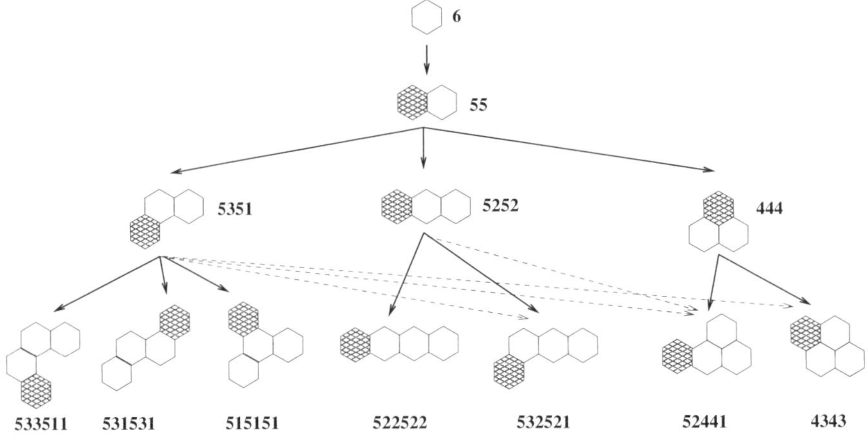

FIGURE 2.7 Isomorph-free search tree for $h = 4$.

removing its first hexagon. This operation in rare cases may disconnect the HS. This occurs precisely when the first hexagon occurs twice rather than once in the code. In such cases, the HS is orphan and cannot be generated via the algorithm's generation tree. A specially designed method for generation of orphan HSs must be devised in these cases. However, Caporossi and Hansen [12] proved that orphan HSs do not occur for $h \leq 28$, so they did not have to deal with the case of orphan HSs in their search.

Next, we describe how hexagons are added to create larger HSs. There are three ways in which a hexagon can be added to an HS, exemplified in Figure 2.8a:

1. A digit $x \geq 3$ in the BEC code corresponding to edges of a hexagon such that one of the edges belong only to this hexagon can be replaced by $a5b$, where $a + b + 1 = x$ and $a \geq 1$ and $b \geq 1$.
2. A sequence xy in the BEC code with $x \geq 2$ and $y \geq 2$ can be replaced by $(x - 1)4(y - 1)$.
3. A sequence $x1y$ with $x \geq 2$ and $y \geq 2$ in the BEC code can be replaced by $(x - 1)3(y - 1)$.

In each of the above cases, we must make sure that the addition of the hexagon does not produce holes. This can be accomplished by checking for the presence of a hexagon in up to three adjacent positions, as shown in Figure 2.8b; if any of these hexagons is present, this addition is not valid.

Procedure GenerateKids that generates, from an HS P with j hexagons, its children in the search with $j + 1$ hexagons is outlined next.

1. *Addition of hexagons:* Any attempt to add a hexagon in the steps below is preceded by a test that guarantees that no holes are created.

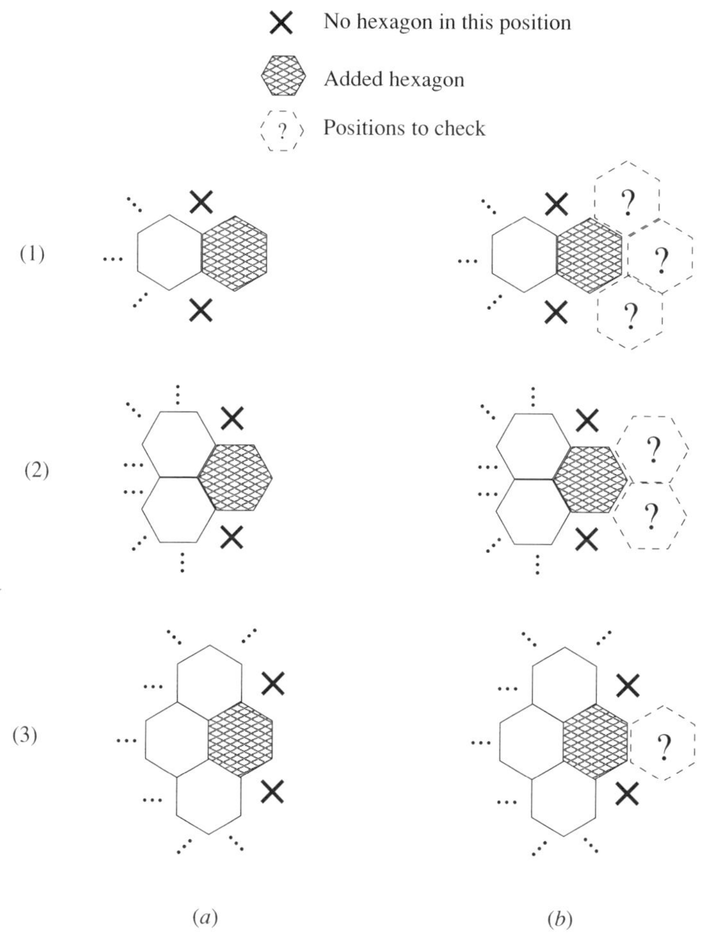

FIGURE 2.8 Ways of adding a hexagon to the boundary of an HS.

- Add a 5 in every possible way to the BEC code of P.
- If the BEC code of P does not begin with a 5, then add a 4 in every possible way to the BEC code of P; otherwise, only consider the addition of a 4 adjacent to the initial 5.
- If the BEC code of P has no 5 and at most two 4s, consider the addition of a 3.

2. *Parenthood validation:* For each HS generated in the previous step, verify that its BEC code can begin on the new hexagon. Reject the ones that cannot.

The correctness of the above procedure comes from the rule used to define who is the parent of an HS, and from the lexicographical properties of the BEC code. Now, putting this into the framework of Algorithm McKayGeneration2, from Section 2.3, gives the final algorithm.

```
Algorithm BECGeneration(P, Pcode, j)
  if (j = h) then output P
  else {
      S=GenerateKids(P, Pcode)
      Remove isomorph copies from S
      for all (P', Pcode') ∈ S do
         BECGeneration(P', Pcode', j + 1)
  }
```

Caporossi and Hansen [12] discuss the possibility of using Algorithm McKay-Generation1, which require computing the symmetries of the parent HS to avoid the isomorphism tests on the fourth line of the above algorithm. However, they report that experiments with this variant gave savings of only approximately 1 percent. Thus, this seem to be a situation in which it is worth using the simpler algorithm given by Algorithm McKayGeneration2.

2.7.2 Generation of Hexagonal Systems and Fusenes Using Labeled Inner Duals

Brinkmann et al. [7,8] exhaustively generate HSs using an algorithm that constructs all fusenes and filters them for HSs. Fusenes are a generalization of polyhexes that allows for irregular hexagons. They only consider simply connected fusenes, of which HSs are therefore a special case. In this section, we shall describe their algorithm for constructing fusenes. Testing whether a fusene fits the hexagonal lattice (checking whether it is an HS) can be easily done, and it is not described here. This algorithm was the first, and so far the only one, to exhaustively generate all HSs with $h = 22$ to $h = 24$.

We first describe the labeled inner dual graph representation of a fusene. The inner dual graph has one vertex for each hexagon, and two vertices are connected if their corresponding hexagons share an edge. This graph does not uniquely describe a fusene, but using an appropriate labeling together with this graph does, see Figure 2.9. Following the boundary cycle of the fusene, associate as many labels with a vertex as the number of times its corresponding hexagon is traversed, so that each label records the number of edges traversed each time. In the cases in which the hexagon occurs only once in the boundary, the label is omitted, as the number of edges in the

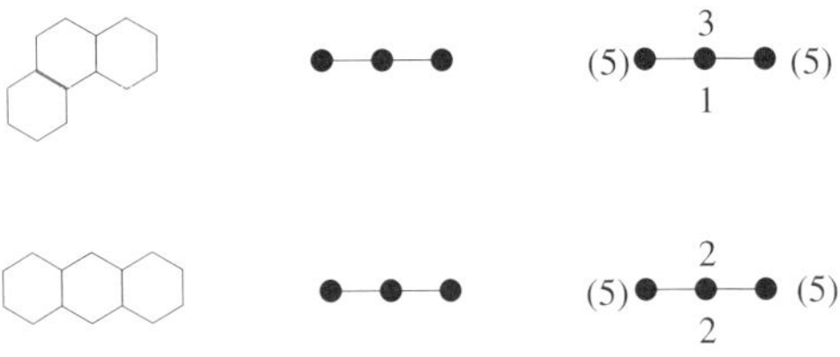

FIGURE 2.9 Hexagonal systems, their inner dual, and labeled inner dual graphs.

boundary is completely determined from $6-\deg(v)$, where $\deg(v)$ is the degree of the corresponding vertex.

Brinkmann et al. characterize the graphs that are inner duals of fusenes, which they call id-fusenes. They show that a planar embedded graph G is an id-fusene if and only if (1) G is connected, (2) all bounded faces of G are triangles, (3) all vertices not on the boundary have degree 6, and (4) for all vertices, the total degree, that is, the degree plus the number of times it occurs in the boundary cycle of the outer face, is at most 6.

Before we describe the algorithm, we need some basic definitions related to graph isomorphisms. Two graphs G_1 and G_2 are *isomorphic* if there exists a bijection (*isomorphism*) from the vertex set of G_1 to the vertex set of G_2 that maps edges to edges (and nonedges to nonedges). An isomorphism from a graph to itself is called an *automorphism* (also called a *symmetry*). The set of all automorphisms of a graph form a permutation group called the *automorphism group* of the graph, denoted $\text{Aut}(G)$. The *orbit* of a vertex v under $\text{Aut}(G)$ is the set of all images of v under automorphisms of G; that is, $\text{orb}(v) = \{g(v) : g \in \text{Aut}(G)\}$. This definition can be naturally extended to a set S of vertices as $\text{orb}(S) = \{g(S) : g \in \text{Aut}(G)\}$, where $g(S) = \{g(x) : x \in S\}$.

In the first step of the algorithm, nonisomorphic inner dual graphs of fusenes (id-fusenes) are constructed via Algorithm McKayGeneration1, described in Section 2.3. This first step is going to be described in more detail later in this section. In the second step, labeled inner duals are generated. We have to assign labels, in every possible way, to the vertices that occur more than once on the boundary, so that the sum of the labels plus the degrees of each vertex equals 6. In this process, we must make sure that we do not construct isomorphic labeled inner dual graphs, which can be accomplished by using some isomorphism testing method. To this end, the authors use the *homomorphism principle* developed by Kerber and Laue (see, for instance, the article by Grüner et al. [28]), which we do not describe here. However, it turns out that isomorphism testing is not needed for the labelings of most inner dual graphs, as discussed in the next paragraph, so the method that we choose for the second step is not so relevant.

One of the reasons for the efficiency of this algorithm is given next. For two labeled inner dual graphs to be isomorphic, we need that their inner dual graphs be isomorphic. Since the first step of the algorithm generates only one representative of each isomorphism class of inner dual graphs, isomorphic labeled inner dual graphs can only result from automorphisms of the same inner dual graph. So, if the inner dual graph has a trivial automorphism group, each of its generated labelings do not have to be tested for isomorphism. It turns out that the majority of fusene inner dual graphs have trivial automorphism group. For instance, for $n = 26$ trivial automorphism groups occur in 99.9994% of the inner dual graphs, each of them with more than 7000 labelings in average. So, this method saves a lot of unnecessary isomorphism tests in the second step of the algorithm.

Now, we give more details on the first step of the algorithm, namely the isomorph-free generation of the inner dual graphs via Algorithm McKayGeneration1, as described by Brinkmann et al. [7]. We need to specify how hexagons are added to

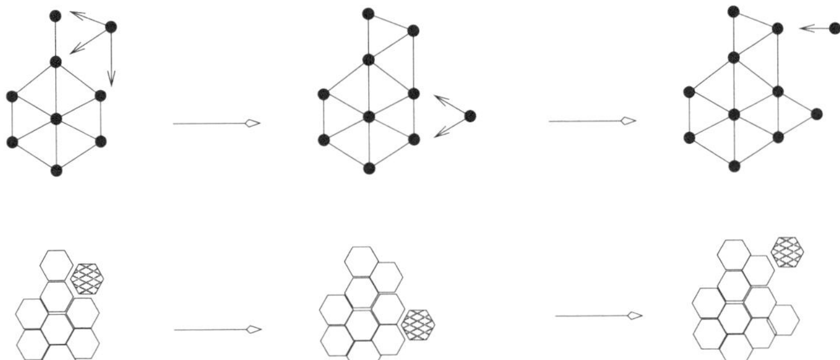

FIGURE 2.10 Valid augmentations of an id-fusene.

existing id-fusenes to create larger ones and how to determine which subgraphs (of order $v-1$) of an id-fusene of order v will be selected to be its parent in the generation tree.

In order to describe how we augment an id-fusene, we need some definitions. A *boundary segment* of an id-fusene is a set of $l-1$ consecutive edges of the boundary cycle. The vertices of the boundary segment are the end vertices of its edges (there are l of them). For convenience, a single vertex in the boundary cycle is a boundary segment with $l=1$. A boundary segment is said to be *augmenting* if the following properties hold: $l \leq 3$, its first and last vertices have total degree at most 5, if $l=1$ its only vertex has total degree at most 4, and if $l=3$ and the middle occurs only once in the boundary, it has total degree 6; see examples of valid augmentations in Figure 2.10. The augmentation algorithm is based on the following lemma.

Lemma 4 *All id-fusenes can be constructed from the inner dual of a single hexagon (a single vertex graph) by adding vertices and connecting them to each vertex of an augmenting boundary segment.*

McKay [46] describes a general way of determining parenthood in Algorithm McKayGeneration1 based on a canonical choice function f. When applied to the case of the current algorithm with the given augmentation, f is chosen to be a function that takes each id-fusene G to an orbit of vertices under the automorphism group of G that satisfy the following conditions:

1. $f(G)$ consists of boundary vertices that occur only once in the boundary cycle and have degree at most 3;
2. $f(G)$ is independent of the vertex numbering of G; that is, if Φ is an isomorphism from G to G', then $\Phi(f(G)) = f(G')$.

Now, as described by McKay [46], graph G is defined to be the parent of graph $G \cup \{v\}$ if and only if $v \in f(G \cup \{v\})$. The specific f used by Brinkmann et al. [7] is a bit technical and would take a page or more to properly explain, so we refer the interested reader to their paper.

Procedure GenerateKidsIDF that generates, from an id-fusene G with v hexagons, its children in the search tree with $v+1$ hexagons is outlined next.

1. *Addition of hexagons*:
   - Compute the orbit of the set of vertices of each augmenting boundary segment of G.
   - Connect the new vertex $n + 1$ to the vertices in one representative of each orbit, creating a new potential child graph G' per orbit.
2. *Parenthood validation*: For each G' created in the previous step, if $n + 1 \in f(G')$ then add G' to S, the set of children of G.

As discussed in the presentation of Algorithm McKayGeneration1, from Section 2.3, no further isomorphism tests are needed between elements of S, unlike the algorithm in Section 2.7.1. Now, putting all these elements into the given framework gives the final algorithm for the isomorph-free generation of id-fusenes.

```
Algorithm IDFGeneration(G, n)
  if (n = h) then output G
  else {
      S=GenerateKidsIDF(G, n)
      for all G' ∈ S do
         IDFGeneration(G', n + 1)
  }
```

For this algorithm and for the one in Section 2.7.1, it is possible and convenient to distribute the generation among several computers, each expanding part of the generation tree. This can be done by having each computer build the generation tree up to certain level and then start the generation starting on a node at that level.

2.7.3 Exercises

1. Draw the edges and vertices in the next level ($h = 5$) of the search tree of the BEC algorithm generation given in Figure 2.7. Recall that it must contain exactly 22 nodes (and edges).
2. Prove that the BEC code of an HS always begins with a digit greater than or equal to 3 [12] .
3. Prove that no HS obtained by the addition of a hexagon sharing more than three consecutive edges with the current HS can be one of its legitimate children in the search tree of Algorithm BECGeneration [12].
4. Consider the three types of addition of hexagons to an HS, given in Figure 2.8a. For each of these cases, prove that the added hexagon creates a polyhex with a hole if and only if at least one of the positions marked with "?" (in the corresponding figure in Fig. 2.8b) contains a hexagon.
5. Prove that any HS with $h \geq 2$ can be obtained from the HS with $h = 2$ by successive additions of hexagons satisfying rules 1–3 in Section 2.7.1 for hexagon additions in the BEC code algorithm.
6. Prove, by induction on n, that a graph with n vertices is an id-fusene if and only if the four properties listed in Section 2.7.2 are satisfied.

7. Give an example of an id-fusene graph that does not correspond to a hexagonal system.
8. Write an algorithm for filtering fusenes for hexagonal systems, that is, an algorithm that verifies whether a labeled inner dual graph of a fusene can be embedded into the hexagonal lattice.
9. Prove Lemma 4 [7].
10. Prove that Algorithm IDFGeneration accepts exactly one member of every isomorphism class of id-fusenes with n vertices [7,46].

2.8 PERFECT MATCHINGS IN HEXAGONAL SYSTEMS

The transformation from molecular structure (e.g., Fig. 2.1a) to an HS (e.g., Fig. 2.1b) leaves out the information about double valences between carbon atoms. Clearly, each carbon atom has a double valence link with exactly one of its neighboring carbon atoms. Thus, double valences correspond to a perfect matching in an HS. Therefore, an HS is the skeleton of a benzenoid hydrocarbon molecule if and only if it has a perfect matching.

An HS that has at least one perfect matching is called *Kekuléan*; otherwise, it is called *non-Kekuléan*. Kekuléan HSs are further classified as either *normal* (if every edge belongs to at least one perfect matching) or *essentially disconnected* (otherwise). Classification of HSs according to the perfect matching property is summarized by Cyvin et al. [14]. An HS with a given perfect matching is called a Kekulé *structure* in chemistry and has great importance. Figure 2.11a and b shows two Kekulé structures that corresponds to the HS in Figure 2.1b.

If the number of vertices of an HS is odd, then clearly there is no perfect matching. We denote by $K(G)$ the number of perfect matchings of a graph G, and refer to it as the

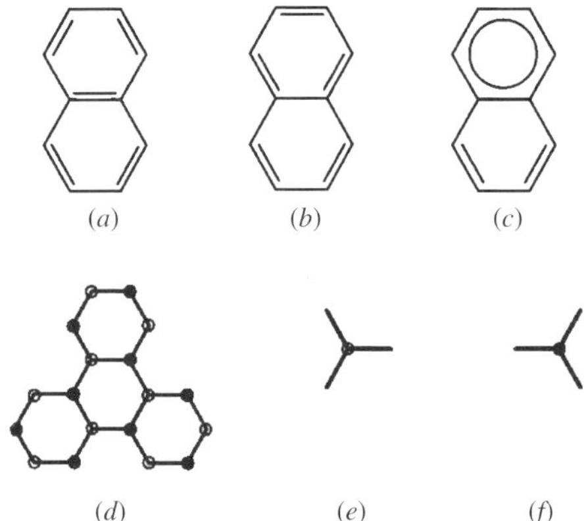

FIGURE 2.11 (*a*–*c*) Kekulé structures and (*d*–*f*) vertex coloring of hexagonal systems.

K number of G. When G is an HS, $K(G)$ is the number of its Kekulé structures. The edges belonging to a given Kekulé structure are *double bonds* while others are *single bonds*. The stability and other properties of HSs have been found to correlate with their K numbers. A whole book [17] is devoted to Kekulé structures in benzenoid hydrocarbons. It contains a list of other references on the problem of finding the "Kekulé structure count" for hydrocarbons.

The vertices of an HS may be divided into two groups, which are conveniently called *black* and *white*. Choose a vertex and color it white, and color all its neighboring vertices black. Continue the process such that all vertices adjacent to a black vertex are white and vice versa. Figure 2.11d shows an example of such coloring. The black and white internal vertices correspond to two different configurations of edges as drawn in Figure 2.11e and f. Every edge joins a black and a white vertex; therefore, HSs are bipartite graphs. Let the number of white and black vertices be nw and nb, respectively, and $\Delta = |nw - nb|$. Clearly, $nw + nb = p + i$ (recall that p is the perimeter and i is the number of internal vertices of an HS). Every edge of a perfect matching of a given HS joins a black and a white vertex. Therefore, if the HS is Kekuléan then $\Delta = 0$. The reverse is not always true. Non-Kekuléan HSs with $\Delta = 0$ exist and are called *concealed*, while for $\Delta > 0$ they are referred to as the *obvious non-Kekuléan*.

2.8.1 *K* Numbers of Hexagonal, Square, and Pentagonal Chains

This section contains a study of the numbers of perfect matchings of square, pentagonal, and hexagonal chains, that is the graphs obtained by concatenating squares, pentagons, and hexagons, respectively. A mapping between square (pentagonal) and hexagonal chains that preserves the number of perfect matchings is established. The results in this section are by Tosic and Stojmenovic [58] (except for the proof of Theorem 1, which is original).

By a *polygonal chain* $P_{k,s}$ we mean a finite graph obtained by concatenating s k-gons in such a way that any two adjacent k-gons (cells) have exactly one edge in common, and each cell is adjacent to exactly two other cells, except the first and last cells (*end cells*) that are adjacent to exactly one other cell each. It is clear that different polygonal chains will result, according to the manner in which the cells are concatenated.

Figure 2.12a shows a hexagonal chain $P_{6,11}$. The LA-sequence of a hexagonal chain is defined by Gutmann [29] as follows. A hexagonal chain $P_{6,s}$ is represented by a word of length s over the alphabet $\{A, L\}$. The ith letter is A (and the corresponding hexagons is called a *kink*) if and only if $1 < i < s$ and the ith hexagon has an edge that does not share a common vertex with any of its two neighbors. Otherwise, the ith letter is L. For instance, the hexagonal chain in Figure 2.12a is represented by the word $LAALALLLALL$, or, in abbreviated form, $LA^2LAL^3AL^2$. The LA-sequence of a hexagonal chain can always be written in the form $P_6\langle x_1, x_2, \ldots, x_n\rangle$ to represent $L^{x_1}AL^{x_2}A\ldots AL^{x_n}$, where $x_1 \geq 1$, $x_n \geq 1$, $x_i \geq 0$ for $i = 2, 3, \ldots, n-1$. For instance, the LA-sequence of the hexagonal chain in Figure 2.12 may be written in the form $P_6\langle 1, 0, 1, 3, 2\rangle$, which represents $LAL^0ALAL^3AL^2$. It is well known that

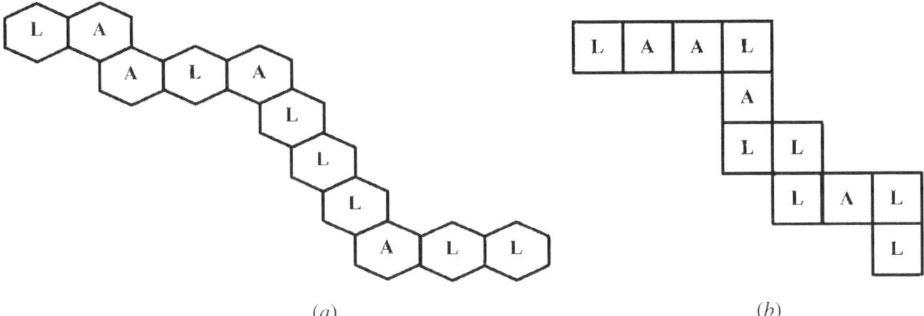

FIGURE 2.12 LA-sequences of (a) hexagonal and (b) square chains.

the K number of a hexagonal chain is entirely determined by its LA-sequence, no matter which way the kinks go [33]. The term *isoarithmicity* for this phenomenon is coined. Thus, $P_6\langle x_1, x_2, \ldots, x_n\rangle$ represents a class of isoarithmic hexagonal chains.

Figure 2.12b shows a square chain $P_{4,11}$. We introduce a representation of square chains in order to establish a mapping between square and hexagonal chains that will enable us to obtain the K numbers for square chains. A square chain $P_{4,s}$ is represented by a word of length s over the alphabet $\{A, L\}$, also called its LA-sequence. The ith letter is A if and only if each vertex of the ith square also belongs to an adjacent square. Otherwise the ith letter is L. For instance, the square chain in Figure 2.12b is represented by the word $LAALALLLALL$, or, in abbreviated form, $LA^2LAL^3AL^2$. Clearly, the LA-sequence of a square chain can always be written in the form $P_4\langle x_1, x_2, \ldots, x_n\rangle$ to represent $L^{x_1}AL^{x_2}A\ldots AL^{x_n}$, where $x_1 \geq 1$, $x_n \geq 1$, $x_i \geq 0$ for $i = 2, 3, \ldots, n-1$. For example, the LA-sequence of the square chain in Figure 2.12 may be written in the form $P_4\langle 1, 0, 1, 3, 2\rangle$ to represent $LAL^0ALAL^3AL^2$. We show below that all square chains of the form $P_4\langle x_1, \ldots, x_n\rangle$ are isoarithmic.

We will draw pentagonal chains so that each pentagon has two vertical edges and a horizontal one that is adjacent to both vertical edges. The common edge of any two adjacent pentagons is drawn vertical. We shall call such way of drawing a pentagonal chain the horizontal representation of that pentagonal chain. From the horizontal representation of a pentagonal chain one can see that it is composed of a certain number (≥ 1) of segments; that is, two adjacent pentagons belong to the same segment if and only if their horizontal edges are adjacent. We denote by $P_5\langle x_1, x_2, \ldots, x_n\rangle$ the pentagonal chain consisting of n segments of lengths $x_1, x_2, \ldots, x_n$, where the segments are taken from left to right. Figure 2.15a shows $P_5\langle 3, 2, 4, 8, 5\rangle$. Notice that one can assume that $x_1 > 1$ and $x_n > 1$.

Among all polygonal chains, the hexagonal chains were studied the most extensively, since they are of great importance in chemistry. We define $P_6\langle\rangle$ as the hexagonal chain with *no hexagons*.

Theorem 1 [58]

$$K(P_6\langle\rangle) = 1,$$
$$K(P_6\langle x_1\rangle) = 1 + x_1,$$
$$K(P_6\langle x_1, \ldots, x_{n-1}, x_n\rangle) = (x_n + 1)K(P_6\langle x_1, \ldots, x_{n-1}\rangle) + K(P_6\langle x_1, \ldots, x_{n-2}\rangle), \quad for\ n \geq 2.$$

Proof. It is easy to deduce the K formula for a single linear chain (polyacene) of x_1 hexagons, $K(P_6\langle x_1\rangle) = 1 + x_1$ [27]. Let H be the last kink (A-mode hexagon) of $\langle x_1, \ldots, x_n\rangle$ and u and v be the vertices belonging only to hexagon H (Fig. 2.13a). We apply the method of fragmentation by attacking the bond uv (Fig. 2.13a). If a perfect matching (Kekulé structure) contains the double bond uv, then the rest of such a perfect matching will be the perfect matching of the graph consisting of two components: $\langle x_n\rangle$ and $\langle x_1, \ldots, x_{n-1}\rangle$ (Fig. 2.13a). The number of such perfect matchings is $K(P_6\langle x_n\rangle)K(P_6\langle x_1, ..., x_{n-1}\rangle)$, that is, $(x_n + l)K(P_6\langle x_1, ..., x_{n-1}\rangle)$. On the contrary, each perfect matching not containing uv (uv is a single bond in the corresponding Kekulé structure) must contain all the double bonds indicated in Figure 2.13b. The rest of such a perfect matching will be a perfect matching of $\langle x_1, x_2, \ldots, x_{n-2}\rangle$ and the number of such perfect matchings is $K(P_6\langle x_1, \ldots, x_{n-2}\rangle)$. The recurrence relation now follows easily. ■

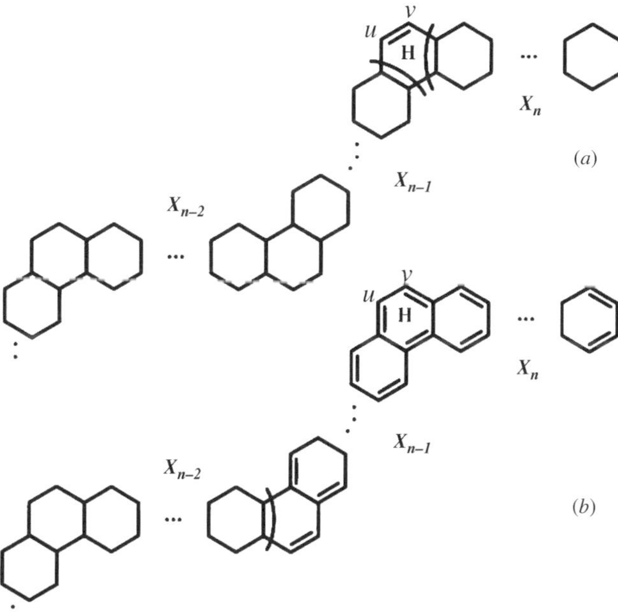

FIGURE 2.13 Recurrence relation for the K number of hexagonal systems.

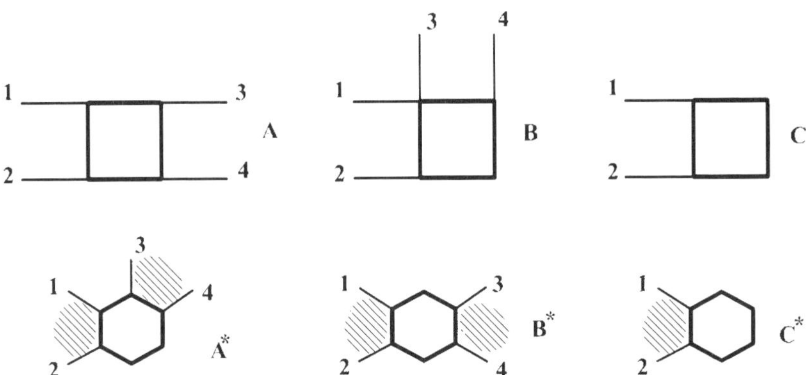

FIGURE 2.14 Transforming square chains into hexagonal chains.

Theorem 2 [58] $K(P_4\langle x_1, x_2, \ldots, x_n\rangle) = K(P_6\langle x_1, x_2, \ldots, x_n\rangle)$.

Proof. Referring to Figure 2.14, it is easy to see that if in a square chain some (or all) structural details of the type A, B, and C are replaced by A*, B*, and C*, respectively, the K number will remain the same. By accomplishing such replacements, each square chain can be transformed into a hexagonal chain with the same LA-sequence. Therefore, a square chain and corresponding hexagonal chain represented by the same LA-sequence have the same K number. For example, the square chain in Figure 2.12b can be transformed into the hexagonal chain in Figure 2.12a. Note that the corner squares of a square chain correspond to the linear hexagons, and vice versa, in this transformation. ■

It is clear that all other properties concerning the K numbers of square chains can be derived from the corresponding results for hexagonal chains and that the investigation of square chains as a separate class from that point of view is of no interest. Let us now study the K number of pentagonal chains. First, let us recall a general result concerning matchings of graphs. Let G be a graph and u, x, y, v distinct vertices, such that ux, xy, yv are edges of G, u and v are not adjacent, and x and y have degree 2. Let the graph H be obtained from G by deleting the vertices x and y and by joining u and v. Conversely, G can be considered as obtained from H by inserting two vertices (x and y) into the edge uv). We say that G can be reduced to H, or that G is reducible to H; clearly, $K(G) = K(H)$.

Theorem 3 [58] *If* $x_1 + x_2 + \cdots + x_n$ *is odd, then* $K(P_5\langle x_1, \ldots, x_n\rangle) = 0$. *Otherwise (i.e., if the sequence* $x_1, x_2, \ldots, x_n$ *contains an even number of odd integers), let* $s(j_1), s(j_2), \ldots, s(j_t)$, $j_1 < j_2 < \cdots < j_t$, *be the odd numbers in the sequence* $s(r) = x_1 + \cdots + x_r, r = 1, 2, \ldots, n$, *and let* $s(j_0) = -1$, *and* $s(j_{t+1}) = s(n) + 1$; *then* $K(P_5\langle x_1, \ldots, x_n\rangle) = K(P_6\langle y_1, y_2, \ldots, y_{t+1}\rangle)$, *where* $y_1 = (s(j_1) - 1)/2 = (s(j_1) - s(j_0) - 2)/2$, $y_{t+1} = (s(n) - s(j_t) - 1)/2 = (s(j_{t+1}) - s(j_t) - 2)/2$, *and, for* $2 \le i \le t$, $y_i = (s(j_i) - s(j_{i-1}) - 2)/2$.

Proof. Clearly, a pentagonal chain consisting of p pentagons has $3p + 2$ vertices. Hence, a pentagonal chain with an odd number of pentagons has no perfect matching. Therefore, we assume that it has an even number of segments of odd length.

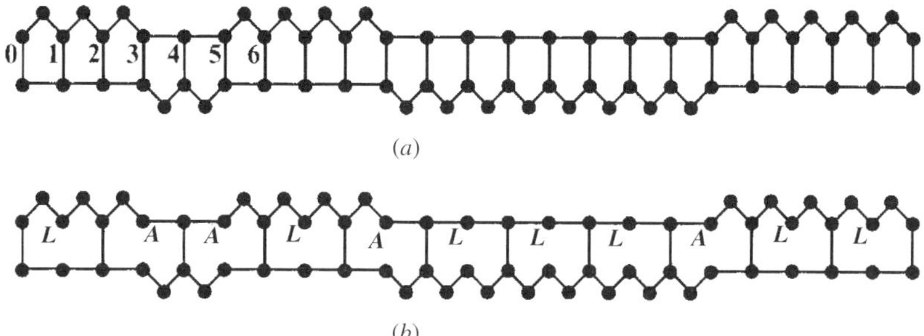

FIGURE 2.15 Transforming (*a*) pentagonal chains into (*b*) octagonal chains.

Consider a horizontal representation of $P_5\langle x_1, x_2, \ldots, x_n\rangle$ (Fig. 2.15a). Label the vertical edges by 0, 1, ..., $s(n)$, from left to right. Clearly, no edge labeled by an odd number can be included in any perfect matching of $P_5\langle x_1, x_2, \ldots, x_n\rangle$, since there are an odd number of vertices on each side of such an edge. By removing all edges labeled with odd numbers we obtain an octagonal chain consisting of $s(n)/2$ octagons (Fig. 2.15b). This octagonal chain can be reduced to a hexagonal chain with $s(n)/2$ hexagons (Fig. 2.12a). It is evident that in the process of reduction, each octagon obtained from the two adjacent pentagons of the same segment becomes an L-mode hexagon, while each octagon obtained from the two adjacent pentagons of different segments becomes a kink. The number of kinks is t, since each kink corresponds to an odd $s(r)$. It means that this hexagonal chain consists of $t+1$ segments. Let y_i be the number of L-mode hexagons in the ith segment. Then the sequence y is defined as given in the theorem. Since reducibility preserves K numbers, it follows that $K(P_5\langle x_1, x_2, ..., x_n\rangle) = K(P_6\langle y_1, y_2, \ldots, y_{t+1}\rangle)$. ■

Corollary 1 [58] *Let* $x_1, x_2, \ldots, x_n$ *be even positive integers,* $n \geq 1$. *Then,* $K(P_5\langle x_1, \ldots, x_n\rangle) = (x_1 + \cdots + x_n)/2 + 1$.

Proof. Since all partial sums $s(r)$ in Theorem 3 are even, no kink is obtained in the process of reduction to a hexagonal chain. Thus, a linear hexagonal chain consisting of $h = (x_1 + x_2 + \cdots + x_n)/2$ hexagons is obtained (i.e. $P_6\langle h\rangle = L^h$). Since $K(P_6\langle h\rangle) = h + 1$, it follows that $K(P_5\langle x_1, \ldots, x_n\rangle) = h + 1$. ■

2.8.2 Clar Formula

A hexagon q in an HS is said to be an *aromatic sextet* when it has exactly three (alternating) single and three double bonds in a given perfect matching. In some references, an aromatic sextet q is called a *resonant hexagon*, defined as a hexagon such that the subgraph of the HS obtained by deleting from it the vertices of q together with their edges has at least one perfect matching. For instance, the upper hexagon in Figure 2.11a is an aromatic sextet. When single and double bonds are exchanged in an aromatic sextet (as in Fig. 2.11b), one obtains another Kekulé structure of the same HS. Aromatic sextets are usually marked by circles inside the hexagon, and such a circle corresponds to two possible matchings of the edges of the hexagon. Figure 2.11c shows an HS with a circle that replaces matchings of Figure 2.11a and b. Clearly, it is not allowed to draw circles in adjacent hexagons. Circles can be drawn in hexagons if the rest of the hexagonal system has at least one perfect matching.

The so-called *Clar formula* of an HS is obtained when the maximal number of circles is drawn such that it leads to a Kekulé structure of the HS. Therefore, not all perfect matchings correspond to a Clar formula (only the maximal ones, when the placement of additional circles is not possible by changing some edges of the matching).

In this section, we shall study Clar formulas of hexagonal chains. We denote by $S(B)$ the number of circles in a Clar formula of a hexagonal chain B. The benzenoid chains with a unique Clar formula (*Clar chains*) are characterized. All the results are taken from the work by Tosic and Stojmenovic [57].

It is clear that the chain with exactly one hexagon ($h = 1$) is a Clar chain. The following theorem describes Clar chains for $h > 1$.

Theorem 4 *A hexagonal chain B is a Clar chain if and only if its LA-sequence is of the form $LA^{m_1}LA^{m_2}L\ldots LA^{m_k}L$, where $k \geq 1$ and all the numbers $m_1, m_2, \ldots, m_k$ are odd.*

Proof. Let B be a benzenoid chain given by its LA-sequence

$$L^{m'_0}A^{m_1}L^{m'_1}A^{m_2}L^{m'_2}\ldots L^{m'_{k-1}}A^{m_k}L^{m'_k},$$

where $m'_0 \geq 1; m'_k \geq 1; m'_i \geq 0$ for $i = 1, \ldots, k-1$; and $m_k \geq 1$, for $i = 1, 2, \ldots, k$.

The part of this chain between the two successive appearances of the A-mode hexagon is said to be an open segment of B. The first m'_0 L-mode hexagons and m'_k last L-mode hexagons also constitute the segments (end segments) of lengths m'_0 and m'_k, respectively. An inner open segment may be without any hexagon: no-hexagon segment. A closed segment is obtained by adding to an open segment two A-mode hexagons that bound it, if it is an inner segment, or one A-mode hexagon that bounds it, if it is an end segment. Two adjoined closed segments always have exactly one common A-mode hexagon.

It easily follows that between any two circles in a Clar-type formula of a benzenoid chain, there must be at least one A-mode hexagon (kink) of that chain. Also, each closed segment of a benzenoid chain contains exactly one circle in any Clar formula of that chain.

Let B be a Clar chain and let H be an A-mode hexagon of B, adjacent to at least one L-mode hexagon of B. Consider a closed segment of B with at least one L-mode hexagon. If any of the two A-mode hexagons of that segment is with circle in a Clar formula of B, then that circle can be replaced by a circle in any of the L-mode hexagon of that segment, producing another Clar formula of B. It is in contradiction with the fact that B is a Clar chain. Thus, H is without circle in any Clar formula of B.

We now show that a Clar chain B does not contain two adjacent L-mode hexagons. Consider a closed segment of B with at least two L-mode hexagons. Neither of the end hexagons of that segment is circled in the Clar formula of B. According to the above two observations, exactly one of the L-mode hexagons of that segment is circled. However, it is clear that each of them can be chosen to be circled. So, the existence of two adjacent L-mode hexagons imply that the Clar formula of B is not unique; that is, B is not a Clar chain. Therefore, each L-mode hexagon of a Clar chain is circled in the Clar formula of that chain.

A benzenoid chain with h hexagons in which all hexagons except the first and the last are A-mode hexagons is called a *zigzag* chain and is denoted by $A(h)$. We show that a zigzag chain $A(h)$ with h hexagons is a Clar chain if and only if h is an odd number. A chain with h hexagons cannot have more than $\lceil h/2 \rceil$ circles in its Clar formula. Now, if $h = 2k + 1$ is odd, then the choice of $\lceil h/2 \rceil = k + 1$ nonadjacent hexagons of $A(h)$ is unique and obviously it determines the unique Clar formula of $A(h)$. Consider now an $A(h)$ with h even. The number of circles in that Clar formula is not greater than $h/2$. However, one can easily draw $h/2$ circles in every second hexagon, thus obtaining two different Clar formulas. Thus, $A(h)$ is not a Clar chain for even h.

The proof proceeds by induction on k. If $k = 1$, then the statement of the theorem follows from the last observation on zigzag chains. Consider the case when B is not a zigzag chain. In that case, B has at least three L-mode hexagons.

($\Rightarrow$) Suppose that B is a Clar chain and for some i, $1 \leq i \leq k$, m_i is even. Consider the part of B corresponding to the subword A^{m_i} (Fig. 2.16), which is a zigzag chain $A(m_i)$. Two L-mode hexagons that bound this zigzag chain in B are with circles in the unique Clar formula of B. It follows that the first and the last hexagons of $A(m_i)$ (numbered by 1 and m_i in Fig. 2.16) are without circles in that formula. The remaining part of $A(m_i)$ is a zigzag chain $A(m_i - 2)$ with an even number of hexagons and it is independent from the rest of B with respect to the distribution of circles in the Clar formula of B. So, $A(m_i - 2)$ itself must be a Clar chain. This is contradiction with the previous observation on zigzag chains. It means that m_i cannot be even. Thus, all m_i, $i = 1, 2, \ldots, k$, are odd.

The number of hexagons of B is $h = m_1 + m_2 + \cdots + m_k + (k + 1)$, where all $m_1, m_2, \cdots, m_k$, are odd numbers; so h must be odd.

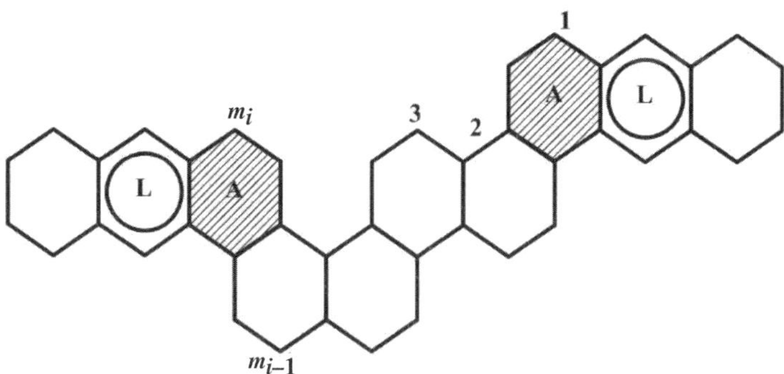

FIGURE 2.16 Clar chain with an even m_i (contradiction).

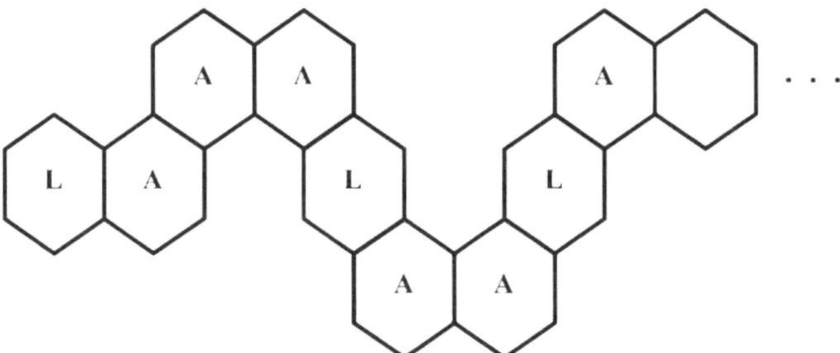

FIGURE 2.17 LA-sequence with odd m_i's.

($\Leftarrow$) Let B be a hexagonal chain with the LA-sequence $LA^{m_1}LA^{m_2}L\ldots LA^{m_k}L$, where all the numbers $m_1, m_2, \ldots, m_k$ are odd, and $k > 1$. Consider B as obtained from two chains B_1 and B_2 with LA-sequences, respectively, $LA^{m_1}L$ and $LA^{m_2}LA^{m_3}L\cdots LA^{m_k}L$, by identifying the last L-mode hexagon of B_1 and the first L-mode hexagon of B_2 (the second L-mode hexagon in Fig. 2.17).

By induction hypothesis, both B_1 and B_2 are Clar chains. The common L-mode hexagon of B_1 and B_2 is with circle in both Clar formulas, for B_1 and B_2. Hence, B is a Clar chain. ■

Let B be a Clar chain with h hexagons. From the discussions in the proof of the previous theorem it follows that, starting from a circled end hexagon, hexagons with and without circle alternate. Thus, the number of circles in the unique Clar formula of B is $S(B) = (h+1)/2$.

We say that two LA-sequences are equivalent if they coincide or can be obtained from each other by reversing. Two benzenoid chains with the same number of hexagons h are isoarithmic if they have equivalent LA-sequences. So, the number of nonisoarithmic chains with h hexagons is equal to the number of nonequivalent LA-sequences of the length h.

We shall determine the number of nonisoarithmic chains with h hexagons and with a unique Clar formula. We denote this number by $N(h)$. Clearly, $N(h) = 0$, if h is an even number, and $N(1) = 1$.

Theorem 5 *Let h be an odd positive integer, $h > 1$. Then*

$$N(h) = 2^{(h-5)/2} + 2^{\lfloor (h-1)/4 \rfloor - 1}.$$

Proof. From Theorem 4, it follows that $N(h)$ is equal to the number of LA-sequences $LA^{m_1}LA^{m_2}L\ldots LA^{m_k}L$, such that $m_1 + m_2 + \cdots + m_k = h - k - 1$, $k \geq 1$, and all the numbers $m_1, m_2, \ldots, m_k$ are odd. Now, the number of such LA-sequences is equal to the number of compositions of $h-1$ into even positive integers, that is, to the number of compositions of $n = (h-1)/2$ into positive integers. This last number is equal to $2^{n-1} = 2^{(h-3)/2}$. Among these compositions there are $2^{\lfloor n/2 \rfloor} = 2^{\lfloor (h-1)/4 \rfloor}$ of those that are symmetric, that is, those that correspond to symmetric (self-reversible) LA-sequences. So, the number of nonequivalent LA-sequences in question is

$$(2^{(h-3)/2} - 2^{\lfloor (h-1)/4 \rfloor})/2 + 2^{\lfloor (h-1)/4 \rfloor} = 2^{(h-5)/2} + 2^{\lfloor (h-1)/4 \rfloor - 1}.$$

That is at the same time the number of nonisoarithmic Clar chains. Among them, $2^{\lfloor (h-1)/4 \rfloor}$ are self-isoarithmic. ■

2.8.3 Exercises

1. Show that every catacondensed HS is normal [33].
2. Assume that an HS is drawn so that some of its edges are vertical. Then, we distinguish peaks and valleys among the vertices on the perimeter. A peak lies above its nearest neighboring vertices, while a valley lies below its nearest neighbors. Let np and nv denote the number of peaks and valleys in a given HS. Prove that $|np - nv| = |nb - nw| = \Delta$ [17].
3. Prove that an HS B is Kekuléan if and only if it has equal numbers of black and white vertices, and if for all edge cuts of B, the fragment F_1 does not have more white vertices than black vertices. An edge cut decomposes HS into two parts F_1 and F_2 (mutually disconnected but each of them is a one-component graph) such that black end vertices of all edges in the cut belong to F_1 [63].
4. Prove that the K number of an HS satisfies $h + 1 \leq K \leq 2^{h-1} + 1$ [32].
5. Let x, y, and z denote the number of double bounds of an HS for each of three edge orientations (i.e., parallel to three alternating edges of a hexagon), respectively. Prove that all Kekulé structures of an HS have the same triplet $\{x, y, z\}$.
6. Prove that a triplet (x, y, z), $x \leq y \leq z$, corresponds to a catacondensed HS if and only if $x + y + z$ is odd and $x + y \geq z + 1$ [65].
7. Prove that every perfect matching of an HS contains three edges, which cover all the six vertices of a hexagon [31].
8. Prove by induction that

$$K(P_6\langle x_1, \ldots, x_{n-1}, x_n \rangle)$$
$$= f_{n+1} + \sum_{0<i_1<\cdots<i_k\leq n, 1\leq k\leq n} f_{n+1-i_k} f_{i_k - i_{k-1}} \cdots f_{i_2 - i_1} f_{i_1} x_{i_1} x_{i_2} \cdots x_{i_k},$$

   where f_n is the nth Fibonacci number [58].
9. Prove that the K number for the chain $LA^{p-1}LA^{q-1}L$ is $f_{p+q+2} + f_{p+1}f_{q+1}$ [35,58].
10. Prove that the K number for the hexagonal chain with n segments of the same length m is [4]

$$K(P_6\langle m, \ldots, m\rangle)$$
$$= \frac{\left(m+1+\sqrt{(m+1)^2+4}\right)^{n+1} - \left(m+1-\sqrt{(m+1)^2+4}\right)^{n+1}}{2^{n+1}\sqrt{(m+1)^2+4}}.$$

11. Prove that the K number for the LA-sequence $L^m AL^{m-1}A \ldots AL^{m-1}AL^m$ (with $n-1$ As) is [2]

$$\frac{1}{\sqrt{m^2+4}} \left[\left(\sqrt{m^2+4}+2\right)\left(\frac{m+\sqrt{m^2+4}}{2}\right)^n + \left(\sqrt{m^2+4}-2\right)\left(\frac{m-\sqrt{m^2+4}}{2}\right)^n\right].$$

12. Prove that the K number for pentagonal chains is [58]

$$K(P_5\langle x_1, \ldots, x_{n-1}, x_n\rangle) = f_{t+2}$$
$$+ \sum_{\substack{0=i_0<i_1<\cdots<i_r\le t+1,\\ 1\le r\le t+1}} (f_{t+2-i_r})/2^r \prod_{l=1}^{r} (s(j_{i_l}) - s(j_{i_l-l}) - 2) f_{i_l-i_{l-1}},$$

where f_k is the kth Fibonacci number and the sequence s is defined in the text.
13. Let m be an odd positive integer > 1. Then, $K(P_5\langle m^2\rangle) = (m^2+2m+5)/4$, and $K(P_5\langle m^4\rangle) = (m^3+2m^2+5m+4)/4$ [25,58].
14. Prove that the K number of the zigzag hexagonal chain with LA-sequence $LA^{k-2}L$ is f_{k+2} [58,61].
15. Prove that the K number of pentagonal zigzag chain with $2k$ pentagons and the K number of hexagonal zigzag chains with k hexagons are the same [58].
16. Prove that $K(P_5\langle 1^{2k}\rangle) = f_{k+2}$ [25,58].
17. Design a general algorithm for the enumeration of Kekulé structures (K numbers) of benzenoid chains and branched catacondensed benzenoids [16,27].
18. Suppose that some edges of an HSs are vertical. Peaks (valleys) are vertices on the perimeter with degree 2 such that both their neighbors are below (above, respectively) them. Prove that the absolute magnitude of the difference between the numbers of peaks and valleys is equal to Δ. Show that the numbers of peaks and valleys in a Kékulean HS are the same.
19. A monotonic path in an HS is a path connecting a pick with a valley, such that starting at the pick we always go downward. Two paths are said to be independent if they do not have common vertices. A monotonic path system of an HS is a collection of independent monotonic paths that involve all the

peaks and all the valleys of the HS. Prove that the number of Kekulé structures of the HS is equal to the number of distinct monotonic path systems of the HS [27,52].

20. Let $p_1, p_2, \ldots, p_k$ be the picks and $v_1, v_2, \ldots, v_k$ the valleys of a given HS. Define a square matrix W of order k such that $(W)_{ij}$ is equal to the number of monotonic paths in the HS starting at p_i and ending at v_j. Prove that the number of Kekulé structures of the HS is $|\det(W)|$ (i.e., the determinant of matrix W) [39].
21. If A is the adjacency matrix of an HS B with n vertices, then prove that $\det(A) = (-1)^{n/2} K(B)^2$ [13,18].
22. The dual graph of an HS is obtained when the centers of all neighboring hexagons are joined by an edge. The outer boundary of the dual graph of a hexagon-shaped HS is a hexagon with parallel edges of size m, n, and k, respectively. Prove that the number of Kekulé structures of such an HS is $\prod_{j=0}^{k-1} \binom{n}{m+n+j} / \binom{n}{n+j}$ [5].
23. Suppose that some edges of an HS are drawn vertically. Prove that in all perfect matchings of the HS a fixed horizontal line, passing through the center of at least one hexagon, intersects an equal number of double bonds [52].
24. Prove that all Kekulé structures of a given HS have an equal number of vertical double bonds (again, some edges are drawn vertically) [64].
25. An edge of an HS is called a single (double) fixed bond if it does not belong (belongs, respectively) to all perfect matchings of the HS. Design an $O(h^2)$ algorithm for the recognition of all fixed bonds in an HS and for determining whether or not a given HS is essentially disconnected [66].
26. A cycle of edges of an HS is called an alternating cycle if there exists a perfect matching of the HS such that edges in the cycle alternatingly belong and do not belong to the perfect matching. Prove that every hexagon of an HS is resonant (i.e., an aromatic sextet) if and only if the perimeter of the HS is an alternating cycle of the HS [62].
27. Determine the number of nonisoarithmic hexagonal chains with h hexagons [17].

ACKNOWLEDGMENTS

The authors would like to thank Gilles Caporossi and Brendan McKay for valuable feedback and suggestions on the presentation and contents of this chapter.

REFERENCES

1. Avis D, Fukuda K. Reverse search for enumeration. Discrete Appl Math 1996;6: 21–46.

2. Balaban AT, Tomescu I. Algebraic expressions for the number of Kekulé structure of isoarithmic catacondensed benzenoid polycyclic hydrocarbons. Match 1983;14:155–182.
3. Balasubramanian K, Kaufman JJ, Koski WS, Balaban AT. Graph theoretical characterisation and computer generation of certain carcinogenic benzenoid hydrocarbons and identification of bay regions. J Comput Chem 1980;1:149–157.
4. Bergan JL, Cyvin BN, Cyvin SJ. The Fibonacci numbers and Kekulé structures of some corona-condensed benzenoids (corannulenes). Acta Chim Hung 1987;124:299.
5. Bodroza O, Gutman I, Cyvin SJ, Tosic R. Number of Kekulé structures of hexagon-shaped benzenoids. J Math Chem 1988;2:287–298.
6. Brinkmann G. Isomorphism rejection in structure generation programs. In: Hansen P, Fowler P, Zheng M, editors. Discrete Mathematical Chemistry. Providence, RI: American Mathematical Society; 2000. p 25–38.
7. Brinkmann G, Caporossi G, Hansen P. A constructive enumeration of fusenes and benzenoids. J Algorithm 2002;45:155–166.
8. Brinkmann G, Caporossi G, Hansen P. A survey and new results on computer enumeration of polyhex and fusene hydrocarbons. J Chem Inform Comput Sci 2003;43:842–851.
9. Brunvoll J, Cyvin BN, Cyvin SJ. Benzenoid chemical isomers and their enumeration. Topics in Current Chemistry. Volume 162. Springer-Verlag; 1992.
10. Brunvoll J, Cyvin SJ, Gutman I, Tosic R, Kovacevic M. Enumeration and classification of coronoid hydrocarbons. J Mol Struct (Theochem) 1989;184:165–177.
11. Brunvoll J, Tosic R, Kovacevic M, Balaban AT, Gutman I, Cyvin SJ. Enumeration of catacondensed benzenoid hydrocarbons and their numbers of Kekulé structures. Rev Roumaine Chim 1990;35:85.
12. Caporossi G, Hansen P. Enumeration of polyhex hydrocarbons to $h = 21$. J Chem Inform Comput Sci 1998;38:610–619.
13. Cvetkovic D, Doob M, Sachs H. Spectra of Graphs, Theory and Applications. New York: Academic Press; 1980.
14. Cyvin BN, Brunvoll J, Cyvin SJ. Enumeration of benzenoid systems and other polyhexes. Topics in Current Chemistry. Volume 162. Springer-Verlag; 1992.
15. Cyvin SJ, Cyvin BN, Brunvoll J. Enumeration of benzenoid chemical isomers with a study of constant-isomer series. Topics in Current Chemistry. Volume 166. Springer-Verlag; 1993.
16. Cyvin SJ, Gutman I. Topological properties of benzenoid systems. Part XXXVI. Algorithm for the number of Kekulé structures in some pericondensed benzenoids. Match 1986;19:229–242.
17. Cyvin SJ, Gutman I. Kekulé Structures in Benzenoid Hydrocarbons. Berlin: Springer-Verlag; 1988.
18. Dewar MJS, Longuet-Higgins HC. The correspondence between the resonance and molecular orbital theories. Proc R Soc Ser A 1952;214:482–493.
19. Dias JR. Handbook of Polycyclic Hydrocarbons. Part A. Benzenoid Hydrocarbons. Amsterdam: Elsevier; 1987.
20. Dias JR. Handbook of Polycyclic Hydrocarbons. Part B. Polycyclic Isomers and Heteroatom Analogs of Benzenoid Hydrocarbons. Amsterdam: Elsevier; 1989.

21. Dias JR. Molecular Orbital Calculations Using Chemical Graph Theory. Berlin: Springer; 1993.
22. Doroslovacki R, Stojmenovic I, Tosic R. Generating and counting triangular systems, BIT 1987;27:18–24.
23. Enting IG. Generating functions for enumerating self-avoiding rings on the square lattice. J Phys A 1980;13:3713–3722.
24. Faradzev IA. Constructive enumeration of combinatorial objects. Problemes Combinatoires et Theorie des Graphes Colloque Internat. CNRS 260. Paris: CNRS; 1978. p 131–135.
25. Farrell EJ. On the occurrences of Fibonacci sequences in the counting of matchings in linear polygonal chains. Fibonacci Quart 1986;24:238–246.
26. Faulon JL, Visco DP, Roe D. Enumerating molecules. In: Lipkowitz K, editor, Reviews in Computational Chemistry. Volume 21. Wiley-VCH; 2005.
27. Gordon M, Davison WHT. Resonance topology of fully aromatic hydrocarbons. J Chem Phys 1952;20:428–435.
28. Grüner T, Laue R, Meringer M. Algorithms for group action applied to graph generation. In: Finkelstein L, Kantor WM, editors. Groups and Computation II, Workshop on Groups and Computation. DIMACS Ser Discrete Math Theor Comput Sci 1997;28: 113–123.
29. Gutman I. Topological properties of benzenoid systems—an identity for the sextet polynomial. Theor Chim Acta 1977;45:309–315.
30. Gutman I. Topological properties of benzenoid molecules. Bull Soc Chim Beograd 1982;47:453–471.
31. Gutman I. Covering hexagonal systems with hexagons. Proceedings of the 4th Yugoslav Seminar on Graph Theory; University of Novi Sad, Novi Sad; 1983. p 151–160.
32. Gutman I. Topological properties of benzenoid systems. Topics in Current Chemistry. Volume 162. Springer-Verlag; 1992. p 1–28.
33. Gutman I, Cyvin SJ. Introduction to the Theory of Benzenoid Hydrocarbons. Springer-Verlag; 1989.
34. Gutman I, Cyvin SJ. Advances in the Theory of Benzenoid Hydrocarbons. Springer-Verlag; 1990.
35. Gutman I, Cyvin SJ. A result on 1-factors related to Fibonacci numbers. Fibonacci Quart 1990; 81–84.
36. Gutman I, Cyvin SJ, Brunvoll J. Advances in the Theory of Benzenoid Hydrocarbons II. Springer-Verlag; 1992.
37. Harary F, Harborth H. Extremal animals. J Comb Inform Syst Sci. 1976;1:1–8.
38. He WJ, He QX, Wang QX, Brunvoll J, Cyvin SJ. Supplements to enumeration of benzenoid and coronoid hydrocarbons. Z Naturforsch. 1988;43a:693–694.
39. John P, Sachs H. Wegesysteme und Linearfaktoren in hexagonalen und quadratischen Systemen (Path systems and linear factors in hexagonal and square systems). Graphen in Forschung und Unterricht. Bad Salzdetfurth, Germany:Verlag Barbara Franzbecker; 1985. p 85–101.
40. Klarner DA. Some results concerning polyominoes. Fibonacci Quart 1965;3:9–20.

41. Knop JV, Müller WP, Szymanski K, Trinajstic N. Use of small computers for large computations: enumeration of polyhex hydrocarbons. J Chem Inform Comput Sci 1990;30:159–160.
42. Knop JV, Szymanski K, Jericevic Z, Trinajstic N. Computer enumeration and generation of benzenoid hydrocarbons and identification of bay regions. J Comput Chem 1983;4:23–32.
43. Kocay W. On writing isomorphism programs. In: Wallis WD editor. Computational and Constructive Design Theory. Kluwer; 1996. p 135–175.
44. McKay BD. Practical graph isomorphism. Congr Numer 1981;30:45–87.
45. McKay BD. Nauty user's guide. Technical Report TR-CS-90-02. Computer Science Department, Australian National University; 1990.
46. McKay BD. Isomorph-free exhaustive generation. J Algorithms 1998;26:306–324.
47. Müller WR, Szymanski K, Knop JV. On counting polyhex hydrocarbons. Croat Chem Acta 1989;62:481–483.
48. Müller WR, Szymanski K, Knop JV, Nikolić S, Trinajstić N. On the enumeration and generation of polyhex hydrocarbons. J Comput Chem 1990;11:223–235.
49. Nikolić S, Trinajstić N, Knop JV, Müller WR, Szymanski K. On the concept of the weighted spanning tree of dualist. J Math Chem 1990;4:357–375.
50. Read RC. Every one a winner. Ann Discrete Math 1978;2:107–120.
51. Redelmeier DH. Counting polyominoes: yet another attack, Discrete Math 1981;36:191–203.
52. Sachs H. Perfect matchings in hexagonal systems. Combinatorica 1984;4:89–99.
53. Stojmenovic I, Tosic R, Doroslovacki R. Generating and counting hexagonal systems. Graph Theory. Proceedings of 6th Yugoslav Seminar on Graph Theory; Dubrovnik, 1985; University of Novi Sad; 1986. p 189–198.
54. Tosic R, Doroslovacki R, Stojmenovic I. Generating and counting square systems. Graph Theory. Proceedings of the 8th Yugoslav Seminar on Graph Theory; University of Novi Sad, Novi Sad; 1987. p 127–136.
55. Tosic R, Kovacevic M. Generating and counting unbranched catacondensed benzenoids. J Chem Inform Comput Sci 1988;28:29–31.
56. Tosic R, Masulovic D, Stojmenovic I, Brunvol J, Cyvin BN, Cyvin SJ. Enumeration of polyhex hydrocarbons to $h = 17$. J Chem Inform Comput Sci 1995;35: 181–187.
57. Tosic R, Stojmenovic I. Benzenoid chains with the unique Clarformula. J Mol Struct (Theochem) 1990;207:285–291.
58. Tosic R, Stojmenovic I. Fibonacci numbers and the numbers of perfect matchings of square, pentagonal, and hexagonal chains. The Fibonacci Quart 1992;30:315–321.
59. Trinajstic N. Chemical Graph Theory. Boca Raton: CRC Press; 1992.
60. Vöge M, Guttman J, Jensen I. On the number of benzenoid hydrocarbons. J Chem Inform Comput Sci 2002;42:456–466.
61. Yen TF. Resonance topology of polynuclear aromatic hydrocarbons. Theor Chim Acta 1971;20:399–404.
62. Zhang F, Chen R. When each hexagon of a hexagonal system covers it. Discrete Appl Math 1991;30:63–75.
63. Zhang FJ, Chen RS, Guo XF. Perfect matchings in hexagonal systems. Graphs Comb 1985;1:383.

64. Zhang FJ, Chen RS, Guo XF, Gutman I. An invariant of the Kekulé structures of benzenoid hydrocarbons. J Serb Chem Soc 1986;51:537.
65. Zhang FJ, Guo XF. Characterization of an invariant for benzenoid systems. Match 1987;22:181–194.
66. Zhang F, Li X, Zhang H. Hexagonal systems with fixed bonds. Discrete Appl Math 1993;47:285–296.

CHAPTER 3

Graph Theoretic Models in Chemistry and Molecular Biology

DEBRA KNISLEY and JEFF KNISLEY

3.1 INTRODUCTION

3.1.1 Graphs as Models

A graph is a mathematical object that is frequently described as a set of points (vertices) and a set of lines (edges) that connect some, possibly all, of the points. If two vertices in the graph are connected by an edge, they are said to be adjacent, otherwise they are nonadjacent. Every edge is incident to exactly two vertices; thus, an edge cannot be drawn unless we identify the two vertices that are to be connected by the edge. The number of edges incident to a vertex is the degree of the vertex. How the edges are drawn, straight, curved, long, or short, is irrelevant, only the connection is relevant. There are many families of graphs and sometimes the same graph can belong to more than one family. For example, a cycle graph is a connected graph where every vertex is of degree 2, meaning every vertex is incident to exactly two edges. A bipartite graph is a graph with the property that there exists a partition of the vertex set into two sets such that there are no edges between any two vertices in the same set. Figure 3.1 shows two drawings of the same graph that can be described both as a cycle on six vertices and as a bipartite graph. The two graphs in Figure 3.1 are said to be isomorphic. Two graphs are isomorphic if there exists a one-to-one correspondence between the vertex sets that preserves adjacencies. In general, it is a difficult problem to determine if two graphs are isomorphic.

An alternate definition of a graph is a set of elements with a well-defined relation. Each element in the set can be represented by a point and if two elements in the set are related by the given relationship, then the corresponding points are connected by an edge. Thus, the common definition of a graph is really a visual representation of a relationship that is defined on a set of elements. In graph theory, one then studies the relational representation as an object in its own right, discerning properties of the object and quantifying the results. These quantities are called graphical invariants

Handbook of Applied Algorithms: Solving Scientific, Engineering and Practical Problems
Edited by Amiya Nayak and Ivan Stojmenović

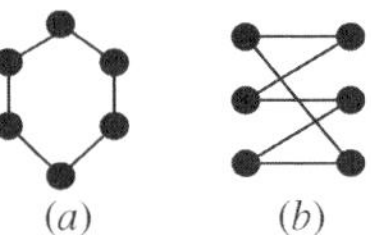

FIGURE 3.1 (*a*) A cycle. (*b*) A bipartite graph.

since their values are the same regardless of how the graph is drawn. The graphical invariants, in turn, tell us about the consequences the relation has on the set. To utilize a graph as a model, we must first determine the set and the relation on the set that we want to study. For example, suppose we want to consider a group of six people, three men and three women. None of the men have ever met each other and none of the women have ever met, but some of the men have met some of the women. Suppose the graph in Figure 3.1b models this set of people where the two people are "related" or associated if they have previously met. Since the two graphs in Figure 3.1 are isomorphic, we immediately know that it is possible to seat the six people around a circular table so that each person is seated next to someone that they have previously met. This illustration shows the usefulness of graphs even with a very simple example. Graphs are frequently used in chemistry to model a molecule. Given the atoms in a molecule as the set, whether or not a bond joins two atoms is well defined and hence the graphical representation of a molecule is the familiar representation.

What is a mathematical model? What is a graph theoretic model? Since graph theory is a field of mathematics, one would assume that a graph theoretic model is a special case or a particular kind of mathematical model. While this is true, the generally accepted definition of a mathematical model among applied mathematicians is somewhat different from the idea of a model in graph theory. In mathematical settings, a model is frequently associated with a set of equations. For example, a biological system is often modeled by a system of equations, and solutions to the equations are used to predict how the biological system responds to stimuli. Molecular biology and biochemistry, however, are more closely aligned with chemistry methodology and literature. Models of molecules in chemistry are often geometric representations of the actual molecule in various formats such as the common ball and stick "model" where balls are used to represent atoms and bonds between the atoms are represented by sticks. As we have seen, this straightforward model of a molecule gives easy rise to a graph where the balls are the vertices and the sticks are the edges. The first appearance of a graph as a model or representation of a molecule appeared in the early nineteenth century. In fact, chemistry and graph theory have been paired since the inception of graph theory and we find that the early work in physical chemistry coincided with the development of graph theory.

As we have seen, a graphical invariant is a measure of some aspect of a graph that is not dependent upon how the graph is drawn. For example, the *girth* of a graph is the length of its shortest cycle. A graph that has no cycle is said to be of infinite girth. The most obvious of invariants are the *order* (number of vertices) and the *size* (number of edges). The minimum number of vertices whose removal will disconnect the graph

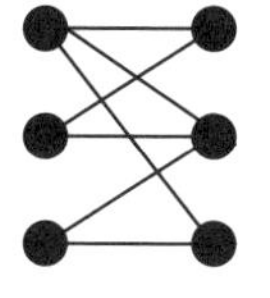

FIGURE 3.2 *G.*

is the (vertex) *connectivity* number. The graph in Figure 3.2 has girth 4, is of order 6, size 7, and connectivity 2.

3.1.2 Early Models in Chemistry

One of the first theorems of graph theory can be stated as follows: The sum of the degrees of a graph is twice the number of edges. Since the sum of the degrees of the vertices of even degree is necessarily an even number, the sum of the degrees of the vertices of odd degree must also be even. As a corollary to the above theorem, we know that the number of vertices of odd degree must be even. As far back as 1843, Laurent [1] and Gerhardt [2] established that the number of atoms of odd valence (degree) in a molecule was always even. What constituted an edge was not well established though. One of the earliest formulations of graphs appeared in 1854 in the work by Couper [3], and in 1861, a chemical bond was represented by a graphical edge following the introduction of the term "molecular structure" by Butlerov [4]. The concept of valence of an atom was later championed by Frankland whose work was published in 1866 [5].

Arthur Cayley, a well-known mathematician from the late 1800s, used combinatorial mathematics to construct chemical graphs [6]. Using mathematics, Cayley enumerated the saturated hydrocarbons by determining the generating function for rooted trees. As an illustration, consider the expansion of the expression $(a+b)^3$. The coefficients of the terms are 1, 3, 3, and 1, respectively, in the expanded form: $1a^3b^0 + 3a^2b^1 + 3a^1b^2 + 1a^0b^3$. Note that the exponents in each term sum to 3 and each term represents a distinct way we can obtain the sum of 3 using two distinct ordered terms. If we let b represent the number of ways we can select to insert an edge (or not to insert an edge), then the corresponding coefficients yield the number of ways this selection can be done. Hence, corresponding to the coefficients, there is one graph with no edges, three graphs with exactly one edge, three graphs with exactly two edges, and one graph with three edges. These are drawn in Figure 3.3. This is the idea behind generating functions. Since the graphical representations of the saturated hydrocarbons are trees, Cayley determined how many such trees are combinatorially possible. At that time, his count exceeded the number of known saturated hydrocarbons by 2. Soon after, two additional hydrocarbons were found. How does one prove that a graphical representation of a saturated hydrocarbon is a tree? First, we must define a tree. A *tree* is a connected graph with no cycles. These two properties, connected and acyclic, imply that any tree with n vertices must contain exactly $n - 1$ edges.

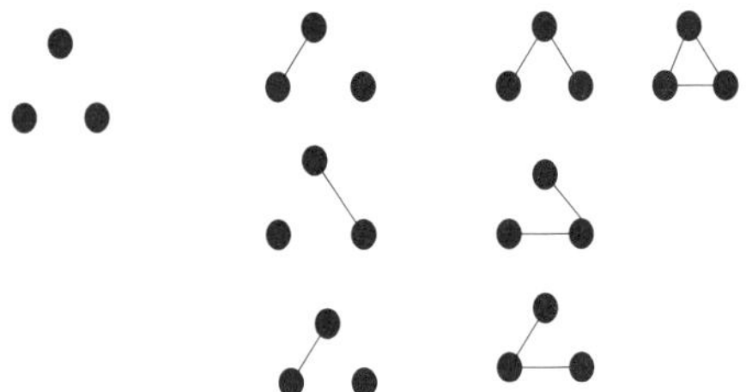

FIGURE 3.3 All possible graphs with three vertices.

A saturated hydrocarbon has the maximum possible number of hydrogen atoms for the number of carbon atoms in a molecule and is denoted by the formula C_mH_{2m+2}. The tree representation of butane, C_4H_{10}, is shown in Figure 3.4.

In order to prove that a graphical representation of a molecule with the above formula will always be represented by a tree, we must conclude that it is connected and acyclic. Since it is molecule, it is inherently connected. Thus, we must show that it will be impossible for a cycle to occur. This is equivalent to showing that there will always be exactly one less edge than the number of vertices. So we proceed with the counting argument. We know that there are $m + 2m + 2$ vertices total by adding the carbon and hydrogen atoms. Thus, there are $3m + 2$ vertices. To count the edges we observe that each carbon atom is incident to exactly four edges and hence there are $4(m)$ edges associated with the carbon atoms. Also, each hydrogen atom is incident to exactly one edge and thus we have $1(2m + 2)$ additional edges. Since each edge is incident to exactly two vertices, each edge has now been counted exactly twice. Thus, the number of edges total is $(1/2)(4m + 2m + 2) = 3m + 1$. Note that $3m + 1$ is exactly one less than the number of vertices.

The mathematician Clifford was first to demonstrate that a saturated hydrocarbon could not possess any cycles and in fact showed that a hydrocarbon with the general formula $C_mH_{2m+2-2x}$ must contain x cycles [7]. In 1878, Sylvester founded the *American Journal of Mathematics*. In its very first issue he wrote a lengthy article on atomic theory and graphical invariants. By labeling the vertices of the graphs, Sylvester was able to devise a method for validating the existence of different types of chemical graphs. This was the first usage of the word graph in the graph theoretic sense [8]. Through the years, chemical graph theory has survived as a little known niche in the field of graph theory. Most textbook applications of graphs have centered on computer networks, logistic problems, optimal assignments strategies, and data structures. Chemical graph theorists persisted and developed a subfield of graph

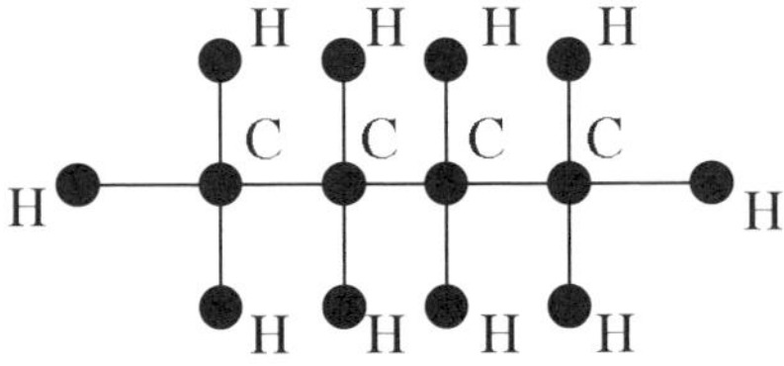

FIGURE 3.4 Butane.

theory built upon molecular graphs. Quantifiers of the molecular graphs are known as "descriptors" or topological indices. These topological indicators are equivalent to graphical invariants in the realm of mathematical graph theory. In the following sections we discuss some of the early graph theoretic models, as well as some of the first graphical invariants and topological indices. For more information on chemical graph theory see the works by Bonchev and Rouvray [9] and Trinajstic [10,11].

3.1.3 New Directions in Chemistry and Molecular Biology

Today graphs are being used extensively to model both chemical molecules and biomolecules. Chemists use molecular descriptors that yield an accurate determination of structural properties to develop algorithms for computer-aided drug designs and computer-based searching algorithms of chemical databases. Just as bioinformatics is the field defined as belonging in the intersection of biology and computer science, cheminformatics lies in the intersection of chemistry and computer science. Cheminformatics can be defined as the application of computational tools to address problems in the efficient storage and retrieval of chemical data. New related fields are emerging, such as chemical genomics and pharmacogenomics. Organic chemicals frequently referred to as "small molecules" are playing a significant part in the discovery of new interacting roles of genes. The completion of the Human Genome Project has changed the way new drugs are being targeted and the expansion of chemical libraries aided by techniques from combinatorial chemistry is seeing more and more graph theoretic applications. While it is generally accepted that graphs are a useful tool for small molecules, graphs are also being utilized for larger biomolecules as well. Graphs are appearing in the literature as DNA structures, RNA structures, and various protein structures. We find that graphs are becoming an invaluable tool for modeling techniques in proteomics and protein homology and thus one could say that chemical graph theory has contributed indirectly to these fields as well. Using graphs to model a molecule has evolved from the early days of chemical graph theory to become an integral part of cheminformatics, combinatorial and computational chemistry, chemical genomics, and pharmacogenomics.

Algorithms that determine maximum common induced subgraphs or other structure similarity searches have played a key role in computational chemistry and cheminformatics. An obvious problem associated with such algorithms is the rapid increase in the number of possible configurations. The exponential growth of the number of graphs with an increasing number of vertices is a difficult challenge that must be addressed. Large graphs result in nonpolynomial time algorithms creating excessive computational expense. In addition, intuition that can often be an aid in determining appropriate molecular descriptors and thus the investigation is greatly hindered by large graphs that cannot be visualized. Methods have been developed for reducing the size of graphs, and such graphs are commonly referred to as reduced graphs. These methods have had a significant impact on the ability to model the relevant biomolecular structures and provide summary representations of chemical and biochemical structures. Reduced graphs offer the ability to represent molecules in terms of their high level features [12,13].

In 2005, in partial fulfillment of the NIH Roadmap stated objectives, NIH announced a plan to fund 10 cheminformatic research centers in response to the identification of critical cheminformatics needs of the biomedical research community. The centers will formulate the strategies to address those needs and will also allow awardees to become familiar with the operation and interactions among the various components of the NIH Molecular Libraries Initiative. These centers are intended to promote multidisciplinary, multiinstitutional collaboration among researchers in computational chemistry, chemical biology, data mining, computer science, and statistics. Stated components of proposed research include the calculation of molecular descriptors, similarity metrics, and specialized methodologies for chemical library design and virtual screening. For example, the Carolina Exploratory Center for Cheminformatics Research plans to establish and maintain an integrated publicly available Cheminformatics Workbench (ChemBench) to support experimental chemists in the Chemical Synthesis centers and quantitative biologists in the Molecular Libraries Screening Centers Network. The Workbench is intended to be a data analytical extension to PubChem.

3.2 GRAPHS AND ALGORITHMS IN CHEMINFORMATICS

3.2.1 Molecular Descriptors

Values calculated from a representation of a molecule that encode some aspect of the chemical or biochemical structure and activities are called molecular descriptors. There are an enormous number of descriptors that have been defined and utilized by researchers in fields such as cheminformatics, computational chemistry, and mathematical chemistry. The *Handbook of Molecular Descriptors* [14] is an encyclopedic collection of more than 3000 descriptors. Molecular descriptors fall into three general categories. Molecular descriptors that quantify some measure of shape and/or volume are called steric descriptors. Electronic descriptors are those that measure electric charge and electrostatic potential, and there are those that measure a molecule's affinity for a lipophilic environment such as log P. log P is calculated as the log ratio of the concentration of the solute in the solvent. Examples of steric descriptors are surface area and bond connectivity. Surface area is calculated by placing a sphere on each atom with the radius given by the Van der Waals radius of the atom. Electronic descriptors include the number of hydrogen bond donors and acceptors and measures of the pi–pi donor–acceptor ability of molecules. With the support of the EU, INTAS (the International Association for the Promotion of Cooperation with Scientists) from the New Independent States (NIS) of the Former Soviet Union created The Virtual Computational Chemistry Laboratory (VCCL) with the aim to promote free molecular properties calculations and data analysis on the Internet [15]. E-Dragon, a program developed by the Milano Chemometrics and QSAR Research Group [16] and a contributor to the VCCL, can calculate more than 1600 molecular descriptors that are divided into 20 categories. Its groups of indices include walk-and-path counts, electronic, connectivity, and information indices. The molecular descriptors that E-Dragon categorizes as topological indices are obtained from molecular graphs (usually H-depleted) that are conformationally independent. E-Dragon is available at VCCL.

All chemical structures can be represented by a simplified linear string using a specific set of conversion and representation rules known as SMILES (Simplified molecular input line entry system). SMILES strings can be converted to representative 3D conformations and 2D representations. While 1D representations are strings and 3D representations are geometric, 2D representations are primarily graphs consisting of vertices (nodes) and their connecting edges. SMILES utilizes the concept of a graph with vertices as atoms and edges as bonds to represent a molecule. The development of SMILES was initiated by the author, David Weininger, at the Environmental Research Laboratory, USEPA, Duluth, MN; the design was completed at Pomona College in Claremont, CA. It was embodied in the Daylight Toolkit with the assistance of Cedar River Software. Parentheses are used to indicate branching points and numeric labels designate ring connection points [17].

Quantities derived from all three representations are considered molecular descriptors. Since we are primarily concerned with graph theoretic models, we will focus on 2D descriptors from graphs and refer to these as topological descriptors or topological indices. Graphs are also useful for 3D models since 3D information can be contained in vertex and edge labeling [18,19]. Descriptors calculated from these types of representations are sometimes called information descriptors. While the 2D graphical model neglects information on bond angles and torsion angles that one finds in 3D models, this can be advantageous since it allows flexibility of the structure to occur without a resulting change in the graph. Methods and tools from computational geometry also often aid in the quantification and simulation of 3D models.

Molecular descriptors are a valuable tool in the retrieval of promising pharmaceuticals from large databases and also in clustering applications. (ADAPT) (Automated Data Analysis Using Pattern Recognition Toolkit) has a large selection of molecular descriptor generation routines (topological, geometrical, electronic, and physicochemical) and the ability to generate hybrid descriptions that combine features. ADAPT was developed by Peter Jurs, the Jurs Research Group at Penn State, and is available over the Internet [20]. The Molecular operating environment (MOE) offered by the Chemical Computing Group [21] has a developed a pedagogical toolkit for educators including a cheminformatics package. This toolkit can calculate approximately 300 descriptors including topological indices, structural keys, and E-state indices.

3.2.2 Graphical Invariants and Topological Indices

A topological index is a number associated with a chemical structure represented by a connected graph. The graph is usually a hydrogen-depleted graph, where atoms are represented by vertices and covalent bonds by edges. On the contrary, many results in graph theory have focused on large graphs and asymptotic results in general. Since chemical graphs are comparatively small, it is not too surprising that graphical invariants and topological indices have evolved separately. However, with the new avenues of research in biochemical modeling of macromolecules, the field of mathematical graph theory may bring new tools to the table. In chemical graph theory, the number of edges, that is, the number of bonds, is an obvious and well-utilized molecular descriptor. Theorems from graph theory or graphical invariants from related fields such as computational complexity and computer architecture may begin to shed new light on the structure and properties of proteins and other large molecules. In recent results by Haynes et al., parameters based on graphical invariants from mathematical graph theory showed promising results in this direction of research [22,23]. It certainly appears that a thorough review of theoretical graphical invariants with an eye toward new applications in biomolecular structures is warranted

Without a doubt, there will be some overlap of concepts and definitions. For example, one of the most highly used topological indices was defined by Hoyosa in 1971 [24]. This index is the sum of the number of ways k disconnected edges can be distributed in a graph G.

$$I(G) = \sum_{k=0}^{n/2} \theta(G, k),$$

where $\theta(G, 0) = 1$ and $\theta(G, 1)$ is the number of edges in G. Let us deviate for a moment and define the graphical invariant, k-factor. To do so, we first define a few other graph theoretic terms. A graph is k-regular if every vertex has degree k. A graph H is a *spanning subgraph* of G if it is a subgraph that has the same vertex set of G. A subgraph H is a k-factor if it is a k-regular spanning subgraph. A 1-factor is a spanning set of edges and a 2-factor of a graph G is a collection of cycle subgraphs that span the vertex set of G. If the collection of spanning cycles consists of a single cycle, then the graph is *Hamiltonian*. Hamiltonian theory is an area that has received substantial attention among graph theorists, as well as the topic of k-factors. We note that $\theta(G, 1)$ is the number of edges in G and that $\theta(G, n/2)$ is equivalent to the number of 1-factors in G [9]. In the following sections, we define selected graphical invariants and topological indices, most of which were utilized in the work by Haynes et al. [22,23].

Domination numbers of graphs have been utilized extensively in fields such as computer network design and fault tolerant computing. The idea of domination is based on sets of vertices that are near (dominate) all the vertices of a graph. A set of vertices *dominate* the vertex set if every vertex in the graph is either in the dominating set or adjacent to at least one vertex in the dominating set. The minimum cardinality among all dominating sets of vertices in the graph is the domination number. For more information on the domination number of graphs see Haynes [25]. If restrictions are placed on the set of vertices that we may select to be in the dominating set, then we obtain variations on the domination number. For example, the independent domination number is the minimum number of nonadjacent vertices that can dominate the graph. Consider Figure 3.5, which contains two trees of order 7, one with independent domination number equal to 3 and the other with independent domination number equal to 2. The vertices in each independent minimum dominating set are labeled $\{u, w, z\}$ and $\{u, z\}$, respectively. Domination numbers have been highly studied in

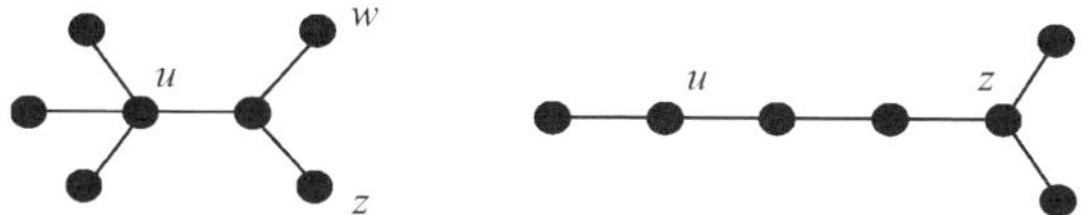

FIGURE 3.5 Dominating vertices $\{u, w, z\}$ and $\{u, z\}$, respectively.

mathematical graph theory and have applications in many fields such as computer networks and data retrieval algorithms.

The *eccentricity* of a vertex is the maximum distance from a vertex v to any other vertex in the graph where distance is defined to be the length of the shortest path and is denoted by $d(u, v)$. The *diameter* of G, diam (G), is the maximum eccentricity where this maximum is taken over all eccentricity values in the graph. That is,

$$\text{diam}(G) = \max_{u,v \in V} d(v, u)$$

and the *radius* of a graph G, denoted by rad (G), is given by the minimum eccentricity value, that is,

$$\text{rad}(G) = \min_{x \in V} \max_{y \in V} \{d(x, y)\}.$$

The diameter and radius are both highly utilized graphical invariants and topological indices.

The *line graph of* G, denoted by $L(G)$, is a graph derived from G so that the edges in G are replaced by vertices in $L(G)$. Two vertices in $L(G)$ are adjacent whenever the corresponding edges in G share a common vertex. Beineke and Zamfirescu [26] studied the kth ordered line graphs and Dix [27] applied the second ordered line graphs to concepts in computational geometry. Figure 3.6 shows a graph G with $L(G)$ and $L^2(G)$, the second iterated line graph. Note that vertex x in $L^2(G)$ corresponds to the edge x in $L(G)$. The edge x in $L(G)$ is defined by the two vertices a and b. These two vertices in $L(G)$ correspond to the two edges a and b in G. Topological indices do not account for angle measures; however, two incident edges represent an angle and thus vertex x in $L^2(G)$ corresponds to the angle, or path of length 2, namely $\{1, 3, 2\}$.

Given that there are over 3000 molecular descriptors defined in the *Handbook of Molecular Descriptors*, we will make no attempt to provide an extensive list of topological indices. Rather we have selected a few representatives that are classical and well known as examples.

The Gordon–Scantlebury index is defined as the number of distinct ways a chain fragment of length 2 can be embedded on the carbon skeleton of a molecule [28]. Thus, if G is the graph in Figure 3.6, then the Gordon–Scantlebury number is 4. The second iterated line graph discussed above not only provides an easy way to determine this index, but also tells us how these paths are related. Notice that the vertices z, w, and y in $L^2(G)$ form a triangle; that is, they are all pairwise adjacent. This is because they are all incident to vertex c in $L(G)$. Since vertex c in $L(G)$ corresponds to edge c in G, we know that the three paths of length 2 corresponding to the vertices in z, w, and y in $L^2(G)$ all share edge c.

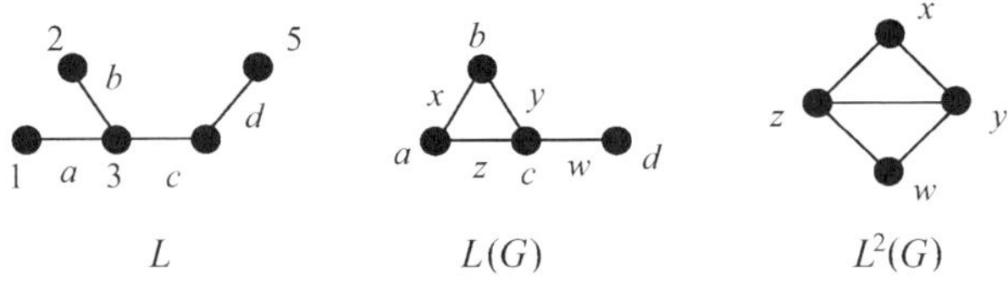

FIGURE 3.6 A graph, its line graph, and the second iterated line graph.

Among the earliest topological indices are the connectivity indices. The classical connectivity index defined by Randic [29] is given by

$$R_0(G) = \sum_{v \in V} \frac{1}{\sqrt{\partial(v)}},$$

$$R_1(G) = \sum_{uv \in E} \frac{1}{\sqrt{\partial(u)\partial(v)}}.$$

The Randic numbers for the graph G in Figure 3.6 are $R_0(G) = 1 + 1 + 1/\sqrt{3} + 1/\sqrt{2} + 1 = 4.28$ and $R_1(G) = 2(1/\sqrt{1 \cdot 3}) + 1/\sqrt{2 \cdot 3} + 1/\sqrt{1 \cdot 2}) = 2.27$. This index can be generalized for paths of length l to define the generalized Randic number $R_l(G)$. One can consider paths as a special type of subgraph. More recently, Bonchev introduced the concept of overall connectivity of a graph G, denoted by TC(G), which is defined to be the sum of vertex degrees of all subgraphs of G [30].

The adjacency matrix is a straightforward way to represent a graph in a computer. Given a graph with n vertices labeled $V = \{v_1, v_2, ..., v_n\}$, the adjacency matrix A is an $n \times n$ matrix with a 1 in the ith row and jth column if vertex v_i is adjacent to vertex v_j and zeros elsewhere. The degree matrix D is the $n \times n$ matrix with $d_{i_j} = \deg(v_i)$ and $d_{i_j} = 0$ if $i \neq j$. The Laplacian matrix is defined as the difference of the adjacency matrix and the degree matrix, $L = D - A$. The *spectrum* of a graph is the set of eigenvalues of the Laplacian matrix. The eigenvalues are related to the density distribution of the edge set, and the pattern of a graph's connectivity is closely related to its spectrum. The second smallest eigenvalue, denoted by λ_2 (often called the Fiedler eigenvalue), is the best measure of the graph's connectivity among all of the eigenvalues. Large values for λ_2 correspond to vertices of high degree that are in close proximity whereas small values for λ_2 correspond to a more equally dispersed edge set.

The *Balaban* index [31], sometimes called the distance sum connectivity index, is considered to be a highly discriminating topological index. The Balaban index $B(G)$ of a graph G is defined as

$$B(G) = \frac{q}{\mu(G) + 1} \sum_{\text{edges}} \frac{1}{\sqrt{s_i s_j}},$$

where s_i is the sum of the distance of the ith vertex to the other vertices in the graph, q is the number of edges, and μ is the minimum number of edges whose removal results in an acyclic graph. The distance matrix T is the $n \times n$ matrix with $d_{i_j} = \text{dist}(v_i, v_j)$.

$d_{ij} = \text{dist}(v_i, v_j)$. The distance matrix and $B(G)$ for G in Figure 3.6 are given below.

$$T = \begin{bmatrix} 0\,2\,1\,2\,3 \\ 2\,0\,1\,2\,3 \\ 1\,1\,0\,1\,2 \\ 2\,2\,1\,0\,1 \\ 3\,3\,2\,1\,0 \end{bmatrix},$$

$$B(G) = 4\left(\frac{1}{\sqrt{8\cdot 5}} + \frac{1}{\sqrt{8\cdot 5}} + \frac{1}{\sqrt{5\cdot 6}} + \frac{1}{\sqrt{6\cdot 9}}\right).$$

The reverse Wiener index was introduced in 2000 [32]. Unlike the distance sums, reverse Wiener indices increase from the periphery toward the center of the graph. As we have seen, there are an enormous number of molecular descriptors utilized in computational chemistry today. These descriptors are frequently used to build what are known as quantitative structure–activity relationships (QSAR). A brief introduction of QSAR is given in the following section.

3.2.3 Quantitative Structure–Activity Relationships

The structure of a molecule facilitates the molecule's properties and its related activities. This is the premise of a QSAR study. QSAR is a method for building models that associate the structure of a molecule with the molecule's corresponding biological activity. QSAR was first developed by Hansch and Fujita in the early 1960s and remains a key player in computational chemistry. The fundamental steps in QSAR are molecular modeling, calculation of molecular descriptors, evaluation and reduction of descriptor set, linear or nonlinear model design, and validation. Researchers at the University of North Carolina at Chapel Hill recently extended the four steps to an approach that employs various combinations of optimization methods and descriptory types. Each descriptor type was used with every QSAR modeling technique, so in total 16 combinations of techniques and descriptor types were considered [33].

A successful QSAR algorithm is predictive. That is, given a molecule and its structure, one can make a reasonable prediction of its biological activity. The ability to predict a molecule's biological activity by computational means has become more important as an ever-increasing amount of biological information is being made available by new technologies. Annotated protein and nucleic databases and vast amounts of chemical data from automated chemical synthesis and high throughput screening require increasingly more sophisticated efforts.

QSAR modeling requires the selection of molecular descriptors that can then be used for either a statistical model or a computational neural network model. Current methods in QSAR development necessarily include feature selection. It is generally accepted that after descriptors have been calculated, this set must be reduced to a set of descriptors that measure the desired structural characteristics. This is obvious, but not always as straightforward as one would hope since the interpretation of a large number of descriptors is not always easy. Since many descriptors may be redundant in the information that they contain, principal component analysis has been the standard tool for descriptor reduction, often reducing the set of calculated invariants. This is accomplished by a vector space description analysis that looks for descriptors that are orthogonal to one another where descriptors that contain essentially the same information are linearly dependent. For example, a QSAR algorithm was developed by Viswanadahn et al. in which a set of 90 graph theoretic and information descriptors representing various structural/topological characteristics of these molecules were calculated. Principal component analysis was used to compress these 90 into the 8 best orthogonal composite descriptors [34]. Often molecular descriptors do not contain molecular information that is relevant to the particular study, which is another drawback one faces in selecting descriptors for a QSAR model. Due to the enormous number of descriptors available, coupled with the lack of interpretation one has for the molecular characteristics they exhibit, very little selection of descriptors is made *a priori*. Randic and Zupan reexamined the structural interpretation of several well-known indices and recommended partitioning indices into bond additive terms [35]. Advances in neural network capabilities may allow for the intermediate steps of molecular descriptor reduction and nonlinear modeling to be combined. Consequently, neural network algorithms are discussed in greater detail in Section 3.4.

Applications of QSAR can be found in the design of chemical libraries, in molecular similarity screening in chemical databases, and in virtual screening in combinatorial libraries. Combinatorial chemistry is the science of synthesizing and testing compound en masse and QSAR predictions have proven to be a valuable tool. The QSAR and Modeling Society Web site is a good source for more information on QSAR and its applications.

3.3 GRAPHS AS BIOMOLECULES

The Randic index is an example of a well-known and highly utilized topological index in cheminformatics. In 2002, Randic and Basak used the term "biodescriptor" when applying a QSAR model for a biomolecular study [36,37]. While graphs have historically been used to model molecules in chemistry, they are beginning to play a fundamental role in the quantification of biomolecules. A new technique for describing the shape and property distribution of proteins, called PPEST (protein property-encoded surface translator) has been developed to help elucidate the mechanism behind protein interactions [38]. The utility of graphs as models of proteins and nucleic acids is fertile ground for the discovery of new and innovative methods for the numerical characterization of biomolecules.

3.3.1 Graphs as RNA

The information contained in DNA must be accessed by the cell in order to be utilized. This is accomplished by what is known as transcription, a process that copies the information contained in a gene for synthesis of genetic products. This copy, RNA, is almost identical to the original DNA, but a letter substitution occurs as thymine (T) is replaced by uracil (U). The other three bases A, C, and G are the same. Since newly produced (synthesized) RNA is single stranded, it is flexible. This allows it to bend back on itself to form weak bonds with another part of the same strand. The initial string is known as the primary structure of RNA and the 2D representation in Figure 3.7 is an example of secondary RNA structure.

While scientists originally believed that the sole function of RNA was to serve as a messenger of DNA to encode proteins, it is now known that there are noncoding or functional RNA sequences. In fact, the widespread conservation of secondary structure points to a very large number of functional RNAs in the human genome [39,40]. Many classes of RNA molecules are characterized by highly conserved secondary structures that have very different primary structure (or primary sequence), which implies that both sequential and structural information is required in order to expand the current RNA databases [41]. RNA was once thought to be the least interesting since it is merely a transcript of DNA. However, since it is now known that RNA is involved in a large variety of processes, including gene regulation, the important task of classifying RNA molecules remains far from complete. Graph theory is quickly becoming one of the fundamental tools used in efforts to determine and identify RNA molecules.

It is assumed that the natural tendency of the RNA molecule is to reach its most energetically stable conformation and this is the premise behind many RNA folding algorithms such as Zucker's well-known folding algorithms [42]. More recently, however, the minimum free energy assumption has been revisited and one potential new player is graph theoretic modeling and biodescriptors. Secondary structure has been represented by various forms in the literature and representations of RNA molecules as graphs is not new. In the classic work of Waterman [43], secondary RNA structure is defined as a graph where each vertex a_i represents a nucleotide base. If a_i pairs with a_j and a_k is paired with a_l where $i < k < j$, then $i < l < j$.

More recently, secondary RNA structures have been represented by various modeling methods as graph theoretic trees. RNA tree graphs were first developed by Le et al. [44] and Benedetti and Morosetti [45] to determine structural similarities in RNA.

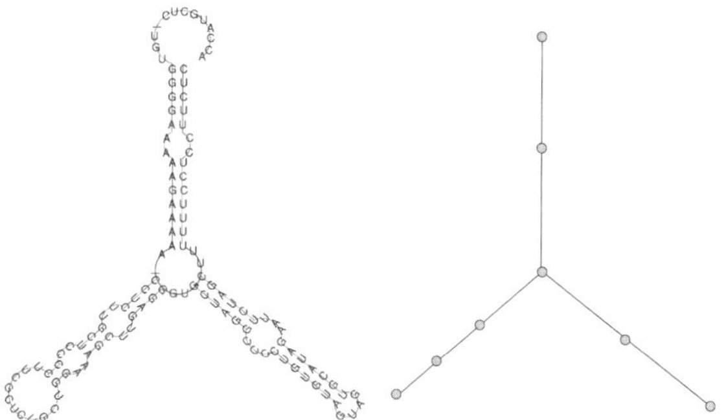

FIGURE 3.7 Secondary RNA structure and its graph.

A modeling concept developed by Barash [46] and Heitsch et al. [47] who noted that the essential arrangement of loops and stems in RNA secondary structure is captured by a tree if one excludes the pseudoknots. A pseudoknot can be conceptualized as switchbacks in the folding of secondary structure. With the exclusion of pseudoknots, the geometric skeleton of secondary RNA structure is easily visualized as a tree as in Figure 3.7. Unlike the classic model developed by Waterman et al. where atoms are represented by vertices and bonds between the atoms by edges in the graph, this model represents stems as edges and breaks in the stems that result in bulges and loops as vertices. A nucleotide bulge, hairpin loop, or internal loop are each represented by a vertex when there is more than one unmatched nucleotide or noncomplementary base pair.

Researchers at New York University in the Computational Biology Group led by Tamar Schlick used this method to create an RNA topology database called RAG (RNA As Graphs) that is published and available at BMC Bioinformatics and Bioinformatics [48,49]. The RNA motifs in RAG are cataloged by their vertex number and Fiedler eigenvalues. This graph theoretic representation provides an alternative approach for classifying all possible RNA structures based on their topological properties. In this work, Schlick et al. find that existing RNA classes represent only a small subset of possible 2D RNA motifs [50,51]. This indicates that there may be a number of additional naturally occuring secondary structures that have not yet been identified. It also points to possible structures that may be utilized in the synthesis of RNA in the laboratory for drug design purposes. The discovery of new RNA structures and motifs is increasing the size of specialized RNA databases. However, a comprehensive method for quantifying and cataloging novel RNAs remains absent. The tree representation utilized by the RAG database provides a useful resource to that end. Other good online resources in addition to the RAG database include the University of Indiana RNA Web site, RNA World, and RNA Base [52].

3.3.2 Graphs as Proteins

Proteins are molecules that consist of amino acids. There are 20 different amino acids; hence, one can think of a chain or sequence from an alphabet of size 20 as the primary structure of a protein. Each amino acid consists of a central carbon atom, an amino group, a carboxyl group, and a unique "side chain" attached to the central carbon. Differences in the side chains distinguish different amino acids. As this string is being produced (synthesized) in the cell, it folds back onto itself creating a 3D object. For several decades or more, biologists have tried to discover how a completely unfolded protein with millions of potential folding outcomes almost instantaneously finds the correct 3D structure. This process is very complex and often occurs with the aid of other proteins known as chaperones that guide the folding protein. The majority of protein structure prediction algorithms are primarily based on dynamic simulations and minimal energy requirements. More recently, it has been suggested that the high mechanical strength of a protein fiber, for example, is due to the folded structural linking rather than thermodynamic stability. This suggest the feasibility and validity of a graph theoretic approach as a model for the molecule.

The 3D structure of the protein is essential for it to carry out its specific function. The 3D structure of a protein has commonly occurring substructures that are referred to as secondary structures. The two most common are alpha helices and beta strands. Bonds between beta strands form beta sheets. We can think of alpha helices and beta sheets as building blocks of the 3D or tertiary structure. As in the case for the secondary RNA trees, graph models can be designed for amino acids, secondary, and tertiary protein structures. In addition to protein modeling, protein structure prediction methods that employ graph theoretic modeling focus on predicting the general protein topology rather than the 3D coordinates. When sequence similarity is poor, but the essential topology is the same, these graph theoretic methods are more advantageous.

The idea of representing a protein structure as a graph is not new and there have been a number of important results on protein structure problems obtained from graphs. Graphs are used for identification of tertiary similarities between proteins by Mitchell et al. [53] and Grindley et al [54]. Koch et al. apply graph theory to the topology of structures in proteins to automate identification of certain motifs [55]. Graph spectral analysis has provided information on protein dynamics, protein motif recognition, and fold. Identification of proteins with similar folds is accomplished using the graph spectra in the work by Patra and Vishveshwara [56]. Clusters important for function, structure, and folding were identified by cluster centers also using the graph's eigenvalues [57]. Fold and pattern identification information was gained by identifying subgraph isomorphisms [58]. For additional information on these results, see the work by Vishveshwara et al. [59]. It is worth noting that all of the above methods relied heavily on spectral graph theory alone.

Some of the early work on amino acid structure by graph theoretic means was accomplished in the QSAR arena. Use of crystal densities and specific rotations of amino acids described by a set of molecular connectivity indices was utilized by Pogliani in a QSAR study [60]. Pogliani also used linear combinations of connectivity indices to model the water solubility and activity of amino acids [61]. Randic et al. utilized a generalized topological index with a multivariate regression analysis QSAR model to determine characteristics of the molar volumes of amino acids [62].

On a larger scale, a vertex can represent an entire amino acid and edges are present if the amino acids are consecutive on the primary sequence or if they are within some specified distance. The graph in the Figure 3.8 shows the modeling of an alpha helix and a beta strand with a total of 24 amino acids.

By applying a frequent subgraph mining algorithm to graph representations of a 3D protein structure, Huan et al. found recurring amino acid residue packing patterns that are characteristic of protein structural families [63]. In their model, vertices represent amino acids, and edges are chosen in one of three ways: first, using a threshold for contact distance between residues; second, using Delaunay tessellation; and third, using the recently developed almost-Delaunay edges. For a set of graphs representing a protein family from the Structural Classification of Proteins (SCOP) database [64], subgraph mining typically identifies several hundred common subgraphs corresponding to the residue packing pattern. They demonstrate that graphs based on almost-Delaunay edges significantly reduced the number of edges in the graph representation and hence presented computational advantage.

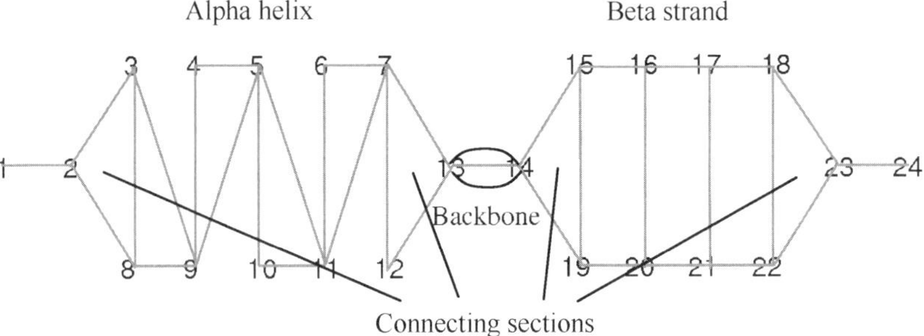

FIGURE 3.8 An alpha helix and a beta strand.

Researchers at the University of California at Berkley and at the Dana Farber Cancer Institute at Harvard Medical School have used aberration multigraphs to model chromosome aberrations [65]. A multigraph is a graph that allows multiple edges between two vertices. Aberration multigraphs characterize and interrelate three basic aberration elements: (1) the initial configuration of a chromosome; (2) the exchange process whose cycle structure helps to describe aberration complexity; and (3) the final configuration of rearranged chromosomes. An aberration multigraph refers in principle to the actual biophysical process of aberration formation. We find that graphical invariants provide information about the processes involved in chromosome aberrations. High diameter for the multigraph corresponds to many different cycles in the exchange process, linked by the fact that they have some chromosomes in common. Girth 2 in a multigraph usually corresponds to a ring formation and girth 3 to inversions. Aberration multigraphs are closely related to cubic multigraphs. An enormous amount is known about cubic multigraphs, mainly because they are related to work on the four-color theorem. Results on cubic multigraphs suggest a mathematical classification of aberration multigraphs. The aberration multigraph models the entire process of DNA damage, beginning with an undamaged chromosome and ending with a damaged one.

A relation is symmetric if "*a* is related to *b*" implies "*b* is related to *a*." Clearly, not all relations are symmetric. If a graph models a relation that is not symmetric, then directions are assigned to the edges. Such graphs are known as digraphs and networks are usually modeled by digraphs. Some network applications exist in chemical graph theory [66]. Since a reaction network in chemistry is a generalization of a graph, the decomposition of the associated graph reflects the submechanisms by closed directed cycles. A reaction mechanism is direct if no distinct mechanisms for the same reaction can be formed from a subset of the steps. Although the decomposition is not unique, the set of all direct mechanisms for a reaction is a unique attribute of a directed graph. Vingron and Waterman [67] utilized the techniques and concepts from electrical networks to explore applications in molecular biology. A variety of novel modeling methods that exploit various areas of mathematical graph theory such as random graph theory are emerging with exciting results. For more examples applications of graphs in molecular biology, see the work by Boncher et al. [68].

3.4 MACHINE LEARNING WITH GRAPHICAL INVARIANTS

Graphical invariants of graph theoretic models of chemical and biological structures can sometimes be used as descriptors [23] in a fashion similar to molecular descriptors in QSPR and QSAR models. Over the past decade, the tools of choice for using descriptors to predict such functional relationships have increasingly been artificial neural networks (ANNs) or algorithms closely related to ANNs [69]. More recently, however, support vector machines (SVMs) have begun to supplant the use of ANNs in QSAR types of applications because of their ability to address issues such as overfitting and hard margins (see, e.g., the works by Xao et al. [70] and Guler and Kocer [71]).

Specifically, the possible properties or activities of a chemical or biological structure define a finite number of specific *classes*. The ANNs and SVMs use descriptors for a given structure to predict the class of the structure, so that properties and activities are predicted via class membership. Algorithms that use descriptors to predict properties and functions of structures are known as *classifiers*. Typically, a collection of structures whose functional relationships have been classified *a priori* are used to *train* the classifier so that the classifier can subsequently be used to predict the classification of a structure whose functional relationships have yet to be identified [72].

3.4.1 Mathematics of Classifiers

Before describing SVMs and ANNs more fully, let us establish a mathematical basis for the study of classification problems. Because a descriptor such as a graphical invariant is real valued, a number n of descriptors of a collection of biological structures form an n-tuple $\mathbf{x} = (x_1, ..., x_n)$ in n-dimensional real space. A classifier is a method that partitions n-dimensional space so that each subset in the partition contains points corresponding to only one class. Training corresponds to using a set of n-tuples for structures with *a priori* classified functional relationships to approximate such a partition. Classification corresponds to using the approximate partition to make predictions about a biological structure whose class is not known [72].

If there are only two classes, as was the case in the work by Haynes et al. [23] where graph theoretic trees were classified as either RNA-like or not RNA-like, the goal is to partition an n-dimensional space into two distinct subsets. If the two subsets can be separated by a hyperplane, then the two classes are said to be *linearly separable*. An algorithm that identifies a suitable separating hyperplane is known as a *linear classifier* (Fig. 3.9).

In a linearly separable classification problem, there are constants $w_1, ..., w_n$ and b such that

$$w_1x_1 + \cdots + w_nx_n + b > 0$$

when $(x_1, ..., x_n)$ is in one class and

$$w_1x_1 + \cdots + w_nx_n + b < 0$$

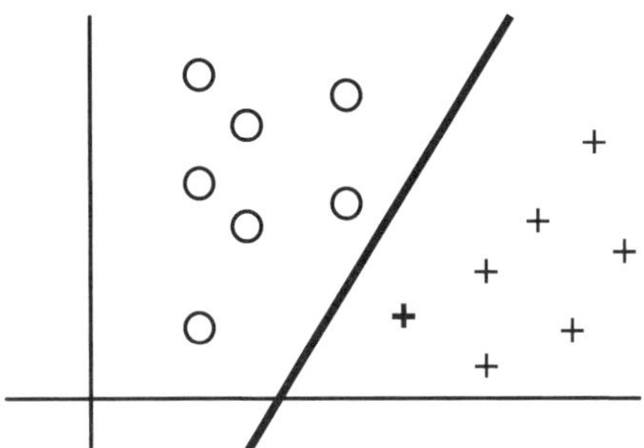

FIGURE 3.9 Linear separability.

when $(x_1, ..., x_n)$ is in the other. Training reduces to choosing the constants so that the distance between the hyperplane and the training data is maximized, and this maximal distance is then known as the margin.

If there are more than two classes and the classes are not linearly separable, then there are at least two different types of classifiers that can be used. An SVM supposes that some mapping $\phi(\mathbf{x})$ from n-space into a larger dimensional vector space known as a *feature space* will lead to linear separability in the larger dimensional space, at which point an optimal hyperplane is computed in the feature space by maximizing the distance between the hyperplane and the closest training patterns. The training patterns that determine the hyperplane are known as *support vectors*.

If $K(\mathbf{x},\mathbf{y})$ is a symmetric, positive definite function, then it can be shown that there exists a feature space with an inner product for which

$$K(\mathbf{x}, \mathbf{y}) = \phi(\mathbf{x}) \cdot \phi(\mathbf{y}).$$

The function $K(\mathbf{x},\mathbf{y})$ is known as the *kernel* of the transformation, and it follows that the implementation of an SVM depends only on the choice of a kernel and does not require the actual specification of the mapping or the feature space. Common kernels include the following:

- *Inner product:* $K(\mathbf{x}, \mathbf{y}) = \mathbf{x} \cdot \mathbf{y}$.
- *Polynomial:* $K(\mathbf{x}, \mathbf{y}) = (\mathbf{x} \cdot \mathbf{y} + 1)^N$, where N is a positive integer.
- *Radial:* $K(\mathbf{x}, \mathbf{y}) = e^{-a\|\mathbf{x}-\mathbf{y}\|^2}$.
- *Neural:* $K(\mathbf{x}, \mathbf{y}) = \tanh(a\mathbf{x} \cdot \mathbf{y} + b)$, where a and b are parameters.

Within the feature space, an SVM is analyzed as a linear classifier [73].

Several implementations of SVMs are readily available. For example, mySVM and YALE, which can be found at http://www.support-vector-machines.org, can be downloaded as windows executables or Java applications [74]. There are also several books, tutorials, and code examples that describe in detail how SVMs are implemented and trained [75].

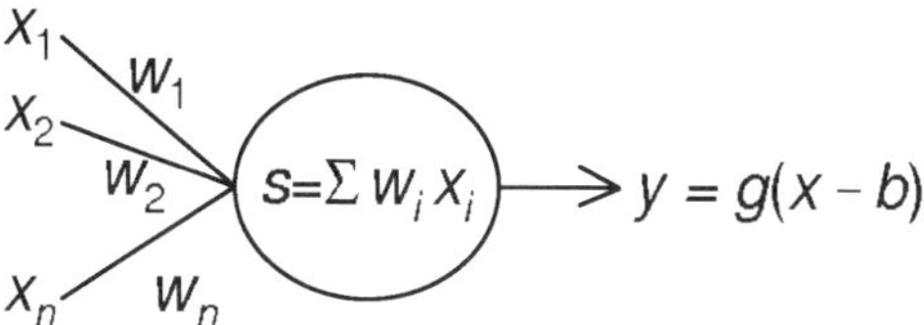

FIGURE 3.10 An artificial neuron.

ANNs are alternatives to SVMs that use networks of linear-like classifiers to predict structure–function classifications. Specifically, let us suppose that the two classes of a linear classifier can be associated with the numbers 1 and 0. If we also define a *firing function* by

$$g(s) = \begin{cases} 1 \text{ if } s > 0, \\ 0 \text{ if } s < 0, \end{cases} \tag{3.1}$$

then the linear classifier can be interpreted to be a single *artificial neuron,* which is shown in Figure 3.10. In this context, $w_1, \ldots, w_n$ are known as *synaptic weights* and b is known as a *bias*. The firing function is also known as the *activation function,* and its output is known as the *activation* of the artificial neuron.

The terminology comes from the fact that artificial neurons began as a caricature of real-world neurons, and indeed, real-world neurons are still used to guide the development of ANNs [76]. The connections with neurobiology also suggest that the activation function $g(s)$ should be sigmoidal, which means that it is differentiable and nondecreasing from 0 up to 1. A commonly used activation function is given by

$$g(s) = \frac{1}{1 + e^{-\kappa s}}, \tag{3.2}$$

where $\kappa > 0$ is a parameter [77], which is related to the hyperbolic tangent via

$$g(s) = \tfrac{1}{2}\tanh(\kappa s) + \tfrac{1}{2}.$$

The choice of a smooth activation function allows two different approaches to training—the synaptic weights can be estimated from a training set either using linear algebra and matrix arithmetic or via optimization with the synaptic weights as dependent variables. The latter is the idea behind the *backpropagation* method, which is discussed in more detail below.

A *multilayer feedforward network* (MLF) is a network of artificial neurons organized into *layers* as shown in Figure 3.11, where a layer is a collection of neurons connected to all the neurons in the previous and next layers, but not to any neurons in the layer itself. The first layer is known as the *input layer*, the last layer is known as the *output layer,* and the intermediate layers are known as *hidden layers*. Figure 3.11 shows a typical three-layer MLF.

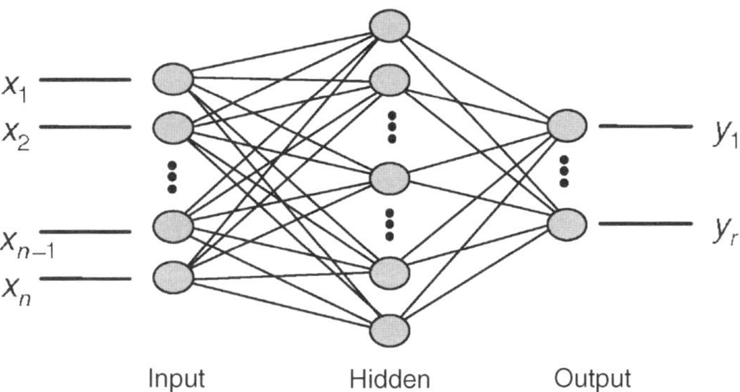

FIGURE 3.11 A three-layer MLP.

In the prediction or *feedforward* stage, the descriptors $x_1, \cdots, x_n$ are presented to the input layer neurons, and their activations are calculated as in Figure 3.10. Those activations are multiplied by the synaptic weights w_{ij} between the ith input neuron and the jth output neuron and used to calculate the activations of the hidden layer neurons. Similarly, the synaptic weights α_{jk} between the kth hidden neurons and the jth output neurons are used to calculate the activations $y_1, \cdots, y_r$ from the output neurons, which are also the predicted classification of the structure that generated the initial descriptors.

If the classification $\mathbf{q} = (q_1, \ldots, q_r)$ for an n-tuple of descriptors $\mathbf{p} = (p_1, \ldots, p_n)$ is known, then the pair $(\mathbf{p}, \mathbf{q})$ is known as a *training pattern.* Training a three-layer MLF using a collection $\left(\mathbf{p}^1, \mathbf{q}^1\right), \ldots, \left(\mathbf{p}^t, \mathbf{q}^t\right)$ of training patterns means using nonlinear optimization to estimate the synaptic weights. In addition, the synaptic weights can be used for *feature selection*, which is to say that a neural network can be used to determine how significant a descriptor is to a classification problem by examining how sensitive the training process is to the values of that descriptor.

3.4.2 Implementation and Training

Both general-purpose and informatics-targeted implementations of MLFs are readily available. For example, the neural network toolbox for MatLab and the modeling kit ADAPT allow the construction of MLFs and other types of neural networks [75,77]. There are also many variations on the MLF ANN structure and training methods, including self-organizing feature maps (SOFM) [78,79] and Bayesian regularized neural network [80]. In addition, several different implementations of neural networks in programming code are also available.

However, it is important not to treat ANNs or SVMs as "canned" routines, because they are similar to other nonlinear regression methods in that they can overfit the data and they can be overtrained to the training set [69]. Overtraining corresponds to the network's "memorizing" of the training set, thus leading to poor predictions for structures not in the training set. This issue is often addressed using cross-validation or "leave-one-out" training methods in which a part of the training set is removed, the network is trained on the remaining training patterns, and then the classification of the removed training patterns is predicted.

Overfitting is a more serious and less avoidable problem [81]. Typically, there is small variation or "noise" in the descriptor values, so that if there are too many parameters—for example, too many neurons in the hidden layer—then training may lead to an "interpolation" of the slightly flawed training set at the expense of poor generalization of the training set. In both overfitting and overtraining, convergence of the nonlinear optimization algorithm is common, but predictions are either meaningless in the case of overfitting or too dependent on the choice of the training.

Because graphical invariants are often discrete valued and highly dependent on the construction of the graphical model, overfitting and overtraining are important issues that cannot be overlooked. For this reason, we conclude with a more mathematical exploration of the ANN algorithm so that their training and predictive properties can be better understood.

To begin with, suppose that $\mathbf{y} = (y_1, ..., y_n)$ denotes the output from a three-layer MLF that has r input neurons connected to m hidden layer neurons that are connected to n neurons in the output layer. It has been shown that with the appropriate selection of synaptic weights, a three-layer MLF can approximate any absolutely integrable mapping of the type

$$f(x_1, \ldots, x_r) = (y_1, \ldots, y_n)$$

to within any $\epsilon > 0$ [82]. That is, a three-layer MLP can theoretically approximation the solution to any classification problem to within any given degree of accuracy, thus leading MLFs to be known as *universal classifiers.* However, in practice the number of hidden layer neurons may necessarily be large, thus contradicting the desire to use small hidden layers to better avoid overfitting and overtraining.

To gain further insights into the innerworkings of a three-layer MLF, let $\mathbf{w}_k = (w_{k1}, \ldots, w_{kr})$ denote the vector of weights between the input layer and the kth-hidden neuron. It follows that $y_j = g\left(s_j - b_j\right)$, where b_j denotes the bias of the jth output neuron, where

$$s_j = \sum_{k=1}^{m} \alpha_{jk} g\left(w_k(x - \theta_k)\right),$$

and where θ_k denotes the bias for the kth hidden neuron. A common method for estimating synaptic weights given a collection $\left(\mathbf{p}^1, \mathbf{q}^1\right), \ldots, \left(\mathbf{p}^t, \mathbf{q}^t\right)$ of training patterns is to define an *energy function*

$$E = \frac{1}{2} \sum_{i=1}^{t} \left(y - q^i\right)\left(y - q^i\right),$$

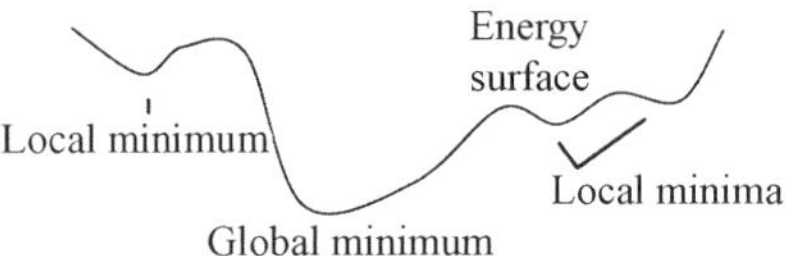

FIGURE 3.12 The energy surface.

and then train the MLP until we have closely approximated

$$\frac{\partial E}{\partial w_{kl}} = 0 \quad \text{and} \quad \frac{\partial E}{\partial \alpha_{jk}} = 0$$

at the inputs $\mathbf{p}^i$ for all $l = 1, \ldots, r$, $k = 1, \ldots, m$, and $j = 1, \ldots, n$. Because these equations cannot be solved directly, a gradient-following method called the *backpropagation* algorithm is used instead.

The backpropagation algorithm is based on the fact that if g is the sigmoidal function defined in equation (3.2), then

$$g' = \kappa g (1 - g).$$

In particular, for each training pattern $\left(\mathbf{p}^i, \mathbf{q}^i\right)$, a three-layer MLP first calculates $\mathbf{y}$ as the output to $\mathbf{p}^i$, which is the *feedforward* step. The weights α_{jk} are subsequently adjusted using

$$\alpha_{jk} \to \alpha_{jk} + \lambda \delta_j \xi_k,$$

where $\xi_k = g(\mathbf{w}_k \cdot \mathbf{x} - \theta_k)$, where $\lambda > 0$ is a fixed parameter called the *learning rate,* and where

$$\delta_j = \kappa y_j \left(1 - y_j\right) \left(q_j^i - y_j\right).$$

The weights w_{kr} are adjusted using

$$w_{kl} \to w_{kl} + \lambda \rho_k \, x_l,$$

where $x_l = g\left(p_l^i - \theta_l\right)$ and where

$$\rho_k = \kappa \xi_k (1 - \xi_k) \sum_{j=1}^{n} \alpha_{jk} \delta_j.$$

Cybenko's theorem implies that the energy E should eventually converge to 0, so training continues until the energy is sufficiently small in magnitude.

However, it is possible that the energy for a given training set does not converge. For example, it is possible for training to converge to a *local minimum* of the energy function, as depicted in Figure 3.12. When this happens, the network can make errant predictions known as *spurious states*. To avoid such local minima, it may be necessary to add small random inputs into each neuron so that training continues beyond any local minima, or it may be necessary to use a process such as *simulated annealing* to avoid such local minima [77].

Similarly, if the synaptic weights are not initialized to small random values, then the network tends to overtrain immediately on the first training pattern presented to it and thus may converge only very slowly. Overtraining can often be avoided by calculating the energy on both the training set and a validation set at each iteration. However, overfitting may not necessarily be revealed by the behavior of the energy during training.

This is because the quantities that define the training process are

$$\delta_j = \kappa y_j \left(1 - y_j\right) \left(q_j^i - y_j\right)$$

and

$$\rho_k = \kappa \xi_k \left(1 - \xi_k\right) \sum_{j=1}^{n} \alpha_{jk} \delta_j,$$

both of which are arbitrarily close to 0 when δ_j is arbitrarily close to 0. In overfitting, this means that once y_j is sufficiently close to q_j^i, the quantities ξ_k can vary greatly without changing the convergence properties of the network. That is, convergence of the output to the training set does not necessarily correspond to convergence of the hidden layer to a definite state. Often this means that two different training sessions with the same training set may lead to different values for the synaptic weights [69].

Careful design and deployment of the network can often avoid many of the issues that may affect ANNs. Large hidden layers are typically not desirable, and often an examination of the synaptic weights over several "test runs" will give some insight into the arbitrariness of the dependent variables ξ_k for the hidden layer, thus indicating when the hidden layer may possibly be too large. In addition, as the network begins to converge, modifying the learning parameter λ as the network converges may "bump" the network out of a local minimum without affecting overall convergence and performance.

3.5 GRAPHICAL INVARIANTS AS PREDICTORS

We conclude with an example of the usefulness of graphical invariants as predictors of biomolecular structures. The RAG database [48] contains all possible unlabeled trees of orders 2 through 10. For the trees of orders 2 through 8, each tree is classified as an RNA tree, an RNA-like tree or not RNA-like tree. For the trees of order 9 and 10, those that represent a known secondary RNA structure are identified as an RNA tree, but no trees are shown to be candidate structures, that is, RNA-like. In the works by Haynes et al. [22,23], the tree modeling method is used to quantify secondary RNA structures with graphical parameters that are defined by variations of the domination number of a graph.

Note that a single graphical invariant may not be sufficient to differentiate between trees that are RNA-like and those that are not. For example, the domination number for trees of order 7, 8, and 9 range from 1 to 4 with no discernable relationship between the value of the invariant and the classification of the tree. However, defining three parameters in terms of graphical invariants does prove to be predictive.

Specifically, an MLP with three input neurons, five hidden neurons, and two output neurons is trained using values of the three parameters

$$P_1 = \frac{\gamma + \gamma_{\mathrm{t}} + \gamma_{\mathrm{a}}}{n},$$

$$P_2 = \frac{\gamma_{\mathrm{L}} + \gamma_{\mathrm{D}}}{n},$$

$$P_3 = \frac{\mathrm{diam}(L(T)) + \mathrm{rad}(L(T)) + |B|}{n},$$

where γ is the domination number, γ_{t} is the total domination number, γ_a is the global alliance number, γ_{L} is the locating domination number of the line graph, and γ_{D} is the differentiating dominating number. For more on variations of the domination numbers of graphs, see the work by Haynes et al. [25]. Additionally, $\mathrm{diam}(L(T))$ is the diameter of the line graph, $\mathrm{rad}(L(T))$ is the radius of the line graph, $|B|$ is the number of blocks in the line graph of the tree, and n is the order of a tree. The use of *leave-one-out* cross-validation during training addresses possible overfitting. We also use the technique of *predicting complements* (also known as leave-v-out cross-validation) with 6, 13, and 20 trees, respectively, in the complement. Table 3.1 shows the average error and standard deviation in predicting either a "1" for a RNA tree or a "0" for a tree that is not RNA-like.

The resulting MLP predicts whether trees of orders 7, 8, and 9 are RNA-like or are not RNA-like. The results are shown in Table 3.2. For the trees of order 7 and 8, the network predictions coincide with the RAG classification with the exception of 2 of the 34 trees. Also, the network was able to predict an additional 28 trees of order 9 as being RNA-like in structure. This information may assist in the development of synthetic RNA molecules for drug design purposes [49].

The use of domination-based parameters as biomolecular descriptors supports the concept of using graphical invariants that are normally utilized in fields such as computer network design to quantify and identify biomolecules. By finding graphical invariants of the trees of orders 7, 8, and using the four additional trees of order 9 in

TABLE 3.1 Accuracy Results for the RNA Classification

	$\lvert\mathrm{Comp}\rvert = 6$	$\lvert\mathrm{Comp}\rvert = 13$	$\lvert\mathrm{Comp}\rvert = 20$
Average error	0.084964905	0.161629391	0.305193489
Standard deviation	0.125919698	0.127051425	0.188008046

TABLE 3.2 RNA Prediction Results

RAG[a]	Class[b]	Error[c]	RAG	Class	Error	RAG	Class	Error
7.4	0	0.00947	9.9	0	0.0554	9.31	1	0.0247
7.5	1	0.0245	9.10	1	2.65E−06	9.32	0	1.99E−06
7.7	1	7.45E−05	9.12	1	5.28E−07	9.33	1	0.0462
7.8	1	1.64E−07	9.14	1	2.32E−07	9.34	1	0.00280
8.1	1	1.05E−06	9.15	0	1.82E−04	9.35	0	2.46E−06
8.2	1	1.24E−06	9.16	1	5.35E−04	9.36	0	7.41E−05
8.4	1	0.0138	9.17	1	6.24E−06	9.37	0	7.41E−05
8.6	1	0.0138	9.18	1	4.87E−07	9.38	1	4.86E−05
8.8	1	5.43E−05	9.19	1	6.06E−07	9.39	0	2.46E−06
8.12	1	3.59E−06	9.20	1	0.0247	9.40	0	4.79E−08
8.13	0	0.0157	9.21	1	6.38E−05	9.41	0	4.79E-08
8.16	1	8.81E−06	9.22	1	0.0247	9.42	1	2.51E−07
9.1	1	1.48E−07	9.23	0	7.41E−05	9.43	1	4.86E−05
9.2	1	0.0151	9.24	1	1.47E−05	9.44	1	0.0247
9.3	1	0.0121	9.25	0	3.85E−07	9.45	0	7.41E−05
9.4	1	4.05E−07	9.26	1	1.48E−04	9.46	0	4.79E−08
9.5	1	5.24E−05	9.28	0	7.41E−05	9.47	0	2.33E−08
9.7	1	6.38E−05	9.29	1	3.61E−07			
9.8	1	6.38E−05	9.30	1	1.47E−05			

[a] Labels from the RAG RNA database [48].

[b] Class = 1 if predicted to be an RNA tree; class = 0 if not RNA-like.

[c] Average deviation from predicted class.

the RAG database, Knisley et al. [23] utilize a neural network to identify novel RNA-like structures from among the unclassified trees of order 9 and thereby illustrate the potential for neural networks coupled with mathematical graphical invariants to predict function and structure of biomolecules.

ACKNOWLEDGMENTS

This work was supported by a grant from the National Science Foundation, grant number DMS-0527311.

REFERENCES

1. Laurent A. Rev Sci 1843;14:314.
2. Gerhardt C. Ann Chim Phys 1843;3(7):129.
3. Russell C. The History of Valency. Leicester: Leicester University Press; 1971.
4. Butlerov A. Zeitschr Chem Pharm 1861;4:549.
5. Frankland E. Lecture Notes for Chemical Students. London: Van Voorst; 1866.
6. Cayley A. Philos Mag 1874;47:444.

7. Lodge O. Philos Mag 1875;50:367.
8. Sylvester J. On an application of the new atomic theory to the graphical representation of the invariants and coinvariants of binary quantics. Am J Math 1878; 1:1.
9. Bonchev D. Rouvray D, editors. Chemical Graph Theory: Introduction and Fundamentals. Abacus Press/Gordon & Breach Science Publishers; 1990.
10. Trinajstic N. Chemical Graph Theory. Volume 1. CRC Press; 1983.
11. Trinajstic N. Chemical Graph Theory. Volume 2. CRC Press; 1983.
12. Barker E, Gardiner E, Gillet V, Ketts P, Morris J. Further development of reduced graphs for identifying bioactive compounds. J Chem Inform Comput Sci 2003;43:346–356.
13. Barker E, Buttar D, Cosgraove D, Gardiner E, Kitts P, Willett P, Gillet V. Scaffold hopping using clique detection applied to reduced graphs. J Chem Inform Model 2006;46:503–511.
14. Todeschini R, Consonni V. In: Mannhold R, Kubinyi H, Timmerman H, editors. Handbook of Molecular Descriptors. Volume 11. Series of Methods and Principles in Medicinal Chemistry. Wiley; 2000.
15. Tetko IV, Gasteiger J, Todeschini R, Mauri A, Livingstone D, Ertl P, Palyulin VA, Radchenko EV, Zefirov NS, Makarenko AS, Tanchuk VY, Prokopenko VV. Virtual computational chemistry laboratory—design and description. J Comput Aided Mol Des 2005;19:453–463. Available at http://www.vcclab.org/
16. Talete: http://www.talete.mi.it/.
17. Weininger D. SMILES, A chemical language and information system. J Chem Inform Comput Sci 1988;28(1):31–36.
18. Schuffenhauer A, Gillet V, Willett P. Similarity searching in files of three-dimensional chemical structures: analysis of the BIOSTER databases using two-dimensional fingerprints and molecular field descriptors. J Chem Inform Comput Sci 2000;40:296–307.
19. Bemis G, Kuntz I. A fast and efficient method for 2D and 3D molecular shape description. J Comput Aided Mol Des 1992;6(6):607–628.
20. Jurs Research Group, http://research.chem.psu.edu/pcjgroup/.
21. The Chemical Computing Group—MOE, http://www.chemcomp.com.
22. Haynes T, Knisley D, Seier E, Zou Y. A quantitative analysis of secondary RNA structure using domination based parameters on trees. BMC Bioinform 2006;7:108, doi:10.1186/1471-2105-7-108.
23. Haynes T, Knisley D, Knisley J, Zoe Y. Using a neural network to identify RNA structures quantified by graphical invariants. Submitted.
24. Hoyosa HB. Chem Soc Jpn 1971;44:2332.
25. Haynes T, Hedetniemi S, Slater P. Fundamentals of Domination in Graphs. Marcel Dekker; 1998.
26. Beineke, L. Zamfirescu C. Connection digraphs and second order line graphs. Discrete Math 1982;39:237–254.
27. Dix D. An application of iterated line graphs to biomolecular conformations. Preprint.
28. Gordon M, Scantlebury G. Trans Faraday Soc 1964;60:604.
29. Randic M. J Am Chem Soc 1975;97:6609.
30. Bonchev D. The overall Weiner index—a new tool for the characterization of molecular topology. J Chem Inform Comput Sci 2001;41(3):582–592.
31. Balaban A. Chem Phys Lett 1982;89:399–404.

32. Balaban A, Mills D, Ivanciuc O, Basak. Reverse wiener indices. CCACAA 2000;73(4):923–941.

33. Lima P, Golbraikh A, Oloff S, Xiao Y, Tropsha. Combinatorial QSAR modeling of P-glycoprotein substrates. J Chem Inform Model 2006;46:1245–1254.

34. Viswanadhan V, Mueller G, Basak S, Weinstein. Comparison of a neural net-based QSAR algorithm with hologram and multiple linear regression-based QSAR approaches: application to 1,4-dihydropyridine-based calcium channel antagonists. J Chem Inform Comput Sci 2001;41:505–511.

35. Randic M, Zupan J. On interpretation of well-known topological indices. J Chem Inform Comput Sci 2001;41:550–560.

36. Randic M, Basak S. A comparative study of proteomic maps using graph theoretical biodescriptors. J Chem Inform Comput Sci 2002;42:983–992.

37. Bajzer Z, Randic M, Plavisic M, Basak S. Novel map descriptors for characterization of toxic effects in proteomics maps. J Mol Graph Model 2003;22(1):1–9.

38. Breneman, CM, Sundling, CM, Sukumar N, Shen L, Katt WP, Embrechts MJ. New developments in PEST—shape/property hybrid descriptors. J Comput Aid Mol Design 2003;17:231–240.

39. Washietl S, Hofacker I, Stadler P. Fast and reliable prediction of noncoding RNAs. Proc Natl Acad Sci USA 2005;101:2454–2459.

40. Washietl S, Hofacker I, Lukasser M, Huttenhofer A, Stadler P. Mapping of conserved RNA secondary structures predicts thousands of functional noncoding RNAs in the human genome. Nat Biotechnol 2005;23(11):1383–1390.

41. Backofen R, Will S. Local sequence–structure motifs in RNA. J Biol Comp Biol 2004; 2(4):681–698.

42. Zuker M, Mathews DH, Turner DH. Algorithms and thermodynamics for RNA secondary structure prediction: a practical guide. In: Barciszewski J, Clark BFC, editors. RNA Biochemistry and Biotechnology. NATO ASI Series. Kluwer Academic Publishers; 1999.

43. Waterman M. An Introduction to Computational Biology: Maps, Sequences and Genomes. Chapman Hall/CRC; 2000.

44. Le S, Nussinov R, Maziel J. Tree graphs of RNA secondary structures and their comparison. Comput Biomed Res 1989;22:461–473.

45. Benedetti G, Morosetti S. A graph-topological approach to recognition of pattern and similarity in RNA secondary structures. Biol Chem 1996;22:179–184.

46. Barash D. Spectral decomposition of the Laplacian matrix applied to RNA folding prediction. Proceedings of the Computational Systems Bioinformatics (CSB); 2003. p 602–6031.

47. Heitsch C, Condon A, Hoos H. From RNA secondary structure to coding theory: a combinatorial approach. In: Hagiya M, Ohuchi A, editors. DNA 8; LNCS; 2003. p 215–228.

48. Fera D, Kim N, Shiffeidrim N, Zorn J. Laserson U, Gan H, Schlick, T. RAG: RNA-As-Graphs web resource. BMC Bioinform 2004;5:88.

49. Gan H, Fera D, Zorn J, Shiffeldrim N, Laserson U, Kim N, Schlick T. RAG: RNA-As-Graphs database—concepts, analysis, and features. Bioinformatics 2004;20:1285–1291.

50. Gan H, Pasquali S, Schlick T. Exploring the repertoire of RNA secondary motifs using graph theory: implications for RNA design. Nucl Acids Res 2003;31(11):2926–2943.

51. Zorn J, Gan HH, Shiffeldrim N, Schlick T. Structural motifs in ribosomal RNAs: implications for RNA design and genomics. Biopolymers 2004;73:340–347.
52. RNA Resources (online): (1) www.indiana.edu/˜tmrna; (2) www. imb-jena.de/RNA.html; (3) www.rnabase.org.
53. Mitchell E, Artymiuk P, Rice D, Willet P. Use of techniques derived from graph theory to compare secondary structure motifs in proteins. J Mol Biol 1989;212(1):151.
54. Grindley H, Artymiuk P, Rice D, Willet. Identification of tertiary structure resemblance in proteins. J Mol Biol 1993;229(3):707.
55. Koch I, Kaden F, Selbig J. Analysis of protein sheet topologies by graph–theoretical techniques. Proteins 1992;12:314–323.
56. Patra S, Vishveshwara S. Backbone cluster identification in proteins by a graph theoretical method. Biophys Chem 2000;84:13–25.
57. Kannan K, Vishveshwara S. Identification of side-chain clusters in protein structures by a graph spectral method. J Mol Biol 1999;292:441–464.
58. Samudrala R, Moult J. A graph–theoretic algorithm for comparative modeling of protein structure. J Mol Biol 1998;279:287–302.
59. Vishveshwara S, Brinda K, Kannan N. Protein structures: insights from graph theory. J Theor Comput Chem 2002;I(1):187–211.
60. Pogliani L. Structure property relationships of amino acids and some dipeptides. Amino Acids 1994;6(2):141–153.
61. Pogliani L. Modeling the solubility and activity of amino acids with the LCCI method. Amino Acids 1995;9(3):217–228.
62. Randic M, Mills D, Basak S. On characterization of physical properties of amino acids. Int J Quantum Chem 2000;80:1199–1209.
63. Huan J, Bandyopadhyay D, Wang W, Snoeyink J, Prins J, Tropsha A. Comparing graph representations of protein structure for mining family-specific residue-based packing motifs. J Comput Biol 2005;12:(6):657–671.
64. Murzin A, Brenner S, Hubbard T, Chothia C. SCOP: a structural classification of proteins database for the investigation of sequences and structures. J Mol Biol 1995;247(4):536–540.
65. Sachs R, Arsuaga J, Vazquez M, Hiatky L, Hahnfeldt P. Using graph theory to describe and model chromosome aberrations. Radiat Res 2002;158:556–567.
66. Gleiss P, Stadler P, Wagner A. Relevant cycles in chemical reaction networks. Adv Complex Syst 2001;1:1–18.
67. Vingron, Waterman M. Alignment networks and electrical networks. Discrete Appl Math: Comput Mol Biol 1996.
68. Bonchev D, Rouvray D. Complexity in Chemistry, Biology and Ecology. Springer; 2005.
69. Winkler D. The role of quantitative structure–activity relationships (QSAR) in biomolecular discovery. Briefings Bioinform 2002;3(1):73–86.
70. Xao XJ, Yao X, Panaye A, Doucet J, Zhang R, Chen H, Liu M, Hu Z, Fan B. Comparative study of QSAR/QSPR correlations using support vector machines, radial basis function neural networks, and multiple linear regression. J Chem Inform Comput Sci 2004;44(4):1257–1266.
71. Guler NF, Kocer S. Use of support vector machines and neural network in diagnosis of neuromuscular disorders. J Med Syst 2005;29(3):271–284.

72. Ivanciuc O. Molecular graph descriptors used in neural network models. In: Devillers J, Balaban AT, editors. Topological Indices and Related Descriptors in QSAR and QSPR. The Netherlands: Gordon and Breach Science Publishers; 1999. p 697–777.

73. Vapnik V. Statistical Learning Theory. New York: Wiley-Interscience; 1998.

74. Rüping S. mySVM, University of Dortmund, http://www-ai.cs.uni-dortmund.de/SOFTWARE/MYSVM/.

75. Kecman V. Learning and Soft Computing: Support Vector Machines, Neural Networks, and Fuzzy Logic Models. Cambridge, MA: The MIT Press; 2001.

76. Knisley J, Glenn L, Joplin K, Carey P. Artificial neural networks for data mining and feature extraction. In: Hong D, Shyr Y, editors. Quantitative Medical Data Analysis Using Mathematical Tools and Statistical Techniques. Singapore: World Scientific; forthcoming.

77. Bose NK, Liang P. Neural Network Fundamentals with Graphs, Algorithms, and Applications. New York: McGraw-Hill; 1996.

78. Tamayo P, Slonim D, Mesirov J, Zhu Q, Kitareewan S, Dmitrovsky E, Lander E, Golub T. Interpreting patterns of gene expression with self-organizing maps methods and application to hematopoietic differentiation. Proc Natl Acad Sci USA 1999;96:2907–2912.

79. Bienfait, B. Applications of high-resolution self-organizing maps to retrosynthetic and QSAR analysis. J Chem Inform Comput Sci 1994;34:890–898.

80. Burden FR, Winkler DA. Robust QSAR models using Bayesian regularized neural networks. J Med Chem 1999;42(16):3183–3187.

81. Lawrence S, Giles C, Tsoi A. Lessons in neural network training: overfitting may be harder than expected. Proceedings of the 14th National Conference on Artificial Intelligence. AAAI-97; 1997. p 540–545.

82. Cybenko G. Approximation by superposition of a sigmoidal function. Math Control Signal Syst 1989;2(4):303–314.

CHAPTER 4

Algorithmic Methods for the Analysis of Gene Expression Data

HONGBO XIE, UROS MIDIC, SLOBODAN VUCETIC, and
ZORAN OBRADOVIC

4.1 INTRODUCTION

The traditional approach to molecular biology consists of studying a small number of genes or proteins that are related to a single biochemical process or pathway. A major paradigm shift recently occurred with the introduction of gene expression microarrays that measure the expression levels of thousands of genes at once. These comprehensive snapshots of gene activity can be used to investigate metabolic pathways, identify drug targets, and improve disease diagnosis. However, the sheer amount of data obtained using the high throughput microarray experiments and the complexity of the existing relevant biological knowledge are beyond the scope of manual analysis. Thus, the bioinformatics algorithms that help to analyze such data are a very valuable tool for biomedical science. This chapter starts with a brief overview of the microarry technology and concepts that are important for understanding the remaining sections. Second, microarray data preprocessing, an important topic that has drawn as much attention from the research community as the data analysis itself, is addressed. Finally, some of the most important methods for microarray data analysis are described and illustrated with examples and case studies.

4.1.1 Biology Background

Most cells within the same living system have identical copies of DNA that store inherited genetic traits. DNA and RNA are the carriers of the genetic information. They are both polymers of nucleotides. There are four different types of nucleotides: adenine (A), thymine/uracil (T/U), guanine (G), and cytosine (C). Thymine is present in DNA, while uracil replaces it in RNA. Genes are fundamental blocks of DNA that encode genetic information and are transcribed into messenger RNA, or mRNA

Handbook of Applied Algorithms: Solving Scientific, Engineering and Practical Problems
Edited by Amiya Nayak and Ivan Stojmenović

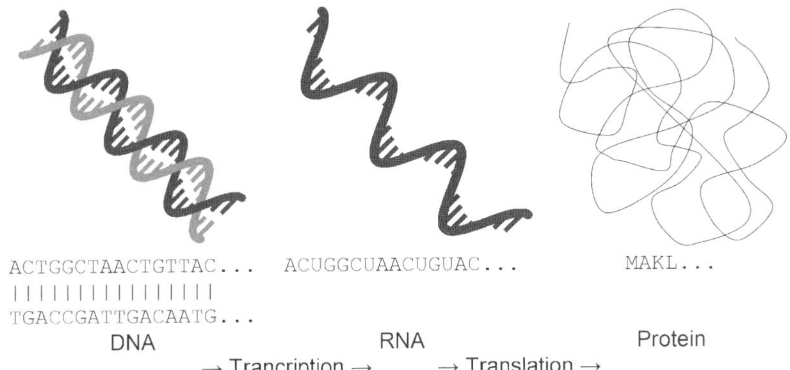

FIGURE 4.1 Central dogma of molecular biology: DNA–RNA–protein relationship.

(hereafter noted simply as "RNA"). RNA sequences are then translated into proteins, which are the primary components of living systems and which regulate most of a cell's biological activities. Activities regulated and/or performed by a protein whose code is contained in the specific gene are also considered functions of that gene. For a gene, the abundance of the respective RNA in a cell (called the "expression level" for that gene) is assumed to correlate with the abundance of the protein into which the RNA translates. Therefore, the measurement of genes' expression levels elucidates the activities of the respective proteins. The relationship between DNA, RNA, and proteins is summarized in the *Central Dogma* of molecular biology as shown in Figure 4.1.

DNA consists of two helical strands; pairs of nucleotides from two strands are connected by hydrogen bonds, creating the so-called base pairs. Due to the chemical and steric properties of nucleotides, adenine can only form a base pair with thymine, while cytosine can only form a base pair with guanine. As a result, if one strand of DNA is identified, the other strand is completely determined. Similarly, the strand of RNA produced during the transcription of one strand of DNA is completely determined by that strand of DNA. The only difference is that uracil replaces thymine as a complement to adenine in RNA. Complementarity of nucleotide pairs is a very important biological feature. Preferential binding—the fact that nucleotide sequences only bind with their complementary nucleotide sequences—is the basis for the microarray technology.

4.1.2 Microarray Technology

Microarray technology evolved from older technologies that are used to measure the expression levels of a small number of genes at a time [1,2]. Microarrays contain a large number—hundreds or thousands—of small spots (hence the term "microarray"), each of them designed to measure the expression level of a single gene. Spots are made up of synthesized short nucleotide sequence segments called probes, which are attached to the chip surface (glass, plastic, or other material). Probes

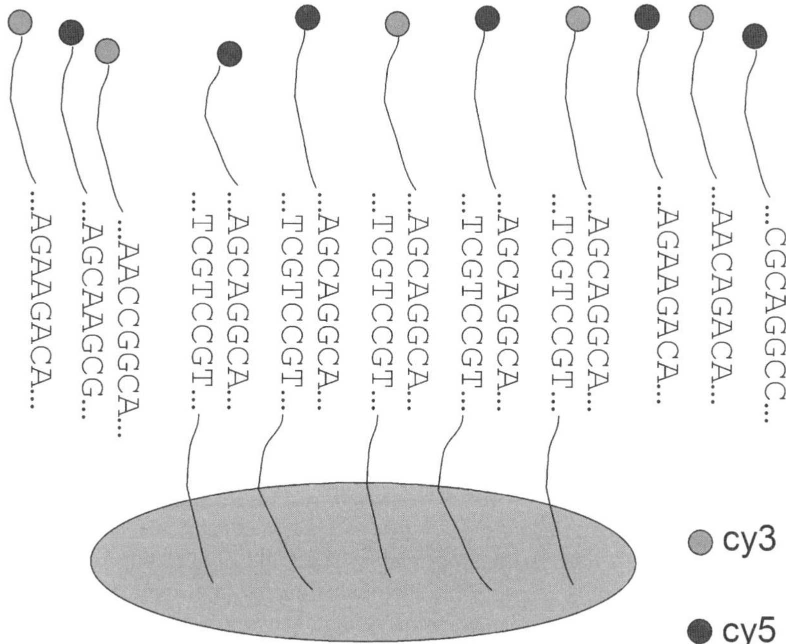

FIGURE 4.2 Binding of probes and nucleotide sequences. Probes in one spot are designed to bind only to one particular type of RNA sequences. This simplified drawing illustrates how only the complementary sequences bind to a probe, while other sequences do not bind to the probe.

in each spot are designed to bind only to the RNA of a single gene through the principle of preferential binding of complementary nucleotide sequences, as illustrated in Figure 4.2. The higher the RNA expression level is for a particular gene, the more of its RNA will bind (or "hybridize") to probes in the corresponding spot.

Single-channel and dual-channel microarrays are the two major types of gene expression microarrays. Single-channel microarrays measure the gene expression levels in a single sample and the readings are reported as absolute (positive) values. Dual-channel microarrays simultaneously measure the gene expression levels in two samples and the readings are reported as relative differences in the expression between the two samples. A sample (or two samples for dual-channel chips) and the microarray chip are processed with a specific laboratory procedure (the technical details of which are beyond the scope of this chapter). Part of the procedure is the attachment of a special fluorescent substrate to all RNA in a sample (this is called the "labeling"). When a finalized microarray chip is scanned with a laser, the substrate attached to sequences excites and emits light. For dual-channel chips, two types of substrates (cy3 and cy5) that emit light at two different wavelengths are used (Fig. 4.3). The intensity of light is proportional to the quantity of RNA bound to a spot, and this intensity correlates to the expression level of the corresponding gene.

FIGURE 4.3 Dual-channel cDNA microarray. A sample of dual-channel microarray chip images, obtained from an image scanner. All images contain only a portion of the chip. From left to right: cy3 channel, cy5 channel, and the computer-generated joint image of cy3 and cy5 channels. A light gray spot in the joint image indicates that the intensity of the cy3 channel spot is higher than intensity of the cy5 channel spot, a dark gray spot indicates a reverse situation, and a white spot indicates similar intensities.

Images obtained from scanning are processed with image processing software. This software transforms an image bitmap into a table of spot intensity levels accompanied by additional information such as estimated spot quality. The focus of this chapter is on the analysis of microarray data starting from this level. The next section describes methods for data preprocessing, including data cleaning, transformation, and normalization. Finally, the last section provides an overview of methods for microarray data analysis and illustrates how these methods are used for knowledge discovery. The overall process of microarray data acquisition and analysis is shown in Figure 4.4.

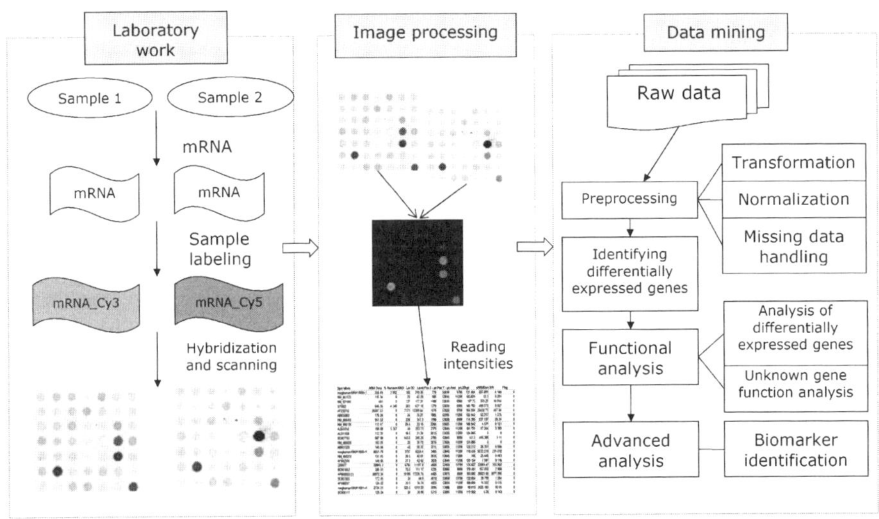

FIGURE 4.4 Data flow schema of microarray data analysis.

4.1.3 Microarray Data Sets

Microarray-based studies consider more than one sample and most often produce several replicates for each sample. The minimum requirement for a useful biological study is to have two samples that can be hybridized on a single dual-channel or on two single-channel microarray chips.

A data set for a single-channel microarray experiment can be described as an $M \times N$ matrix in which each column represents gene expression levels for one of the N chips (arrays), and each row is a vector containing expression levels of one of the M genes in different arrays (called "expression profile"). A data set for a dual-channel microarray experiment can be observed as a similar matrix in which each chip is represented by a single column of expression ratios between the two channels ($cy3$ and $cy5$), or by two columns of absolute expression values of the two channels. A typical microarray data table has a fairly small number of arrays and a large number of genes ($M \gg N$); for example, while microarrays can measure the expression of thousands of genes, the number of arrays is usually in the range from less than 10 (in small-scale studies) to several hundred (in large-scale studies).

Methods described in this chapter are demonstrated by case studies on acute leukemia, *Plasmodium falciparum* intraerythrocytic developmental cycle, and chronic fatigue syndrome microarray data sets. *Acute leukemia data set* [3] contains 7129 human genes with 47 arrays of acute lymphoblastic leukemia (ALL) samples and 25 arrays of acute myeloid leukemia (AML) samples. The data set is used to demonstrate a generic approach to separating two types of human acute leukemia (AML versus ALL) based on their gene expression patterns. This data set is available at http://www.broad.mit.edu/cgi-bin/cancer/publications/pub_paper.cgi?mode=view&paper_id=43. *Plasmodium falciparum data set* [4] contains 46 arrays with samples taken during 48 h of intraerythrocytic developmental cycle of *Plasmodium falciparum* to provide the comprehensive overview of the timing of transcription throughout the cycle. Each array consists of 5080 spots, related to 3532 unique genes. This data set is available at http://biology.plosjournals.org/archive/1545-7885/1/1/supinfo/10.1371_journal.pbio.0000005.sd002.txt. *Chronic fatigue syndrome (CFS) data set* contains 79 arrays from 39 clinically identified CFS patients and 40 non-CFS (NF) patients [5]. Each chip measures expression levels of 20,160 genes. This data set was used as a benchmark at the 2006 Critical Assessment of Microarray Data Analysis (CAMDA) contest and is available at http://www.camda.duke.edu/camda06/datasets.

4.2 MICROARRAY DATA PREPROCESSING

Images obtained by scanning microarray chips are preprocessed to identify the spots, estimate their intensities, and flag the spots that cannot be read reliably. Data obtained from a scanner are usually very noisy; the use of raw unprocessed data would likely bias the study and possibly lead to false conclusions. In order to reduce these problems, several preprocessing steps are typically performed and are described in this section.

4.2.1 Data Cleaning and Transformation

4.2.1.1 Reduction of Background Noise in Microarray Images The background area outside of the spots in a scanned microarray image should ideally be dark (indicating no level of intensity), but in practice, the microarray image background has a certain level of intensity known as *background noise*. It is an indicator of the systematic error introduced by the laboratory procedure and microarray image scanning. This noise can often effectively be reduced by estimating and subtracting the mean background intensity from spot intensities. A straightforward approach that uses the mean background intensity of the whole chip is not appropriate when noise intensity is not uniform in all parts of the chip. In such situations, *local estimation* methods are used to estimate the background intensity individually for each spot from a small area surrounding the spot.

4.2.1.2 Identification of Low Quality Gene Spots Chip scratching, poor washing, bad hybridization, robot injection leaking, bad spot shape, and other reasons can result in microarray chips containing many damaged spots. Some of these gene spot problems are illustrated in Figure 4.5. Low quality gene spots are typically identified by comparing the spot signal and its background noise [6,7]. Although statistical techniques can provide a rough identification of problematic gene spots, it is important to carefully manually evaluate the microarray image to discover the source of the problem and to determine how to address problematic spots. The most simplistic method is to remove all data for the corresponding genes from further analysis. However, when the spots in question are the primary focus of the biological study, it is preferable to process microarray images using specialized procedures [8]. Unfortunately, such a process demands intensive manual and computational work. To reduce the data uncertainty due to damaged spots, it is sometimes necessary to repeat the hybridization of arrays with a large area or fraction of problematic spots.

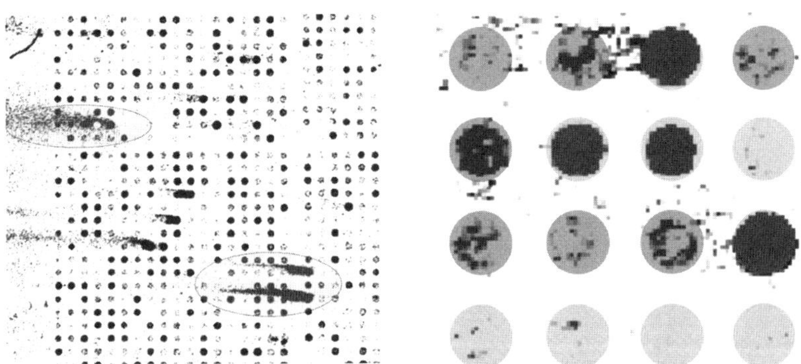

FIGURE 4.5 Examples of problematic spots. The light gray ovals in the left image are examples of poor washing and scratching. The black circle spots in the right image are good-quality spots. The light gray circles indicate empty (missing) spots. The dark gray circles mark badly shaped spots.

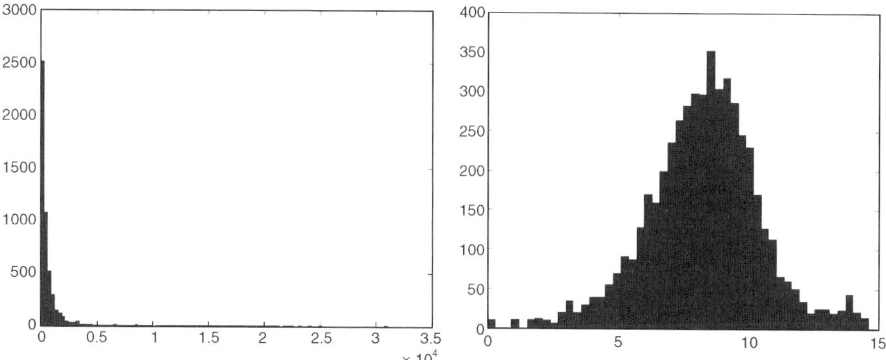

FIGURE 4.6 Data distribution before and after logarithmic transformation. Histograms show gene expression data distribution for patient sample #1 from acute lymphoblastic leukemia data set (X-axis represents the gene expression levels and Y-axis represents the amount of genes with given expression level). The distribution of raw data on the left is extremely skewed. The log-2 transformed data have a bell-shaped, approximately normal distribution, shown on the right.

4.2.1.3 Microarray Data Transformation After the numerical readings are obtained from the image, the objective of microarray data transformation is to identify outliers in the data and to adjust the data to meet the distribution assumptions implied by statistical analysis methods. A simple *logarithmic transformation* illustrated in Figure 4.6 is commonly used. It reshapes the data distribution into a bell shape that resembles normal distribution. This transformation is especially beneficial for data from dual-channel arrays, since data from these arrays are often expressed as ratios of signal intensities of pairs of samples. Alternative transformations used in practice include *arcsinh* function, *linlog* transformation, *curve-fitting* transformations, and *shift* transformation [9]; among them, the *linlog* transformation was demonstrated to be the most beneficial.

4.2.2 Handling Missing Values

Typical data sets generated by microarray experiments contain large fractions of missing values caused by low quality spots. Techniques for handling missing values have to be chosen carefully, since they involve certain assumptions. When these assumptions are not correct, artifacts can be added into the data set that may substantially bias the evaluation of biological hypotheses.

The straightforward approach is to completely discard genes with at least one missing value. However, if a large fraction of genes are eliminated because of missing values, then this approach is not appropriate.

A straightforward imputation method consists of replacing all missing values for a given gene with the mean of its valid expression values among all available arrays. This assumes that the data for estimating the most probable value of a missing gene expression were derived under similar biological conditions; for instance, they could be derived from replicate arrays. Most microarray experiments lack replicates due to the experimental costs. When there are no replicates available, a better choice for imputation is to replace all of the missing data in an array with the average of valid expression values within the array.

The *k-nearest-neighbor based method (KNN)* does not demand experimental replicates. Given a gene with missing gene expression readings, k genes with the most similar expression patterns (i.e., its k neighbors) are found. The given gene's missing values are imputed as the average expression values of its k neighbors [10], or predicted with the *local least squares (LLS)* method [11]. Recent research has demonstrated that the weighted nearest-neighbors imputation method (WeNNI), in which both spot quality and correlations between genes were used in the imputation, is more effective than the traditional KNN method [12].

Domain knowledge can help estimate missing values based on the assumption that genes with similar biological functions have similar expression patterns. Therefore, a missing value for a given gene can be estimated by evaluating the expression values of all genes that have the same or similar functions [13]. Although such an approach is reasonable in terms of biology, its applicability is limited when the function is unknown for a large number of the genes.

In addition to the problems that are related to poor sample preparation, such as chip scratching or poor washing, a major source of problematic gene spots is relatively low signal intensity compared to background noise. It is important to check the reasons for low signal intensity. Gene expression might be very low, for instance, if the biological condition successfully blocks the gene expression. In this case, the low gene expression signal intensity is correct and the imputation of values estimated by the above-mentioned methods would probably produce a value that is too high. An alternative is to replace such missing data with the lowest obtained intensity value within the same chip or with an arbitrary small number.

4.2.3 Normalization

Microarray experiments are prone to systematic errors that cause changes in the data distribution and make statistical inference unreliable. The objective of normalization is to eliminate the variation in data caused by errors of the experimental methods, making further analysis based only on the real variation in gene expression levels. All normalization methods may introduce artifacts and should be used with care. Most methods are sensitive to outliers, so outlier removal is crucial for the success of normalization.

There are two major types of normalization methods: *within-chip normalization* uses only the data within the same chip and is performed individually on each chip, while *between-chip normalization* involves microarray data from all chips simultaneously. Reviews on microarray data normalization methods are provided in [14–16].

4.2.3.1 Within-Chip Normalization Several within-chip normalization methods are based on linear transformations of the form *new_value* =(*original_value*–

$a)/b$, where parameters a and b are fixed for one chip. *Standardization normalization* assumes that the gene expression levels in one chip follow the standard normal distribution. Parameter a is set to the mean, while parameter b is set to the standard deviation of gene expression levels in a chip. This method can be applied to both dual-channel and single-channel microarray data.

Linear regression normalization [15] is another linear transformation that uses a different way to choose parameters a and b. The basic assumption for dual-channel arrays is that for a majority of genes, the intensity for the cy3 channel is similar to intensity for the cy5 channel. As a result, the two intensities should be highly correlated, and the fitted regression line should be very close to the main diagonal of the scatterplot. Parameters a and b in linear transformation are chosen so that the regression line for transformed data points aligns with the main diagonal.

A more advanced normalization alternative is the *loess transformation*. It uses a scatterplot of log ratio of two channel intensities ($\log(cy3/cy5)$) against average value of two channel intensities ($(cy3 + cy5)/2$). A locally weighted polynomial regression is used on this scatterplot to form a smooth regression curve. Original data are then transformed using the obtained regression curve. Loess normalization can also be used with single-channel microarrays where two arrays are observed as two channels and normalized together. For data from more than two arrays, loess normalization can be iteratively applied on all distinct pairs of arrays, but this process has larger computational cost. Some other forms of loess normalization are *local loess* [17], *global loess*, and *two-dimensional loess* [18].

Several normalization methods make use of domain knowledge. All organisms have a subset of genes—called housekeeping genes—that maintain necessary cell activities, and, as a result, their expression levels are nearly constant under most biological conditions. All the above-mentioned methods can be modified so that all transformation parameters are calculated based only on the expression levels of housekeeping genes.

4.2.3.2 Between-Chip Normalization

Row–column normalization [19] is applied to a data set comprised of several arrays, observed as a matrix with M rows (representing genes) and N columns (representing separate arrays and array channels). In one iteration, the mean value of a selected row (or column) is subtracted from all of the elements in that row (or column). This is iteratively repeated for all rows and columns of the matrix, until the mean values of all rows and columns approach zero. This method fixes variability among both genes and arrays. A major problem with this method is its sensitivity to outliers, a problem that can significantly increase computation time. Outlier removal is thus crucial for the performance of this method. The computation time can also be improved if standardization is first applied to all individual arrays.

Distribution (quantile) normalization [20] is based on the idea that a quantile–quantile plot is a straight diagonal line if two sample vectors come from the same distribution. Data samples can be forced to have the same distribution by projecting data points onto the diagonal line. For microarray data matrix with m rows and n columns, each column is separately sorted in descending order, and the mean values are calculated for all rows in the new matrix. Each value in the original matrix is then replaced with the mean value of the row in the sorted matrix where that value was placed during sorting. Distribution normalization may improve the reliability of statistical inference. However, it may also introduce artifacts; after normalization, low intensity genes may have the same (very low) intensity across all arrays.

Statistical model-fitting normalization involves the fitting of gene expression level data using a statistical model. The fitting residues can then be treated as bias-free transformation of expression data. For example, for a given microarray data set with genes g $(g = 1, \ldots, n)$, biological conditions $T_i (i = 1, \ldots, m)$, and arrays $A_j (j = 1, \ldots, k)$, the intensity I of gene g at biological condition i and array j can be fitted using a model [21]

$$I_{gij} = u + T_i + A_j + (TA)_{ij} + \varepsilon_{gij}.$$

The fitting residues ε_{gij} for this model can be treated as bias-free data for gene g at biological condition i and array j after normalization.

In experiments with dual-channel arrays, it is possible to distribute (possibly multiple) samples representing m biological conditions over k arrays in many different ways. Many statistical models have recently been proposed for model-fitting normalization [22,23]. The normalization approaches of this type have been demonstrated to be very effective in many applications, especially in the identification of differentially expressed genes [21,24].

4.2.4 Data Summary Report

The data summary report is used to examine preprocessed data in order to find and correct inconsistencies in the data that can reduce the validity of statistical inference. Unlike other procedures, there are no golden standards for this step. It is a good practice to evaluate the data summary report before and after data preprocessing. Approaches used to inspect the data include the evaluation of a histogram to provide information about data distribution in one microarray, a boxplot of the whole data set to check the similarities of all data distributions, and the evaluation of correlation coefficient maps (see Fig. 4.7) to check consistency among arrays. Correlation coefficient heat maps plot the values of correlation coefficients between pairs of arrays. For a given pair of arrays, #i and #j, their expression profiles are observed as vectors and the correlation coefficient between the two vectors is plotted as two pixels—in symmetrical positions (ij) and (ji)—in the heat map (the magnitude of correlation coefficient is indicated by the color of the pixel). Correlation coefficients are normally expected to be high, since we assume that the majority of gene expression levels are similar in different arrays. A horizontal (and the corresponding vertical) line in a heat map represents all of the correlation coefficients between a given array and all other arrays. If a line has a near-constant color representing a very low value, we should suspect a problem with the corresponding array.

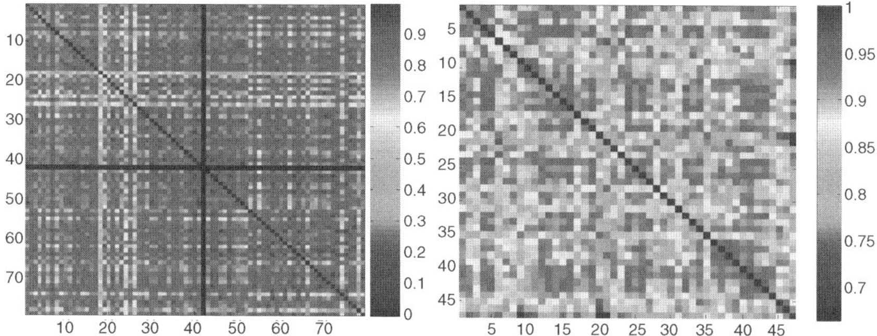

FIGURE 4.7 Correlation coefficient heat maps. The left heat map shows the correlation coefficients among the 79 samples of the CFS data set. The first 40 samples are from the nonfatigue (control) group. The remaining 39 samples are from the group of CFS patients. The shade of a pixel represents the magnitude of the correlation coefficient (as shown in the shaded bar on the right). The correlation coefficients on the diagonal line are 1, since they compare each sample to itself. There are two clearly visible horizontal and vertical lines in the heat map on the left, corresponding to the sample #42. This indicates that this sample is different from the others; its correlation coefficients with all other samples are near zero. Therefore, we need to inspect this sample's chip image. Another sample that draws our attention is sample #18, which also has near-uniform correlation coefficients (around 0.5) with other samples. After inspecting the sample's chip image, we found that these correlation coefficients reflected sample variation and that we should not exclude sample #18 from our study. A similar heat map on the right shows the correlation coefficients among the 47 ALL samples from the acute leukemia data set. Overall data consistency is fairly high with an average correlation coefficient over 0.89.

4.3 MICROARRAY DATA ANALYSIS

This section provides a brief outline of methods for the analysis of preprocessed microarray data that include the identification of differentially expressed genes, discovery of gene expression patterns, characterization of gene functions, pathways analysis, and discovery of diagnostic biomarkers. All methods described in this section assume that the data have been preprocessed; see Section 4.2 for more details on microarray data preprocessing methods.

4.3.1 Identification of Differentially Expressed Genes

A gene is differentially expressed if its expression level differs significantly for two or more biological conditions. A straightforward approach for the identification of differentially expressed genes is based on the selection of genes with absolute values of log-2 ratio of expression levels larger than a prespecified threshold (such as 1). This simple approach does not require replicates, but is subject to high error rate (both false positive and false negative) due to the large variability in microarray data.

More reliable identification is possible by using statistical tests. However, these methods typically assume that the gene expression data follow a certain distribution, and require sufficiently large sample size that often cannot be achieved due to microarray experimental conditions or budget constraints. Alternative techniques, such as bootstrapping, impose less rigorous requirements on the sample size and distribution while still providing reliable identification of differentially expressed genes.

Given the data, a statistical test explores whether a *null hypothesis* is valid and calculates the p-value, which refers to the probability that the observed statistics are generated by the null model. If the p-value is smaller than some fixed threshold (e.g., 0.05), the null hypothesis is rejected. If the p-value is above the threshold, however, it should not be concluded that the original hypothesis is confirmed; the result of the test is that the observed events do not provide a reason to overturn it [25]. The most common null hypothesis in microarray data analysis is that there is no difference between two groups of expression values for a given gene. In this section, we briefly introduce the assumptions and requirements for several statistical tests that are often used for the identification of differentially expressed genes.

4.3.1.1 Parametric Statistical Approaches The *Student's t-test* examines the null hypothesis that the means of distributions from which two samples are obtained are equal. The assumptions required for t-test are that the two distributions are normal and that their variances are equal. The null hypothesis is rejected if the p-value for the t-statistics is below some fixed threshold (e.g., 0.05). The t-test is used in microarray data analysis to test—for each individual gene—the equality of the means of expression levels under two different biological conditions. Genes for which a t-test rejects the null hypothesis are considered differentially expressed.

The t-test has two forms: *dependent sample t-test* and *independent sample t-test. Dependent sample t-test* assumes that each member in one sample is related to a specific member of the other sample; for example, this test can be used to evaluate the drug effects by comparing the gene expression levels of a group of patients before and after they are given a certain type of drug. *Independent sample t-test* is used when the samples are independent of each other; for example, this test can be used to evaluate the drug effects by comparing gene expression levels for a group of patients treated with the drug to the gene expression levels of another group of patients treated with a placebo. The problem with using the t-test in microarray data analysis is that the distribution normality requirement is often violated in microarray data.

One-way analysis of variance (ANOVA) is a generalization of the t-test to samples from more than two distributions. ANOVA also requires that the observed distributions are normal and that their variances are approximately equal. ANOVA is used in microarray data analysis when gene expression levels are compared under two or more biological conditions, such as for a comparison of gene expression levels for a group of patients treated with drug A, a group of patients treated with drug B, and a group of patients treated with placebo.

The *volcano plot* (see Fig. 4.8) is often used in practice for the identification of differentially expressed genes; in this case, it is required that a gene both

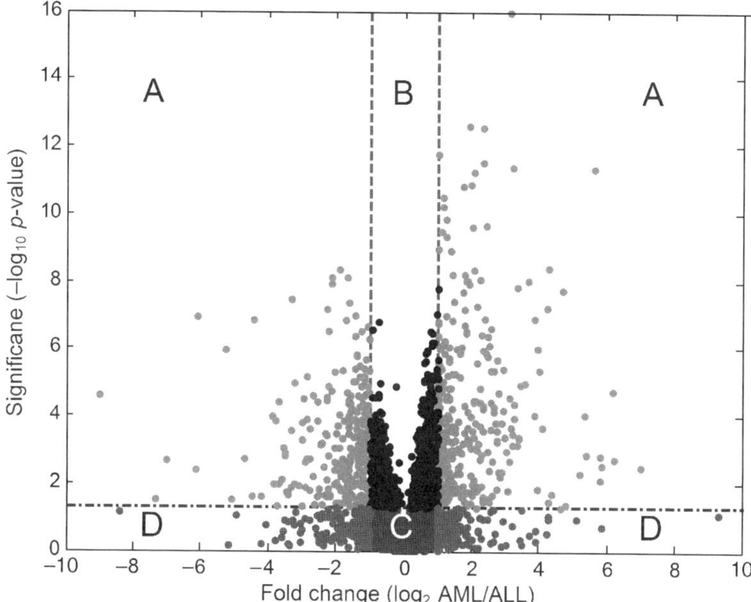

FIGURE 4.8 The volcano plot of significance versus fold change. This figure is a plot of the significance (p-value from ANOVA test, on a –log-10 scale) against fold change (log-2 ratio), for testing the hypothesis on the differences in gene expression levels between the AML group and the ALL group in the acute leukemia data set. The horizontal line represents a significance level threshold of 0.05. The two vertical lines represent the absolute fold-change threshold of 2. The genes plotted in the two "A" regions are detected as significant by both methods, while the genes plotted in region "C" are detected as insignificant by both methods. This type of plot demonstrates two types of errors that occur with the ratio-based method: false positive errors plotted in the two "D" regions, and false negative errors plotted in the "B" region. A common practice is to identify only the genes plotted in the two "A" regions as differentially expressed and discard the genes plotted in the "B" region.

passes the significance test and that its expression level log ratio is above the threshold.

4.3.1.2 Nonparametric Statistical Approaches Nonparametric tests relax the assumptions posed by the parametric tests. Two popular nonparametric tests are the Wilcoxon rank-sum test for equal median and the Kruskal–Wallis nonparametric one-way analysis of variance test.

The Wilcoxon rank-sum test (also known as Mann–Whitney U-test) tests the hypothesis that two independent samples come from distributions with equal medians. This is a nonparametric version of the t-test. It replaces real data values with their sorted ranks and uses the sum of ranks to obtain a p-value. Kruskal–Wallis test compares the medians of the samples. It is a nonparametric version of the one-way ANOVA, and an extension of the Wilcoxon rank-sum test to more than two groups.

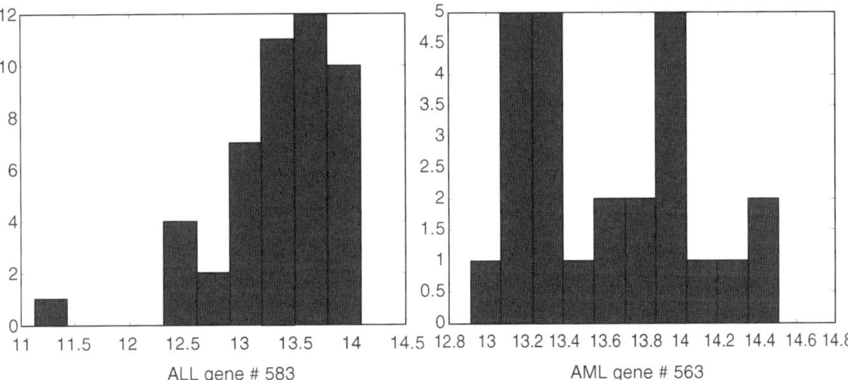

FIGURE 4.9 Importance of data distribution type for the choice of statistical test. Two histograms show the distribution of expression levels for gene #563 in two groups of samples in the acute leukemia data set: ALL on the left and AML on the right. The two distributions are clearly different. When testing the equality of means of two groups, the Kruskal–Wallis test gives us the p-value of 0.16, and the ANOVA test gives us the p-value of 0.05. Since the data distribution in the right panel has two major peaks, it is not close to normal distribution; therefore, it is preferable to choose the Kruskal–Wallis test.

Nonparametric tests tend to reject less null hypotheses than the related parametric tests and have lower sensitivity, which leads to an increased rate of false negative errors. They are more appropriate when the assumptions for parametric tests are not satisfied, as is often the case with microarray data (see Fig. 4.9). However, this does not imply that nonparametric tests will necessarily identify a smaller number of genes as differentially expressed than the parametric test, or that the sets of genes identified by one parametric test and one nonparametric test will necessarily be in a subset relationship. To illustrate the difference in results we used both ANOVA and the Kruskal–Wallis test to identify differentially expressed genes in the acute leukemia data set. Out of 7129 genes, 1030 genes were identified as differentially expressed by both methods. In addition to that, 155 genes were identified only by ANOVA, while 210 genes were identified only by the Kruskal–Wallis test.

4.3.1.3 Advanced Statistical Models Recently, more sophisticated models and methods for the identification of differentially expressed genes have been proposed [26,27]. For example, when considering the factors of array (A), gene (G), and biological condition (T), a *two-step mix-model* [21] first fits the variance of arrays, biological conditions, and interactions between arrays and biological conditions using one model, and then uses the residues from fitting the first model to fit the second model. An overview of mix-model methods is provided in the work by Wolfinger et al. [28]. Other advanced statistical approaches with demonstrated good results in identifying differentially expressed genes include the significance analysis of microarray (SAM) [29], regression model approaches [30], empirical Bayes analysis [31], and the bootstrap approach to gene selection (see the case study below).

Case Study 4.1: Bootstrapping Procedure for Identification of Differentially Expressed Genes

We illustrate the bootstrapping procedure for the identification of differentially expressed genes on an acute leukemia data set. The objective is to identify the genes that are differentially expressed between 47 ALL and 25 AML arrays. For each gene, we first calculate the p-value p_0 of two-sample t-test on the gene's expression levels in AML group versus ALL group. Next, the set of samples is randomly split into two subsets with 47 and 25 elements, and a similar t-test is performed with these random subsets and p-value p_1 is obtained. This step is repeated a large number of times ($n > 1000$), and as a result we obtain p-values $p_1, p_2, p_3, \ldots, p_n$. These p-values are then compared to the original p_0. We define the bootstrap p-value as $p_b = c/n$, where c is the number of times when values $p_i (i = 1, \ldots, n)$ are smaller than p_0. If p_b is smaller than some threshold (e.g., 0.05), then we consider the gene to be differentially expressed.

For the 88th gene in the data set, the expression levels are

ALL	AML
759, 1656, 1130, 1062,	1801, 1024, 3084, 1974,
822, 1020, 1068, 1455,	1084, 1090, 908, 2474,
1099, 1164, 662, 753,	1635, 1591, 1323, 857,
728, 918, 943, 644,	1872, 1593, 1981, 2668,
2703, 916, 677, 1251,	1128, 3601, 2153, 1603,
138, 1557, 750, 814,	769, 893, 2513, 2903,
667, 616, 1187, 1214,	2147
1080, 1053, 674, 708,	
1260, 1051, 1747, 1320,	
730, 825, 1072, 774,	
690, 1119, 866, 564,	
958, 1377, 1357	

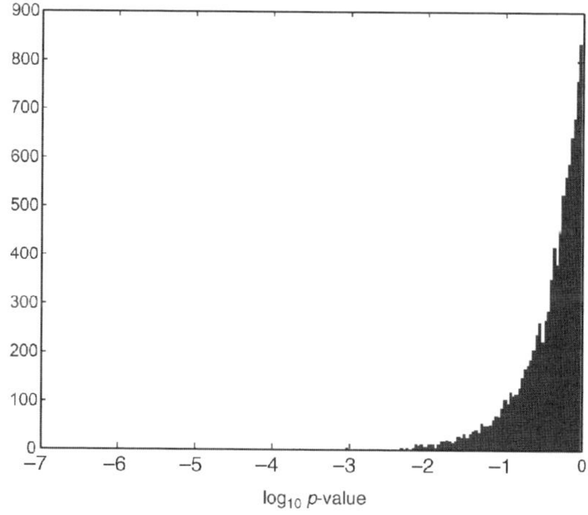

Figure 4.10

The p-value of the t-test for this gene is $p_0 = 3.4\text{E} - 007$, which is smaller than the threshold 0.05. The distribution of p-values obtained on randomly selected subsets $(p_1, \ldots, p_{1000})$ is shown in Figure 4.10. The bootstrap p-value is $p_b = 0$, so the bootstrapping procedure confirms the result of the t-test, that is, the 88th gene is differentially expressed.

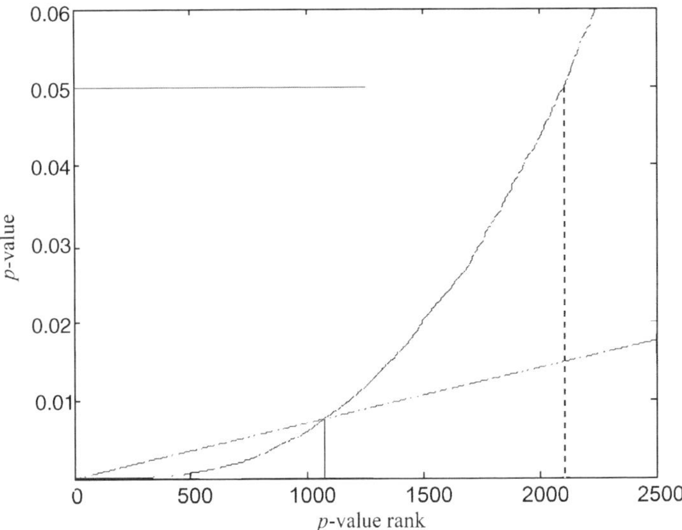

FIGURE 4.11 Benjamini–Hochberg FDR control. This figure compares the use of constant p-value threshold (in this case 0.05) and the use of Benjamini–Hochberg (BH) FDR control method for the two-sample t-test on acute leukemia data set. The curve is the plot of the original p-values obtained from the t-tests for individual genes, sorted in an increasing order. The horizontal line represents the constant p-value threshold of 0.05. There are 2106 genes with a p-value smaller than this threshold. The slanted line represents the p-value thresholds $p_i = \alpha_0 \cdot i/N$ that BH method uses to control the FDR at level of $\alpha_0 = 0.05$ (N is the total number of genes). It intersects with the curve at p-value 0.0075. Only the 1071 genes whose p-values are smaller than 0.0075 are considered to be significantly differentially expressed. The remaining 935 genes are considered to be false positive discoveries made by individual t-tests.

4.3.1.4 False Discovery Rate (FDR) Control Statistical procedures for the identification of differentially expressed genes can be treated as multiple hypothesis testing. A p-value threshold that is appropriate for a single test does not provide good control on false positive discovery for the overall procedure. For example, testing of 10,000 genes with p-value threshold of 0.05 is expected to identify $10,000 \times 0.05 = 500$ genes as differentially expressed even if none of the genes are actually differentially expressed. The false positive rate can be controlled by evaluating the expected proportion of true rejected null hypotheses out of the total number of rejected null hypothesis. An example of FDR control is shown in Figure 4.11.

If N is the total number of genes, α_0 is the p-value threshold, and $p_i (i = 1, \ldots, N)$ are p-values in ascending order, then the ith ranked gene is selected if $p_i \leq \alpha_0 \cdot i/N$ [32]. A comprehensive review of this statistical FDR control is presented in the work by Qian and Huang [33]. It is worth noting that a bootstrap procedure for FDR control has also been introduced [29] and was shown to be suitable for gene selecting when data distribution deviates from normal distribution.

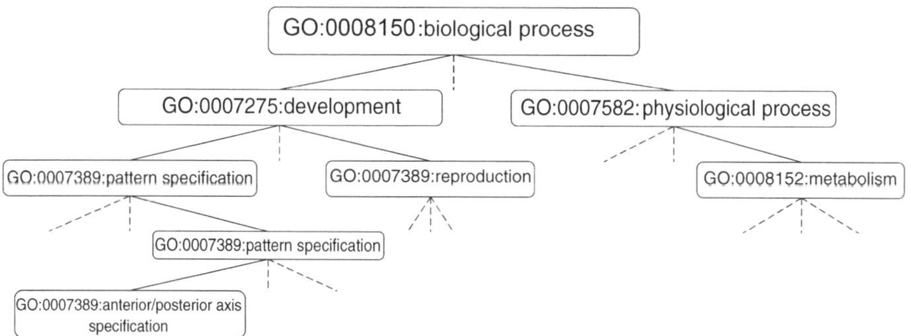

FIGURE 4.12 Part of the Gene Ontology direct acyclic graph. The shortest path between GO:0007275:*development* and GO:0009948:*anterior/posterior axis specification* is 3 (the nearest common ancestor for the two terms is GO:0007275:*development*). The shortest path between the terms GO:0007275:*development* and GO:0008152:*metabolism* is 3 but the only ancestor for them is GO:0008150:*biological processes*, so the distance between them is 3 + 23, where 23 is the added penalty distance, which is the maximum distance in Biological Process part of Gene Ontology DAG.

4.3.2 Functional Annotation of Genes

One of the goals of microarray data analysis is to aid in discovering biological functions of genes. One of the most important sources of domain knowledge on gene functions is Gene Ontology (GO), developed and maintained by the Gene Ontology Consortium [34,35]. Using a controlled and limited vocabulary of terms describing gene functions, each term in Gene Ontology consists of a unique identifier, a name, and a definition that describes its biological characteristic. GO terms are split into three major groups: biological processes, molecular functions, and cellular component categories. Within each category, GO terms are organized in a direct acyclic graph (DAG) structure, where each term is a node in the DAG, and each node can have several child and parent nodes. The GO hierarchy is organized with a general-to-specific relation between higher and lower level GO terms (see Fig. 4.12).

Sometimes, it is useful to compare several GO terms and determine if they are similar. Although there is no commonly accepted similarity measure between different GO terms, various distance measures were proposed for measuring the similarity between GO terms [36,37]. For example, the distance between nodes X and Y in a DAG can be measured as the length of the shortest path between X and Y within the GO hierarchy normalized by the length of maximal chain from the top to the bottom of the DAG [38]. One possible modification, illustrated in Figure 4.12, is to add a large penalty for paths that cross the root of a DAG to account for unrelated terms.

4.3.3 Characterizing Functions of Differentially Expressed Genes

After identifying differentially expressed genes, the next step in analysis is often to explore the functional properties of these genes. This information can be extremely useful to domain scientists for the understanding of biological properties of different sample groups. Commonly used methods for such analysis are described in this section. The *chi-square* and the *Fisher's exact* tests are used to test whether the selected genes are overannotated with a GO term F, as compared to the set of remaining genes spotted on a microarray [39,40]. For instance, the following 2×2 contingency table contains the data that can be used to test whether the frequency of genes annotated with a GO term F among the selected genes is different than the same frequency among the remaining genes:

	Number of genes		
	Selected genes	Remaining genes	Total
Annotated with a GO term F	f_{11}	f_{12}	r_1
Not annotated with a GO term F	f_{21}	f_{22}	r_2
Total	c_1	c_2	S

Chi-square test uses a χ^2 statistic with formula

$$\chi^2 = \sum_{i=1}^{2}\sum_{i=1}^{2}\frac{(f_{ij} - r_i c_j/S)^2}{r_i c_j/S}.$$

The chi-square test is not suitable when any of the expected values $r_i c_j/S$ are smaller than 10. *Fisher's exact* test is more appropriate in such cases. In practice, all genes annotated with term F and all terms in the subtree of term F are considered to be annotated with F.

4.3.4 Functional Annotation of Uncharacterized Genes

The functional characterization of genes involves a considerable amount of biological laboratory work. Therefore, only a small fraction of known genes and proteins is functionally characterized. An important microarray application is the prediction of gene functions in a cost-effective manner. Numerous approaches use microarray gene expression patterns to identify unknown gene functions [41–43]. In the following section, we outline some of the most promising ones.

4.3.4.1 Unsupervised Methods for Functional Annotation Gene expression profiles can be used to measure distances among genes. The basic assumption in functional annotation is that genes with similar biological functions are likely to have similar expression profiles. The functions of a given gene could be inferred by considering the known functions of genes with similar expression profiles. A similar approach is to group all gene expression profiles using clustering methods and to find the overrepresented functions within each cluster [44,45]. Then, all genes within a cluster are annotated with the overrepresented functions of that cluster. An alternative is to first cluster only the genes with known functions. An averaged expression profile of all genes within the cluster can then be used as the representative of a cluster [4]. The gene with the unknown function can be assigned functions based on its distance to the representative expression profiles. Conclusions from these procedures are often unreliable: a gene may have multiple functions that may be quite distinctive; also, genes with the same function can have quite different expression profiles. Therefore, it is often very difficult to select representative functions from a cluster of genes.

Many unsupervised methods for functional annotation face the issue of model selection in clustering, such as choosing the proper number of clusters, so that the genes within the cluster have similar functions. Domain knowledge is often very helpful in the model selection [46].

As we already mentioned, nearest-neighbor and clustering methods for assigning functions to genes are based on assumptions that genes with similar functions will have similar expression profiles [47]. However, this assumption is violated for more than half of the GO terms [48]. A more appropriate approach, therefore, is to first determine a subset of GO terms for which the assumption is valid, and use only these GO terms in gene function annotation.

4.3.4.2 Supervised Methods for Functional Annotation Supervised methods for functional characterization involve building classification models that predict gene functions based on gene expression profiles. A predictor for a given function is trained to predict whether a given gene has that function or not [49]. Such a predictor is trained and tested on a collection of genes with known functions. If testing shows that the accuracy of the predictor is significantly higher than that for a trivial predictor, the predictor can then be used on the uncharacterized genes to annotate them. Previous research shows that the support-vector machines (SVM) model achieves the best overall accuracy when compared to other competing prediction methods [50]. The SVM-based predictor can overcome some of the difficulties that are present with the unsupervised methods. It can flexibly select the expression profile similarity measure

Case Study 4.2: Identification of GO Terms with Conserved Expression Profiles

We applied a bootstrapping procedure to identify GO terms that have conserved gene expression profiles in the *Plasmodium* data set that contains 46 arrays. Each of the 46 arrays in the *Plasmodium* data set measures expression levels of 3532 genes at a specific time point over the 48-h *Plasmodium falciparum* intraerythrocytic developmental cycle (IDC). The bootstrap procedure was applied to 884 GO terms that are associated with at least two genes. For a given GO term with l associated genes, we collected their expression profiles and calculated the average pairwise correlation coefficients ρ_0. We compared ρ_0 to average expression profile correlation coefficients of randomly selected pairs of genes. In each step of the bootstrap procedure, we randomly selected l genes and computed their average correlation coefficient ρ_i. This was repeated 10,000 times to obtain $\rho_1, \rho_2, \ldots, \rho_{10{,}000}$. We counted the number c of ρ_i that are greater than ρ_0 and calculated the bootstrap p-value as $p_b = c/n$. If p_b is smaller than 0.05, the expression profiles of the GO term are considered to be conserved.

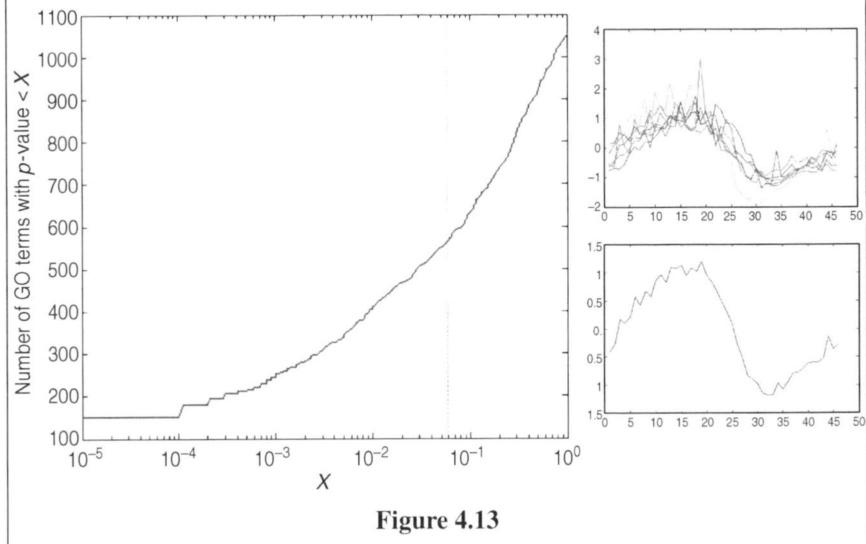

Figure 4.13

The plot in the left part of Figure 4.13 shows the cumulative number of GO terms with p-value smaller than **x**. Four hundred and twenty-eight (48.4 percent) of the 884 GO terms have p-value smaller than 0.05; 199 of these are molecular function and 229 are biological process GO terms. This result validates to a large extent the hypothesis that genes with identical functions have similar expression profiles. However, it also reveals that for a given microarray experiment, a large fraction of functions do not follow this hypothesis.

Figure 4.13 also contains expression profiles of genes annotated with GO term GO:0006206 (pyrimidine base metabolism; bootstrap p-value 0) and its representative expression profile.

and handle a large feature space. The unresolved problem of the supervised approach is the presence of multiple classes and class imbalance; a function can be associated with only a few genes, and there are several thousand functions describing genes in a given microarray data set.

4.3.5 Correlations Among Gene Expression Profiles

A major challenge in biological research is to understand the metabolic pathways and mechanisms of biological systems. The identification of correlated gene expressions in a microarray experiment is aimed at facilitating this objective. Several methods for this task are described in this section.

4.3.5.1 Main Methods for Clustering of Gene Expression Profiles Hierarchical clustering and K-means clustering are two of the most popular approaches for the clustering of microarray data. The *hierarchical clustering* approach used with microarray data is the *bottom-up approach*. This approach begins with single-member clusters, and small clusters are iteratively grouped together to form larger clusters,

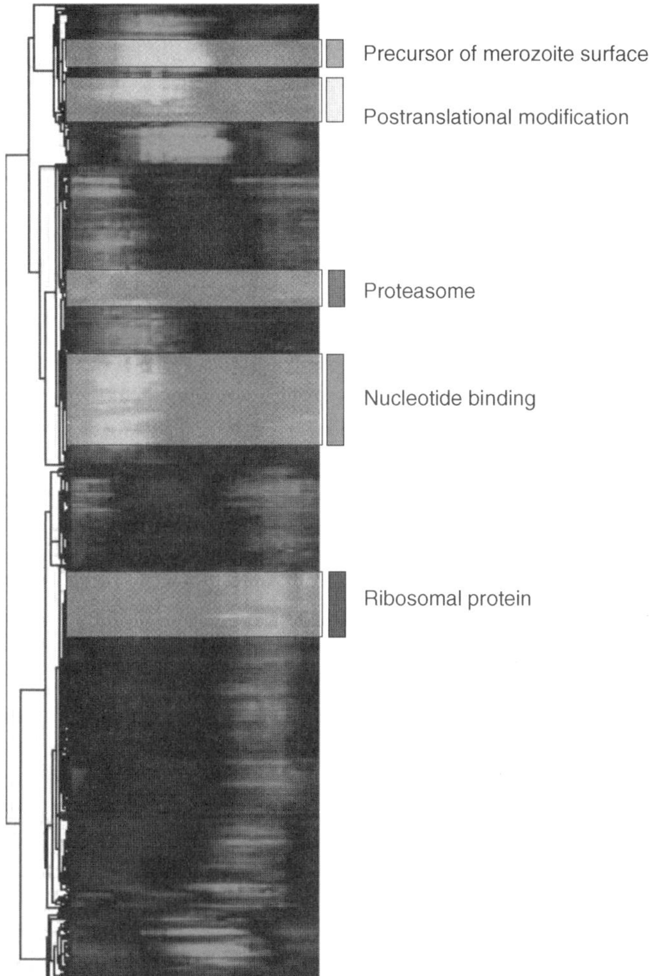

FIGURE 4.14 Visualization of hierarchically clustered data with identified functional correlation. The *Plasmodium* data set was clustered using hierarchical clustering. Rows of pixels represent genes' expression levels at different time points. Columns of pixels represent the expression level of all genes in one chip at one given time point in the IDC process, and their order corresponds to the order of points in time. The cluster hierarchy tree is on the left side. The image contains clearly visible patterns of dark gray and light gray pixels that correspond to upregulated and downregulated expression levels, respectively. A domain expert investigated the higher level nodes in the clustering tree, examining the similarity of functions in each cluster for genes with known functions. Five examples of clusters for which the majority of genes are annotated with a common function are marked using the shaded bars and the names of the common functions. These clusters can be used to infer the functions of the genes within the same cluster whose function is unknown or unclear.

until a single cluster containing the whole set is obtained. In each iteration, the two clusters that are chosen for joining are two clusters with the closest distance to each other. The result of hierarchical clustering is a binary tree; descendants of each cluster in that tree are the two subclusters of which the cluster consists. The distance between two clusters in the tree reflects their correlation distance. Hierarchical clustering provides a visualization of the relationships between gene expression profiles (see Fig. 4.14).

K-means clustering groups genes into a prespecified number of clusters by minimizing the distances within each cluster and maximizing the distances between clusters. The K-means clustering method first chooses k genes called centroids (which can be done randomly or by making sure that their expression profiles are very different). It then examines all gene expression profiles and assigns each of these to the cluster with the closest centroid. The position of a centroid is recalculated each time a gene expression profile is added to the cluster by averaging all profiles within the cluster. This procedure is iteratively repeated until stable clusters are obtained, and no gene expression profiles switch clusters between iterations. The K-means method is computationally less demanding than hierarchical clustering. However, an obvious disadvantage is the need for the selection of parameter k, which is generally not a trivial task.

4.3.5.2 Alternative Clustering Methods for Gene Expression Profiles

Alternative clustering methods that are used with gene expression data include the self-organizing map (SOM) and random forest (RF) clustering.

An SOM is a clustering method implemented with a neural network and a special training procedure. The comparison of SOM with hierarchical clustering methods shows that an SOM is superior in both robustness and accuracy [51]. However, as K-means clusters, an SOM requires the value of parameter k to be prespecified.

RF clustering is based on an RF predictor that is a collection of individual classification trees. After an RF is constructed, the similarity measure between two samples can be defined as the number of times a tree predictor places the two samples in the same terminal node. This similarity measure can be used to cluster gene expression data [52]. It was demonstrated that the RF-based clustering of gene profiles is superior compared to the standard Euclidean distance measure [53].

Other advanced techniques proposed for clustering gene expression data include the mixture model approach [54], the shrinkage-based similarity procedure [55], the kernel method [56], and bootstrapping analysis [57].

4.3.5.3 Distance of Gene Expression Profile Clusters There are many ways to measure the distance between gene expression profiles and clusters of gene expression profiles. The Pearson correlation coefficient and the Euclidean distance are often used for well-normalized microarray data sets. However, microarray gene expression profiles contain noise and outliers. Nonparametric distance measures provide a way to avoid these problems. For instance, the Spearman correlation replaces gene expression values with their ranks before measuring the distance.

Average linkage, single linkage, and complete linkage are commonly used to measure the distances between clusters of gene expression profiles. Average linkage

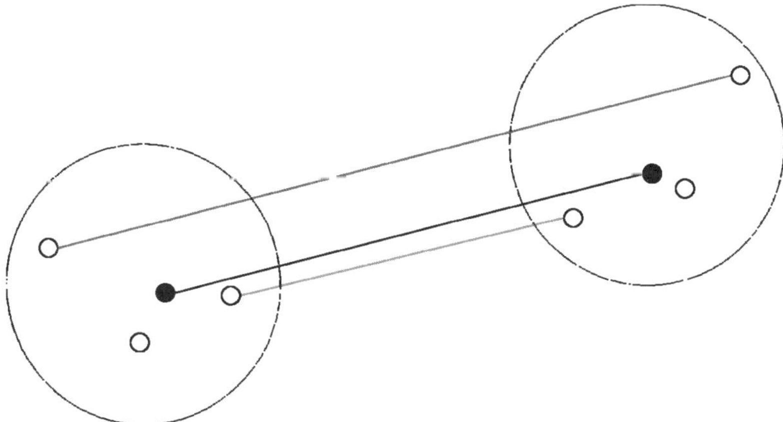

FIGURE 4.15 Cluster distance definitions. Hollow dots represent data points, and the two circles represent two distinct clusters of data points, while black dots are weighted centers of data points in each cluster. The bottom line illustrates the single linkage method of cluster distance, the top line illustrates the complete linkage method, and the middle line represents the average linkage method.

computes the distances between all pairs of gene expression profiles from two clusters and the average of these distances becomes the distance between the clusters. Single linkage defines the distance between two clusters as the distance between the two closest representatives of these clusters. Complete linkage defines the distance between two clusters as the distance between the two farthest representatives. The difference between these three definitions is illustrated in Figure 4.15.

4.3.5.4 Cluster Validation Regardless of the type of clustering, all obtained clusters need to be evaluated for biological validity before proceeding to further analysis. Visual validation is aimed at determining whether there are outliers in clusters or whether the gene expression profiles within each cluster are correlated to each other. If a problem is detected by validation, clusters are often refined by adjusting the number of clusters (parameter k), the distance measuring method, or even by repeating the clustering with a different clustering method. Microarray data sets are highly dimensional. It is often difficult to provide a clear view of gene expression profile types within each cluster. By reducing the dimension of the microarray data set to two or three dimensions, analysis can be simplified and a visual overview of the data can be generated, which may provide useful information on gene expression profile clustering. Such a dimensionality reduction is typically achieved with principal component analysis (PCA). This technique finds the orthogonal components (also called *principal components*) of the input vectors and retains two or three orthogonal components with the highest variance. A visual examination of the projected clusters can help determine an appropriate number of distinct clusters for clustering as illustrated in Figure 4.16.

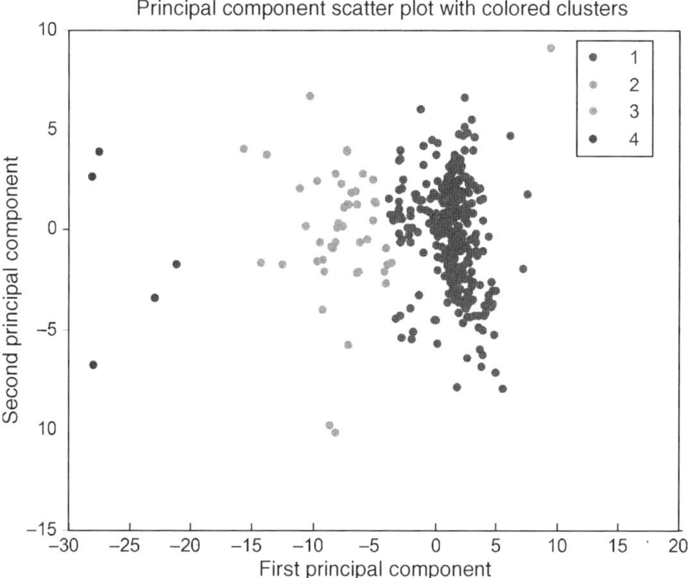

FIGURE 4.16 Principal component analysis. This scatterplot was obtained by plotting the first and the second principal component of the first 100 genes in an acute leukemia data set. It illustrates the benefit of PCA for visualizing data. There are apparently two to four clusters (depending on the criteria of separation of clusters), which is valuable information for the choice of parameter k in many clustering algorithms. A possible clustering to two groups of genes is shown as light gray and dark gray points, while black and lighter gray (top right) points can be discarded as outliers.

4.3.6 Biomarker Identification

One major challenge of microarray data analysis is sample classification. Examples of classification include the separation of people with and without CFS, or the classification of cancer patients into prespecified subcategories. Classifier construction includes the selection of the appropriate prediction model and the selection of features. Feature selection is a technique whereby genes with the most useful expression levels for classification are selected. Such genes can also be useful as *biomarkers* that in turn can be used for practical and cost-effective classification systems.

4.3.6.1 Classical Feature Selection Methods *Forward feature selection* is an iterative process. It starts with an empty set of genes and at each iteration step adds the most informative of the remaining genes based on their ability to discriminate different classes of samples. This process is repeated until no further significant improvement of classification accuracy can be achieved. A reverse procedure, *backward feature elimination*, is also widely applied. It begins by using all the available genes and continues by dropping the least important genes until no significant improvement can be achieved.

In the *filter feature selection methods*, various statistical measures are used to rank genes by their discriminative powers. Successful measures include using the t-test, the chi-square test, information gain, and the Kruskal–Wallis test.

A recently proposed biomarker identification approach involves clustering gene expression profiles [58]. In such an approach, genes are clustered based on their microarray expression profiles. Then, within each cluster, the most representative gene is selected (the representative gene could be the gene closest to the mean or median expression value within the cluster). The representative genes are collected and used as selected features to build a predictor for classification of unknown samples. However, selected sets of genes often lack biological justification and their size is usually too large for experimental validation.

4.3.6.2 Domain Knowledge-Based Feature Selection

A recently proposed feature selection approach exploits the biological knowledge of gene functions as a criterion for selection [59]. The underlying hypothesis for this approach is that the difference between samples lies in a few key gene functions. Genes annotated with those key functions are likely to be very useful for classification. To use this observation, a statistical test is applied to microarray data in order to rank genes by their p-values and generate a subset of significant genes. Selected genes are compared to the overall population in order to identify the most significant function. Only genes associated with the most significant function are selected for classification. This approach results in a small set of genes that provide high accuracy (see the case study below).

Case Study 4.3: Feature Selection for Classification

The CFS data set contains 39 test samples from patients clinically diagnosed with CFS and 40 control samples from subjects without CFS (nonfatigue, NF). The objective is to develop a predictor that classifies new subjects either as CFS or NF based on their gene expressions. Each microarray measures 20,160 genes.

We first used the Kruskal–Wallis test with p-value threshold of 0.05 for the initial gene selection. For each GO term, we count how many genes in the original set of 20,160 genes, as well as how many of the selected, are annotated with it. We then use the hypergeometric test to evaluate whether the representation of this GO term in the selected subset of genes is significantly greater than that in the original set of genes. We rank GO terms by their p-values and find the most overrepresented (those with smallest p-value) GO term. We narrow the selection of genes to include only the genes that are the most overrepresented GO term. We then select these genes as features for classification. Feature selection methods were tested using a leave-one-out cross-validation procedure. The prediction model used in all experiments was an SVM with quadratic kernel $k(x, y) = (C + x^{\mathrm{T}}y)^2$.

The Kruskal–Wallis test with a threshold of 0.05 produced the initial selection of 1296 genes. The overall accuracy of prediction with this feature selection method was 53 percent, which is barely better than the 50 percent accuracy of a random predictor. The proposed procedure narrowed the selection down to 17 genes. Although the number of features was reduced by almost two orders of magnitude, the overall accuracy of prediction with this smaller feature set improved to 72 percent. The GO term that was most often selected was GO:0006397 (mRNA processing). Interestingly, mRNA processing was verified by unrelated biological research as very important for CFS diagnosis [60]. We can compare the accuracy of the obtained predictor (72 percent) to the accuracy of a predictor with 17 features with the smallest p-values selected by the Kruskal–Wallis test, which was close to 50 percent; in other words, the predictor was not better than a trivial random predictor.

4.3.7 Conclusions

Microarray data analysis is a significant and broad field with many unresolved problems. This chapter briefly introduces some of the most commonly used methods for the analysis of microarray data, but many topics still remain. For example, microarray data can be used to construct gene networks, which are made up of links that represent relationships between genes, such as coregulation. Computational models for gene networks include Bayesian networks [61], Boolean networks [62], Petri nets [63], graphical Gaussian models [64], and stochastic process calculi [65].

Microarrays can also be studied in conjunction with other topics, such as microarray-related text mining, microarray resources and database construction, drug discovery, drug response study, and design clinical trials.

Several other types of microarrays are used in addition to gene expression microarrays: protein microarrays (including antibody microarrays), single-nucleotide polymorphism (SNP) microarrays, and chemical compound microarrays. Other experimental technologies, such as mass spectrometry, also produce results at a high throughput rate. Methods for the analysis of these various types of biological data have a certain degree of similarity with microarray data analysis. For example, methods used for the identification of differentially expressed genes are similar to the methods used for the identification of biomarkers in mass spectrometry data. Overall, there are many challenging open topics on analyzing high throughput biological data that can provide research opportunities for the data mining and machine learning community. Progress toward solving these challenges and the future directions of research in this area are discussed at various bioinformatics meetings; these include a specialized *International Conference for the Critical Assessment of Microarray Data Analysis (CAMDA)* that was established in 2000, and that was aimed at the assessment of the state-of-the-art methods in large-scale biological data mining. CAMDA provided standard data sets and put an emphasis on various challenges of analyzing large-scale biological data: time series cell cycle data analysis [45] and cancer sample classification using microarray data [3], functional discovery [42] and drug response [66], microarray data sample variance [67], integration of information from different microarray lung cancer data sets [68–71], the malaria transcriptome monitored by microarray data [4], and integration of different types of high throughput biological data related to CFS.

ACKNOWLEDGMENTS

This project is funded in part under a grant with the Pennsylvania Department of Health. The Department specifically disclaims responsibility for any analyses, interpretations, or conclusions. We thank Samidh Chatterjee, Omkarnath Prabhu, Vladan Radosavljević, Lining Yu, and Jingting Zeng at our laboratory for carefully reading and reviewing this text. In addition, we would like to express special thanks to the external reviewers for their valuable comments on a preliminary manuscript.

REFERENCES

1. Schena M, Shalon D, Davis RW, Brown PO. Quantitative monitoring of gene expression patterns with a complementary DNA microarray. Science 1995;270:467–470.
2. Lockhart DJ, Dong H, Byrne MC, Follettie MT, Gallo MV, Chee MS, Mittmann M, Wang C, Kobayashi M, Horton H, Brown EL. Expression monitoring by hybridization to high-density oligonucleotide arrays. Nat Biotechnol 1996;14:1675–1680.
3. Golub TR, Slonim DK, Tamayo P, Huard C, Gaasenbeek M, Mesirov JP, Coller H, Loh ML, Downing JR, Caligiuri MA, Bloomfield CD, Lander ES. Molecular classification of cancer: class discovery and class prediction by gene expression monitoring. Science 1999;286:531–537.
4. Bozdech Z, Llinas M, Pulliam BL, Wong ED, Zhu J, DeRisi JL. The transcriptome of the intraerythrocytic developmental cycle of *Plasmodium falciparum*. PLoS Biol 2003;1:E5.
5. Vernon SD, Reeves WC. The challenge of integrating disparate high-content data: epidemiological, clinical and laboratory data collected during an in-hospital study of chronic fatigue syndrome. Pharmacogenomics 2006;7:345–354.
6. Yang YH, Buckley MJ, Speed TP. Analysis of cDNA microarray images. Brief Bioinform 2001;2:341–349.
7. Yap G. Affymetrix, Inc. Pharmacogenomics 2002;3:709–711.
9. Kooperberg C, Fazzio TG, Delrow JJ, Tsukiyama T. Improved background correction for spotted DNA microarrays. J Comput Biol 2002;9:55–66.
9. Cui X, KM, Churchill GA. Transformations for cDNA microarray data. Stat Appl Genet Mol Biol 2003;2:article 4.
10. Troyanskaya O, Cantor M, Sherlock G, Brown P, Hastie T, Tibshirani R, Botstein D, Altman RB. Missing value estimation methods for DNA microarrays. Bioinformatics 2001;17:520–525.
11. Kim H, Golub GH, Park H. Missing value estimation for DNA microarray gene expression data: local least squares imputation. Bioinformatics 2005;21:187–198.
12. Johansson P, Hakkinen J. Improving missing value imputation of microarray data by using spot quality weights. BMC Bioinform 2006;7:306.
13. Tuikkala J, Elo L, Nevalainen OS, Aittokallio T. Improving missing value estimation in microarray data with gene ontology. Bioinformatics 2006;22:566–572.
14. Quackenbush J. Microarray data normalization and transformation. Nat Genet 2002; 32(Suppl):496–501.

15. Yang YH, Dudoit S, Luu P, Lin DM, Peng V, Ngai J, Speed TP. Normalization for cDNA microarray data: a robust composite method addressing single and multiple slide systematic variation. Nucleic Acids Res 2002;30:e15.
16. Smyth GK, Speed T. Normalization of cDNA microarray data. Methods 2003;31: 265–273.
17. Berger JA, Hautaniemi S, Jarvinen AK, Edgren H, Mitra SK, Astola J. Optimized LOWESS normalization parameter selection for DNA microarray data. BMC Bioinform 2004;5: 194.
18. Colantuoni CHG, Zeger S, Pevsner J. Local mean normalization of microarray element signal intensities across an array surface: quality control and correction of spatially systematic artifacts. Biotechniques 2002;32:1316–1320.
19. Holter NS, Mitra M, Maritan A, Cieplak M, Banavar JR, Fedoroff NV. Fundamental patterns underlying gene expression profiles: simplicity from complexity. Proc Natl Acad Sci USA 2000;97:8409–8414.
20. Bolstad BM, Irizarry RA, Astrand M, Speed TP, A comparison of normalization methods for high density oligonucleotide array data based on variance and bias. Bioinformatics 2003;19:185–193.
21. Wolfinger RD, Gibson G, Wolfinger ED, Bennett L, Hamadeh H, Bushel P, Afshari C, Paules RS. Assessing gene significance from cDNA microarray expression data via mixed models. J Comput Biol 2001;8:625–637.
22. Schadt EE, Li C, Ellis B, Wong WH, Feature extraction and normalization algorithms for high-density oligonucleotide gene expression array data. J Cell Biochem Suppl 2001;37:120–125.
23. Irizarry RA, Hobbs B, Collin F, Beazer-Barclay YD, Antonellis KJ, Scherf U, Speed TP. Exploration, normalization, and summaries of high density oligonucleotide array probe level data. Biostatistics 2003;4:249–264.
24. Yu X, Chu TM, Gibson G, Wolfinger RD, A mixed model approach to identify yeast transcriptional regulatory motifs via microarray experiments. Stat Appl Genet Mol Biol 2004;3:article22.
25. Ramsey FL, Shafer DW. The Statistical Sleuth: A Course in Methods of Data Analysis. Belmont, CA: Duxbury Press; 1996.
26. Kerr MK, Martin M, Churchill GA, Analysis of variance for gene expression microarray data. J Comput Biol 2000;7:819–837.
27. Pan WA. Comparative review of statistical methods for discovering differentially expressed genes in replicated microarray experiments. Bioinformatics 2002;18:546–554.
28. Singer JD. Using SAS PROC MIXED to fit multilevel models, hierarchical models, and individual growth models. J Educ Behav Stat 1998;24:323–355.
29. Tusher VG, Tibshirani R, Chu G. Significance analysis of microarrays applied to the ionizing radiation response. Proc Natl Acad Sci USA 2001;98:5116–5121.
30. Thomas JG, Olson JM, Tapscott SJ, Zhao LP. An efficient and robust statistical modeling approach to discover differentially expressed genes using genomic expression profiles. Genome Res 2001;11:1227–1236.
31. Efron B, Tibshirani R. Empirical Bayes methods and false discovery rates for microarrays. Genet Epidemiol 2002;23:70–86.

32. Benjamini Y, Hochberg Y. Controlling the false discovery rate: a practical and powerful approach to multiple testing. J R Stat Soc Ser B 1995;57:289–300.

33. Qian HR, Huang S. Comparison of false discovery rate methods in identifying genes with differential expression. Genomics 2005;86:495–503.

34. Ashburner M, Ball CA, Blake JA, Botstein D, Butler H, Cherry JM, Davis AP, Dolinski K, Dwight SS, Eppig JT, Harris MA, Hill DP, Issel-Tarver L, Kasarskis A, Lewis S, Matese JC, Richardson JE, Ringwald M, Rubin GM, Sherlock G. Gene ontology: tool for the unification of biology. The Gene Ontology Consortium. Nat Genet 2000;25:25–29.

35. Gene Ontology Consortium. Creating the gene ontology resource: design and implementation. Genome Res 2001;11:1425–1433.

36. Lord PW, Stevens RD, Brass A, Goble CA. Investigating semantic similarity measures across the Gene Ontology: the relationship between sequence and annotation. Bioinformatics 2003;19:1275–1283.

37. Schlicker A, Domingues FS, Rahnenfuhrer J, Lengauer T. A new measure for functional similarity of gene products based on Gene Ontology. BMC Bioinform 2006; 7:302.

38. Rada R, Mili H, Bicknell E, Blettner M. development and application of a metric on semantic nets. IEEE Trans Syst Man Cybernet 1989;19:17–30.

39. Beissbarth T, Speed TP. GOstat: find statistically overrepresented Gene Ontologies within a group of genes. Bioinformatics 2004;20:1464–1465.

40. Dennis G, Jr, Sherman BT, Hosack DA, Yang J, Gao W, Lane HC, Lempicki RA. DAVID: Database for annotation, visualization, and integrated discovery. Genome Biol 2003;4:P3.

41. Chu S, DeRisi J, Eisen M, Mulholland J, Botstein D, Brown PO, Herskowitz I. The transcriptional program of sporulation in budding yeast. Science 1998;282:699–705.

42. Hughes TR, Marton MJ, Jones AR, Roberts CJ, Stoughton R, Armour CD, Bennett HA, Coffey E, Dai H, He YD, Kidd MJ, King AM, Meyer MR, Slade D, Lum PY, Stepaniants SB, Shoemaker DD, Gachotte D, Chakraburtty K, Simon J, Bard M, Friend SH. Functional discovery via a compendium of expression profiles. Cell 2000;102:109–126.

43. Karaoz U, Murali TM, Letovsky S, Zheng Y, Ding C, Cantor CR, Kasif S. Whole-genome annotation by using evidence integration in functional-linkage networks. Proc Natl Acad Sci USA 2004;101:2888–2893.

44. Eisen MB, Spellman PT, Brown PO, Botstein D. Cluster analysis and display of genome-wide expression patterns. Proc Natl Acad Sci USA 1998;95:14863–14868.

45. Spellman PT, Sherlock G, Zhang MQ, Iyer VR, Anders K, Eisen MB, Brown PO, Botstein D, Futcher B. Comprehensive identification of cell cycle-regulated genes of the yeast *Saccharomyces cerevisiae* by microarray hybridization. Mol Biol Cell 1998;9: 3273–3297.

46. Whitfield ML, Sherlock G, Saldanha AJ, Murray JI, Ball CA, Alexander KE, Matese JC, Perou CM, Hurt MM, Brown PO, Botstein D. Identification of genes periodically expressed in the human cell cycle and their expression in tumors. Mol Biol Cell 2002;13:1977–2000.

47. Zhou X, Kao MC, Wong WH. Transitive functional annotation by shortest-path analysis of gene expression data. Proc Natl Acad Sci USA 2002;99:12783–12788.

48. Xic H, Vucetic S, Sun H, Hedge P, Obradovic Z. Characterization of gene functional expression profiles of *Plasmodium falciparum*. Proceedings of the 5th Conference on Critical Assessment of Microarray Data Analysis; 2004.

49. Barutcuoglu Z, Schapire RE, Troyanskaya OG. Hierarchical multi-label prediction of gene function. Bioinformatics 2006;22:830–836.
50. Brown MP, Grundy WN, Lin D, Cristianini N, Sugnet CW, Furey TS, Ares M Jr, Haussler D. Knowledge-based analysis of microarray gene expression data by using support vector machines. Proc Natl Acad Sci USA 2000;97:262–267.
51. Mangiameli P, Chen SK, West D. A comparison of SOM of neural network and hierarchical methods. Eur J Oper Res 1996;93:402–417.
52. Breiman L. Random forests. Mach Learning 2001;45:5–32.
53. Shi T, S D, Belldegrun AS, Palotie A, Horvath S. Tumor classification by tissue microarray profiling: random forest clustering applied to renal cell carcinoma. Mod Pathol 2005;18:547–557.
54. McLachlan GJ, Bean RW, Peel D. A mixture model-based approach to the clustering of microarray expression data. Bioinformatics 2002;18:413–422.
55. Cherepinsky V, Feng J, Rejali M, Mishra B. Shrinkage-based similarity metric for cluster analysis of microarray data. Proc Natl Acad Sci USA 2003;100:9668–9673.
56. Verri A. A novel kernel method for clustering. IEEE Trans Pattern Anal Mach Intell 2005;27:801–805.
57. Kerr K, Churchill GA. Bootstrapping cluster analysis: access the reliable of conclusions from microarray experiments. Proc Natl Acad Sci USA 2001;98:8961–8965.
58. Au W, Chan K, Wong A, Wang Y. Attribute clustering for grouping, selection, and classification of gene expression data. IEEE/ACM Trans Comput Biol Bioinform 2005;2:83–101.
59. Xie H, Obradovic Z, Vucetic S. Mining of microarray, proteomics, and clinical data for improved identification of chronic fatigue syndrome. In: Proceedings of the Sixth International Conference for the Critical Assessment of Microarray Data Analysis; 2006.
60. Whistler T, Unger ER, Nisenbaum R, Vernon SD. Integration of gene expression, clinical, and epidemiologic data to characterize chronic fatigue syndrome. J Transl Med 2003;1: 10.
61. Hartemink AJ, Gifford DK, Jaakkola TS, Young RA. Combining location and expression data for principled discovery of genetic regulatory network models. Pac Symp Biocomput 2002;437–449.
62. Akutsu T, Miyano S, Kuhara S. Identification of genetic networks from a small number of gene expression patterns under the Boolean network model. Pac Symp Biocomput 1999;17–28.
63. Gambin A, Lasota S, Rutkowski M. Analyzing stationary states of gene regulatory network using Petri nets. In Silico Biol 2006;6:0010.
64. Toh H, Horimoto K. Inference of a genetic network by a combined approach of cluster analysis and graphical Gaussian modeling. Bioinformatics 2002;18:287–297.
65. Golightly A, Wilkinson DJ. Bayesian inference for stochastic kinetic models using a diffusion approximation. Biometrics 2005;61:781–788.
66. Scherf U, Ross DT, Waltham M, Smith LH, Lee JK, Tanabe L, Kohn KW, Reinhold WC, Myers TG, Andrews DT, Scudiero DA, Eisen MB, Sausville EA, Pommier Y, Botstein D, Brown PO, Weinstein JN. A gene expression database for the molecular pharmacology of cancer. Nat Genet 2000;24:236–244.
67. Pritchard CC, Hsu L, Delrow J, Nelson PS. Project normal: defining normal variance in mouse gene expression. Proc Natl Acad Sci USA 2001;98:13266–13271.

68. Wigle DA, Jurisica I, Radulovich N, Pintilie M, Rossant J, Liu N, Lu C, Woodgett J, Seiden I, Johnston M, Keshavjee S, Darling G, Winton T, Breitkreutz BJ, Jorgenson P, Tyers M, Shepherd FA, Tsao MS. Molecular profiling of non-small cell lung cancer and correlation with disease-free survival. Cancer Res 2002;62:3005–3008.

69. Beer DG, et al. Gene-expression profiles predict survival of patients with lung adenocarcinoma. Nat Med 2002;8:816–824

70. Garber ME, Troyanskaya OG, Schluens K, Petersen S, Thaesler Z, Pacyna-Gengelbach M, van de Rijn M, Rosen GD, Perou CM, Whyte RI, Altman RB, Brown PO, Botstein D, Petersen I. Diversity of gene expression in adenocarcinoma of the lung. Proc Natl Acad Sci USA 2001;98:13784–13789.

71. Bhattacharjee A, Richards WG, Staunton J, Li C, Monti S, Vasa P, Ladd C, Beheshti J, Bueno R, Gillette M, Loda M, Weber G, Mark EJ, Lander ES, Wong W, Johnson BE, Golub TR, Sugarbaker DJ, Meyerson M. Classification of human lung carcinomas by mRNA expression profiling reveals distinct adenocarcinoma subclasses. Proc Natl Acad Sci USA 2001;98: 13790–13795.

CHAPTER 5

Algorithms of Reaction–Diffusion Computing

ANDREW ADAMATZKY

We give a case study introduction to the novel paradigm of wave-based computing in chemical systems. We show how selected problems and tasks of computational geometry, robotics, and logics can be solved by encoding data in configuration of chemical medium's disturbances and programming wave dynamics and interaction.

5.1 INTRODUCTION

> It is usually very difficult, and sometimes impossible, to solve variational problems explicitly in terms of formulas or geometric constructions involving known simple elements. Instead, one is often satisfied with merely proving the existence of a solution under certain conditions and afterward investigating properties of the solution. In many cases, when such an existence proof turns to be more or less difficult, it is stimulating to realize the mathematical conditions of the problem by corresponding physical devices, or rather, to consider mathematical problem as an interpretation of a physical phenomenon. The existence of the physical phenomenon then represents the solution of the mathematical problem [16].

In 1941, in their timeless treatise Courant and Robbins [16] discussed one of the "classical examples of nonclassical computing"— an idea of physics-based computation, traced back to 1800s where Plateau experimented with the problem on calculation of the surface of smallest area bounded by a given closed contour in space. We will rephrase this as follows. Given a set of planar points, connect the points by a graph with minimal sum of edge lengths (it is allowed to add more points; however, a number of additional points should be minimal). The solution offered is extraordinarily simple and hence nontrivial. Mark given planar points on a flat surface. Insert pins in the points. Place another sheet on top of the pins. Briefly immerse the device in soap solution. Wait till the soap film dries. Record (draw, make a photo) topology of dried soap film. This represents minimal Steiner tree spanning given planar points.

Handbook of Applied Algorithms: Solving Scientific, Engineering and Practical Problems
Edited by Amiya Nayak and Ivan Stojmenović

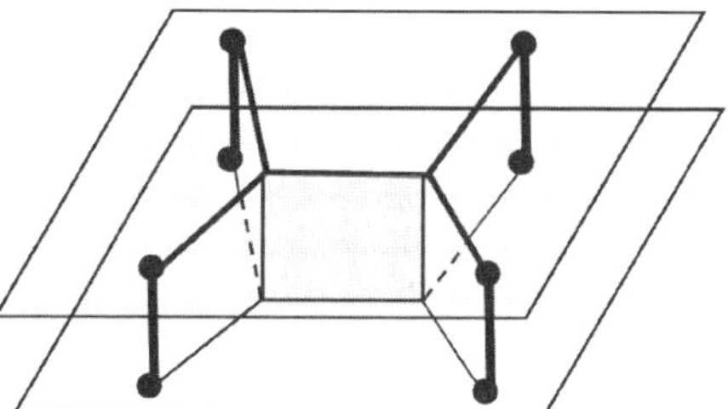

FIGURE 5.1 Soap computer constructs spanning tree of four points [16].

An example of the computing device is shown in Figure 5.1. Owing to surface tension the soap film between the pins, representing points, will try to minimize total surface area. The shrinking can be constrained by a fixed pressure, assuming that the foam film is a cross section of a three-dimensional foam. A length-minimizing curve enclosing a fixed-area region consists of circular arcs of positive outward curvature and line segments [41]. Curvature of the arcs is inversely proportional to pressure. By gradually increasing pressure (Fig. 5.2) we transform arcs to straight lines, and thus spanning tree is calculated.

> In the nineteenth century many of the fundamental theorems of function theory were discovered by Riemann by thinking of simple experiments concerning the flow of electricity in thin metallic sheets [16].

At that time ideas on unconventional, or nature-inspired, computing were flourishing as ever, and Lord Kelvin made his famous differential analyzer, a typical example of a general-purpose analog computer generating functions of the time measure in volts [37]. He wrote in 1876

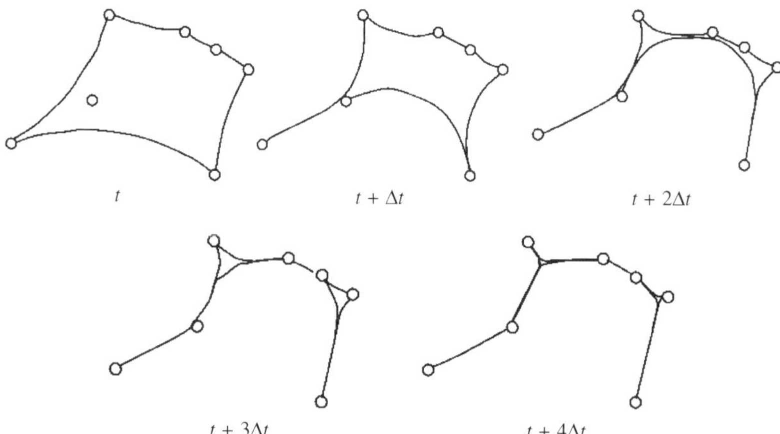

FIGURE 5.2 Several steps of spanning tree constructions by soap film [41].

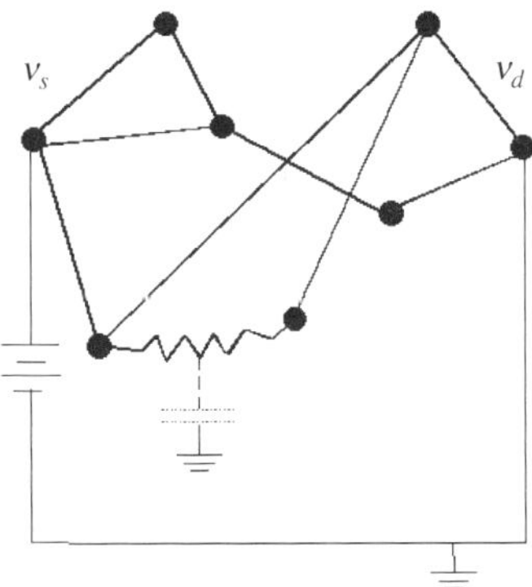

FIGURE 5.3 An electrical machine that computes connectivity of graph edges [50].

> It may be possible to conceive that nature generates a computable function of a real variable directly and not necessarily by approximation as in the traditional approach [37].

The main idea of a field computing on graphs and networks lies in the application of a voltage to a graph, where edges and nodes are assumed to have certain resistance, and measuring resistance or capacities of the networks. This technique was used, at least implicitly, from the beginning of the century or even early but the earliest publication with the emphasis on the algorithmic part is the paper by Vergis et al. [50]. They solve a well-known (s, t)-connectivity problem by constructing a virtual electrical model of the given graph (Fig. 5.3): Given two vertexes s and t of a graph, decide whether there is a path from s to t. This is solved as follows. Put wires instead of edges and connect them at the nodes. Apply a voltage between the nodes s and t. Measure the current. If near null current is recorded, there is no path between s and t.

The method works on the assumption that resistance is proportional only to the length of a wire; therefore, if there is no path between s and t then resistance is nearly infinite high resistance, if there is no path between v_s and v_t. If lengths of wires grow linearly with the number of graph nodes, the total capacity of the voltage source and total resistance have the upper bound $O(|\mathbf{E}^2|)$, which leads to the total size and power consumption $O(|\mathbf{E}^4|)$; that is, the electric machine operates polynomial resources [50].

Surface tension, propagating waves, and electricity have been principal "engines" of nature-inspired computers for over two centuries; even so they never were combined together till Kuhnert's pioneer work on image transformations in light-sensitive Belousov–Zhabotinsky system [27]. A reaction–diffusion computer is a spatially extended chemical system, which processes information using interacting growing patterns, and excitable and diffusive waves. In reaction–diffusion processors, both the data and the results of the computation are encoded as concentration profiles of the reagents. The computation is performed via the spreading and interaction of wave fronts.

The reaction–diffusion computers are parallel because myriads of their microvolumes update their states simultaneously, and molecules diffuse and react in parallel. Liquid-phase chemical media are wet analogs of massive parallel (millions of elementary processors in a small chemical reactor) and locally connected (every microvolume of the medium changes its state depending on the states of its closest neighbors) processors. They have parallel inputs and outputs; for example, optical input is parallel because of the control of initial excitation dynamics by illumination masks while, output is parallel because concentration profile representing results of computation is visualized by indicators. The reaction–diffusion computers are fault tolerant and capable of automatic reconfiguration, namely if we remove some quantity of the computing substrate, the topology is restored almost immediately.

Reaction–diffusion computers are based on three principles of physics-inspired computing. First, physical action measures amount of information: we exploit active processes in nonlinear systems and interpret dynamics of the systems as computation. Second, physical information travels only finite distance: this means that computation is local and we can assume that the nonlinear medium is a spatial arrangement of elementary processing units connected locally; that is, each unit interacts with closest neighbors. Third, nature is governed by waves and spreading patterns: computation is therefore spatial.

Reaction–diffusion computers give us best examples of unconventional computers; their features follow Jonathan Mills' classification of convention versus unconventional [32]: wetware, nonsilicon computing substrate; parallel processing; computation occurring everywhere in substrate space; computation is based on analogies; spatial increase in precision; holistic and spatial programming; visual structure; and implicit error correcting.

A theory of reaction–diffusion computing was established and a range of practical applications are outlined in the work by Adamatzky [1]; recent discoveries are published in a collective monograph [5]. The chapter in no way serves as a substitute for these books but rather an introduction to the field and a case study of several characteristic examples.

The chapter is populated with cellular automaton examples of reaction–diffusion processes. We have chosen cellular automatons to study computation in reaction–diffusion media because cellular automatons can provide just the right fast prototypes of reaction–diffusion models. The examples of "best practice" include models of BZ reactions and other excitable systems [21,31], chemical systems exhibiting Turing patterns [54,56,58], precipitating systems [5], calcium wave dynamics [55], and chemical turbulence [23]. We therefore consider it reasonable to interpret the cellular automaton local update rules in terms of reaction–diffusion chemical systems and reinterpret the cellular automaton rules in novel designs of the chemical laboratory reaction–diffusion computers.

Cellular automaton models of reaction–diffusion and excitable media capture essential aspects of the natural media in a computationally tractable form. A cellular automaton is a—in our case two-dimensional—lattice of finite automatons, or an array of cells. The automatons evolve in a discrete time and take their states from a finite set. All automatons of the lattice update their states simultaneously. Every automaton calculates its next state depending on the states of its closest neighbors (throughout the chapter we assume every nonedge cell x of a cellular automaton updates its state depending on the states of its eight closest neighbors).

The best way to learn riding bicycle is to ride a bicycle. Therefore, instead of wasting time on pointless theoretical constructions, we immediately describe and analyze working reaction–diffusion algorithms for image processing, computational geometry, logical and arithmetical circuits, memory devices, path planning and robot navigation, and control of massive parallel actuators.

Just few words of warning—when thinking about chemical algorithms some of you may realize that diffusive and phase waves are pretty slow in physical time. The sluggishness of computation is the only point that may attract criticism to reaction–diffusion chemical computers. There is however a solution—to speed up we are implementing the chemical medium in silicon, microprocessor LSI analogs of reaction–diffusion computers [11]. Further miniaturization of the reaction–diffusion computers can be reached when the system is implemented as a two-dimensional array of single-electron nonlinear oscillators diffusively coupled to each other [12]. Yet another point of developing reaction–diffusion computers is to design embedded controllers for soft-bodied robots, where usage of conventional silicon materials seem to be inappropriate.

5.2 COMPUTATIONAL GEOMETRY

In this section we discuss "mechanics" of reaction–diffusion computing on example of plane subdivision. Let **P** be a nonempty finite set of planar points. A planar Voronoi diagram of the set **P** is a partition of the plane into such regions that for any element of **P**, a region corresponding to a unique point p contains all those points of the plane that are closer to p than to any other node of **P**. A unique region $\text{vor}(p) = \{z \in \mathbf{R}^2 : d(p, z) < d(p, m) \forall m \in \mathbf{R}^2,\ m \neq z\}$ assigned to point p is called a Voronoi cell of the point p. The boundary of the Voronoi cell of a point p is built of segments of bisectors separating pairs of geographically closest points of the given planar set **P**. A union of all boundaries of the Voronoi cells determines the planar Voronoi diagram: $VD(\mathbf{P}) = \cup_{p \in \mathbf{P}} \partial\, \text{vor}(p)$. A variety of Voronoi diagrams and algorithms of their construction can be found in the work by Klein [26].

The basic concept of constructing Voronoi diagrams with reaction-diffusion systems is based on a very simple intuitive technique for detecting the bisector points separating two given points of the set **P**. If we drop reagents at the two data points, the diffusive waves, or phase waves if computing substrate is active, spread outward from the drops with the same speed. The waves travel the same distance from the sites of origination before they meet one another. The points, where the waves meet, are the bisector points. This idea of a Voronoi diagram computation was originally implemented in cellular automaton models and in experimental parallel chemical processors (see extensive bibliography in the works by Adamatzky et al. [1,5]).

Assuming that the computational space is homogeneous and locally connected, and every site (microvolume of the chemical medium or cell of the automaton array) is coupled to its closest neighbors by the same diffusive links, we can easily draw a parallel between distance and time, and thus put our wave-based approach into action. In cellular automaton representation of physical reality, cell neighborhood u determines that all processes in the cellular automaton model are constrained to the discrete metric L_∞. So, when studying automaton models we should think rather about discrete Voronoi diagram than its Euclidean representation. Chemical laboratory prototypes of reaction–diffusion computers do approximate continuous Voronoi diagram as we will see further.

A discrete Voronoi diagram can be defined on lattices or arrays of cells, for example, a two-dimensional lattice $\mathbf{Z}^2$. The distance $d(\cdot, \cdot)$ is calculated not in Euclidean but in one of the discrete metrics, for example, L_1 and L_∞. A discrete bisector of nodes x and y of $\mathbf{Z}^2$ is determined as $B(x, y) = \{z \in \mathbf{Z}^2 : d(x, z) = d(y, z)\}$. However, following such definition we sometimes generate bisectors that fill a quarter of the lattices or produce no bisector at all [1]. If we want the constructed diagrams be closer to the real world, then we could redefine discrete bisector as follows: $B(x, y) = \{z \in \mathbf{Z}^2 : |d(x, z) - d(y, z)| \leq 1\}$. The redefined bisector will comprise edges of Voronoi diagrams constructed in discrete, cellular automaton models of reaction–diffusion and excitable media.

Now we will discuss several versions of reaction–diffusion wave-based construction of Voronoi diagrams, from a naïve model, where the number of reagents grow proportionally to the number of data points, to a minimalist implementation with just one reagent and one substrate [1].

Let us start with $O(n)$-reagent model. In a naïve version of reaction–diffusion computation of a Voronoi diagram, one needs two reagents and a precipitate to mark a bisector separating two points. Therefore, $n + 2$ reagents, including precipitate and substrate, are required to approximate a Voronoi diagram of n points. When place n unique reagents on n points of the given data set $\mathbf{P}$, waves of these reagents spread around the space and interact with each other where they meet. When at least two different reagents meet at the same or adjacent sites of the space, they react and form a precipitate—sites that contain the precipitate represent edges of the Voronoi cell, and therefore constitute the Voronoi diagram. In "chemical reaction" equations, the idea looks as follows: α and β are different reagents and # is a precipitate: $\alpha + \beta \rightarrow \#$. This can be converted to cellular automaton interpretation as follows:

$$x^{t+1} = \begin{cases} \rho, \text{ if } x^t = \bullet \text{ and } \Psi(x)^t \subset \{\rho, \bullet\}, \\ \#, \text{ if } x^t \neq \# \text{ and } |\Psi(x)^t/\#| > 1, \\ x^t, \text{ otherwise}, \end{cases}$$

where $\bullet$ is a resting state (cell in this state does not contain any reagents), $\rho \in \mathbf{R}$ is a reagent from the set $\mathbf{R}$ of n reagents, and $\Psi(x)^t = \{y^t : y \in u(x)\}$ characterizes the reagents that are present in the local neighborhood $u(x)$ of the cell x at time step t.

The first transition of the above rule symbolizes diffusion. A resting cell takes the state ρ if only this reagent is present in the cell's neighborhood. If there are two different reagents in the cell's neighborhood, then the cell takes the precipitate state #. Diffusing reagents halt because the formation of precipitate reduces the number of "vacant" resting cells. Precipitate does not diffuse. Cell in state # remains in this

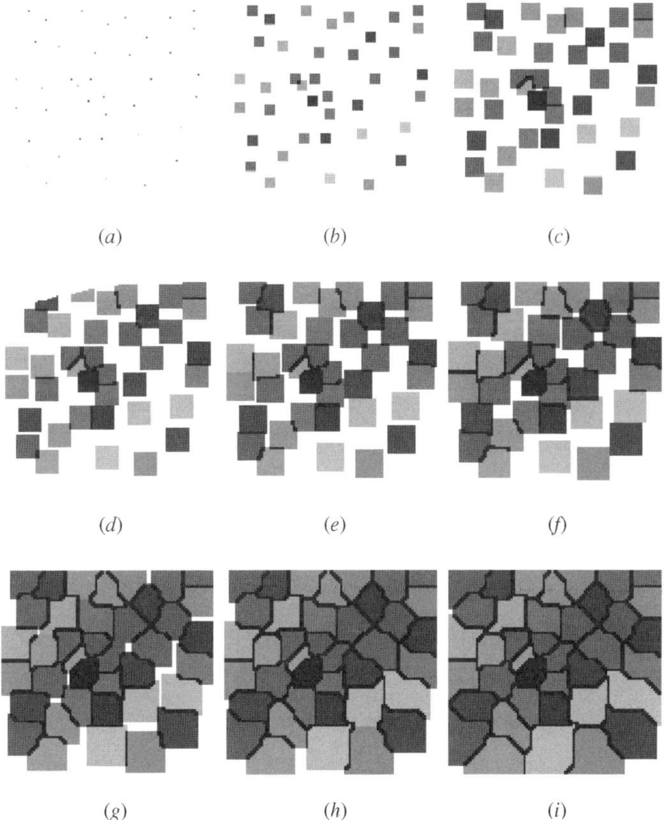

FIGURE 5.4 Computation of a Voronoi diagram in a cellular automaton model of a chemical processor with $O(n)$ reagents. Precipitate is shown in black $(a)\, t = 1; (b)\, t = 3; (c)\, t = 5; (d)\, t = 6; (e)\, t = 7; (f)\, t = 8; (g)\, t = 10; (h)\, t = 12; (i)\, t = 15$.

indefinitely. An example of a cellular automaton simulation of $O(n)$-reagent chemical processor is shown in Figure 5.4.

The $O(n)$-reagent model is demonstrative; however, it is computationally inefficient. Clearly, we can reduce number of reagents to four—using map coloring theorems—but preprocessing time will be unfeasibly high. The number of participating reagents can be sufficiently reduced to $O(1)$ when the topology of the spreading waves is taken into account [1].

Now we go from one extreme to another and consider a model with just one reagent and a substrate. The reagent α diffuses from sites corresponding two point of a data planar set **P**. When two diffusing wave fronts meet a superthreshold concentration of reagents, they do not spread further. A cellular automaton model represents this as follows.

Every cell has two possible states: resting or substrate state $\bullet$ and reagent state α. If the cell is in state α, it remains in this state indefinitely. If the cell is in state

• and between one and four of its neighbors are in state α, then the cell takes the state α. Otherwise, the cell remains in the state • — this reflects the "superthreshold inhibition" idea. A cell state transition rule is follows:

$$x^{t+1} = \begin{cases} \alpha, \text{ if } x^t = \bullet \text{ and } 1 \leq \sigma(x)^t \leq 4, \\ x^t, \text{ otherwise}, \end{cases}$$

where $\sigma(x)^t = |y \in u(x) : y^t = \alpha|$.

Increasing number of reagents to two (one reagent and one precipitate) would make life easy. A reagent β diffuses on a substrate, from the initial points (drop of reagent) of **P**, and forms a precipitate in the reaction $m\beta \rightarrow \alpha$, where $1 \leq m \leq 4$.

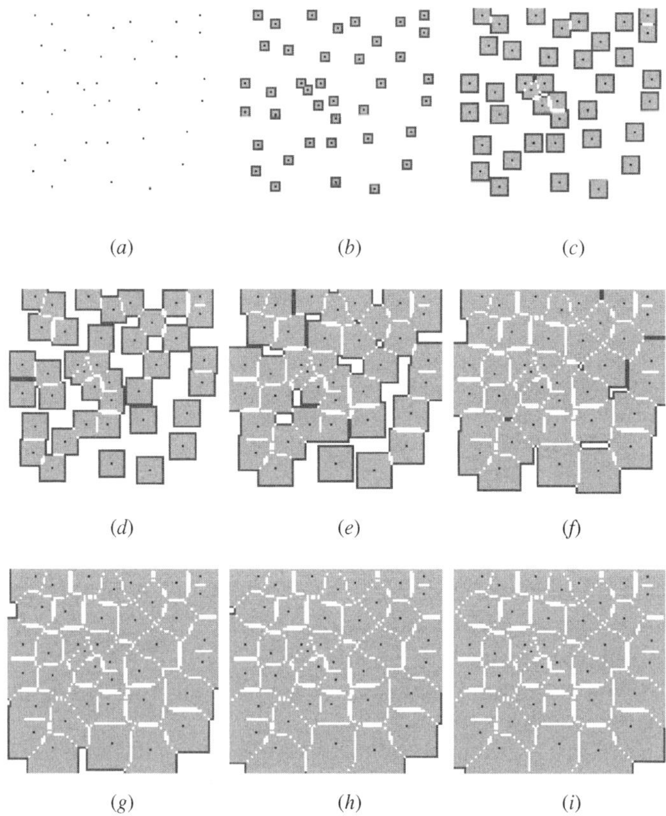

FIGURE 5.5 An example of Voronoi diagram computing in an automaton model of reaction–diffusion medium with one reagent and one substrate. Reactive parts of wave fronts are shown in black. Precipitate is gray and edges of Voronoi diagram are white (*a*) $t = 1$; (*b*) $t = 3$; (*c*) $t = 5$; (*d*) $t = 7$; (*e*) $t = 9$; (*f*) $t = 11$; (*g*) $t = 13$; (*h*) $t = 15$; (*i*) $t = 17$.

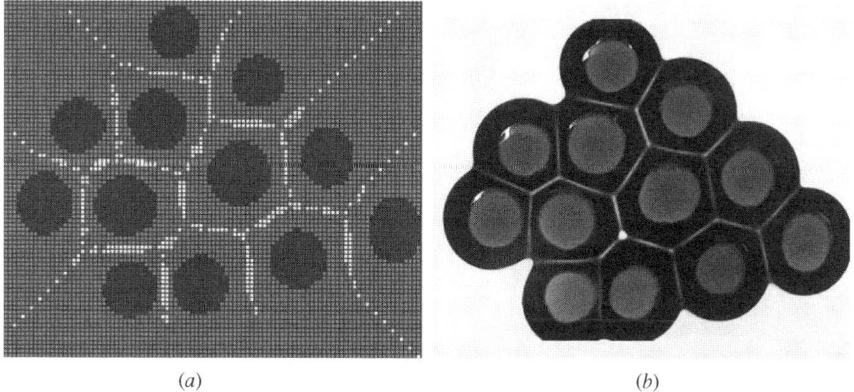

(*a*) (*b*)

FIGURE 5.6 Planar Voronoi diagram computed in (*a*) cellular automaton and (*b*) palladium reaction–diffusion chemical processor [5].

Every cell takes three states: • (resting cell, no reagents), α (e.g., colored precipitate), and β (reagent). The cell updates its states by the rule:

$$x^{t+1} = \begin{cases} \beta, \text{ if } x^t = \bullet \text{ and } 1 \leq \sigma(x)^t \leq 4, \\ \alpha, \text{ if } x^t = \beta \text{ and } 1 \leq \sigma(x)^t \leq 4, \\ x^t, \text{ otherwise,} \end{cases}$$

where $\sigma(x)^t = |y \in u(x) : y^t = \beta|$.

An example of a Voronoi diagram computed in an automaton model of a reaction–diffusion medium with one reagent and one substrate is shown in Figure 5.5.

By increasing number of cell state and enlarging cell neighborhood in cellular automaton model we can produce more realistic—almost perfectly matching outcomes of chemical laboratory experiments—Voronoi diagrams (Fig. 5.6).

Let us consider the following model. Cells of the automaton take state from interval $[\rho, \alpha]$, where ρ is a minimum refractory value and α is maximum excitation value; $\rho = -2$ and $\alpha = 5$ in our experiments. Cell x's state transitions are strongly determined by normalized local excitation $\sigma_x^t = \sum_{y \in u_x}(y^t/\sqrt{(}|u_x|))$. Every cell x updates its state at time $t+1$, depending on its state x^t and state u_x^t of its neighborhood u_x—in experiments we used 15×15 cell neighborhood—as follows:

$$x^{t+1} = \begin{cases} \alpha, \text{ if } x^t = 0 \text{ and } \sigma_x^t \geq \alpha, \\ 0, \text{ if } x^t = 0 \text{ and } \sigma_x^t < \alpha, \\ x^t + 1, \text{ if } x^t < 0, \\ x^t - 1, \text{ if } x^t > 1, \\ \rho, \text{ if } x^t = 1. \end{cases}$$

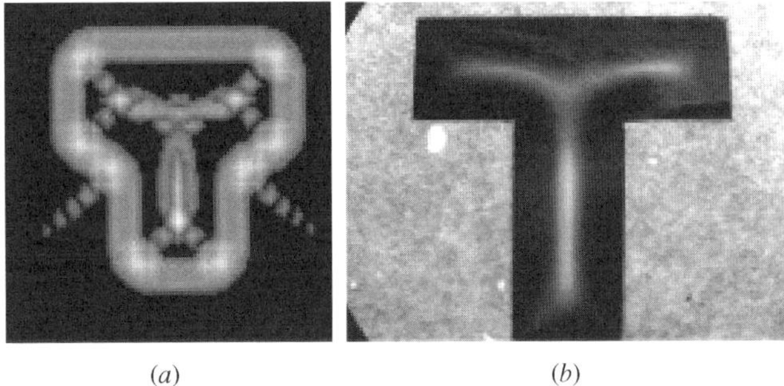

(*a*) (*b*)

FIGURE 5.7 Skeleton—internal Voronoi diagram—of planar T-shape constructed in multi-state cellular automaton model (*a*) and chemical laboratory Prussian blue reaction–diffusion processor (*b*) [10].

This rule represents spreading of "excitation," or simply phase wave fronts, in computational space, interaction, and annihilation of the wave fronts. To allow the reaction–diffusion computer "memorize" sites of wave collision, we add a precipitate state p_x^t. Concentration p_x^t of precipitate at site x at moment t is calculated as $p_x^{t+1} \sim |\{y \in u_x : y^t = \alpha\}|$.

As shown in Figure 5.7, the model represents cellular automaton Voronoi diagrams in "unlike phase" with experimental chemical representation of the diagram. Sites of higher concentration of precipitate in cellular automaton configurations correspond to sites with lowest precipitate concentration in experimental processors.

5.3 LOGICAL UNIVERSALITY

Certain families of thin-layer reaction–diffusion chemical media can implement sensible transformation of initial (data) spatial distribution of chemical species concentrations to final (result) concentration profile [1,45]. In these reaction–diffusion computers, a computation is realized via spreading and interaction of diffusive or phase waves. Specialized, intended to solve a particular problem, experimental chemical processors implement basic operations of image processing [5,28,39,40], computation of optimal paths [5,9,46], and control of mobile robots [5].

A device is called computationally universal if it implements a functionally complete system of logical gates, for example, a tuple of negation and conjunction, in its space–time dynamics.

A number of computationally universal reaction–diffusion devices were implemented: the findings include logical gates [42,48] and diodes [17,29,34] in Belousov-Zhabotinsky (BZ) medium, and xor gate in palladium processor [2]. All the known so far experimental prototypes of reaction–diffusion processors exploit interaction of wave fronts in a geometrically constrained chemical medium; that is, the computation is based on a stationary architecture of medium's inhomogeneities. Constrained by stationary wires and gates, chemical universal processors pose a little computational novelty and none dynamical reconfiguration ability because they simply imitate architectures of silicon computing devices.

Experimental prototypes of reaction–diffusion processors exploit interaction of wave fronts in a geometrically constrained chemical medium; that is, the computation is based on a stationary architecture of medium's inhomogeneities. Constrained by stationary wires and gates reaction–diffusion chemical universal processors pose a little computational novelty and no dynamic reconfiguration ability because they simply imitate architectures of conventional silicon computing devices. To appreciate in full massive parallelism of thin-layer chemical media and to free the chemical processors from limitations of fixed computing architectures, we adopt an unconventional paradigm of architectureless, or collision-based, computing. An architecture-based, or stationary, computation implies that a logical circuit is embedded into the system in such a manner that all elements of the circuit are represented by the system's stationary states. The architecture is static. If there is any kind of "artificial" or "natural" compartmentalization, the medium is classified as an architecture-based computing device. Personal computers, living neural networks, cells, and networks of chemical reactors are typical examples of architecture-based computers.

A collision-based, or dynamical, computation employs mobile compact finite patterns, mobile self-localized excitations or simply localizations, in active nonlinear medium. Essentials of collision-based computing are the following. Information values (e.g., truth values of logical variables) are given by either absence or presence of the localizations or other parameters of the localizations. The localizations travel in space and do computation when they collide with each other. There are no predetermined stationary wires; a trajectory of the traveling pattern is a momentary wire. Almost any part of the medium space can be used as a wire. Localizations can collide anywhere within a space sample; there are no fixed positions at which specific operations occur, nor location specified gates with fixed operations. The localizations undergo transformations, form bound states, annihilate, or fuse when they interact with other mobile patterns. Information values of localizations are transformed as a result of collision and thus a computation is implemented [3].

The paradigm of collision-based computing originates from the technique of proving computational universality of game of life [14], conservative logic and billiard ball model [20], and their cellular automaton implementations [30].

Solitons, defects in tubulin microtubules, excitons in Scheibe aggregates, and breather in polymer chains are most frequently considered candidates for a role of information carrier in nature-inspired collision-based computers (see overview in the work by Adamatzky [1]). It is experimentally difficult to reproduce all these artifacts in natural systems; therefore, existence of mobile localizations in an experiment-friendly chemical media would open new horizons for fabrication of collision-based computers.

The basis for material implementation of collision-based universality of reaction–diffusion chemical media is discovered by Sendina-Nadal et al. [44]. They experimentally proved the existence of localized excitations—traveling wave fragments that behave like quasiparticles—in photosensitive subexcitable Belousov–Zhabotinsky medium.

We show how logical circuits can be fabricated in a subexcitable BZ medium via collisions between traveling wave fragments. While implementation of collision-based logical operations is relatively straightforward [5], more attention should be paid to control of signal propagation in the homogeneous medium. It has been demonstrated that applying light of varying intensity we can control excitation dynamics in Belousov–Zhabotinsky medium [13,22,36], wave velocity [47], and pattern formation [51]. Of particular interest are experimental evidences of light-induced back-propagating waves, wave front splitting, and phase shifting [59]; we can also manipulate medium's excitability by varying intensity of the medium's illumination [15]. On the basis of these facts we show how to control signal wave fragments by varying geometric configuration of excitatory and inhibitory segments of impurity reflectors.

We built our model on a two-variable Oregonator equation [19,49] adapted to a light-sensitive BZ reaction with applied illumination [13]:

$$\frac{\partial u}{\partial t} = \frac{1}{\epsilon}\left(u - u^2 - (fv + \phi)\frac{u-q}{u+q}\right) + D_u \nabla^2 u,$$

$$\frac{\partial v}{\partial t} = u - v,$$

where variables u and v represent local concentrations of bromous acid ($HBrO_2$) and the oxidized form of the catalyst ruthenium (Ru(III)), respectively, ϵ sets up a ratio of timescale of variables u and v, q is a scaling parameter depending on reaction rates, f is a stoichiometric coefficient, and ϕ is a light-induced bromide production rate proportional to intensity of illumination (an excitability parameter—moderate intensity of light will facilitate excitation process, higher intensity will produce excessive quantities of bromide which suppresses the reaction). We assumed that the catalyst is immobilized in a thin layer of gel; therefore, there is no diffusion term for v. To integrate the system we used the Euler method with five-node Laplacian operator, time step $\Delta t = 10^{-3}$, and grid point spacing $\Delta x = 0.15$, with the following parameters: $\phi = \phi_0 + A/2$, $A = 0.0011109$, $\phi_0 = 0.0766$, $\epsilon = 0.03$, $f = 1.4$, and $q = 0.002$. Chosen parameters correspond to a region of "higher excitability of the subexcitability regime" outlined in the work by Sedina-Nadal et al. [44] (see also how to adjust f and q in the work by Qian and Murray [38]) that supports propagation of sustained wave fragments (Fig. 5.8a). These wave fragments are used as quanta of information in our design of collision-based logical circuits. The waves were initiated by locally disturbing initial concentrations of species; for example, 10 grid sites in a chain are given value $u = 1.0$ each; this generated two or more localized wave fragments, similarly to counterpropagating waves induced by temporary illumination in experiments [59]. The traveling wave fragments keep their shape for around 4×10^3–10^4 steps of simulation (4–10 time units), then decrease in size and vanish. The wave's lifetime is sufficient, however, to implement logical gates; this also allows us not to worry about "garbage collection" in the computational medium.

We model signals by traveling wave fragments [13,44]: a sustainably propagating wave fragment (Fig. 5.8a) represents TRUE value of a logical variable corresponding to the wave's trajectory (momentarily wire).

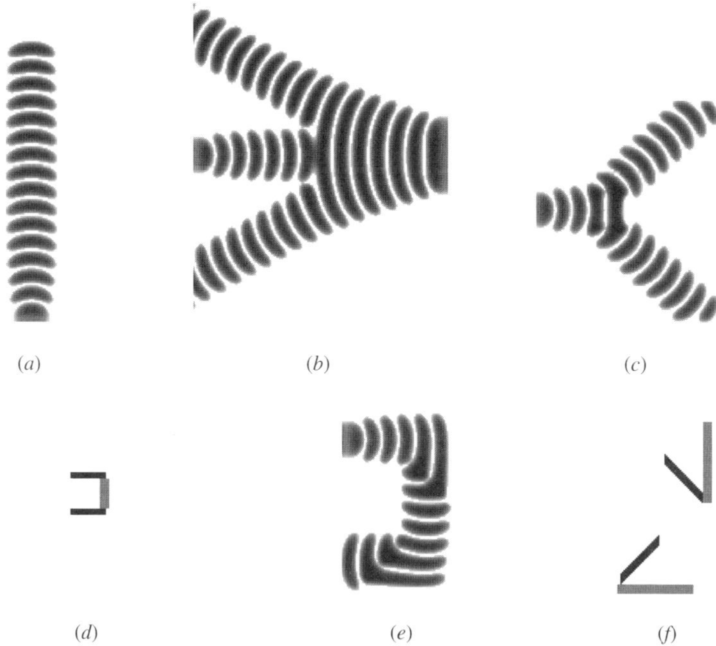

FIGURE 5.8 Basic operations with signals. Overlay of images taken every 0.5 time units. Exciting domains of impurities are shown in black; inhibiting domains of impurities are shown in gray. (*a*) Wave fragment traveling north. (*b*) Signal branching without impurities: a wave fragment traveling east splits into two wave fragments (traveling southeast and northeast) when it collides with a smaller wave fragment traveling west. (*c*) Signal branching with impurity: wave fragment traveling west is split by impurity (*d*) into two waves traveling northwest and southwest. (*e*) Signal routing (U-turn) with impurities: a wave fragment traveling east is routed north and then west by two impurities. (*f*) An impurity reflector consists of inhibitory (gray) and excitatory (black) chains of grid sites.

To demonstrate that a physical system is logically universal, it is enough to implement negation and conjunction or disjunction in spatiotemporal dynamics of the system. To realize a fully functional logical circuit, we must also know how to operate input and output signals in the system's dynamics, namely to implement signal branching and routing; delay can be realized via appropriate routing.

We can branch a signal using two techniques. First, we can collide a smaller auxiliary wave to a wave fragment representing the signal, the signal wave will split then into two signals (these daughter waves shrink slightly down to stable size and then travel with constant shape further 4×10^3 time steps of the simulation) and the auxiliary wave will annihilate (Fig. 5.8b).

Second, we can temporarily and locally apply illumination impurities on a signal's way to change properties of the medium and thus cause the signal to split (Fig. 5.8c and d). We must mention, it was already demonstrated in the work by Yoneyama [59], that wave front influenced by strong illumination (inhibitory segments of the impurity) splits and its ends do not form spirals, as in typical situations of excitable media.

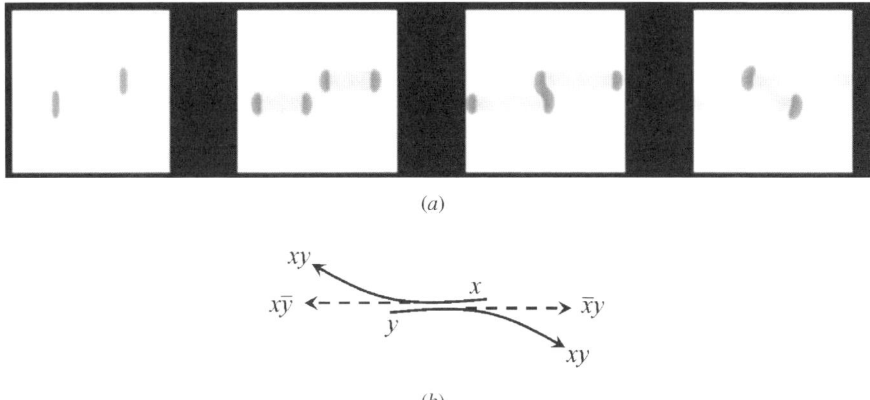

FIGURE 5.9 Implementation of conservative gate in Belousov–Zhabotinsky system. (*a*) Elastic co-collision of two wave fragments, one traveling west and the other east. The fragments change directions of their motion to northwest and southeast, respectively, as a result of the collision. (*b*) Scheme of the gate. In (*a*), logical variables are represented as $x = 1$ and $y = 1$.

A control impurity, or reflector, consists of a few segments of sites whose illumination level is slightly above or below overall illumination level of the medium. Combining excitatory and inhibitory segments we can precisely control wave's trajectory, for example, realize U-turn of a signal (Fig. 5.8e and f).

A typical billiard ball model interaction gate [20,30] has two inputs—x and y, and four outputs—$x\overline{y}$ (ball x moves undisturbed in absence of ball y), $\overline{x}y$ (ball y moves undisturbed in absence of ball x), and twice xy (balls x and y change their trajectories when collided with each other). Such conservative interaction gate can be implemented via elastic collision of wave fragment see Fig. 5.9.

The elastic collision is not particularly common in laboratory prototypes of chemical systems; more often interacting waves either fuse or one of the waves annihilates as a result of the collision with another wave. This leads to nonconservative version

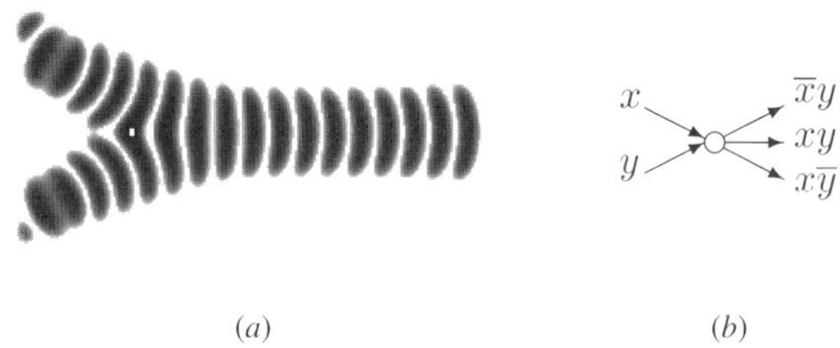

(*a*) (*b*)

FIGURE 5.10 Two wave fragments undergo angle collision and implement interaction gate $\langle x, y\rangle \rightarrow \langle x\overline{y}, xy, \overline{x}y\rangle$. (*a*) In this example $x = 1$ and $y = 1$, both wave fragments are present initially. Overlay of images taken every 0.5 time units. (*b*) Scheme of the gate. In upper-left and bottom-left corners of (*a*) we see domains of wave generation two echo wave fragments are also generated; they travel outward gate area and thus do not interfere with computation.

of the interaction gate with two inputs and three outputs, that is, just one xy output instead of two. Such collision gate is shown in Figure 5.10.

Rich dynamics of subexcitable Belousov-Zhabotinsky medium allows us also to implement complicated logical operations just in a single interaction event (see details in the work by Adamatzky et al. [5]).

5.4 MEMORY

Memory in chemical computers can be represented in several following ways. In precipitating systems, any site with precipitate is a memory element. However, they are not rewritable. In "classical" excitable chemical systems, like Belousov–Zhabotinsky dynamics, one can construct memory as a configuration of sources of spiral or target ways. We used this technique to program movement of wheeled robot controlled by onboard chemical reactor with Belouso–Zhabotinsky system [5]. The method has the same drawback as precipitating memory—as soon as reaction space is divided by spiral or target waves, it is quite difficult if not impossible to sensibly move source of the waves. This is only possible with external inhibition or complete reset of the medium.

In geometrically constrained excitable chemical medium, as demonstrated in the work by Motoike et al. [33], we can employ old-time techniques of storing information in induction coils and other types of electrical circuits, that is, dynamical memory. A ring with an input channel is prepared from reaction substrate. The ring is broken by a small gap and the input is also separated from the ring with a gap of similar width [33]; the gaps play a role of one-way gates to prevent excitation from spreading backwards. The waves enter the ring via input channel and travel along the ring "indefinitely" (till substrate lasts) [33]. The approach aims to split reaction–diffusion system into many compartments, and thus does not fit our paradigm of computing in uniform medium.

In our search for real-life chemical systems exhibiting both mobile and stationary localizations, we discovered a cellular automaton model [53] of an abstract activator–inhibitor reaction diffusion system, which ideally fits the framework of the collision-based computing paradigm and reaction–diffusion computing. The phenomenology of the automaton was discussed in detail in our previous work [53]; therefore, in the present paper we draw together the computational properties of the reaction–diffusion cellular hexagonal automaton. The automaton imitates spatiotemporal dynamics of the following reaction equations:

$$
\begin{aligned}
A + 6S \rightarrow A \qquad\qquad A + I \rightarrow I \qquad\qquad A + 3I \rightarrow I \\
A + 2I \rightarrow S \qquad\qquad 2A \rightarrow I \\
3A \rightarrow A \qquad\qquad \beta A \rightarrow I \\
I \rightarrow S.
\end{aligned}
$$

Each cell of the automaton takes three states—substrate S, activator A, and inhibitor I. Adopting formalism from [7], we represent the cell state transition rule as a matrix $\mathbf{M} = (m_{ij})$, where $0 \leq i \leq j \leq 7$, $0 \leq i + j \leq 7$, and $m_{ij} \in \{I, A, S\}$. The output state of each neighborhood is given by the row index i, the number of neighbors in cell state I, and column index j (the number of neighbors in cell state A). We do not have to count the number of neighbors in cell state S, because it is given by $7 - (i + j)$. A cell with a neighborhood represented by indexes i and j will update to cell state M_{ij} that can be read off the matrix. In terms of the cell state transition function, this can be presented as follows: $x^{t+1} = M_{\sigma_2(x)^t \sigma_1(x)^t}$, where $\sigma_i(x)^t$ is a sum of cell x's neighbors in state i, $i = 1, 2$, at time step t. The exact matrix structure, which corresponds to matrix M_3 in the work by Wuensche and Adamatzky [53], is as follows:

$$M = \left\{ \begin{array}{cccccccc} S & A & I & A & I & I & I & I \\ S & I & I & A & I & I & I & \\ S & S & I & A & I & I & & \\ S & I & I & A & I & & & \\ S & S & I & A & & & & \\ S & S & I & & & & & \\ S & S & & & & & & \\ S & & & & & & & \end{array} \right\}.$$

The cell state transition rule reflects the nonlinearity of activator–inhibitor interactions for subthreshold concentrations of the activator. Namely, for small concentration of the inhibitor and for threshold concentrations, the activator is suppressed by the inhibitor, while for critical concentrations of the inhibitor both inhibitor and activator dissociate producing the substrate. In exact words, $M_{01} = A$ symbolizes the diffusion of activator A, $M_{11} = I$ represents the suppression of activator A by the inhibitor I, and $M_{z2} = I$ $(z = 0, \cdots, 5)$ can be interpreted as self-inhibition of the activator in particular concentrations. $M_{z3} = A$ $(z = 0, \ldots, 4)$ means a sustained excitation under particular concentrations of the activator. $M_{z0} = S$ $(z = 1, \ldots, 7)$ means that the inhibitor is dissociated in absence of the activator, and that the activator does not diffuse in subthreshold concentrations. And, finally, $M_{zp} = I, p \geq 4$ is an upper-threshold self-inhibition.

Among nontrivial localizations, see full "catalog" in the work by Adamatzky and Wuensche Study [8], found in the medium we selected eaters gliders G_4 and G_{34}, mobile localizations with activator head and inhibitor tail, and eaters E_6, stationary localizations transforming gliders colliding into them, as components of the memory unit.

The eater E_6 can play the role of a six-bit flip-flop memory device. The substrate sites (bit-down) between inhibitor sites (Fig. 5.11) can be switched to an inhibitor state (bit-up) by a colliding glider. An example of writing one bit of information in E_6 is shown in Figure 5.12. Initially, E_6 stores no information. We aim to write one bit in the substrate site between the northern and northwestern inhibitor sites (Fig. 5.12a). We

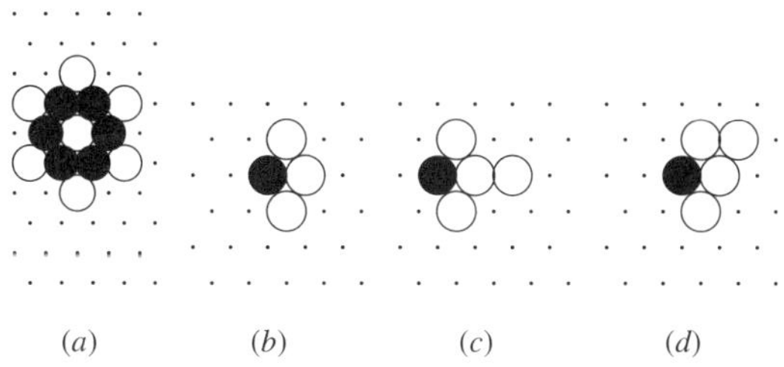

FIGURE 5.11 Localizations in reaction–diffusion hexagonal cellular automaton. Cell with inhibitor I are empty circles, and cells with activator A are black disks. (*a*) Stationary localization eater E_6, (*b*), (*c*) two forms of glider G_{34}, and (*d*) glider G_4 [8].

generate a glider G_{34} (Fig. 5.12b and c) traveling west. G_{34} collides with (or brushes past) the north edge of E_6, resulting in G_{34} being transformed to a different type of glider, G_4 (Fig. 5.12g and h). There is now a record of the collision—evidence that writing was successful. The structure of E_6 now has one site (between the northern and northwestern inhibitor sites) changed to an inhibitor state (Fig. 5.12j)—a bit was saved [8].

To read a bit from the E_6 memory device with one bit-up (Fig. 5.13a), we collide (or brush past) with glider G_{34} (Fig. 5.13b). Following the collision, the glider G_{34} is transformed into a different type of basic glider, G_{34} (Fig. 5.13g), and the bit is erased (Fig. 5.13j).

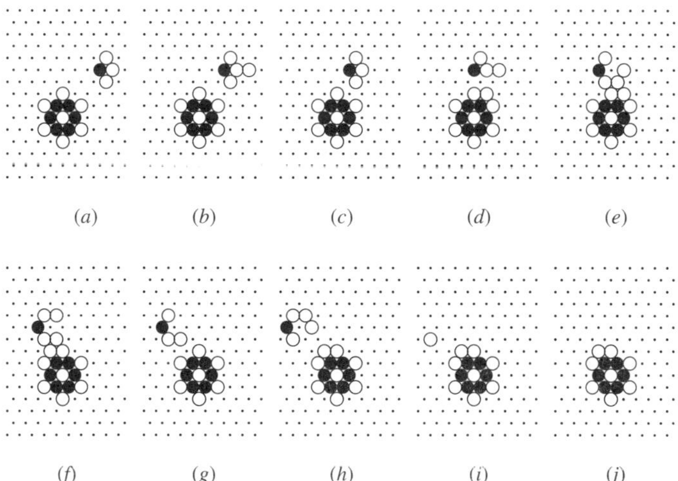

FIGURE 5.12 Write bit [8]. (*a*) t; (*b*) $t+1$; (*c*) $t+2$; (*d*) $t+3$; (*e*) $t+4$; (*f*) $t+5$; (*g*) $t+6$; (*h*) $t+7$; (*i*) $t+8$; (*j*) $t+9$.

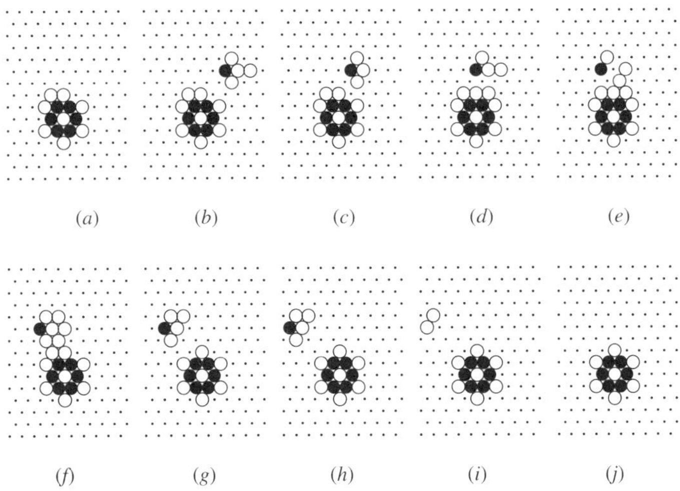

FIGURE 5.13 Read and erase bit [8]. $(a)\, t$; $(b)\, t+5$; $(c)\, t+7$; $(d)\, t+8$; $(e)\, t+9$; $(f)\, t+10$; $(g)\, t+11$; $(h)\, t+12$; $(i)\, t+13$; $(j)\, t+14$.

5.5 PROGRAMMABILITY

When developing a coherent theoretical foundation of reaction–diffusion computing in chemical media, one should pay particular attention to issues of programmability. In chemical laboratory, the term programmability means controllability.

How real chemical systems can be controlled? The majority of the literature, related to theoretical and experimental studies concerning the controllability of reaction–diffusion medium, deals with the application of an electric field. For example, in a thin-layer Belousov–Zhabotinsky reactor stimulated by an electric field the following phenomena are observed. The velocity of excitation waves is increased by a negative and decreased by a positive electric field. Very high electric field, applied across the medium, splits a wave into two waves that move in opposite directions; stabilization and destabilization of wave fronts are also observed (see [5]).

The other control parameters may include temperature (e.g., program transitions between periodic and chaotic oscillations), substrate's structure (controlling formation, annihilation, and propagation of waves), and illumination (inputting data and routing signals in light-sensitive chemical systems).

Let us demonstrate a concept of control-based programmability in models of reaction–diffusion processors. First, we show how to adjust reaction rates in chemical medium to make it perform computation of Voronoi diagram over a set of given points. Second, we show how to switch excitable system between specialized-processor and universal-processor modes (see the work by Adamatzky et al. [5] for additional examples and details).

Let a cell x of a two-dimensional lattice take four states: resting $\circ$, excited $(+)$, refractory $(-)$ and precipitate $\star$, and update their states in discrete time t depending

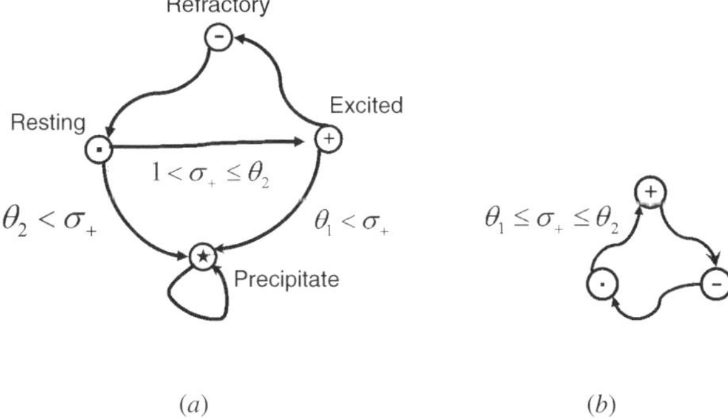

FIGURE 5.14 Cell state transition diagrams: (*a*) model of precipitating reaction–diffusion medium and (*b*) model of excitable system.

on the number $\sigma^t(x)$ of excited neighbors in its eight-cell neighborhood as follows (Fig. 5.14a):

- A resting cell x becomes excited if $0 < \sigma^t(x) \leq \theta_2$ and precipitates if $\theta_2 < \sigma^t(x)$.
- An excited cell "precipitates" if $\theta_1 < \sigma^t(x)$ or otherwise becomes refractory.
- A refractory cell recovers to the resting state unconditionally, and the precipitate cell does not change its state.

Initially, we perturb the medium, excite it in several sites, thus inputting data. Waves of excitation are generated, they grow, collide with each other, and annihilate as a result of the collision. They may form a stationary inactive concentration profile of a precipitate, which represents the result of the computation. Thus, we can only be concerned with reactions of precipitation: $+ \xrightarrow{k_1} \star$ and $\circ\boxplus + \xrightarrow{k_2} \star$, where k_1 and k_2 are inversely proportional to θ_1 and θ_2, respectively. Varying θ_1 and θ_2 from 1 to 8, and thus changing precipitation rates from the maximum possible to the minimum, we obtain various kinds of precipitate patterns, as shown in Figure 5.15.

Precipitate patterns developed for relatively high ranges of reaction rates ($3 \leq \theta_1, \theta_2 \leq 4$) represent discrete Voronoi diagrams (a given "planar" set, represented by sites of initial excitation, is visible in pattern $\theta_1 = \theta_2 = 3$ as white dots inside the Voronoi cells) derived from the set of initially excited sites (see Fig. 5.16a and b). This example demonstrates that by externally controlling precipitation rates we can force the reaction–diffusion medium to compute a Voronoi diagram.

When dealing with excitable media excitability is the key parameter for tuning spatiotemporal dynamics. We demonstrated that by varying excitability we can force the medium to exhibit almost all possible types of excitation dynamics [1].

Let each cell of 2D automaton take three states: resting ($\cdot$), exciting ($+$), and refractory ($-$), and update its state depending on number σ_+ of excited neighbors in its eight-cell neighborhood (Fig. 5.14a). A cell goes from excited to refractory and from

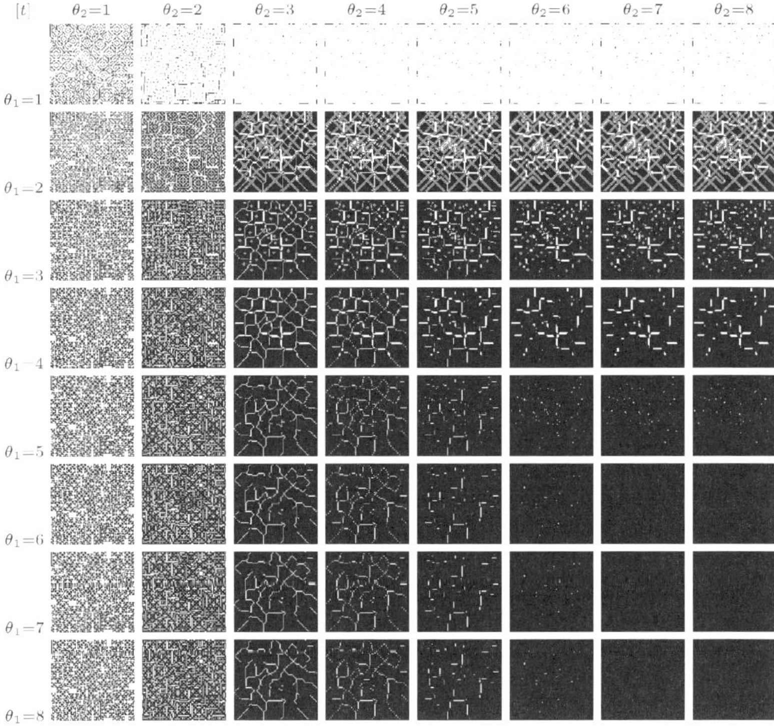

FIGURE 5.15 Final configurations of reaction–diffusion medium for $1 \leq \theta_1 \leq \theta_2 \leq 2$. Resting sites are black, precipitate is white [4].

refractory to resting states unconditionally, and resting cell excites if $\sigma_+ \in [\theta_1, \theta_2]$, $1 \leq \theta_1 \leq \theta_2 \leq 8$. By changing θ_1 and θ_2 we can move the medium dynamics in a domain of "conventional" excitation waves, useful for image processing and robot navigation [5] (Fig. 5.17a), as well as make it exhibit mobile localized excitations

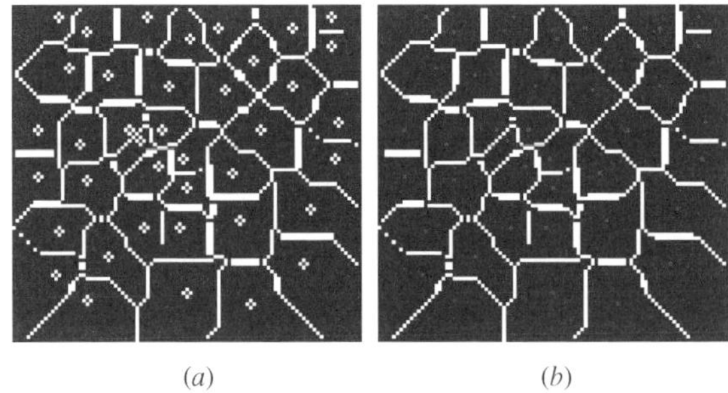

FIGURE 5.16 Exemplary configurations of reaction–diffusion medium for (*a*) $\theta_1 = 3$ and $\theta_2 = 3$, and (*b*) $\theta_1 = 4$ and $\theta_2 = 3$. Resting sites are black, precipitate is white [5].

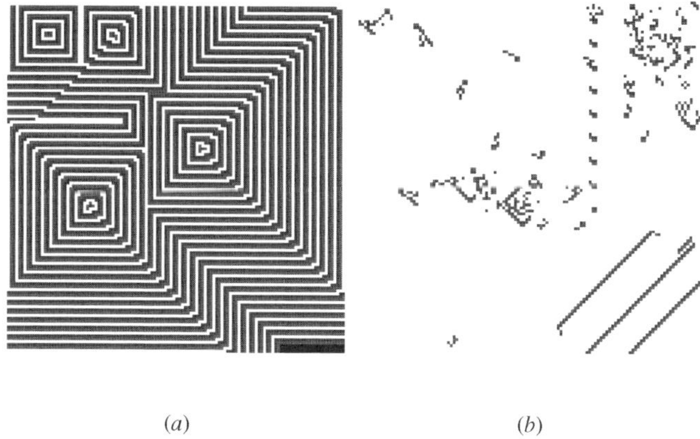

(*a*) (*b*)

FIGURE 5.17 Snapshots of space–time excitation dynamics for excitability $\sigma_+ \in [1, 8]$ (*a*) and $\sigma_+ \in [2, 2]$ (*b*).

(Fig. 5.17b), quasiparticles, and discrete analogs of dissipative solitons, employed in collision-based computing [1].

5.6 ROBOT NAVIGATION AND MASSIVE MANIPULATION

As we have seen in previous sections, reaction–diffusion chemical systems can solve complex problems and implement logical circuits. Embedded controllers for nontraditional robotics architectures would be yet another potentially huge field of application of reaction–diffusion computers. The physicochemical artifacts are well known to be capable of sensible motion. Most famous are Belousov–Zhabotinsky vesicles [24], self-propulsive chemosensitive drops [25,35], and ciliar arrays. Their motion is directional but somewhere lacks sophisticated control mechanisms.

At the present stage of reaction–diffusion computing research, it seems to be difficult to provide effective solutions for experimental prototyping of combined sensing, decision making, and actuating. However, as a proof-of-concept we can always consider hybrid "wetware + hardware" systems. For example, to fabricate a chemical controller for robot, we can place a reactor with Belousov–Zhabotinsky solution onboard of a wheeled robot and allow the robot to observer excitation wave dynamics in the reactor. When the medium is stimulated at one point, target waves are formed. The robot becomes aware of the direction toward source of stimulation from the topology of the wave fronts [2,5].

A set of remarkable experiments were undertaken by Hiroshi Yokoi and Ben De Lacy Costello. They built interface between robotic hand and Belousov–Zhabotinsky chemical reactor [57]. Excitation waves propagating in the reactor were sensed by photodiodes, which triggered finger motion. When the bending fingers touched the chemical medium with their glass nails filled with colloid silver, circular waves were triggered in the medium [5]. Starting from any initial configuration, the chemical robotic system does always reach a coherent activity mode, where fingers move in regular, somewhat melodic patterns, and few generators of target waves govern dynamics of excitation in the reactor [57].

The chemical processors for navigating wheeled robot and for controlling, and actively interacting with, a robotic hand are well discussed in our recent monograph [5]; therefore, we do not go into details in the present chapter. Instead, we concentrate on rather novel findings on coupling of reaction–diffusion system with massive parallel array of virtual actuators.

How a reaction–diffusion medium can manipulate objects? To find out we couple a simulated abstract parallel manipulator with an experimental Belousov–Zhabotinsky (BZ) chemical medium, so the excitation dynamics in the chemical system are reflected in changing the OFF–ON mode of elementary actuating units. In this case, we convert experimental snapshots of the spatially distributed chemical system to a force vector field and then simulate the motion of manipulated objects in the force field, thus achieving reaction–diffusion medium controlled actuation. To build an interface between the recordings of space–time snapshots of the excitation dynamics in BZ medium and simulated physical objects, we calculate force fields generated by mobile excitation patterns and then simulate the behavior of an object in this force field.

Chemical medium to perform actuation is prepared following the typical receipt[1] (see the works by Adamatzky et al. [6] and Field and Winfee [18]), based on a ferroin-catalyzed BZ reaction. A silica gel plate is cut and soaked in a ferroin solution. The gel sheet is placed in a Petri dish and BZ solution is added. Dynamics of the chemical system is recorded at 30-s intervals using a digital camera.

The cross-section profile of the BZ wave front recorded on a digital snapshot shows a steep rise of red color values in the pixels at the wave front's head and a gradual descent in the pixels along the wave front's tail. Assuming that excitation waves push the object, local force vectors generated at each site—pixel of the digitized image—of the medium should be oriented along local gradients of the red color values. From the digitized snapshot of the BZ medium we extract an array of red components from the snapshot's pixels and then calculate the projection of a virtual vector force at the pixel. Force fields generated by the excitation patterns in a BZ system (Fig. 5.18) result in tangential forces being applied to a manipulated object, thus causing translational and rotational motions of the object [6].

Nonlinear medium controlled actuators can be used for sorting and manipulating both small objects, comparable in size to the elementary actuating unit, and larger objects, with lengths of tens or hundreds of actuating units. Therefore, we demonstrate here two types of experiments with BZ-based manipulation of pixel-sized objects and of planar convex shapes.

Pixel objects, due to their small size, are subjected to random forces, caused by impurities of the physical medium and imprecision of the actuating units. In this case, no averaging of forces is allowed and the pixel objects themselves sensitively react to a single force vector. Therefore, we adopt the following model of manipulating a

[1] Chemical laboratory experiments are undertaken by Dr. Ben De Lacy Costello (UWE, Bristol, UK).

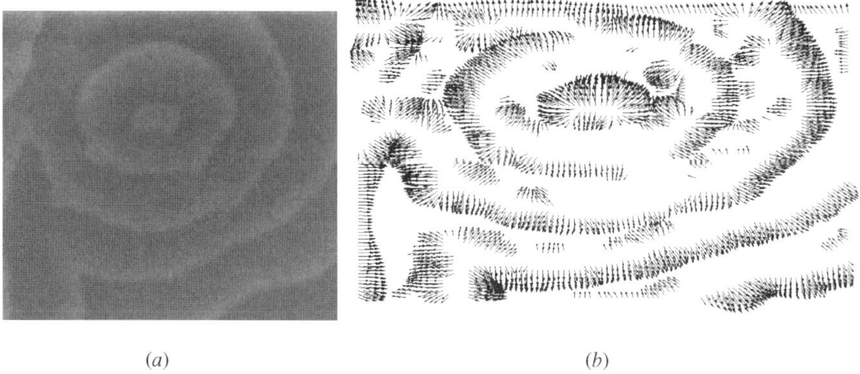

(*a*) (*b*)

FIGURE 5.18 Force vector field (*b*) calculated from BZ medium's image (*a*) [6].

pixel object: if all force vectors at the eight-pixel neighborhood of the current site of the pixel object are nil, then the pixel object jumps to a randomly chosen neighboring pixel of its neighborhood, otherwise the pixel object is translated by the maximum force vector in its neighborhood.

When placed on the simulated manipulating surface, pixel objects move at random in the domains of the resting medium; however, by randomly drifting each pixel object does eventually encounter a domain of coaligned vectors (representing excitation wave front in BZ medium) and is translated along the vectors. An example of several pixel objects transported on a "frozen" snapshot of the chemical medium is shown in Figure 5.19. Trajectories of pixel objects (Fig. 5.19a) show distinctive intermittent modes of random motion separated by modes of directed "jumps" guided by traveling wave fronts. Smoothed trajectories of pixel objects (Fig. 5.19b) demonstrate that despite a very strong chaotic component in manipulation, pixel objects are transported to the sites of the medium where two or more excitation wave fronts meet.

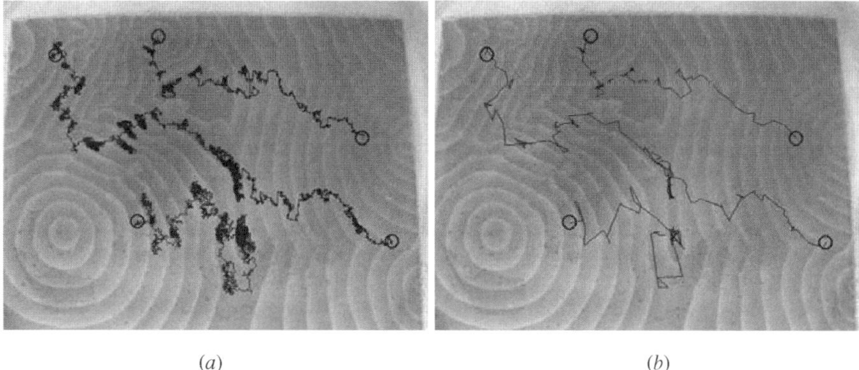

(*a*) (*b*)

FIGURE 5.19 Examples of manipulating five pixel objects using the BZ medium: (*a*) trajectories of pixel objects, (*b*) jump trajectories of pixel objects recorded every 100th time step. Initial positions of the pixel objects are shown by circles [6].

The overall speed of pixel object transportation depends on the frequency of wave generations by sources of target waves. As a rule, the higher the frequency, the faster the objects are transported. This is because in parts of the medium spanned by low frequency target waves there are lengthy domains of resting system, where no force vectors are formed. Therefore, pixel-sized object can wander randomly for a long time till climbing next wave front [6].

To calculate the contribution of each force we partitioned the object into fragments, using a square grid, in which each cell of the grid corresponds to one pixel of the image. We assume that the magnitude of the force applied to each fragment above given pixel is proportional to the area of the fragment and is codirectional with a force vector. A momentum of inertia of the whole object with respect to axis normal to the object and passing through the object's center of mass is calculated from the position of the center of mass and the mass of every fragment. Since the object's shape and size are constant, it is enough to calculate the moment of inertia only at the beginning of simulation. We are also taking into account principal rotational momentum created by forces and angular acceleration of the object around its center of mass. Therefore, object motion in our case can be sufficiently described by coordinates of its center of mass and its rotation at every moment of time [6].

Spatially extended objects follow the general pattern of motion observed for the pixel-sized objects. However, due to integration of many force vectors the motion of planar objects is smoother and less sensitive to the orientation of any particular force vector.

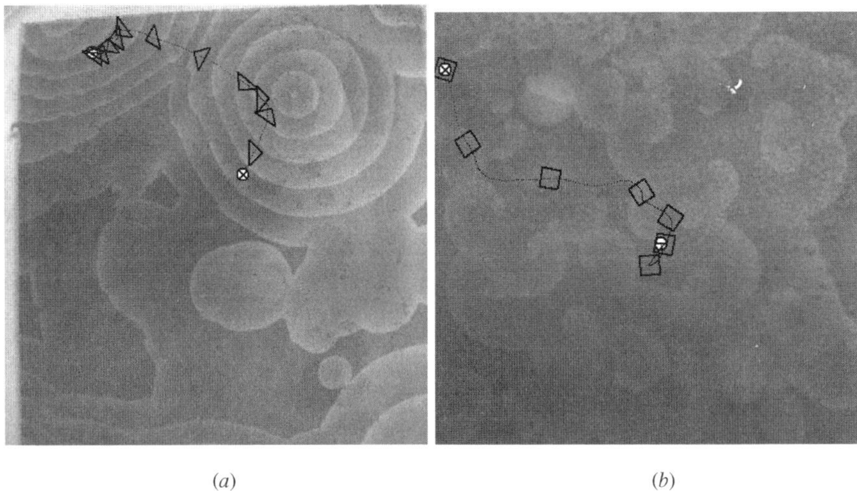

FIGURE 5.20 Manipulating planar object in BZ medium. (*a*) Right-angled triangle moved by fronts of target waves. (*b*) Square object moved by fronts of fragmented waves in subexcitable BZ medium. Trajectories of center of mass of the square are shown by the dotted line. Exact orientation of the objects is displayed every 20 steps. Initial position of the object is shown by $\ominus$ and the final position by $\otimes$ [6].

Outcome of manipulation depends on the size of the object; with increasing size of the object—due to larger numbers of local vector forces acting on the object—the objects become more controllable by the excitation wave fronts (Fig. 5.20).

5.7 SUMMARY

The field of reaction–diffusion computing started 20 years ago [27] as a subfield of physics and chemistry dealing with image processing operations in uniform thin-layer excitable chemical media. The basic idea was to apply input data as two-dimensional profile of heterogeneous illumination, then allow excitation waves spread and interact with each other, and then optically record result of the computation. The first even reaction–diffusion computers were already massively parallel, with parallel optical inputs and outputs. Later computer engineers entered the field and started to exploit traditional techniques—wires were implemented by channels where wave pulses travel, and specifically shaped junctions acted as logical valves. In this manner, most "famous" chemical computing devices were implemented, including Boolean gates, coincidence detectors, memory units, and more. The upmost idea of reaction–diffusion computation was if not ruined then forced into cul-de-sac of nonclassical computation. The breakthrough happened when paradigms and solutions from the field of dynamical, collision-based computing and conservative logic were mapped onto realms of spatially extended chemical systems. The computers became uniform and homogeneous.

In several examples we demonstrated that reaction–diffusion chemical systems are capable of solving combinatorial problems with natural parallelism. In spatially distributed chemical processors, the data and the results of the computation are encoded as concentration profiles of the chemical species. The computation per se is performed via the spreading and interaction of wave fronts.

The reaction–diffusion computers are parallel because the chemical medium's microvolumes update their states simultaneously, and molecules diffuse and react in parallel. During the last decades, a wide range of experimental prototypes of reaction–diffusion computing devices have been fabricated and applied to solve various problems of computer science, including image processing, pattern recognition, path planning, robot navigation, computational geometry, logical gates in spatially distributed chemical media, and arithmetical and memory units.

These important, but scattered across many scientific fields, results convince us that reaction–diffusion systems can do a lot. Are they capable enough to be intelligent? Yes, reaction–diffusion systems are smart—showing a state of readiness to respond, able to cope with difficult situations, capable for determining something by mathematical and logical methods—and endowed with capacity to reason. Reaction–diffusion computers allow for massive parallel input of data. Equivalently, reaction–diffusion robots would need no dedicated sensors, each microvolume of the medium, each site of the matrix gel, is sensitive to changes in one or another physical characteristic of the environment. Electric field, temperature, and illumination are "sensed" by reaction–diffusion devices, and these are three principal parameters in controlling and programming reaction–diffusion robots.

Hard computational problems of geometry, image processing, and optimization on graphs are resource efficiently solved in reaction–diffusion media due to intrinsic natural parallelism of the problems [1]. In this chapter we demonstrated efficiency of reaction–diffusion computers on example of construction of Voronoi diagram. The Voronoi diagram is a subdivision of plane by data planar set. Each point of the data set is represented by a drop of a reagent. The reagent diffuses and produces a color precipitate when reacting with the substrate. When two or more diffusive fronts of the "data" chemical species meet, no precipitate is produced (due to concentration-dependent inhibition). Thus, uncolored domains of the computing medium represent bisectors of the Voronoi diagram. The precipitating chemical processor can also compute a skeleton. The skeleton of a planar shape is computed in the similar manner. A contour of the shape is applied to computing substrate as a disturbance in reagent concentrations. The contour concentration profile induces diffusive waves. A reagent diffusing from the data contour reacts with the substrate and the precipitate is formed. Precipitate is not produced at the sites of diffusive waves' collision. The uncolored domains correspond to the skeleton of the data shape. To compute a collision-free shortest path in a space with obstacles, we can couple two reaction–diffusion media. Obstacles are represented by local disturbances of concentration profiles in one of the media. The disturbances induce circular waves traveling in the medium and approximating a scalar distance-to-obstacle field. This field is mapped onto the second medium, which calculates a tree of "many-sources-one-destination" shortest paths by spreading wave fronts [5].

There is still no rigorous theory of reaction–diffusion computing, and God knows if one will ever be developed; however, algorithms are intuitively convincing and range of applications is wide, and after all the whole field of nature-inspired computing is built on interpretations:

> Of course, this is only a plausible consideration and not a mathematical proof, since the question still remains whether the mathematical interpretation of the physical event is adequate in a strict sense, or whether it gives only an adequate image of physical reality. Sometimes such experiments, even if performed only in imagination, are convincing even to mathematicians [16].

5.8 ACKNOWLEDGEMENTS

Many thanks to Ben De Lacy Costello, who implemented chemical laboratory prototypes of reaction–diffusion computers discussed in the chapter. I am grateful to Andy Wuensche (hexagonal cellular automatons), Hiroshi Yokoi (robotic hand controlled by Belousov–Zhabotinsky reaction), Chris Melhuish (control of robot navigation), Sergey Skachek (massive parallel manipulation), Tetsuya Asai (LSI prototypes of reaction–diffusion computers) and Genaro Martinez (binary-state cellular automatons) for their cooperation. Some pictures, where indicated, where adopted from our coauthored publications. Special thanks to Ikuko Motoike for correcting the original version of the chapter.

REFERENCES

1. Adamatzky A. Computing in Nonlinear Media and Automata Collectives. Institute of Physics Publishing; 2001.
2. Adamatzky A, De Lacy Costello BPJ. Experimental logical gates in a reaction–diffusion medium: the XOR gate and beyond. Phys Rev E 2002;66:046112.
3. Adamatzky A, editor. Collision Based Computing. Springer; 2003.
4. Adamatzky A. Programming reaction–diffusion computers. In: Unconventional Programming Paradigms. Springer; 2005.
5. Adamatzky A, De Lacy Costello B, Asai T. Reaction-Diffusion Computers. Elsevier; 2005.
6. Adamatzky A, De Lacy Costello B, Skachek S, Melhuish C. Manipulating objects with chemical waves: open loop case of experimental Belousov–Zhabotinsky medium. Phys Lett A 2005.
7. Adamatzky A, Wuensche A, De Lacy Costello B. Glider-based computation in reaction–diffusion hexagonal cellular automata. Chaos, Solitons Fract 2006;27:287–295.
8. Adamatzky A, Wuensche A. Computing in 'spiral rule' reaction–diffusion hexagonal cellular automaton. Complex Syst. 2007;16:1–27.
9. Agladze K, Magome N, Aliev R, Yamaguchi T, Yoshikawa K. Finding the optimal path with the aid of chemical wave. Physica D 1997;106:247–254.
10. Asai T, De Lacy Costello B, Adamatzky A. Silicon implementation of a chemical reaction-diffusion processor for computation of Voronoi diagram. Int J Bifurcation Chaos 2005;15(1).
11. Asai T, Kanazawa Y, Hirose T, Amemiya Y. Analog reaction–diffusion chip imitating Belousov–Zhabotinsky reaction with hardware oregonator model. Int J Unconven Comput 2005;1:123–147.
12. Oya T, Asai T, Fukui T, Amemiya Y. Reaction–diffusion systems consisting of single-electron oscillators. Int J Unconvent Comput 2005;1:179–196.
13. Beato V, Engel H. Pulse propagation in a model for the photosensitive Belousov–Zhabotinsky reaction with external noise. In: Schimansky-Geier L, Abbott D, Neiman A, Van den Broeck C, editors. Noise in Complex Systems and Stochastic Dynamics. Proc SPIE 2003;5114:353–62.
14. Berlekamp ER, Conway JH, Guy RL. Winning Ways for Your Mathematical Plays. Volume 2. Academic Press; 1982.
15. Brandtstädter H, Braune M, Schebesch I, Engel H. Experimental study of the dynamics of spiral pairs in light-sensitive Belousov–Zhabotinskii media using an open-gel reactor. Chem Phys Lett 2000;323:145–154.
16. Courant R, Robbins H. What is Mathematics? Oxford University Press; 1941.
17. Dupont C, Agladze K, Krinsky V. Excitable medium with left–right symmetry breaking. Physica A 1998;249:47–52.
18. Field R, Winfree AT. Travelling waves of chemical activity in the Zaikin–Zhabotinsky–Winfree reagent. J Chem Educ 1979; 56:754.

19. Field RJ, Noyes RM. Oscillations in chemical systems. IV. Limit cycle behavior in a model of a real chemical reaction. J Chem Phys 1974;60:1877–1884.
20. Fredkin F, Toffoli T. Conservative logic. Int J Theor Phys 1982;21:219–253.
21. Gerhardt M, Schuster H, Tyson JJ. A cellular excitable media. Physica D 1990;46:392–415.
22. Grill S, Zykov VS, Müller SC. Spiral wave dynamics under pulsatory modulation of excitability. J Phys Chem 1996;100:19082–19088.
23. Hartman H, Tamayo P. Reversible cellular automata and chemical turbulence. Physica D 1990;45:293–306.
24. KItahata H, Aihara R, Magome N, Yoshikawa K. Convective and periodic motion driven by a chemical wave. J Chem Phys 2002;116:5666.
25. Kitahata H, Yoshikawa K. Chemo-mechanical energy transduction through interfacial instability. Physica D 2005;205:283–291.
26. Klein R. Concrete and abstract Voronoi diagrams. Berlin: Springer-Verlag; 1990.
27. Kuhnert L. A new photochemical memory device in a light sensitive active medium. Nature 1986;319:393.
28. Kuhnert L, Agladze KL, Krinsky VI. Image processing using light-sensitive chemical waves. Nature 1989;337:244–247.
29. Kusumi T, Yamaguchi T, Aliev R, Amemiya T, Ohmori T, Hashimoto H, Yoshikawa K. Numerical study on time delay for chemical wave transmission via an inactive gap. Chem Phys Lett 1997;271:355–360.
30. Margolus N. Physics-like models of computation. Physica D 1984;10:81–95.
31. Markus M, Hess B. Isotropic cellular automata for modelling excitable media. Nature 1990;347:56–58.
32. Mills J. The new computer science and its unifying principle: complementarity and unconventional computing. Position Papers. International Workshop on the Grand Challenge in Nonclassical Computation; New York; 2005 Apr 18–19.
33. Motoike IN, Yoshikawa K, Iguchi Y, Nakata S. Real-time memory on an excitable field. Phys Rev E 2001;63:036220.
34. Motoike IN, Yoshikawa K. Information operations with multiple pulses on an excitable field. Chaos Solitons Fract 2003;17:455–461.
35. Nagai K, Sumino Y, Kitahata H, Yoshikawa K. Mode selection in the spontaneous motion of an alcohol droplets. Phys Rev E 2005;71:065301.
36. Petrov V, Ouyang Q, Swinney HL. Resonant pattern formation in a chemical system. Nature 1997;388:655–657.
37. Pour–El MB. Abstract computability and its relation to the general purpose analog computer (some connections between logic, differential equations and analog computers). Trans Am Math Soc 1974;199:1–28.
38. Qian H, Murray JD. A simple method of parameter space determination for diffusion-driven instability with three species. Appl Math Lett 2001;14:405–411.
39. Rambidi NG. Neural network devices based on reaction–diffusion media: an approach to artificial retina. Supramol Sci 1998;5:765–767.
40. Rambidi NG, Shamayaev KR, Peshkov GY. Image processing using light-sensitive chemical waves. Phys Lett A 2002;298:375–382.
41. Saltenis V. Simulation of wet film evolution and the Euclidean Steiner problem. Informatica 1999;10:457–466.

42. Sielewiesiuk J, Gorecki J. Logical functions of a cross junction of excitable chemical media. J Phys Chem A 2001;105:8189–8195.
43. Schenk CP, Or-Guil M, Bode M, Purwins HG. Interacting pulses in three-component reaction–diffusion systems on two-dimensional domains. Phys Rev Lett 1997;78:3781–3784.
44. Sedina-Nadal I, Mihaliuk E, Wang J, Pérez-Munuzuri V, Showalter K. Wave propagation in subexcitable media with periodically modulated excitability. Phys Rev Lett 2001;86:1646–1649.
45. Sienko T, Adamatzky A, Rambidi N, Conrad M, editors. Molecular Computing. The MIT Press; 2003.
46. Steinbock O, Toth A, Showalter K. Navigating complex labyrinths: optimal paths from chemical waves. Science 1995;267:868–871.
47. Schebesch I, Engel H. Wave propagation in heterogeneous excitable media. Phys Rev E 1998;57:3905–3910.
48. Tóth A, Showalter K. Logic gates in excitable media. J Chem Phys 1995;103:2058–2066.
49. Tyson JJ, Fife PC. Target patterns in a realistic model of the Belousov–Zhabotinskii reaction. J Chem Phys 1980;73:2224–2237.
50. Vergis A, Steiglitz K, Dickinson B. The complexity of analog computation. Math Comput Simulat 1986;28:91–113.
51. Wang J. Light-induced pattern formation in the excitable Belousov–Zhabotinsky medium. Chem Phys Lett 2001;339:357–361.
52. Weaire D, Hutzler S, Cox S, Kern N, Alonso MD Drenckhan W. The fluid dynamics of foams. J Phys: Condens Matter 2003;15:S65–S73.
53. Wuensche A, Adamatzky A. On spiral glider-guns in hexagonal cellular automata: activator-inhibitor paradigm. Int J Modern Phys C 2006;17.
54. Yaguma S, Odagiri K, Takatsuka K. Coupled-cellular-automata study on stochastic and pattern-formation dynamics under spatiotemporal fluctuation of temperature. Physica D 2004;197:34–62.
55. Yang X. Computational modelling of nonlinear calcium waves. Appl Math Model 2006;30:200–208.
56. Yang X. Pattern formation in enzyme inhibition and cooperativity with parallel cellular automata. Parallel Comput 2004;30:741–751.
57. Yokoi H, Adamatzky A, De Lacy Costello B, Melhuish C. Excitable chemical medium controlled for a robotic hand: closed loop experiments. Int J Bifurcation Chaos 2004.
58. Young D. A local activator–inhibitor model of vertebrate skin patterns. Math Biosci 1984;72:51.
59. Yoneyama M. Optical modification of wave dynamics in a surface layer of the Mn-catalyzed Belousov–Zhabotinsky reaction. Chem Phys Lett 1996;254:191–196.

CHAPTER 6

Data Mining Algorithms I: Clustering

DAN A. SIMOVICI

6.1 INTRODUCTION

Activities of contemporary society generate enormous amounts of data that are used in decision support processes. Many databases have current volumes in the hundreds of terabytes. An academic estimate [4] puts the volume of data created in 2002 alone at 5 hexabytes (the equivalent of 5 million terabytes). The difficulty of analyzing these kinds of data volumes by human operators is clearly insurmountable. This lead to a rather new area of computer science, data mining, whose aim is to develop automatic means of data analysis for discovering new and useful patterns embedded in data.

Data mining builds on several disciplines, statistics, artificial intelligence, databases, visualization techniques, and others, and has crystallized as a distinct discipline in the last decade of the past century.

The range of subjects in data mining is very broad. Among the main directions of this branch of computer science, one should mention identification of associations between data items, clustering, classification, summarization, outlier detection, and so on. The diversity of these preoccupations makes impossible an exhaustive presentation of data mining algorithms in a very limited space. In this chapter, we concentrate on clustering algorithms. This choice will allow us a presentation that is as self-contained as possible and gives a quite accurate image of the challenges posed by data mining.

6.2 CLUSTERING ALGORITHMS

Clustering is the process of grouping together objects that are similar. The groups formed by clustering are referred to as *clusters*. Similarity between objects that belong to a set S is usually measured using a dissimilarity $d : S \times S \longrightarrow \mathbb{R}_{\geq 0}$ that is definite (see Section 6.3), this means that $d(x, y) = 0$ if and only if $x = y$ and $d(x, y) = d(y, x)$

Handbook of Applied Algorithms: Solving Scientific, Engineering and Practical Problems
Edited by Amiya Nayak and Ivan Stojmenović

for every $x, y \in S$. Two objects x, y are similar if the value of $d(x, y)$ is small; what "small" means depends on the context of the problem.

Clustering can be regarded as a special type of classification, where the clusters serve as classes of objects. It is a widely used data mining activity with multiple applications in a variety of scientific activities ranging from biology and astronomy to economics and sociology.

There are several points of view for examining clustering techniques. We follow here the taxonomy of clustering presented in the work by Jain et al. [5].

Clustering may or may not be *exclusive*, where an exclusive clustering technique yields clusters that are disjoint, while a nonexclusive technique produces overlapping clusters. From an algebraic point of view, an exclusive clustering generates a partition of the set of objects, and most clustering algorithms fit in this category.

Clustering may be *intrinsic* or *extrinsic*. Intrinsic clustering is an unsupervised activity that is based only on the dissimilarities between the objects to be clustered. Most clustering algorithms fall into this category. Extrinsic clustering relies on information provided by an external source that prescribes, for example, which objects should be clustered together and which should not.

Finally, clustering may be *hierarchical* or *partitional*.

In hierarchical clustering algorithms, a sequence of partitions is constructed. In *hierarchical agglomerative algorithms*, this sequence is increasing and it begins with the least partition of the set of objects whose blocks consist of single objects; as the clustering progresses, certain clusters are fused together. As a result, an agglomerative clustering is a chain of partitions on the set of objects that begins with the least partition α_S of the set of objects S and ends with the largest partition ω_S. In a *hierarchical divisive algorithm*, the sequence of partitions is decreasing. Its first member is the one-block partition ω_S and each partition is built by subdividing the blocks of the previous partition.

A partitional clustering creates a partition of the set of objects whose blocks are the clusters such that objects in a cluster are more similar to each other than to objects that belong to different clusters. A typical representative algorithm is the k-means algorithm and its many extensions.

Our presentation is organized around the last dichotomy. We start with a class of hierarchical agglomerative algorithms. This is continued with a discussion of the k-means algorithm, a representative of partitional algorithms. Then, we continue with a discussion of certain limitations of clustering centered around Kleinberg's impossibility theorem. We conclude with an evaluation of clustering quality.

6.3 BASIC NOTIONS: PARTITIONS AND DISSIMILARITIES

Definition 1 Let S be a nonempty set. A partition of S is a nonempty collection of nonempty subsets of S, $\pi = \{B_i | i \in I\}$ such that $i \neq j$ implies $B_i \cap B_j = \emptyset$ and $\bigcup\{B_i | i \in I\} = S$.

The members of the collection π are the blocks of the partition π. The collection of partitions of a set S is denoted by $\mathrm{PART}(S)$.

Example 1 Let $S = \{a, b, c, d, e\}$ be a set. The following collections of subsets of S are partitions of S:

$$\pi_0 = \{\{a\}, \{b\}, \{c\}, \{d\}, \{e\}\},$$

$$\pi_1 = \{\{a, b\}, \{c\}, \{d, e\}\},$$

$$\pi_2 = \{\{a, c\}, \{b\}, \{d, e\}\},$$

$$\pi_3 = \{\{a, b, c\}\{d, e\}\},$$

$$\pi_4 = \{\{a, b, c, d, e\}\}.$$

□

A partial order relation can be defined on PART(S) by taking $\pi \leq \sigma$ if every block of π is included in some block of σ. It is easy to see that for the partitions defined in Example 1, we have $\pi_0 \leq \pi_1 \leq \pi_3 \leq \pi_4$ and $\pi_0 \leq \pi_2 \leq \pi_3 \leq \pi_4$; however, we have neither $\pi_1 \leq \pi_2$ nor $\pi_2 \leq \pi_1$.

The partially ordered set (PART(S), $\leq$) has as its least element the partition whose blocks are singletons of the form $\{x\}$,

$$\alpha_S = \{\{x\}|x \in S\},$$

and as its largest element the one-block partition $\omega_S = \{S\}$. For the partitions defined in Example 1 we have $\pi_0 = \alpha_S$ and $\pi_4 = \omega_S$.

We refer the reader to the work by Birkhoff [1] for a detailed discussion of the properties of this partial ordered set.

To obtain a quantitative expression of the differences that exist between objects we use the notion of dissimilarity.

Definition 2 A dissimilarity on a set S is a function $d : S^2 \longrightarrow \mathbb{R}_{\geq 0}$ satisfying the following conditions:

(i) $d(x, x) = 0$ for all $x \in S$;
(ii) $d(x, y) = d(y, x)$ for all $x, y \in S$.

The pair (S, d) is a dissimilarity space.

The set of dissimilarities defined on a set S is denoted by $\mathcal{D}_S$.

The notion of dissimilarity can be strengthened in several ways by imposing certain supplementary conditions. A nonexhaustive list of these conditions is given next.

1. $d(x, y) = 0$ implies $d(x, z) = d(y, z)$ for every $x, y, z \in S$ (evenness);
2. $d(x, y) = 0$ implies $x = y$ for every x, y (definiteness);
3. $d(x, y) \leq d(x, z) + d(z, y)$ for every x, y, z (triangular inequality);
4. $d(x, y) \leq \max\{d(x, z), d(z, y)\}$ for every x, y, z (the ultrametric inequality).

The set of definite dissimilarities on a set S is denoted by $\mathcal{D}'_S$.

Example 2 Consider the mapping $d : (\mathbf{Seq}_n(S))^2 \longrightarrow \mathbb{R}_{\geq 0}$ defined by

$$d(\boldsymbol{p}, \boldsymbol{q}) = |\{i | 0 \leq i \leq n-1 \text{ and } \boldsymbol{p}(i) \neq \boldsymbol{q}(i)\}|,$$

for every sequences $\boldsymbol{p}, \boldsymbol{q}$ of length n on the set S.

Clearly, d is a dissimilarity that is both even and definite. Moreover, it satisfies the triangular inequality. Indeed, let $\boldsymbol{p}, \boldsymbol{q}, \boldsymbol{r}$ be three sequences of length n on the set S. If $\boldsymbol{p}(i) \neq \boldsymbol{q}(i)$, then $\boldsymbol{r}(i)$ must be distinct from at least one of $\boldsymbol{p}(i)$ and $\boldsymbol{q}(i)$. Therefore,

$$\begin{aligned}&\{i | 0 \leq i \leq n-1 \text{ and } \boldsymbol{p}(i) \neq \boldsymbol{q}(i)\}\\ &\subseteq \{i | 0 \leq i \leq n-1 \text{ and } \boldsymbol{p}(i) \neq \boldsymbol{r}(i)\} \cup \{i | 0 \leq i \leq n-1 \text{ and } \boldsymbol{r}(i) \neq \boldsymbol{q}(i)\},\end{aligned}$$

which implies the triangular inequality. □

The ultrametric inequality implies the triangular inequality; both the triangular inequality and definiteness imply evenness (see Exercise 10).

Definition 3 A dissimilarity $d \in \mathcal{D}_S$ is

1. a metric, if it satisfies the definiteness property and the triangular inequality;
2. an ultrametric, if it satisfies the definiteness property and the ultrametric inequality.

The set of metrics and the set of ultrametrics on a set S are denoted by $\mathcal{M}_S$ and $\mathcal{U}_S$, respectively.

If d is a metric or an ultrametric on a set S, then (S, d) is a metric space or an ultrametric space, respectively.

Definition 4 The diameter of a finite metric space (S, d) is the number $\text{diam}_{S,d} = \max\{d(x, y) | x, y \in S\}$.

Exercise 10 implies that $\mathcal{U}_S \subseteq \mathcal{M}_S \subseteq \mathcal{D}_S$.

Example 3 Let $\mathcal{G} = (V, E)$ be a connected graph. Define the mapping $d : V^2 \longrightarrow \mathbb{R}_{\geq 0}$ by $d(x, y) = m$, where m is the length of the shortest path that connects x and y. Then, d is a metric.

Indeed, we have $d(x, y) = 0$ if and only if $x = y$. The symmetry of d is obvious.

If $\boldsymbol{p}$ is a shortest path that connects x to z and $\boldsymbol{q}$ is a shortest path that connects z to y, then $\boldsymbol{pq}$ is a path of length $d(x, z) + d(z, y)$ that connects x to y. Therefore, $d(x, y) \leq d(x, z) + d(z, y)$. □

In this chapter, we shall use frequently the notion of sphere in a metric space.

Definition 5 Let (S, d) be a metric space. The closed sphere centered in $x \in S$ of radius r is the set

$$B_d(x, r) = \{y \in S | d(x, y) \leq r\}.$$

The open sphere centered in $x \in S$ of radius r is the set

$$C_d(x, r) = \{y \in S | d(x, y) < r\}.$$

Let d be a dissimilarity and let $S(x, y)$ be the set of all nonnull sequences $\mathbf{s} = (s_1, \ldots, s_n) \in \mathbf{Seq}(S)$ such that $s_1 = x$ and $s_n = y$. The d-amplitude of $\mathbf{s}$ is the number $\text{amp}_d(\mathbf{s}) = \max\{d(s_i, s_{i+1}) | 1 \leq i \leq n - 1\}$.

If d is a ultrametric we have $d(x, y) \leq \min\{\text{amp}_d(\mathbf{s}) | \mathbf{s} \in S(x, y)\}$ (Exercise 1).

Dissimilarities defined on finite sets can be represented by matrices. If $S = \{x_1, \ldots, x_n\}$ is a finite set and $d : S \times S \longrightarrow \mathbb{R}_{\geq 0}$ is a dissimilarity, let $D_d \in (\mathbb{R}_{\geq 0})^{n \times n}$ be the matrix defined by $(D_d)_{ij} = d(x_i, x_j)$ for $1 \leq i, j \leq n$. Clearly, all main diagonal elements of D_d are 0 and the matrix D is symmetric.

6.4 ULTRAMETRIC SPACES

Ultrametrics represent a strengthening of the notion of metric, where the triangular inequality is replaced by the stronger ultrametric inequality. They play an important role in studying hierarchical clustering algorithm, which we discuss in Section 6.5.

A simple, interesting property of triangles in ultrametric spaces is given next.

Theorem 1 *Let (S, d) be an ultrametric space. For every $x, y, z \in S$, two of the numbers $d(x, y), d(x, z), d(y, z)$ are equal and the third is not larger than the other two equal numbers.*

Proof. Let $d(x, y)$ be the least of the numbers $d(x, y), d(x, z), d(y, z)$. We have $d(x, z) \leq \max\{d(x, y), d(y, z\} = d(y, z)$ and $d(y, z) \leq \max\{d(x, y), d(x, z)\} = d(x, z)$. Therefore, $d(y, z) = d(x, z)$ and $d(x, y)$ is not larger than the other two. ■

Theorem 1 can be paraphrased by saying that in an ultrametric space any triangle is isosceles and the side that is not equal to the other two cannot be longer than these.

In an ultrametric space, a closed sphere has all its points as centers.

Theorem 2 *Let $B(x, r)$ be a closed sphere in the ultrametric space (S, d). If $z \in B(x, d)$, then $B(x, r) = B(z, r)$. Moreover, if two closed spheres $B(x, r)$, $B(y, r')$ space have a point in common, they one of the closed spheres is included in the other.*

Proof. See Exercise 7. ■

Theorem 2 implies $S = B(x, \text{diam}_{S,d})$ for any point $x \in S$.

6.4.1 Construction of Ultrametrics

There is a strong link between ultrametrics defined on a finite set S and chains of equivalence relations on S (or chains of partitions on S). This is shown in the next statement.

Theorem 3 *Let S be a finite set and let $d : S \times S \longrightarrow \mathbb{R}_{\geq 0}$ be a function whose range is $\text{Ran}(d) = \{r_1, \ldots, r_m\}$, where $r_1 = 0$ such that $d(x, y) = 0$ if and only if $x = y$. For $u \in S$ and $r \in \mathbb{R}_{\geq 0}$ define the set $D_{u,r} = \{x \in S | d(u, x) \leq r\}$. Define the collection of sets $\pi_{r_i} = \{D(u, r_i) | u \in S\}$ for $1 \leq i \leq m$.*

The function d is an ultrametric on S if and only if the sequence of collections $\pi_{r_1}, \ldots, \pi_{r_m}$ is an increasing sequence of partitions on S such that $\pi_{r_1} = \alpha_S$ and $\pi_{r_m} = \omega_S$.

Proof. Suppose that d is an ultrametric on S. Then, the sets of the form $D(x, r)$ are precisely the closed spheres $B(x, r)$. Since $x \in B(x, r)$ for $x \in S$, it follows that none of these sets is empty and that $\bigcup_{x \in S} B(x, r) = S$. Any two distinct spheres $B(x, r)$, $B(y, r)$ are disjoint by Theorem 2.

It is straightforward to see that $\pi^{r_1} \preceq \pi^{r_2} \preceq \cdots \preceq \pi^{r_m}$; that is, this sequence of relations is indeed a chain of equivalences.

Conversely, suppose that $\pi^{r_1}, \ldots, \pi^{r_m}$ is an increasing sequence of partitions on S such that $\pi^{r_1} = \alpha_S$ and $\pi^{r_m} = \omega_S$, where π^{r_i} consists of the sets of the form D_{u,r_i} for $u \in S$.

Since $D_{x,0} = \{x\}$, it follows that $d(x, y) = 0$ if and only if $x = y$.

We claim that

$$d(x, y) = \min\{r | \{x, y\} \subseteq B \in \pi^r\}. \tag{6.1}$$

Indeed, since $\pi^{r_m} = \omega_S$, it is clear that there is a partition π^{r_i} such that $\{x, y\} \subseteq B \in \pi^{r_i}$. If x and y belong to the same block of π^{r_i}, the definition of π^{r_i} implies $d(x, y) \leq r_i$, so $d(x, y) \leq \min\{r | \{x, y\} \subseteq B \in \pi^r\}$. This inequality can be easily seen to become an equality since $x, y \subseteq B \in \pi^{d(x,y)}$. This implies immediately that d is symmetric.

To prove that d satisfies the ultrametric inequality, let x, y, z be three members of the set S. Let $p = \max\{d(x, z), d(z, y)\}$. Since $\{x, z\} \subseteq b \in \pi^{d(x,z)} \leq \pi^p$ and $\{z, y\} \subseteq B' \in \pi^{d(z,y)} \leq \pi^p$, it follows that x, y belong to the same block of the partition π^p. Thus, $d(x, y) \leq p = \max\{d(x, z), d(z, y)\}$, which proves the triangular inequality for d. ■

6.4.2 Hierarchies and Ultrametrics

Definition 6 Let S be a set. A hierarchy on the set S is a collection of sets $\mathcal{H} \subseteq \mathcal{P}(S)$ that satisfies the following conditions:

(i) the members of $\mathcal{H}$ are nonempty sets;

(ii) $S \in \mathcal{H}$;

(iii) for every $x \in S$ we have $\{x\} \in \mathcal{H}$;
(iv) if $H, H' \in \mathcal{H}$ and $H \cap H' \neq \emptyset$, then we have either $H \subseteq H'$ or $H' \subseteq H$.

Example 4 Let $S = \{s, t, u, v, w, x, y\}$ be a finite set. It is easy to verify that the family of subsets of S defined by

$$\mathcal{H} = \{\{s\}, \{t\}, \{u\}, \{v\}, \{w\}, \{x\}, \{y\},$$
$$\{s, t, u\}, \{w, x\}, \{s, t, u, v\}, \{w, x, y\}, \{s, t, u, v, w, x, y\}\}$$

is a hierarchy on the set S. □

Chains of partitions defined on a set generate hierarchies as we show next.

Theorem 4 *Let S be a set and let $C = (\pi_1, \pi_2, \ldots, \pi_n)$ be an increasing chain of partitions* $(\mathrm{PART}(S), \leq)$ *such that $\pi_1 = \alpha_S$ and $\pi_n = \omega_S$. Then, the collection $\mathcal{H}_C = \bigcup_{i=1}^{n} \pi_i$ that consists of the blocks of all partitions in the chain is a hierarchy on S.*

Proof. The blocks of any of the partitions are nonempty sets, so $\mathcal{H}_C$ satisfies the first condition of Definition 6.

Note that $S \in \mathcal{H}_C$ because S is the unique block of $\pi_n = \omega_S$. Also, since all singletons $\{x\}$ are blocks of $\alpha_S = \pi_1$ it follows that $\mathcal{H}_C$ satisfies the second and the third conditions of Definition 6. Finally, let H, H' be two sets of $\mathcal{H}_C$ such that $H \cap H' \neq \emptyset$. Because of this condition it is clear that these two sets cannot be blocks of the same partition. Thus, there exist two partitions π_i and π_j in the chain such that $H \in \pi_i$ and $H' \in \pi_j$. Suppose that $i < j$. Since every block of π_j is a union of blocks of π_i, H' is a union of blocks of π_i and $H \cap H' \neq \emptyset$ means that H is one of these blocks. Thus, $H \subseteq H'$. If $j > i$, we obtain the reverse inclusion. This allows us to conclude that $\mathcal{H}_C$ is indeed a hierarchy. ■

Of course, Theorem 4 could be stated in terms of chains of equivalences; we give this alternative formulation for convenience.

Theorem 5 *Let S be a finite set and let $(\rho_1, \ldots, \rho_n)$ be a chain of equivalence relations on S such that $\rho_1 = \iota_S$ and $\rho_n = \theta_S$. Then, the collection of blocks of the equivalence relations ρ_r, that is, the set $\bigcup_{1 \leq r \leq n} S/\rho_r$, is a hierarchy on S.*

Proof. The proof is a mere restatement of the proof of Theorem 4. ■

Define the relation "$\prec$" on a hierarchy $\mathcal{H}$ on S by $H \prec K$ if $H, K \in \mathcal{H}$, $H \subset K$, and there is no set $L \in \mathcal{H}$ such that $H \subset L \subset K$.

Lemma 1 *Let $\mathcal{H}$ be a hierarchy on a finite set S and let $L \in \mathcal{H}$. The collection $\mathcal{P}_L = \{H \in \mathcal{H} | H \prec L\}$ is a partition of the set L.*

Proof. We claim that $L = \bigcup \mathcal{P}_L$. Indeed, it is clear that $\bigcup \mathcal{P}_L \subseteq L$.

Conversely, suppose that $z \in L$ but $z \notin \bigcup \mathcal{P}_L$. Since $\{z\} \in \mathcal{H}$ and there is no $K \in \mathcal{P}_L$ such that $z \in K$, it follows that $\{z\} \in \mathcal{P}_L$, which contradicts the assumption that $z \notin \bigcup \mathcal{P}_L$. This means that $L = \bigcup \mathcal{P}_L$.

Let $K_0, K_1 \in \mathcal{P}_L$ be two distinct sets. These sets are disjoint since otherwise we would have either $K_0 \subset K_1$, or $K_1 \subset K_0$, and this would contradict the definition of $\mathcal{P}_L$. ■

Theorem 6 *Let $\mathcal{H}$ be a hierarchy on a set S. The graph of the relation $\prec$ on $\mathcal{H}$ is a tree whose root is S; its leaves are the singletons $\{x\}$ for every $x \in S$.*

Proof. Since $\prec$ is an antisymmetric relation on $\mathcal{H}$ it is clear that the graph $(\mathcal{H}, \prec)$ is acyclic. Moreover, for each set $K \in \mathcal{H}$ there is a unique path that joins K to S, so the graph is indeed a rooted tree. ■

Definition 7 Let $\mathcal{H}$ be a hierarchy on a set S. A *grading function* for $\mathcal{H}$ is a function $h : \mathcal{H} \longrightarrow \mathbb{R}$ that satisfies the following conditions:

(i) $h(\{x\}) = 0$ for every $x \subset S$, and
(ii) if $H, K \in \mathcal{H}$ and $H \subset K$, then $h(H) < h(K)$.

If h is a grading function for a hierarchy $\mathcal{H}$, the pair $(\mathcal{H}, h)$ is a graded hierarchy.

Example 5 For the hierarchy $\mathcal{H}$ defined in Example 4 on the set $S = \{s, t, u, v, w, x, y\}$, the function $h : \mathcal{H} \longrightarrow \mathbb{R}$ given by

$$h(\{s\}) = h(\{t\}) = h(\{u\}) = h(\{v\}) = h(\{w\}) = h(\{x\}) = h(\{y\}) = 0,$$
$$h(\{s, t, u\}) = 3, h(\{w, x\}) = 4, h(\{s, t, u, v\}) = 5, h(\{w, x, y\}) = 6,$$
$$h(\{s, t, u, v, w, x, y\}) = 7$$

is a grading function and the pair $(\mathcal{H}, h)$ is a graded hierarchy on S. □

Theorem 4 can be extended to graded hierarchies.

Theorem 7 *Let S be a finite set and let $C = (\pi_1, \pi_2, \ldots, \pi_n)$ be an increasing chain of partitions* (PART(S), $\leq$) *such that $\pi_1 = \alpha_S$ and $\pi_n = \omega_S$.*

Consider a function $f : \{1, \ldots, n\} \longrightarrow \mathbb{R}_{\geq 0}$ such that $f(1) = 0$. The function $h : \mathcal{H}_C \longrightarrow \mathbb{R}_{\geq 0}$ given by $h(K) = f\left(\min\{j | K \in \pi_j\}\right)$ for $K \in \mathcal{H}_C$ is a grading function for the hierarchy $\mathcal{H}_C$.

Proof. Since $\{x\} \in \pi_1 = \alpha_S$ it follows that $h(\{x\}) = 0$, so h satisfies the first condition of Definition 7.

Suppose that $H, K \in \mathcal{H}_C$ and $H \subset K$. If $\ell = \min\{j | H \in \pi_j\}$, it is impossible for K to be a block of a partition that precedes π_ℓ. Therefore, $\ell < \min\{j | K \in \pi_j\}$, so $h(H) < h(K)$, so $(\mathcal{H}_C, h)$ is indeed a graded hierarchy. ■

A graded hierarchy defines an ultrametric as shown next.

Theorem 8 *Let $(\mathcal{H}, h)$ be a graded hierarchy on a finite set S. Define the function $d : S^2 \longrightarrow \mathbb{R}$ as*

$$d(x, y) = \min\{h(U) | U \in \mathcal{H} \text{ and } \{x, y\} \subseteq U\}$$

for $x, y \in S$. The mapping d is an ultrametric on S.

Proof. Note that for every $x, y \in S$ there exists a set $H \in \mathcal{H}$ such that $\{x, y\} \subseteq H$ because $S \in \mathcal{H}$.

It is immediate that $d(x, x) = 0$. Conversely, suppose that $d(x, y) = 0$. Then, there exists $H \in \mathcal{H}$ such that $\{x, y\} \subseteq H$ and $h(H) = 0$. If $x \neq y$, then $\{x\} \subset H$; hence $0 = h(\{x\}) < h(H)$, which contradicts the fact that $h(H) = 0$. Thus, $x = y$.

The symmetry of d is immediate.

To prove the ultrametric inequality, let $x, y, z \in S$ and suppose that $d(x, y) = p$, $d(x, z) = q$, and $d(z, y) = r$. There exist $H, K, L \in \mathcal{H}$ such that $\{x, y\} \subseteq H$, $h(H) = p$, $\{x, z\} \subseteq K$, $h(K) = q$, and $\{z, y\} \subseteq L$, $h(L) = r$. Since $K \cap L \neq \emptyset$ (because both sets contain z), we have either $K \subseteq L$ or $L \subseteq K$, so $K \cup L$ equals either K or L, and in either case, $K \cup L \in \mathcal{H}$. Since $\{x, y\} \subseteq K \cup L$, it follows that

$$d(x, y) \leq h(K \cup L) = \max\{h(K), H(L)\} = \max\{d(x, z), d(z, y)\},$$

which is the ultrametric inequality. ■

We refer to the ultrametric d whose existence is shown in Theorem 8 as the ultrametric generated by the graded hierarchy $(\mathcal{H}, h)$.

Example 6 The values of the ultrametric generated by the graded hierarchy $(\mathcal{H}, h)$ on the set S, introduced in Example 5, are given in the following table.

d	s	t	u	v	w	x	y
s	0	3	3	5	7	7	7
t	3	0	3	5	7	7	7
u	3	3	0	5	7	7	7
v	5	5	5	0	7	7	7
w	7	7	7	7	0	4	6
x	7	7	7	7	4	0	6
y	7	7	7	7	6	6	0

□

The hierarchy introduced in Theorem 5 that is associated with an ultrametric space can be naturally equipped with a grading function, as shown next.

Theorem 9 *Let (S, d) be a finite ultrametric space. There exists a graded hierarchy $(\mathcal{H}, h)$ on S such that d is the ultrametric associated to $(\mathcal{H}, h)$.*

Proof. Let $\mathcal{H}$ be the collection of equivalence classes of the equivalences $\eta_r = \{(x, y) \in S^2 | d(x, y) \leq r\}$ defined by the ultrametric d on the finite set S, where the index r takes its values in the range R_d of the ultrametric d. Define $h(E) = \min\{r \in R_d | E \in S/\eta_r\}$ for every equivalence class E.

It is clear that $h(\{x\}) = 0$ because $\{x\}$ is an η_0-equivalence class for every $x \in S$. Let $[x]_t$ be the equivalence class of x relative to the equivalence η_t.

Suppose that E, E' belong to the hierarchy and $E \subset E'$. We have $E = [x]_r$ and $E' = [x]_s$ for some $x \in X$. Since E is strictly included in E', there exists $z \in E' - E$ such that $d(x, z) \leq s$ and $d(x, z) > r$. This implies $r < s$. Therefore,

$$h(E) = \min\{r \in R_d | E \in S/\eta_r\} \leq \min\{s \in R_d | E' \in S/\eta_s\} = h(E'),$$

which proves that $(\mathcal{H}, h)$ is a graded hierarchy.

The ultrametric e generated by the graded hierarchy $(\mathcal{H}, h)$ is given by

$$\begin{aligned} e(x, y) &= \min\{h(B) | B \in \mathcal{H} \text{ and } \{x, y\} \subseteq B\} \\ &= \min\{r | (x, y) \in \eta_r\} \\ &- \min\{r | d(x, y) \leq r\} \\ &= d(x, y), \end{aligned}$$

for $x, y \in S$; in other words, we have $e = d$. ■

Example 7 Starting from the ultrametric on the set $S = \{s, t, u, v, w, x, y\}$ defined by the table given in Example 6, we obtain the following quotient sets:

Values of r	S/η_r
$[0, 3)$	$\{s\}, \{t\}, \{u\}, \{v\}, \{w\}, \{x\}, \{y\}$
$[3, 4)$	$\{s, t, u\}, \{v\}, \{w\}, \{x\}, \{y\}$
$[4, 5)$	$\{s, t, u\}, \{v\}, \{w, x\}, \{y\}$
$[5, 6)$	$\{s, t, u, v\}, \{w, x\}, \{y\}$
$[6, 7)$	$\{s, t, u, v\}, \{w, x, y\}$
$[7, \infty)$	$\{s, t, u, v, w, x, y\}$

□

We shall draw the tree of a graded hierarchy $(\mathcal{H}, h)$ using a special representation known as a *dendrogram*. In a dendrogram, an interior vertex K of the tree is represented by a horizontal line drawn at the height $h(K)$. For example, the dendrogram of the graded hierarchy of Example 5 is shown in Figure 6.1.

As we saw in Theorem 8, the value $d(x, y)$ of the ultrametric d generated by a hierarchy $\mathcal{H}$ is the smallest height of a set of a hierarchy that contains both x and y. This allows us to "read" the value of the ultrametric generated by $\mathcal{H}$ directly from the dendrogram of the hierarchy.

Example 8 For the graded hierarchy of Example 5, the ultrametric extracted from Figure 6.1 is clearly the same as the one that was obtained in Example 6. □

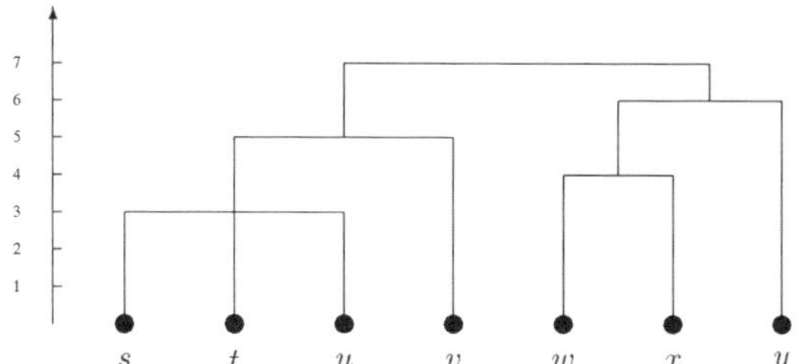

FIGURE 6.1 Dendrogram of graded hierarchy of Example 5.

6.4.3 The Poset of Ultrametrics

Let S be a set. Recall that we denoted the set of dissimilarities by $\mathcal{D}_S$. Define a partial order $\leq$ on $\mathcal{D}_S$ by $d \leq d'$ if $d(x, y) \leq d'(x, y)$ for every $x, y \in S$. It is easy to verify that $(\mathcal{D}_S, \leq)$ is a poset.

Note that $\mathcal{U}_S$, the set of ultrametrics on S, is a subset of $\mathcal{D}_S$.

Theorem 10 *Let d be a dissimilarity on a set S and let U_d be the set of ultrametrics:*

$$U_d = \{e \in \mathcal{U}_S | e \leq d\}.$$

The set U_d has a largest element in the poset $(\mathcal{D}_S, \leq)$.

Proof. Note that the set U_d is nonempty because the zero dissimilarity d_0 given by $d_0(x, y) = 0$ for every $x, y \in S$ is an ultrametric and $d_0 \leq d$.

Since the set $\{e(x, y) | e \in U_d\}$ has $d(x, y)$ as an upper bound, it is possible to define the mapping $e_1 : S^2 \longrightarrow \mathbb{R}_{\geq 0}$ as

$$e_1(x, y) = \sup\{e(x, y) | e \in U_d\}$$

for $x, y \in S$. It is clear that $e \leq e_1$ for every ultrametric e. We claim that e_1 is an ultrametric on S.

We prove only that e_1 satisfies the ultrametric inequality. Suppose that there exist $x, y, z \in S$ such that e_1 violates the ultrametric inequality, that is

$$\max\{e_1(x, z), e_1(z, y)\} < e_1(x, y).$$

This is equivalent to

$$\sup\{e(x, y) | e \in U_d\} > \max\{\sup\{e(x, z) | e \in U_d\}, \sup\{e(z, y) | e \in U_d\}\}.$$

Thus, there exists $\hat{e} \in U_d$ such that

$$\hat{e}(x, y) > \sup\{e(x, z) | e \in U_d\},$$
$$\hat{e}(x, y) > \sup\{e(z, y) | e \in U_d\}.$$

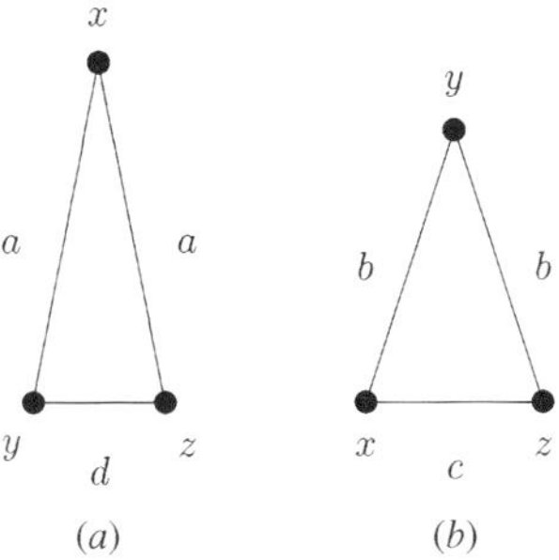

FIGURE 6.2 Two ultrametrics on the set $\{x, y, z\}$.

In particular, $\hat{e}(x, y) > \hat{e}(x, z)$ and $\hat{e}(x, y) > \hat{e}(z, y)$, which contradicts the fact that $\hat{e}$ is an ultrametric. ■

The ultrametric defined by Theorem 10 is known as the maximal subdominant ultrametric for the dissimilarity d.

The situation is not symmetric with respect to the infimum of a set of ultrametrics because, in general, the infimum of a set of ultrametrics is not necessarily an ultrametric.

For example, consider a three-element set $S = \{x, y, z\}$, four distinct nonnegative numbers a, b, c, d such that $a > b > c > d$, and the ultrametrics d and d' defined by the triangles shown in Figure 6.2a and b, respectively. The dissimilarity d_0 defined by $d_0(u, v) = \min\{d(u, v), d'(u, v)\}$ for $u, v \in S$ is given by

$$d_0(x, y) = b,\ d_0(y, z) = d, \text{ and } d_0(x, z) = c,$$

and d_0 is clearly not an ultrametric because the triangle xyz is not isosceles.

In the sequel, we give an algorithm for computing the maximal subdominant ultrametric for a dissimilarity defined on a finite set S.

We will define inductively an increasing sequence of partitions $\pi_1 \prec \pi_2 \prec \cdots$ and a sequence of dissimilarities $d_1, d_2, \ldots$ on the sets of blocks of $\pi_1, \pi_2, \ldots$, respectively.

For the initial phase, $\pi_1 = \alpha_S$ and $d_1(\{x\}, \{y\}) = d(x, y)$ for $x, y \in S$.

Suppose that d_i is defined on π_i. If $B, C \in \pi_i$ is a pair of blocks such that $d_i(B, C)$ has the smallest value, define the partition π_{i+1} by

$$\pi_{i+1} = (\pi_i - \{B, C\}) \cup \{B \cup C\}.$$

In other words, to obtain π_{i+1} we replace two of the closest blocks B, C of π_i (in terms of d_i) with new block $B \cup C$. Clearly, $\pi_i \prec \pi_{i+1}$ in PART(S) for $i \geq 1$. Note that the collection of blocks of the partitions π_i form a hierarchy $\mathcal{H}_d$ on the set S. The dissimilarity d_{i+1} is given by

$$d_{i+1}(U, V) = \min\{d(x, y) | x \in U, y \in V\} \tag{6.2}$$

for $U, V \in \pi_{i+1}$.

We introduce a grading function h_d on the hierarchy defined by this chain of partitions starting from the dissimilarity d. The definition is done for the blocks of the partitions π_i by induction on i.

For $i = 1$ the blocks of the partition π_1 are singletons; in this case we define $h_d(\{x\}) = 0$ for $x \in S$.

Suppose that h_d is defined on the blocks of π_i and let D be the block of π_{i+1} that is generated by fusing the blocks B, C of π_i. All other blocks of π_{i+1} coincide with the blocks of π_i. The value of the function h_d for the new block D is given by

$$h_d(D) = \min\{d(x, y) | x \in B, y \in C\}.$$

It is clear that h_d satisfies the first condition of Definition 7.

For a set U of $\mathcal{H}_d$ define $p_U = \min\{i | U \in \pi_i\}$ and $q_U = \max\{i | U \in \pi_i\}$. To verify the second condition of Definition 7, let $H, K \in \mathcal{H}_d$ such that $H \subset K$. It is clear that $q_H \leq p_K$. The construction of the sequence of partitions implies that there are $H_0, H_1 \in \pi_{p_H - 1}$ and $K_0, K_1 \in \pi_{p_K - 1}$ such that $H = H_0 \cup H_1$ and $K = K_0 \cup K_1$. Therefore,

$$h_d(H) = \min\{d(x, y) | x \in H_0, y \in H_1\},$$

$$h_d(K) = \min\{d(x, y) | x \in K_0, y \in K_1\}.$$

Since H_0, H_1 have been fused (to produce the partition π_{p_H}) before K_0, K_1 (to produce the partition π_{p_K}), it follows that $h_d(H) < h_d(K)$.

By Theorem 8 the graded hierarchy $(\mathcal{H}_d, h_d)$ defines an ultrametric; we denote this ultrametric by e and we will prove that e is the maximal subdominant ultrametric for d. Recall that e is given by

$$e(x, y) = \min\{h_d(W) | \{x, y\} \subseteq W\},$$

and that $h_d(W)$ is the least value of $d(u, v)$ such that $u \in U, v \in V$ if $W \in \pi_{p_W}$ is obtained by fusing the blocks U and V of $\pi_{p_W - 1}$. The definition of $e(x, y)$ implies that we have neither $\{x, y\} \subseteq U$ nor $\{x, y\} \subseteq V$. Thus, we have either $x \in U$ and $y \in V$ or $x \in V$ and $y \in U$. Thus, $e(x, y) \leq d(x, y)$.

We now prove that

$$e(x, y) = \min\{\text{amp}_d(\mathbf{s}) | \mathbf{s} \in S(x, y)\},$$

for $x, y \in S$.

Let D be the minimal set in $\mathcal{H}_d$ that includes $\{x, y\}$. Then, $D = B \cup C$, where B, C are two disjoint sets of $\mathcal{H}_d$ such that $x \in B$ and $y \in C$. If $\mathbf{s}$ is a sequence included in D, then there are two consecutive components of $\mathbf{s}$, s_k, s_{k+1} such that $s_k \in B$ and $s_{k+1} \in C$. This implies

$$\begin{aligned} e(x, y) &= \min\{d(u, v) | u \in B, v \in C\} \\ &\leq d(s_k, s_{k+1}) \\ &\leq \text{amp}_d(\mathbf{s}). \end{aligned}$$

If $\mathbf{s}$ is not included in D, let s_q, s_{q+1} be two consecutive components of $\mathbf{s}$ such that $s_q \in D$ and $s_{q+1} \notin D$. Let E be the smallest set of $\mathcal{H}_d$ that includes $\{s_q, s_{q+1}\}$. Note that $D \subseteq E$ (because $s_k \in D \cap E$), and therefore, $h_d(D) \leq h_d(E)$. If E is obtained as the union of two disjoint sets E', E'' of $\mathcal{H}_d$ such that $s_k \in E'$ and $s_{k+1} \in E''$, we have $D \subseteq E'$. Consequently,

$$h_d(E) = \min\{d(u, v) | u \in E', v \in E''\} \leq d(s_k, s_{k+1}),$$

which implies

$$e(x, y) = h_d(D) \leq h_d(E) \leq d(s_k, s_{k+1}) \leq \text{amp}_d(\mathbf{s}).$$

Therefore, we conclude that $e(x, y) \leq \text{amp}_d(\mathbf{s})$ for every $\mathbf{s} \in S(x, y)$.

We show now that there is a sequence $\mathbf{w} \in S(x, y)$ such that $e(x, y) \geq \text{amp}_d(\mathbf{w})$, which implies the equality $e(x, y) = \text{amp}_d(\mathbf{w})$. To this end, we prove that for every $D \in \pi_k \subseteq \mathcal{H}_d$ there exists $\mathbf{w} \in S(x, y)$ such that $\text{amp}_d(\mathbf{w}) \leq h_d(D)$. The argument is by induction on k.

For $k = 1$, the statement obviously holds. Suppose that it holds for $1, \ldots, k-1$ and let $D \in \pi_k$. The set D belongs to π_{k-1} or D is obtained by fusing the blocks B, C of π_{k-1}. In the first case, the statement holds by inductive hypothesis. The second case has several subcases:

(i) If $\{x, y\} \subseteq B$, then by inductive hypothesis, there exists a sequence $\mathbf{u} \in S(x, y)$ such that $\text{amp}_d(\mathbf{u}) \leq h_d(B) \leq h_d(D) = e(x, y)$.

(ii) The case $\{x, y\} \subseteq C$ is similar to the first case.

(iii) If $x \in B$ and $y \in C$, there exist $u, v \in D$ such that $d(u, v) = h_d(D)$. By the inductive hypothesis, there is a sequence $\mathbf{u} \in S(x, u)$ such that $\text{amp}_d(\mathbf{u}) \leq h_d(B)$ and there is a sequence $\mathbf{v} \in S(v, y)$ such that $\text{amp}_d(\mathbf{v}) \leq h_d(C)$. This allows us to consider the sequence $\mathbf{w}$ obtained by concatenating the sequences $\mathbf{u}$, (u, v), $\mathbf{v}$; clearly, we have $\mathbf{w} \in S(x, y)$ and

$$\text{amp}_d(\mathbf{w}) = \max\{\text{amp}_d(\mathbf{u}), d(u, v), \text{amp}_d(\mathbf{v})\} \leq h_d(D).$$

To complete the argument we need to show that if e' is an other ultrametric such that $e(x, y) \leq e'(x, y) \leq d(x, y)$, then $e(x, y) = e'(x, y)$ for every $x, y \in S$. By the previous argument there exists a sequence $\mathbf{s} = (s_0, \ldots, s_n) \in S(x, y)$ such that $\text{amp}_d(\mathbf{s}) = e(x, y)$. Since $e'(x, y) \leq d(x, y)$ for every $x, y \in S$, it follows that $e'(x, y) \leq \text{amp}_d(\mathbf{s}) = e(x, y)$. Thus, $e(x, y) = e'(x, y)$ for every $x, y \in S$, which means that $e = e'$. This concludes our argument.

6.5 HIERARCHICAL CLUSTERING

Hierarchical clustering is a recursive process that begins with a metric space of objects (S, d) and results in a chain of partitions of the set of objects. In each of the partitions, similar objects belong to the same block and objects that belong to distinct blocks tend to be dissimilar.

In the agglomerative hierarchical clustering, the construction of this chain begins with the unit partition $\pi^1 = \alpha_S$. If the partition constructed at step k is

$$\pi^k = \{U_1^k, \ldots, U_{m_k}^k\},$$

then two distinct blocks U_p^k and U_q^k of this partition are selected using a *selection criterion*. These blocks are fused and a new partition

$$\pi^{k+1} = \{U_1^k, \ldots, U_{p-1}^k, U_{p+1}^k, \ldots, U_{q-1}^k, U_{q+1}^k, \ldots, U_p^k \cup U_q^k\}$$

is formed. Clearly, we have $\pi^k \prec \pi^{k+1}$. The process must end because the poset $(\text{PART}(S), \leq)$ is of finite height. The algorithm halts when the one-block partition ω_S is reached.

As we saw in Theorem 4, the chain of partitions $\pi^1, \pi^2, \ldots$ generates a hierarchy on the set S. Therefore, all tools developed for hierarchies, including the notion of dendrogram, can be used for hierarchical algorithms.

When data to be clustered is numerical, that is, when $S \subseteq \mathbb{R}^n$, we can define the *centroid* of a nonempty subset U of S as

$$\boldsymbol{c}_U = \frac{1}{|U|} \sum \{\boldsymbol{o} | \boldsymbol{o} \in U\}.$$

If $\pi = \{U_1, \ldots, U_m\}$ is a partition of S, then the *sum of the squared errors* of π is the number

$$\text{sse}(\pi) = \sum_{i=1}^{m} \sum \{d^2(\boldsymbol{o}, \boldsymbol{c}_{U_i}) | \boldsymbol{o} \in U_i\}, \tag{6.3}$$

where d is the Euclidean distance in $\mathbb{R}^n$.

If two blocks U, V of a partition π are fused into a new block W to yield a new partition π' that covers π, then the variation of the sum of squared errors is given by

$$\begin{aligned}\text{sse}(\pi') - \text{sse}(\pi) &= \sum \{d^2(\boldsymbol{o}, \boldsymbol{c}_W) | \boldsymbol{o} \in U \cap V\} \\ &\quad - \sum \{d^2(\boldsymbol{o}, \boldsymbol{c}_U) | \boldsymbol{o} \in U\} - \sum \{d^2(\boldsymbol{o}, \boldsymbol{c}_V) | \boldsymbol{o} \in V\}.\end{aligned}$$

The centroid of the new cluster W is given by

$$\begin{aligned}\boldsymbol{c}_W &= \frac{1}{|W|} \sum \{\boldsymbol{o} | \boldsymbol{o} \in W\} \\ &= \frac{|U|}{|W|} \boldsymbol{c}_U + \frac{|V|}{|W|} \boldsymbol{c}_V.\end{aligned}$$

This allows us to evaluate the increase in the sum of squared errors:

$$\begin{aligned}\mathrm{sse}(\pi') - \mathrm{sse}(\pi) &= \sum\{d^2(\boldsymbol{o}, \boldsymbol{c}_W) | \boldsymbol{o} \in U \cup V\} \\ &\quad - \sum\{d^2(\boldsymbol{o}, \boldsymbol{c}_U) | \boldsymbol{o} \in U\} - \sum\{d^2(\boldsymbol{o}, \boldsymbol{c}_V) | \boldsymbol{o} \in V\} \\ &= \sum\{d^2(\boldsymbol{o}, \boldsymbol{c}_W) - d^2(\boldsymbol{o}, \boldsymbol{c}_U) | \boldsymbol{o} \in U\} \\ &\quad + \sum\{d^2(\boldsymbol{o}, \boldsymbol{c}_W) - d^2(\boldsymbol{o}, \boldsymbol{c}_V) | \boldsymbol{o} \in V\}.\end{aligned}$$

Observe that

$$\begin{aligned}&\sum\{d^2(\boldsymbol{o}, \boldsymbol{c}_W) - d^2(\boldsymbol{o}, \boldsymbol{c}_U) | \boldsymbol{o} \in U\} \\ &= \sum_{o \in U} ((\boldsymbol{o} - \boldsymbol{c}_W)(\boldsymbol{o} - \boldsymbol{c}_W) - (\boldsymbol{o} - \boldsymbol{c}_U)(\boldsymbol{o} - \boldsymbol{c}_U)) \\ &= |U|(\boldsymbol{c}_W^2 - \boldsymbol{c}_U^2) + 2(\boldsymbol{c}_U - \boldsymbol{c}_W) \sum_{o \in U} \boldsymbol{o} \\ &= |U|(\boldsymbol{c}_W^2 - \boldsymbol{c}_U^2) + 2|U|(\boldsymbol{c}_U - \boldsymbol{c}_W)\boldsymbol{c}_U \\ &= (\boldsymbol{c}_W - \boldsymbol{c}_U)(|U|(\boldsymbol{c}_W + \boldsymbol{c}_U) - 2|U|\boldsymbol{c}_U) \\ &= |U|(\boldsymbol{c}_W - \boldsymbol{c}_U)^2.\end{aligned}$$

Using the equality $\boldsymbol{c}_W - \boldsymbol{c}_U = |U|/|W|\boldsymbol{c}_U + |V|/|W|\boldsymbol{c}_V - \boldsymbol{c}_U = |V|/|W|(\boldsymbol{c}_V - \boldsymbol{c}_U)$, we obtain $\sum\{d^2(\boldsymbol{o}, \boldsymbol{c}_W) - d^2(\boldsymbol{o}, \boldsymbol{c}_U) | \boldsymbol{o} \in U\} = |U||V|^2/|W|^2\,(\boldsymbol{c}_V - \boldsymbol{c}_U)^2$. Similarly, we have

$$\sum\{d^2(\boldsymbol{o}, \boldsymbol{c}_W) - d^2(\boldsymbol{o}, \boldsymbol{c}_V) | \boldsymbol{o} \in V\} = \frac{|U|^2|V|}{|W|^2} (\boldsymbol{c}_V - \boldsymbol{c}_U)^2,$$

so

$$\mathrm{sse}(\pi') - \mathrm{sse}(\pi) = \frac{|U||V|}{|W|} (\boldsymbol{c}_V - \boldsymbol{c}_U)^2. \tag{6.4}$$

The dissimilarity between two clusters U, V can be defined using one of the following real-valued, two-argument functions defined on the set of subsets of S:

$$\begin{aligned}\mathrm{sl}(U, V) &= \min\{d(u, v) | u \in U, v \in V\}; \\ \mathrm{cl}(U, V) &= \max\{d(u, v) | u \in U, v \in V\}; \\ \mathrm{gav}(U, V) &= \frac{\sum\{d(u, v) | u \in U, v \in V\}}{|U| \cdot |V|};\end{aligned}$$

$$\text{cen}(U, V) = (\boldsymbol{c}_U - \boldsymbol{c}_V)^2;$$
$$\text{ward}(U, V) = \frac{|U||V|}{|U| + |V|} (\boldsymbol{c}_V - \boldsymbol{c}_U)^2 .$$

The names of the functions sl, cl, gav, and cen defined above are acronyms of the terms "single link," "complete link," "group average," and "centroid," respectively. They are linked to variants of the hierarchical clustering algorithms that we discuss in later. Note that in the case of the ward function the value equals the increase in the sum of the square errors when the clusters U, V are replaced with their union.

The specific selection criterion for fusing blocks defines the clustering algorithm. All algorithms store the dissimilarities between the current clusters $\pi^k = \{U_1^k, \dots, U_{m_k}^k\}$ in a $m_k \times m_k$ matrix $D^k = (d_{ij}^k)$, where d_{ij}^k is the dissimilarity between the clusters U_i^k and U_j^k. As new clusters are created by merging two existing clusters, the distance matrix must be adjusted to reflect the dissimilarities between the new cluster and existing clusters.

The general form of the algorithm is

matrix_agglomerative_clustering {
  compute the initial dissimilarity matrix D^1;
  $k = 1$;
  while (π^k contains more than one block) **do**
    merge a pair of two of the closest clusters;
    $k++$;
    compute the dissimilarity matrix D^k;
  endwhile;
}

Next, we show the computation of the dissimilarity between a new cluster and existing clusters.

Theorem 11 *Let U, V be two clusters of the clustering π that are joined into a new cluster W. Then, if $Q \in \pi - \{U, V\}$ we have*

$$\text{sl}(W, Q) = \tfrac{1}{2}\text{sl}(U, Q) + \tfrac{1}{2}\text{sl}(V, Q) - \tfrac{1}{2}\left|\text{sl}(U, Q) - \text{sl}(V, Q)\right|;$$

$$\text{cl}(W, Q) = \tfrac{1}{2}\text{cl}(U, Q) + \tfrac{1}{2}\text{cl}(V, Q) + \tfrac{1}{2}\left|\text{cl}(U, Q) - \text{cl}(V, Q)\right|;$$

$$\text{gav}(W, Q) = \frac{|U|}{|U| + |V|}\text{gav}(U, Q) + \frac{|V|}{|U| + |V|}\text{gav}(V, Q);$$

$$\text{cen}(W, Q) = \frac{|U|}{|U| + |V|}\text{cen}(U, Q) + \frac{|V|}{|U| + |V|}\text{cen}(V, Q) - \frac{|U||V|}{(|U| + |V|)^2}\text{cen}(U, V);$$

$$\begin{aligned}\text{ward}(W, Q) = {} & \frac{|U| + |Q|}{|U| + |V| + |Q|}\text{ward}(U, Q) + \frac{|V| + |Q|}{|U| + |V| + |Q|}\text{ward}(V, Q) \\ & - \frac{|Q|}{|U| + |V| + |Q|}\text{ward}(U, V).\end{aligned}$$

Proof. The first two equalities follow from the fact that

$$\begin{aligned}\min\{a, b\} &= \tfrac{1}{2}(a + b) - \tfrac{1}{2}|a - b|, \\ \max\{a, b\} &= \tfrac{1}{2}(a + b) + \tfrac{1}{2}|a - b|,\end{aligned}$$

for every $a, b \in \mathbb{R}$.

For the third equality, we have

$$\begin{aligned}\text{gav}(W, Q) &= \frac{\sum\{d(w, q)|w \in W, q \in Q\}}{|W| \cdot |Q|} \\ &= \frac{\sum\{d(u, q)|u \in U, q \in Q\}}{|W| \cdot |Q|} + \frac{\sum\{d(v, q)|v \in V, q \in Q\}}{|W| \cdot |Q|} \\ &= \frac{|U|}{|W|}\frac{\sum\{d(u, q)|u \in U, q \in Q\}}{|U| \cdot |Q|} + \frac{|V|}{|W|}\frac{\sum\{d(v, q)|v \in V, q \in Q\}}{|V| \cdot |Q|} \\ &= \frac{|U|}{|U| + |V|}\text{gav}(U, Q) + \frac{|V|}{|U| + |V|}\text{gav}(V, Q).\end{aligned}$$

The equality involving the function cen is immediate. The last equality can be easily translated into

$$\begin{aligned}&\frac{|Q||W|}{|Q| + |W|}\left(\boldsymbol{c}_Q - \boldsymbol{c}_W\right)^2 \\ &= \frac{|U| + |Q|}{|U| + |V| + |Q|}\frac{|U||Q|}{|U| + |Q|}\left(\boldsymbol{c}_Q - \boldsymbol{c}_U\right)^2 \\ &\quad + \frac{|V| + |Q|}{|U| + |V| + |Q|}\frac{|V||Q|}{|V| + |Q|}\left(\boldsymbol{c}_Q - \boldsymbol{c}_V\right)^2 \\ &\quad - \frac{|Q|}{|U| + |V| + |Q|}\frac{|U||V|}{|U| + |V|}\left(\boldsymbol{c}_V - \boldsymbol{c}_U\right)^2,\end{aligned}$$

which can be verified replacing $|W| = |U| + |V|$ and $\boldsymbol{c}_W = |U|/|W|\boldsymbol{c}_U + |V|/|W|\boldsymbol{c}_V$. ■.

The equalities contained by Theorem 11 are often presented as a single equality involving several coefficients.

Corollary 1 (The Lance–Williams formula) *Let U, V be two clusters of the clustering π that are joined into a new cluster W. Then, if $Q \in \pi - \{U, V\}$ the dissimilarity* *between W and Q can be expressed as*

$$d(W, Q) = a_U d(U, Q) + a_V d(V, Q) + bd(U, V) + c|d(U, Q) - d(V, Q)|,$$

where the coefficients a_U, a_V, b, c are given by the following table.

Function	a_U	a_V	b	c
sl	$1/2$	$1/2$	0	$-(1/2)$
cl	$1/2$	$1/2$	0	$1/2$
gav	$\|U\|/(\|U\| + \|V\|)$	$\|V\|/\|U\| + \|V\|$	0	0
cen	$\|U\|/(\|U\| + \|V\|)$	$\|V\|/\|U\| + \|V\|$	$-(\|U\|\|V\|(\|U\| + \|V\|)^2)$	0
ward	$\|U\| + \|Q\|(\|U\| + \|V\| + \|Q\|)$	$\|V\| + \|Q\|/\|U\| + \|V\| + \|Q\|$	$-(\|Q\|\|U\| + \|V\| + \|Q\|)$	0

Proof. This statement is an immediate consequence of Theorem 9. ■

The variant of the algorithm that makes use of the function sl is known as the single-link clustering. It tends to favor elongated clusters.

Example 9 We use single-link clustering for the data set shown in Figure 6.3, $S = \{\mathbf{o}_1, \ldots, \mathbf{o}_7\}$, that consists of seven objects.

The distances between the objects of S are specified by the 7×7 matrix

$$D^1 = \begin{pmatrix} 0 & 1 & \sqrt{5} & \sqrt{20} & \sqrt{32} & \sqrt{61} & \sqrt{58} \\ 1 & 0 & \sqrt{2} & \sqrt{13} & 5 & \sqrt{50} & \sqrt{45} \\ \sqrt{5} & \sqrt{2} & 0 & \sqrt{5} & \sqrt{13} & \sqrt{32} & \sqrt{29} \\ \sqrt{20} & \sqrt{13} & \sqrt{5} & 0 & 2 & \sqrt{13} & \sqrt{10} \\ \sqrt{32} & \sqrt{5} & \sqrt{13} & 2 & 0 & \sqrt{5} & \sqrt{10} \\ \sqrt{61} & \sqrt{50} & \sqrt{32} & \sqrt{13} & \sqrt{5} & 0 & \sqrt{5} \\ \sqrt{58} & \sqrt{45} & \sqrt{29} & \sqrt{10} & \sqrt{10} & \sqrt{5} & 0 \end{pmatrix}.$$

Let us apply the hierarchical clustering algorithm using the single-link variant to the set S. Initially, the clustering is

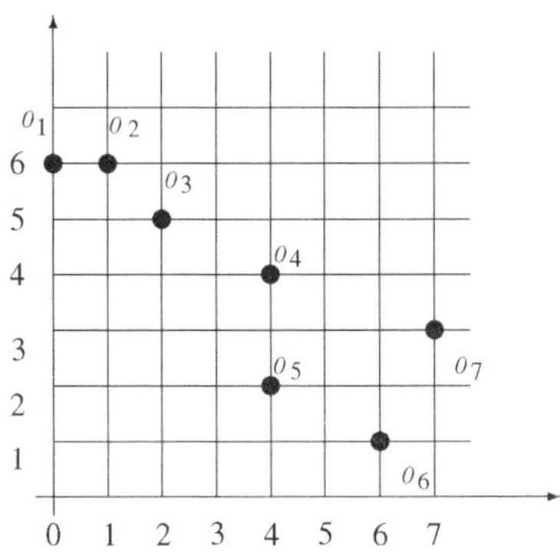

FIGURE 6.3 Set of seven points in $\mathbb{R}^2$.

$$\pi^1 = \{\{o_1\}, \{o_2\}, \{o_3\}, \{o_4\}, \{o_5\}, \{o_6\}, \{o_7\}\}.$$

The closest clusters are $\{o_1\}, \{o_2\}$; these clusters are fused into the cluster $\{o_1, o_2\}$, the new partition is

$$\pi^2 = \{\{o_1, o_2\}, \{o_3\}, \{o_4\}, \{o_5\}, \{o_6\}, \{o_7\}\},$$

and the matrix of dissimilarities becomes the 6×6 matrix

$$D^2 = \begin{pmatrix} 0 & \sqrt{2} & \sqrt{13} & 5 & \sqrt{50} & \sqrt{45} \\ \sqrt{2} & 0 & \sqrt{5} & \sqrt{13} & \sqrt{32} & \sqrt{29} \\ \sqrt{13} & \sqrt{5} & 0 & 2 & \sqrt{13} & \sqrt{10} \\ 5 & \sqrt{13} & 2 & 0 & \sqrt{5} & \sqrt{10} \\ \sqrt{50} & \sqrt{32} & \sqrt{13} & \sqrt{5} & 0 & \sqrt{5} \\ \sqrt{45} & \sqrt{29} & \sqrt{10} & \sqrt{10} & \sqrt{5} & 0 \end{pmatrix}.$$

Next, the closest clusters are $\{o_1, o_2\}$ and $\{o_3\}$. These clusters are fused into the cluster $\{o_1, o_2, o_3\}$ and the new 5×5 matrix is

$$D^3 = \begin{pmatrix} 0 & \sqrt{5} & \sqrt{13} & \sqrt{32} & \sqrt{29} \\ \sqrt{5} & 0 & 2 & \sqrt{13} & \sqrt{10} \\ \sqrt{13} & 2 & 0 & \sqrt{5} & \sqrt{10} \\ \sqrt{32} & \sqrt{13} & \sqrt{5} & 0 & \sqrt{5} \\ \sqrt{29} & \sqrt{10} & \sqrt{10} & \sqrt{5} & 0 \end{pmatrix},$$

which corresponds to the partition

$$\pi^3 = \{\{o_1, o_2, o_3\}, \{o_4\}, \{o_5\}, \{o_6\}, \{o_7\}\}.$$

Next, the closest clusters are $\{o_4\}$ and $\{o_5\}$. Fusing these yields the partition

$$\pi^4 = \{\{o_1, o_2, o_3\}, \{o_4, o_5\}, \{o_6\}, \{o_7\}\}$$

and the 4×4 matrix

$$D^4 = \begin{pmatrix} 0 & \sqrt{5} & \sqrt{32} & \sqrt{29} \\ \sqrt{5} & 0 & \sqrt{5} & \sqrt{10} \\ \sqrt{32} & \sqrt{5} & 0 & \sqrt{5} \\ \sqrt{29} & \sqrt{10} & \sqrt{5} & 0 \end{pmatrix}.$$

We have two choices now: we could fuse $\{o_1, o_2, o_3\}$ with $\{o_4, o_5\}$, or $\{o_4, o_5\}$ with $\{o_6\}$ since in either case the intercluster dissimilarity is $\sqrt{5}$. We choose the first option and we form the cluster $\{o_1, o_2, o_3, o_4, o_5\}$. Now the partition is

$$\pi^5 = \{\{o_1, o_2, o_3, o_4, o_5\}, \{o_6\}, \{o_7\}\}$$

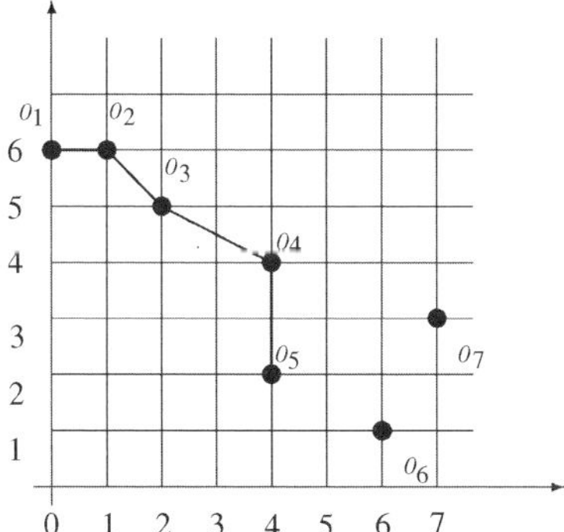

FIGURE 6.4 Elongated cluster produced by the single-link algorithm.

and the matrix is

$$D^5 = \begin{pmatrix} 0 & \sqrt{5} & \sqrt{10} \\ \sqrt{5} & 0 & \sqrt{5} \\ \sqrt{10} & \sqrt{5} & 0 \end{pmatrix}.$$

Observe that the large cluster formed so far has an elongated shape (see Fig. 6.4); this is typical for single-link variant of the algorithm. Fusing now $\{\boldsymbol{o}_1, \boldsymbol{o}_2, \boldsymbol{o}_3, \boldsymbol{o}_4, \boldsymbol{o}_5\}$ with $\{\boldsymbol{o}_6\}$ gives the two-block partition

$$\pi^6 = \{\{\boldsymbol{o}_1, \boldsymbol{o}_2, \boldsymbol{o}_3, \boldsymbol{o}_4, \boldsymbol{o}_5, \boldsymbol{o}_6\}, \{\boldsymbol{o}_7\}\}$$

and the 2×2 matrix

$$D^6 = \begin{pmatrix} 0 & \sqrt{5} \\ \sqrt{5} & 0 \end{pmatrix}.$$

In the final step, the two clusters are fused and the algorithm stops.

The dendrogram of the hierarchy produced by the algorithm is given in Figure 6.5. □

The variant of the algorithm that uses the function cl is known as the complete-link clustering. It tends to favor globular clusters.

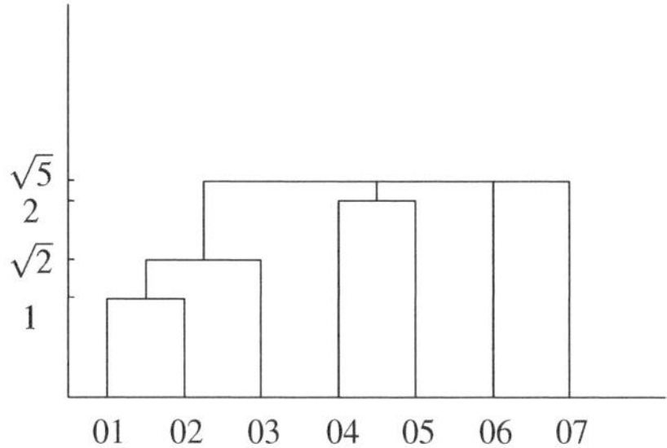

FIGURE 6.5 Dendrogram of single-link clustering.

Example 10 Now we apply the complete-link algorithm to the set S considered in Example 9. It is easy to see that the initial two partitions and the initial matrix are the same as for the single-link algorithm.

However, after creating the first cluster $\{o_1, o_2\}$, the distance matrices begin to differ. The next matrix is

$$D^2 = \begin{pmatrix} 0 & \sqrt{5} & \sqrt{20} & \sqrt{32} & \sqrt{61} & \sqrt{58} \\ \sqrt{5} & 0 & \sqrt{5} & \sqrt{13} & \sqrt{32} & \sqrt{29} \\ \sqrt{20} & \sqrt{5} & 0 & 2 & \sqrt{13} & \sqrt{10} \\ \sqrt{32} & \sqrt{13} & 2 & 0 & \sqrt{5} & \sqrt{10} \\ \sqrt{61} & \sqrt{32} & \sqrt{13} & \sqrt{5} & 0 & \sqrt{5} \\ \sqrt{58} & \sqrt{29} & \sqrt{10} & \sqrt{10} & \sqrt{5} & 0 \end{pmatrix},$$

which shows that the closest clusters are now $\{o_4\}$ and $\{o_5\}$. Thus,

$$\pi^3 = \{\{o_1, o_2\}, \{o_3\}, \{o_4, o_5\}, \{o_6\}, \{o_7\}\}$$

and the new matrix is

$$D^3 = \begin{pmatrix} 0 & \sqrt{5} & \sqrt{32} & \sqrt{61} & \sqrt{58} \\ \sqrt{5} & 0 & \sqrt{13} & \sqrt{32} & \sqrt{29} \\ \sqrt{32} & \sqrt{13} & 0 & \sqrt{10} & \sqrt{10} \\ \sqrt{61} & \sqrt{32} & \sqrt{13} & 0 & \sqrt{5} \\ \sqrt{58} & \sqrt{29} & \sqrt{10} & \sqrt{5} & 0 \end{pmatrix}.$$

Now there are two pairs of clusters that correspond to the minimal value in D^3: $\{o_1, o_2\}, \{o_3\}$ and $\{o_6\}, \{o_7\}$; if we merge the last pair we get the partition $\pi^4 = \{\{o_1, o_2\}, \{o_3\}, \{o_4, o_5\}, \{o_6, o_7\}\}$ and the matrix

$$D^4 = \begin{pmatrix} 0 & \sqrt{32} & \sqrt{61} & \sqrt{58} \\ \sqrt{32} & 0 & \sqrt{13} & \sqrt{10} \\ \sqrt{61} & \sqrt{13} & 0 & \sqrt{5} \\ \sqrt{58} & \sqrt{10} & \sqrt{5} & 0 \end{pmatrix}.$$

Next, the closest clusters are $\{o_1, o_2\}, \{o_3\}$. Merging those clusters will result in the partition $\pi^5 = \{\{o_1, o_2, o_3\}, \{o_4, o_5\}, \{o_6, o_7\}\}$ and the matrix

$$D^5 = \begin{pmatrix} 0 & \sqrt{32} & \sqrt{61} \\ \sqrt{32} & 0 & \sqrt{13} \\ \sqrt{61} & \sqrt{13} & 0 \end{pmatrix}.$$

The current clustering is shown in Figure 6.6. Observe that in the case of the complete-link method clusters that appear early tend to enclose objects that are closed in the sense of the distance.

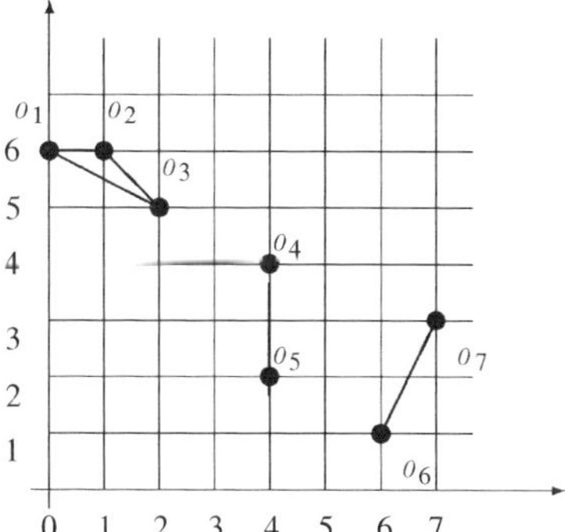

FIGURE 6.6 Partial clustering obtained by complete-link method.

Now the closest clusters are $\{o_4, o_5\}$ and $\{o_6, o_7\}$. Merging those clusters will give the partition $\pi^5 = \{\{o_1, o_2, o_3\}, \{o_4, o_5, o_6, o_7\}\}$ and the matrix

$$D^6 = \begin{pmatrix} 0 & \sqrt{61} \\ \sqrt{61} & 0 \end{pmatrix}.$$

The dendrogram of the resulting clustering is given in Figure 6.7. □

The *group average method* that makes use of the gav function is an intermediate approach between the single-link and the complete-link method. What the methods mentioned so far have in common is the *monotonicity property* expressed by the following statement.

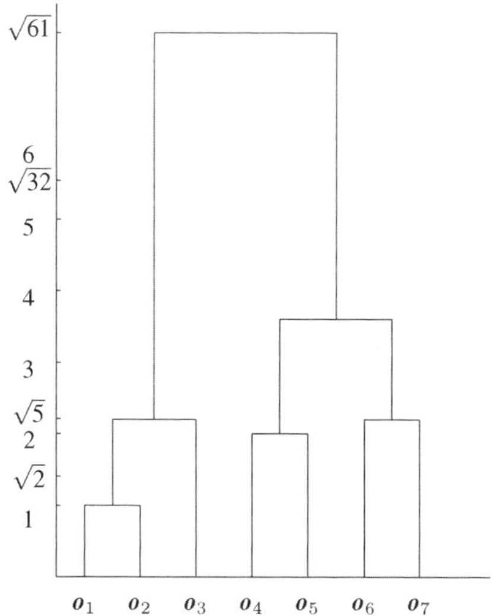

FIGURE 6.7 Dendrogram of complete-link clustering.

Theorem 12 *Let (S, d) be finite metric space and let $D^1, \ldots, D^m$ be the sequence of matrices constructed by any of the first three hierarchical methods (single, complete, or average link), where $m = |S|$. If μ_i is the smallest entry of the matrix D^i for $1 \leq i \leq m$, then $\mu_1 \leq \mu_2 \leq \cdots \leq \mu_m$. In other words, the dissimilarity between clusters that are merged at each step is nondecreasing.*

Proof. Suppose that the matrix D^{j+1} is obtained from the matrix D^j by merging the clusters C_p and C_q that correspond to the lines p, q and to columns p, q of D^j. This happens because $d_{pq} = d_{qp}$ is one of the minimal elements of the matrix D^j. Then, these lines and columns are replaced with a line and column that corresponds to the new cluster C_r and to the dissimilarities between this new cluster and the previous clusters C_i, where $i \neq p, q$. The elements d_{rh}^{j+1} of the new line (and column) are obtained either as $\min\{d_{ph}^j, d_{qh}^j\}$, $\max\{d_{ph}^j, d_{qh}^j\}$, or as $(|C_p|/|C_r|)d_{ph}^j + (|C_q|/|C_r|)d_{qh}^j$, for the single-link, complete-link, or group average methods, respectively. In any of these case, it is not possible to obtain a value for d_{rh}^{j+1} that is less than the minimal value of an element of D^j. ■

The last two methods captured by the Lance–Williams formula are, respectively, the centroid method and the Ward method of clustering. As we observed before, formula (6.4) shows that the dissimilarity of two cluster in the case of Ward's method equals the increase in the sum of the squared errors that results when the clusters are merged. The centroid method adopts the distance between the centroids as the distance between the corresponding clusters. Either method lacks the monotonicity properties.

To evaluate the space and time complexity of hierarchical clustering note that the algorithm must handle the matrix of the dissimilarities between objects and this is a symmetric $n \times n$ matrix having all elements on its main diagonal equal to 0; in other words, the algorithm needs to store $(n(n-1)/2)$ numbers. To keep track of the clusters, an extra space that does not exceed $n - 1$ is required. Thus, the total space required is $O(n^2)$.

The time complexity of agglomerative clustering algorithms has been evaluated in the work by Kurita [9]; the proposed implementation requires a heap that contains the pairwise distances between clusters and therefore has a size of n^2. The pseudocode of this algorithm is

generic_agglomerative algorithm {
  construct a heap H of size n^2
    for inter-cluster dissimilarities;
  while the number of clusters is larger than 1 **do**
    get the nearest pairs of clusters C_p, C_q that correspond to $H[0]$;
    reduce the number of clusters by 1 through merging C_p and C_q;
    update the heap to reflect the revised distances and
      remove unnecessary elements;

endwhile;

}

Note that the while loop is performed n times as each execution reduces the number of clusters by 1. The initial construction of the heap requires a time of $O(n^2 \log n^2) = O(n^2 \log n)$. Then, each of operations inside the loop requires no more than $O(\log n^2) = O(\log n)$ (because the heap has size n^2). Thus, we conclude that the time complexity is $O(n^2 \log n)$.

There exists an interesting link between the single-link clustering algorithm and the subdominant ultrametric of a dissimilarity, which we examined in Section 6.4.3.

To construct the subdominant ultrametric for a dissimilarity dissimilarity space (S, d), we built an increasing chain of partitions $\pi_1, \pi_2, \ldots$ of S (where $\pi_1 = \alpha_S$) and a sequence of dissimilarities $d_1, d_2, \ldots$ (where $d_1 = d$) on the sets of blocks of $\pi_1, \pi_2, \ldots$, respectively. We claim that this sequence of partitions $\pi_1, \pi_2, \ldots$ coincides with the sequence of partitions $\pi^1, \pi^2, \ldots$, and that the sequence of dissimilarities $d_1, d_2, \ldots$ coincides with the sequences of dissimilarities $d^1, d^2, \ldots$ defined by the matrices D^i constructed by the single-link algorithm. This is clearly the case for $i = 1$.

Suppose that the statement is true for i. The partition π_{i+1} is obtained from π_i by fusing the blocks B, C of π such that $d_i(B, C)$ has the smallest value, that is,

$$\pi_{i+1} = (\pi_i - \{B, C\}) \cup \{B \cup C\}.$$

Since this is exactly how the partition π^{i+1} is constructed from π^i, it follows that $\pi_{i+1} = \pi^{i+1}$. The inductive hypothesis implies that

$$d^i(U, V) = d_i(U, V) = \min\{d(u, v) | u \in U, v \in V\}$$

for all $U, V \in \pi_i$. Since the dissimilarity d_{i+1} is $d_{i+1}(U, V) = \min\{d(u, v) | u \in U, u \in V\}$ for every pair of blocks U, V of π_{i+1}, it is clear that $d_{i+1}(U, V) = d_i(U, V) = d^i(U, V) = d^{i+1}(U, V)$ when neither U nor V equal the block $B \cup C$. Then,

$$\begin{aligned}
&d_{i+1}(B \cup C, W) \\
&= \min\{d(t, w) | t \subset B \cup C, w \in W\} \\
&= \min\{\min\{d(b, w) | b \in B, w \in W\}, \min\{d(c, w) | c \in C, w \in W\}\} \\
&= \min\{d_i(B, W), d_i(C, W)\} \\
&= \min\{d^i(B, W), d^i(C, W)\} \\
&= d^{i+1}(B \cup C, W).
\end{aligned}$$

Thus, $d_{i+1} = d^{i+1}$.

Let x, y be a pair of elements of S. The value of the subdominant ultrametric is given by

$$e(x, y) = \min\{h_d(W) | W \in \mathcal{H}_d \text{ and } \{x, y\} \subseteq W\}.$$

This is the height of W in the dendrogram of the single-link clustering, and therefore, the subdominant ultrametric can be read directly from this dendrogram.

Example 11 The subdominant ultrametric of the Euclidean metric considered in Example 9 is given by the following table.

$e(\boldsymbol{o}_i, \boldsymbol{o}_j)$	$\boldsymbol{o}_1$	$\boldsymbol{o}_2$	$\boldsymbol{o}_3$	$\boldsymbol{o}_4$	$\boldsymbol{o}_5$	$\boldsymbol{o}_6$	$\boldsymbol{o}_7$
$\boldsymbol{o}_1$	0	1	$\sqrt{2}$	2	$\sqrt{5}$	$\sqrt{5}$	$\sqrt{5}$
$\boldsymbol{o}_2$	1	0	$\sqrt{2}$	$\sqrt{5}$	$\sqrt{5}$	$\sqrt{5}$	$\sqrt{5}$
$\boldsymbol{o}_3$	$\sqrt{2}$	$\sqrt{2}$	0	$\sqrt{5}$	$\sqrt{5}$	$\sqrt{5}$	$\sqrt{5}$
$\boldsymbol{o}_4$	2	$\sqrt{5}$	$\sqrt{5}$	0	$\sqrt{5}$	$\sqrt{5}$	$\sqrt{5}$
$\boldsymbol{o}_5$	$\sqrt{5}$	$\sqrt{5}$	$\sqrt{5}$	$\sqrt{5}$	0	$\sqrt{5}$	$\sqrt{5}$
$\boldsymbol{o}_6$	$\sqrt{5}$	$\sqrt{5}$	$\sqrt{5}$	$\sqrt{5}$	$\sqrt{5}$	0	$\sqrt{5}$
$\boldsymbol{o}_7$	$\sqrt{5}$	$\sqrt{5}$	$\sqrt{5}$	$\sqrt{5}$	$\sqrt{5}$	$\sqrt{5}$	0

□

6.6 THE *k*-MEANS ALGORITHM

The k-means algorithm is a partitional algorithm that requires the specification of the number of clusters k as an input. The set of objects to be clustered $S = \{\boldsymbol{o}^1, \ldots, \boldsymbol{o}^n\}$ is a subset of $\mathbb{R}^m$. Due to its simplicity and its many implementations, it is a very popular algorithm despite this requirement.

The k-means algorithm begins with a randomly chosen collection of k points $\boldsymbol{c}^1, \ldots, \boldsymbol{c}^k$ in $\mathbb{R}^m$ called centroids. An initial partition of the set S of objects is computed by assigning each object $\boldsymbol{o}^i$ to its closest centroid $\boldsymbol{c}^j$. Let U_j be the set of points assigned to the centroid $\boldsymbol{c}^j$.

The assignments of objects to centroids are expressed by a matrix (b_{ij}), where

$$b_{ij} = \begin{cases} 1 \text{ if } \boldsymbol{o}^i \in U_j, \\ 0 \text{ otherwise.} \end{cases}$$

Since each object is assigned to exactly one cluster, we have $\sum_{j=1}^{k} b_{ij} = 1$. On the contrary, $\sum_{i=1}^{n} b_{ij}$ equals the number of objects assigned to the centroid $\boldsymbol{c}^j$.

After these assignments, expressed by the matrix (b_{ij}), the centroids $\boldsymbol{c}^j$ must be recomputed using the formula

$$\boldsymbol{c}^j = \frac{\sum_{i=1}^{n} b_{ij}\boldsymbol{o}^i}{\sum_{i=1}^{n} b_{ij}} \tag{6.5}$$

for $1 \leq j \leq k$.

The sum of squared errors of a partition $\pi = \{U_1, \ldots, U_k\}$ of a set of objects S was defined in equality (6.3) as

$$\text{sse}(\pi) = \sum_{j=1}^{k} \sum_{\boldsymbol{o} \in U_j} d^2(\boldsymbol{o}, \boldsymbol{c}^j),$$

where $\boldsymbol{c}^j$ is the centroid of U_j for $1 \le j \le k$. The error of such an assignment is the sum of squared errors of the partition $\pi = \{U_1, \ldots, U_k\}$ defined as

$$\begin{aligned}\text{sse}(\pi) &= \sum_{i=1}^{n} \sum_{j=1}^{k} b_{ij} ||\boldsymbol{o}^i - \boldsymbol{c}^j||^2 \\ &= \sum_{i=1}^{n} \sum_{j=1}^{k} b_{ij} \sum_{p=1}^{m} \left(o_p^i - c_p^j\right)^2.\end{aligned}$$

The mk necessary conditions for a local minimum of this function,

$$\frac{\partial \text{sse}(\pi)}{\partial c_p^j} = \sum_{i=1}^{n} b_{ij} \left(-2(o_p^i - c_p^j)\right) = 0$$

for $1 \le p \le m$ and $1 \le j \le k$, can be written as

$$\sum_{i=1}^{n} b_{ij} o_p^i = \sum_{i=1}^{n} b_{ij} c_p^j = c_p^j \sum_{i=1}^{n} b_{ij},$$

or as

$$c_p^j = \frac{\sum_{i=1}^{n} b_{ij} o_p^i}{\sum_{i=1}^{n} b_{ij}}$$

for $1 \le p \le m$. In vectorial form, these conditions amount to

$$\boldsymbol{c}^j = \frac{\sum_{i=1}^{n} b_{ij} \boldsymbol{o}^i}{\sum_{i=1}^{n} b_{ij}},$$

which is exactly formula (6.5) that is used to update the centroids. Thus, the choice of the centroids can be justified by the goal of obtaining local minima of the sum of squared errors of the clusterings.

Since we have new centroids, objects must be reassigned, which means that the values of b_{ij} must be recomputed, which, in turn, will affect the values of the centroids, and so on.

The halting criterion of the algorithm depends on particular implementations and it may involve

(i) performing a certain number of iterations;
(ii) lowering the sum of squared errors sse(π) below a certain limit;
(iii) the current partition coincides with the previous partition.

This variant of the k-means algorithm is known as *Forgy's* algorithm:

```
k_means_forgy{
        obtain a randomly chosen collection of
        k points c_1, ..., c_k in R^n;
        assign each object o^i to the closest centroid c^j;
        let π = {U_1, ..., U_k} be the partition defined by
            c^1, ..., c^k;
        recompute the centroids of the clusters U_1, ..., U_k;
        while (halting criterion is not met) do
                compute the new value of the partition π
                using the current centroids;
        recompute the centroids of the blocks of π;
        endwhile
}
```

The popularity of the k-means algorithm stems on its simplicity and its low time complexity that is $O(kn\ell)$, where n is the number of objects to be clustered and ℓ is the number of iterations that the algorithm is performing.

Another variant of the k-means algorithm redistributes objects to clusters based on the effect of such a reassignment on the objective function. If sse(π) decreases, the object is moved and the two centroids of the affected clusters are recomputed. This variant is carefully analyzed in the work by Berkin and Becher [3].

6.7 THE PAM ALGORITHM

Another algorithm named PAM (an acronym of partition around medoids) developed by Kaufman and Rousseeuw [7] also requires as an input parameter the number k of clusters to be extracted.

The k clusters are determined based on a representative object from each cluster called the *medoid* of the cluster. The medoid is intended to have the most central position in the cluster relative to all other members of the cluster. Once medoids are selected, each remaining object o is assigned to a cluster represented by a medoid o_i if the dissimilarity $d(o, o_i)$ is minimal.

In the second phase, swapping objects and existing medoids are considered. The cost of a swap is defined with the intention of penalizing swaps that diminish the centrality of the medoids in the clusters. Swapping continues as long as useful swaps (i.e., swaps with negative costs) can be found.

PAM begins with a set of objects S, where $|S| = n$, a dissimilarity $n \times n$ matrix D, and a prescribed number of clusters k. The d_{ij} entry of the matrix D is the dissimilarity

$d(o_i, o_j)$ between the objects o_i and o_j. PAM is more robust than Forgy's variant of k-clustering because it minimizes the sum of the dissimilarities instead of the sum of the squared errors.

The algorithm has two distinct phases: the *building phase* and the *swapping phase*.

The building phase aims to construct a set L of selected objects, $L \subseteq S$. The set or remaining objects is denoted by R; clearly, $R = S - L$. We begin by determining the most centrally located object.

The quantities $Q_i = \sum_{j=1}^{n} d_{ij}$ are computed starting from the matrix D. The most central object o_q is the determined by

$$q = \arg\min_i Q_i.$$

The set L is initialized as $L = \{o_q\}$.

Suppose now that we have constructed a set of L of selected objects and $|L| < k$. We need to add a new selected object to the set L. To do this, we need to examine all objects that have not been included in L so far, that is, all objects in R. The selection is determined by a merit function $M : R \longrightarrow \mathbb{N}$.

To compute the merit $M(o)$ of an object $o \in R$, we scan all objects in R distinct from o. Let $o' \in R - \{o\}$ be such an object. If $d(o, o') < d(L, o')$, then adding o to L could benefit the clustering (from the point of view of o') because $d(L, o')$ will diminish. The potential benefit is $d(o', L) - d(o, o')$. Of course, if $d(o, o') \geq d(L, o')$ no such benefit exists (from the point of view of o'). Thus, we compute the merit of o as

$$M(o) = \sum_{o' \in R - \{o\}} \max\{D(L, o') - d(o, o'), 0\}.$$

We add to L the unselected object o that has the largest merit value. The building phase halts when $|L| = k$.

The objects in set L are the potential medoids of the k clusters that we seek to build. The second phase of the algorithm aims to improve the clustering by considering the merit of swaps between selected and unselected objects. So, assume now that o_i is a selected object, $o_i \in L$, and o_h is an unselected object, $o_h \in R = S - L$. We need to determine the cost $C(o_i, o_h)$ of swapping o_i and o_h. Let o_j be an arbitrary unselected object. The contribution c_{ihj} of o_j to the cost of the swap between o_i and o_h is defined as follows:

1. If $d(o_i, o_j)$ and $d(o_h, o_j)$ are greater than $d(o, o_j)$ for any $o \in L - \{o_i\}$, then $c_{ihj} = 0$.
2. If $d(o_i, o_j) = d(L, o_j)$, then two cases must be considered depending on the distance $e(o_j)$ from e_j to the second closest object of S.
   (a) If $d(o_h, o_j) < e(o_j)$, then $c_{ihj} = d(o_h, o_j) - d(S, o_j)$.
   (b) If $d(o_h, o_j) \geq e(o_j)$, then $c_{ihj} = e(o_j) - d(S, o_j)$.

   In either of these two subcases, we have

$$c_{ihj} = \min\{d(o_h, o_j), e_j\} - d(o_i, o_j).$$

3. If $d(o_i, o_j) > d(L, o_j)$ (i.e., o_j is more distant from o_i than from at least one other selected object) and $d(o_h, o_j) < d(L, o_j)$ (which means that o_j is closer to o_h than to any selected object), then $c_{ihj} = d(o_h, o_j) - d(S, o_j)$.

The cost of the swap is $C(o_i, o_h) = \sum_{o_j \in R} c_{ihj}$. The pair that minimizes $C(o_i, o_j)$ is selected. If $C(o_i, o_j) < 0$, then the swap is carried out. All potential swaps are considered.

The algorithm halts when no useful swap exists, that is, no swap with negative cost can be found.

The pseudocode of the algorithm is

k_means_PAM{
construct the set L of k medoids;
repeat
compute the costs $C(o_i, o_h)$ for $o_i \in L$ and $o_h \in R$;
select the pair (o_i, o_h) that corresponds to the minimum $m = C(o_i, o_h)$;
until $(m > 0)$;
}

Note that inside the loop **repeat**... **until** there are $l(n-l)$ pairs of objects to be examined and for each pair we need to involve $n - l$ nonselected objects. Thus, one execution of the loop requires $O(l(n-l)^2)$ and the total execution may require up to $O\left(\sum_{l=1}^{n-l} l(n-l)^2\right)$, which is $O(n^4)$. Thus, the usefulness of PAM is limited to rather small data set (no more than a few hundred objects).

6.8 LIMITATIONS OF CLUSTERING

As we stated before, an exclusive clustering of a set of objects S is a partition of S whose blocks are the *clusters*. A clustering method starts with a definite dissimilarity on S and generates a clustering. This is formalized in the next definition.

Definition 8 Let S be a set of objects and let $\mathcal{D}'_S$ be the set of definite dissimilarities that can be defined on S.

A clustering function on S is a mapping $f : \mathcal{D}'_S \longrightarrow \text{PART}(S)$.

Example 12 Let $g : \mathbb{R}_{\geq 0} \longrightarrow \mathbb{R}_{\geq 0}$ be a continuous, nondecreasing, and unbounded function and let $S \subseteq \mathbb{R}^n$ be a finite subset of $\mathbb{R}^n$. For $k \in \mathbb{N}$ and $k \geq 2$, define a (g, k)-clustering function as follows.

Begin by selecting a set T of k points from S such that the function $\Lambda_d^g(T) = \sum_{x \in S} g(d(x, T))$ is minimized. Here $d(x, T) = \min\{d(x, t) | t \in T\}$. Then, define a partition of S into k clusters by assigning each point to the point in T that is the closest and breaking the ties using a fixed (but otherwise arbitrary) order on the set of points. The clustering function defined by (d, g), denoted by f^g maps d to this partition.

The k-median clustering function is obtained by choosing $g(x) = x$ for $x \in \mathbb{R}_{\geq 0}$; the k-means clustering function is obtained by taking $g(x) = x^2$ for $x \in \mathbb{R}_{\geq 0}$. □

Definition 9 Let κ be a partition of S and let $d, d' \in \mathcal{D}'_S$. The definite dissimilarity d' is a κ-transformation of d if the following conditions are satisfied:

(i) If $x \equiv_\kappa y$, then $d'(x, y) \leq d(x, y)$;
(ii) If $x \not\equiv_\kappa y$, then $d'(x, y) > d(x, y)$.

In other words, d' is a κ-transformation of d if for two objects that belong to the same κ-cluster $d'(x, y)$ is smaller than $d(x, y)$, while for two objects that belong to two distinct clusters $d'(x, y)$ is larger than $d(x, y)$.

Next, we consider three desirable properties of a clustering function.

Definition 10 Let S be a set and let $f : \mathcal{D}'_S \longrightarrow \text{PART}(S)$ be a clustering function. The function f is

(i) scale invariant, if for every $d \in \mathcal{D}'_S$ and every $\alpha > 0$ we have $f(d) = f(\alpha d)$;
(ii) rich, if $\text{Ran}(f) = \text{PART}(S)$;
(iii) consistent, if for every $d, d' \in \mathcal{D}'_S$ and $\kappa \in \text{PART}(S)$ such that $f(d) = \kappa$ and d' is a κ-transformation of d we have $f(d') = \kappa$,

Unfortunately, as we shall see in Theorem 14, established in the work by Kleinburg [8], there is no clustering function that enjoys all three properties.

The following definition will be used in the proof of Lemma 2.

Definition 11 A dissimilarity $d \in \mathcal{D}'_S$ is (a, b)-conformant to a clustering κ if $x \equiv_\kappa y$ implies $d(x, y) \leq a$ and $x \not\equiv_\kappa y$ implies $d(x, y) \geq b$.

A dissimilarity is *conformant* to a clustering κ if it is (a, b)-conformant to κ for some pair of numbers (a, b).

Note that if d' is a κ-transformation of d, and d is (a, b)-conformant to κ, then d' is also (a, b)-conformant to κ.

Definition 12 Let $\kappa \in \text{PART}(S)$ be a partition on S and f be a clustering function on S. A pair of positive numbers (a, b) is *κ-forcing with respect to f* if for every $d \in \mathcal{D}'_S$ that is (a, b)-conformant to κ we have $f(d) = \kappa$.

Lemma 2 *If f is a consistent clustering function on a set S, then for any partition $\kappa \in \text{Ran}(f)$ there exist $a, b \in \mathbb{R}_{>0}$ such that the pair (a, b) is κ-forcing.*

Proof. For $\kappa \in \text{Ran}(f)$ there exists $d \in \mathcal{D}'_S$ such that $f(d) = \kappa$. Define the numbers

$$a_{\kappa,d} = \min\{d(x, y) | x \neq y, x \equiv_\kappa y\},$$
$$b_{\kappa,d} = \max\{d(x, y) | x \not\equiv_\kappa y\}.$$

In other words, $a_{\kappa,d}$ is the smallest d value for two distinct objects that belong to the same κ-cluster, and $b_{\kappa,d}$ is the largest d value for two objects that belong to different κ-clusters.

Let (a, b) a pair of positive numbers such that $a \leq a_{\kappa,d}$ and $b \geq b_{\kappa,d}$. If d' is a definite dissimilarity that is (a, b)-conformant to κ, then $x \equiv_\kappa y$ implies $d'(x, y) \leq a \leq a_{\kappa,d} \leq d(x, y)$ and $x \not\equiv_\kappa y$ implies $d'(x, y) \geq b > b_{\kappa,d} > d(x, y)$, so d' is a κ-transformation of d. By the consistency property of f, we have $f(d') = \kappa$. This implies that (a, b) is κ-forcing. ■

Theorem 13 *If f is a scale-invariant and consistent clustering function on a set S, then its range is an antichain in poset* $(\text{PART}(S), \leq)$.

Proof. This statement is equivalent to saying that for any scale invariant and consistent clustering function no two distinct partitions of S that are values of f are comparable.

Suppose that there are two clusterings, κ_0 and κ_1, in the range of a scale-invariant and consistent clustering such that $\kappa_0 < \kappa_1$.

Let (a_i, b_i) be a κ_i-forcing pair for $i = 0, 1$, where $a_0 < b_0$ and $a_1 < b_1$. Let a_2 be a number such that $a_2 \leq a_1$ and choose ϵ such that

$$0 < \epsilon < \frac{a_0 a_2}{b_0}.$$

By Exercise 3 construct a distance d such that:

1. for any points x, y that belong to the same block of π_0, $d(x, y) \leq \epsilon$;
2. for points that belong to the same cluster of π_1, but not to the same cluster of π_0, $a_2 \leq d(x, y) \leq a_1$;
3. for points that do not belong to the same cluster of π_1, $d(x, y) \geq b_1$.

The distance d is (a_1, b_1)-conformant to π_1 and so we have $f(d) = \pi_1$. Take $\alpha = b_0/a_2$, and define $d' = \alpha d$. Since f is scale invariant, we have $f(d') = f(d) = \pi_1$. Note that for points x, y that belong to the same cluster of κ_0 we have

$$d'(x, y) \leq \frac{\epsilon b_0}{a_2} < a_0,$$

while for points x, y that do not belong to the same cluster of κ_0 we have

$$d'(x, y) \geq \frac{a_2 b_0}{a_2} \geq b_0.$$

Thus, d' is (a_0, b_0)-conformant to κ_0, and so we must have $f(d') = \kappa_0$. Since $\kappa_0 \neq \kappa_1$, this is a contradiction. ■

Theorem 14 (Kleinberg's impossibility theorem) *If $|S| \geq 2$, there is no clustering function that is scale invariant, rich, and consistent.*

Proof. If S contains at least two elements than the poset $(\mathrm{PART}(S), \leq)$ is not an antichain. Therefore, this statement is a direct consequence of Theorem 13. ■

Theorem 15 *For every antichain A of the poset* $(\mathrm{PART}(S), \leq)$ *there exists a clustering function f that is scale invariant and consistent such that* $\mathrm{Ran}(f) = A$.

Proof. Suppose that A contains more than one partition. We define $f(d)$ as the first partition $\pi \in A$ (in some arbitrary but fixed order) that minimizes the quantity:

$$\Phi_d(\pi) = \sum_{x \equiv_\pi y} d(x, y).$$

Note that $\Phi_{\alpha d} = \alpha \Phi_d$. Therefore, f is scale invariant.

We need to prove that every partition of A is in the range of f.

For a partition $\rho \in A$ define d such that $d(x, y) < 1/|S|^3$ if $x \equiv_\rho y$ and $d(x, y) \geq 1$ otherwise. Observe that $\Phi_d(\rho) < 1$. Suppose that $\Phi_d(\theta) < 1$. The definition of d means that

$$\Phi_d(\theta) = \sum_{x \equiv_\theta y} d(x, y) < 1,$$

so for all pairs $(x, y) \in \equiv_\theta$ we have $d(x, y) < 1/|S|^3$, which means that $x \equiv_\rho y$. Therefore, we have $\pi < \rho$. Since A is an antichain, it follows that ρ must minimize Φ_d over all partitions of A, and consequently, $f(d) = \rho$.

To verify the consistency of f suppose that $f(d) = \pi$ and let d' be a π-transformation of d. For $\sigma \in \mathrm{PART}(S)$ define $\delta(\sigma)$ as $\Phi_d(\sigma) - \Phi_{d'}(\sigma)$. For $\sigma \in A$ we have

$$\begin{aligned}
\delta(\sigma) &= \sum_{x \equiv_\sigma y} (d(x, y) - d'(x, y)) \\
&\leq \sum_{\substack{x \equiv_\sigma y \\ \text{and } x \equiv_\pi y}} (d(x, y) - d'(x, y)) \\
&\quad \text{(only terms corresponding to pairs in the same cluster are nonnegative)} \\
&\leq \delta(\pi) \\
&\quad \text{(every term corresponding to a pair in the same cluster is nonnegative).}
\end{aligned}$$

Consequently,

$$\Phi_d(\sigma) - \Phi_{d'}(\sigma) \leq \Phi_d(\pi) - \Phi_{d'}(\pi),$$

or $\Phi_d(\sigma) - \Phi_d(\pi) \leq \Phi_{d'}(\sigma) - \Phi_{d'}(\pi)$. Thus, if π minimizes $\Phi_d(\pi)$, then $\Phi_d(\sigma) - \Phi_d(\pi) \geq 0$ for every $\sigma \in A$, and therefore, $\Phi_{d'}(\sigma) - \Phi_{d'}(\pi) \geq 0$, which means that π also minimizes $\Phi_{d'}(\pi)$. This implies $f(d') = \pi$, which shows that f is consistent. ■

Example 13 It is possible to show that for $k \geq 2$ and for sufficiently large sets of objects the clustering function f^g introduced in Example 12 is not consistent.

Suppose that $\kappa = \{C_1, C_2, \ldots, C_k\}$ is a partition of S and d is a definite dissimilarity on S such that $d(x, y) = r_i$ if $x \neq y$ and $\{x, y\} \subseteq C_i$ for some $1 \leq i \leq k$ and $d(x, y) = r + a$ if x and y belong to two distinct blocks of κ, where $r = \max\{r_i | 1 \leq i \leq k\}$ and $a > 0$.

Suppose that T is a set of k members of S. Then, the value of $g(d(x, T))$ is $g(r)$ if the closest member of T is in the same block as x and is $g(r + a)$ otherwise. This means that the smallest value of $\Lambda_d^g(T) = \sum_{x \in C_i} g(d(x, T))$ is obtained when each block C_i contains a member t_i of T for $1 \leq i \leq k$ and the actual value is $\Lambda_d^g(T) = \sum_{i=1}^{k}(|C_i| - 1)r^2 = (|S| - k)r^2$.

Consider now a partition $\kappa' = \{C_1', C_1'', C_2, \ldots, C_k\}$, where $C_1 = C_1' \cup C_1''$, so $\kappa' < \kappa$. Choose r' to be a positive number such that $r' < r$ and define the dissimilarity d' on S such that $d'(x, y) = r'$ if $x \neq y$ and $x \equiv_{\kappa'} y$ and $d'(x, y) = d(x, y)$ otherwise. Clearly, d' is a κ-transformation of d. The minimal value for $\Lambda_d^g(T')$ will be achieved when T' consists of $k + 1$ points, one in each of the block of κ'; as a result, the value of the clustering function for d' will be $\kappa' \neq \kappa$, which shows that no clustering function obtained by this technique is consistent. □

6.9 CLUSTERING QUALITY

There are two general approaches for evaluating the quality of a clustering: *unsupervised evaluation* that measures the cluster cohesion and the separation between clusters and *supervised evaluation* that measures the extent to which the clustering we analyze matches a partition of the set of objects that is specified by an external labeling of the objects.

6.9.1 Object Silhouettes

The *silhouette method* is an unsupervised method for evaluation of clusterings that computes certain coefficients for each object. The set of these coefficients allows an evaluation of the quality of the clustering.

Let $O = \{u_1, \ldots, u_n\}$ be a collection of objects, $d : O \times O \longrightarrow \mathbb{R}_+$ be a dissimilarity on O, and let $\kappa : O \longrightarrow \{C_1, \ldots, C_k\}$ be a clustering function.

Suppose that $\kappa(u_i) = C_\ell$. The (κ, d)-average dissimilarity is the function $a_{k,d} : O \longrightarrow \mathbb{R}$ given by

$$a_{\kappa,d}(u_i) = \frac{\sum\{d(u_i, u) | \kappa(u) = \kappa(u_i) \text{ and } u \neq u_i\}}{|\kappa(u_i)|},$$

that is, the average dissimilarity of u_i to all objects of $\kappa(u_i)$, the cluster to which u_i is assigned.

For a cluster C and an object u_i let

$$d(u_i, C) = \frac{\sum\{d(u_i, u) | \kappa(u) = C\}}{|C|}$$

be the average dissimilarity between u_i and the objects of the cluster C.

Definition 13 Let $\kappa : O \longrightarrow \{C_1, \ldots, C_k\}$ be a clustering function. A *neighbor of* u_i is a cluster $C \neq \kappa(u_i)$ for which $d(u_i, C)$ is minimal.

In other words, a neighbor of an object u_i is "the second best choice" for a cluster for u_i. Let $b : O \longrightarrow \mathbb{R}$ be the function defined by

$$b_{\kappa,d}(u_i) = \min\{d(u_i, C) | C \neq \kappa(u_i)\}.$$

If κ and d are clear from context, we shall simply write $a(u_i)$ and $b(u_i)$ instead of $a_{\kappa,d}(u_i)$ and $b_{\kappa,d}(u_i)$, respectively.

Definition 14 The *silhouette* of the object u_i for which $|\kappa(u_i)| \geq 2$ is the number $\text{sil}(u_i)$ given by

$$\text{sil}(u_i) = \begin{cases} 1 - \dfrac{a(u_i)}{b(u_i)} & \text{if } a(u_i) < b(u_i) \\ 0 & \text{if } a(u_i) = b(u_i) \\ \dfrac{b(u_i)}{a(u_i)} - 1 & \text{if } a(u_i) > b(u_i). \end{cases}$$

Equivalently, we have

$$\text{sil}(u_i) = \frac{b(u_i) - a(u_i)}{\max\{a(u_i), b(u_i)\}}$$

for $u_i \in O$.

If $\kappa(u_i) = 1$, then $s(u_i) = 0$.

Observe that $-1 \leq \text{sil}(u_i) \leq 1$. When $\text{sil}(u_i)$ is close to 1, this means that $a(u_i)$ is much smaller than $b(u_i)$ and we may conclude that u_i is well classified. When $\text{sil}(u_i)$ is near 0, it is not clear which is the best cluster for u_i. Finally, if $\text{sil}(u_i)$ is close to -1, the average distance from u to its neighbor(s) is much smaller than the average distance between u_i and other objects that belong to the same cluster $\kappa(u_i)$. In this case, it is clear that u_i is poorly classified.

Definition 15 Let average silhouette width of a cluster C is

$$\text{sil}(C) = \frac{\sum\{\text{sil}(u) | u \in C\}}{|C|}.$$

The average silhouette width of a clustering κ is

$$\text{sil}(\kappa) = \frac{\sum\{\text{sil}(u) | u \in O\}}{|O|}.$$

The silhouette of a clustering can be used for determining the "optimal" number of clusters. If the silhouette of the clustering is above 0.7, we have a strong clustering.

6.9.2 Supervised Evaluation

Suppose that we intend to evaluate the accuracy of a clustering algorithm $\mathcal{A}$ on a set of objects S relative to a collection of classes on S that forms a partition σ of S. In other words, we wish to determine the extent to which the clustering produced by $\mathcal{A}$ coincides with the partition determined by the classes.

If the set S is large, the evaluation can be performed by extracting a random sample T from S, applying $\mathcal{A}$ to T, and then comparing the clustering partition of T computed by $\mathcal{A}$ and the partition of T into the preexisting classes.

Let $\kappa = \{C_1, \ldots, C_m\}$ be the clustering partition of T and let $\sigma = \{K_1, \ldots, K_n\}$ be the partition of T of classes. The evaluation is helped by $n \times m$ matrix $\mathbf{Q}$, where $q_{ij} = |C_i \cap K_j|$ named the confusion matrix.

We can use distances associated with the generalized entropy, $d_\beta(\kappa, \sigma)$, to evaluate the distinction between these partitions. This was already observed by Rand [11], who proposed as a measure the cardinality of the symmetric difference of the sets of pairs of objects that belong to the equivalences that correspond to the two partitions.

Frequently, one uses the conditional entropy

$$\mathcal{H}(\sigma|\kappa) = \sum_{i=1}^{m} \frac{|C_i|}{|T|} \mathcal{H}(\sigma_{C_i}) = \sum_{i=1}^{m} \frac{|C_i|}{|T|} \sum_{j=1}^{n} \frac{|C_i \cap K_j|}{|C_i|} \log_2 \frac{|C_i \cap K_j|}{|C_i|}$$

to evaluate the "purity" of the clusters C_i relative to the classes $K_1, \ldots, K_n$. Low values of this number indicate a high degree of purity.

Some authors [14] define the *purity* of a cluster C_i as a as $\text{pur}_\sigma(C_i) = \max_j |C_i \cap K_j|/|C_i|$ and the purity of the clustering κ relative to σ as

$$\text{pur}_\sigma(\kappa) = \sum_{i=1}^{n} \frac{|C_i|}{|T|} \text{pur}_\sigma(C_i).$$

Larger values of the purity indicate better clusterings (from the point of view of the matching with the class partition of the set of objects).

Example 14 Suppose that a set of 1000 objects consists of three classes of objects K_1, K_2, K_3, where $|K_1| = 500$, $|K_2| = 300$, and $|K_1| = 200$. Two clustering algorithms $\mathcal{A}$ and $\mathcal{A}'$ yield the clusterings $\kappa = \{C_1, C_2, C_3\}$ and $\kappa' = \{C'_1, C'_2, C'_3\}$ and the confusion matrices Q and Q', respectively:

	K_1	K_2	K_3
C_1	400	0	25
C_2	60	200	75
C_3	40	100	100

and

	K_1	K_2	K_3
C'_1	60	0	180
C'_2	400	50	0
C'_3	40	250	20

The distances $d_2(\kappa, \sigma)$ and $d_2(\kappa', \sigma)$ are 0.5218 and 0.4204 suggesting that the clustering κ' produced by the second algorithm is closer to the partition in classes.

As expected, the purity of the first clustering, 0.7, is smaller than the purity of the second clustering, 0.83. □

Another measure of clustering quality proposed in the work by Ray and Turi [12] applies to objects in $\mathbb{R}^n$ and can be applied, for example, to the clustering that results from the k-means method, the validity of clustering. Let $\pi = \{U_1, \ldots, U_k\}$ be a clustering of N objects, $\boldsymbol{c}_1, \ldots, \boldsymbol{c}_k$ the centroids of the clusters, then the clustering validity is

$$\text{val}(\pi) = \frac{\text{sse}(\pi)}{N \min_{i<j} d^2(\boldsymbol{c}_i, \boldsymbol{c}_j)}.$$

The variety of clustering algorithms is very impressive and it is very helpful to the reader to consult two excellent surveys of clustering algorithms [2,5] before exploring in depth this domain.

6.10 FURTHER READINGS

Several general introductions in data mining [13,14] provide excellent references for clustering algorithms. Basic reference books for clustering algorithms are authored by Jain and Dubes [6] and Kaufmann and Rousseeuw [7]. Recent surveys such as those by Berkhin [2] and Jain et al. [5] allow the reader to get familiar with current issues in clustering.

6.11 EXERCISES

1. Let d be a ultrametric and let $S(x, y)$ be the set of all non-null sequences $\mathbf{s} = (s_1, \ldots, s_n) \in \mathbf{Seq}(S)$ such that $s_1 = x$ and $s_n = y$. If d is a ultrametric prove that $d(x, y) \leq \min\{\text{amp}_d(\mathbf{s}) | \mathbf{s} \in S(x, y)\}$ (Exercise 1).
2. Let S be a set, π be a partition of S, and let a, b be two numbers such that $a < b$. Prove that the mapping $d : S^2 \longrightarrow \mathbb{R}_{\geq 0}$ given by $d(x, x) = 0$ for $x \in S$, $d(x, y) = a$ if $x \neq y$ and $\{x, y\} \subseteq B$ for some block B of π and $d(x, y) = b$, otherwise is an ultrametric on S.
3. Prove the following extension of the statement from Exercise 2.

   Let S be a set, $\pi_0 < \pi_1 < \cdots < \pi_{k-1}$ be a chain of partitions on S, and let $a_0 < a_1 \ldots < a_{k-1} < a_k$ be a chain of positive reals.

   Prove that the mapping $d : S^2 \longrightarrow \mathbb{R}_{\geq 0}$ given by

$$d(x, y) = \begin{cases} 0 & \text{if } x = y \\ a_0 & \text{if } x \neq y \text{ and } x \equiv_{\pi_0} y \\ \vdots & \qquad \vdots \\ a_{k-1} & \text{if } x \not\equiv_{\pi_{k-2}} y \text{ and } x \equiv_{\pi_{k-1}} y \\ a_k & \text{if } x \not\equiv_{\pi_{k-1}} y \end{cases}$$

   is an ultrametric on S.
4. Let $f : \mathbb{R}_{\geq 0} \longrightarrow \mathbb{R}_{\geq 0}$ be a function that satisfies the following conditions:
   (a) $f(x) = 0$ if and only if $x = 0$;
   (b) f is monotonic on $\mathbb{R}_{\geq 0}$, that is, $x \leq y$ implies $f(x) \leq f(y)$ for $x, y \in \mathbb{R}_{\geq 0}$;
   (c) f is subadditive on $\mathbb{R}_{\geq 0}$, that is, $f(x + y) \leq f(x) + f(y)$ for $x, y \in \mathbb{R}_{\geq 0}$.

   (c) Prove that if d is a metric on a set S, then fd is also a metric on S.
   (d) Prove that if d is a metric on S, the $\sqrt{d}$ and $d/1 + d$ are also metrics on S; what can be said about d^2?
5. A function $F : \mathbb{R} \geq 0 \longrightarrow \mathbb{R}$ is *convex* if for every $s, t \in \mathbb{R}_{\geq 0}$ and $a \in [0, 1]$ we have $F(as + (1 - a)t) \leq aF(s) + (1 - 1)F(t)$.
   (a) Prove that if $F(0) = 0$, F is monotonic and convex, then F is subadditive.
   (b) Prove that if f is a metric on the set S, then the function given by

$$d'(x, y) = 1 - e^{-kd(x,y)},$$

   where k is a positive constant and $x, y \in S$ is also a metric on S. This metric is known as the Schoenberg transform of d.
6. Let S be a finite set and let $d : S^2 \longrightarrow \mathbb{R}_{\geq 0}$ be a dissimilarity. Prove that there exists $a \in \mathbb{R}_{\geq 0}$ such that the dissimilarity d_a defined by $d_a(x, y) = (d(x, y))^a$ satisfies the triangular inequality.

Hint: Observe that $\lim_{a \to 0} d_a(x, y)$ is a dissimilarity that satisfies the triangular inequality.

7. Prove Theorem 2.
8. Let (S, d) be a finite metric space. Prove that the functions $D, E : \mathcal{P}(S)^2 \longrightarrow \mathbb{R}$ defined by

$$D(U, V) = \max\{d(u, v) | u \in U, v \in V\}$$

$$E(U, V) = \frac{1}{|U| \cdot |V|} \sum \{d(u, v) | u \in U, v \in V\}$$

for $U, V \in \mathcal{P}(S)$ are metrics on $\mathcal{P}(S)$.

9. Prove that if we replace max by min in Exercise 8, then the resulting function $F : \mathcal{P}(S)^2 \longrightarrow \mathbb{R}$ defined by

$$D(U, V) = \min\{d(u, v) | u \in U, v \in V\}$$

for $U, V \in \mathcal{P}(S)$ is not a metric on $\mathcal{P}(S)$, in general.

10. Prove that the ultrametric inequality implies the triangular inequality; also, show that both the triangular inequality and definiteness imply evenness for an ultrametric.
11. Let $(\mathcal{T}, v_0)$ be a finite rooted tree, V be the set of vertices of the tree $\mathcal{T}$, and let S be a finite, nonempty set such that the rooted tree $(\mathcal{T}, v_0)$ has $|S|$ leaves. Consider a function $M : V \longrightarrow \mathcal{P}(S)$ defined as follows:

    (a) the tree $\mathcal{T}$ has $|S|$ leaves and each for each leaf v the set $M(v)$ is a distinct singleton of S;

    (b) if an interior vertex v of the tree has the descendants $v_1, v_2, \ldots, v_n$, then $M(v) = \bigcup_{i=1}^{n} M(v_i)$.

    Prove that the collection of sets $\{M(v) | v \in V\}$ is a hierarchy on S.

12. Apply hierarchical clustering to the data set given in Example 9 using the average-link method, the centroid method and the Ward method. Compare the shapes of the clusters that are formed during the aggregation process. Draw the dendrograms of the clusterings.
13. Using a random number generator produce h sets of points in $\mathbb{R}^n$ normally distributed around h given points in $\mathbb{R}^n$. Use k-means to cluster these points with several values for k and compare the quality of the resulting clusterings.
14. A variant of the k-means clustering introduced in the work by Stainbach [13] is the *bisecting k-means algorithm* described below. The parameters are S, the set of objects to be clustered; k, the desired number of clusters; and *nt*, the number of trial bisections.

bisecting k-means{
  set_of_clusters = $\{S\}$;
  while (|*set_of_clusters*| < k)

```
        extract a cluster C from the set_of_clusters;
        k = 0;
        for i = 1 to nt do
            let C0i, C1i be the two clusters obtained from C by bisecting C
                using standard k-means (k = 2);
            if (i = 1) then s = sse({C0i, C1i});
            if (sse({C0i, C1i}) ≤ s) then
                k = i;
                s = sse({C0i, C1i});
            endif;
        endfor;
        add C0k, C1k to set_of_clusters;
    endwhile
}
```

The cluster C that is bisected may be the largest cluster, or the cluster having the largest sse.

Evaluate the time performance of bisecting k-means compared with the standard k-means and with some variant of a hierarchical clustering.

15. One of the issues that the k-means algorithm must confront is that the number of clusters k must be provided as an input parameter. Using clustering validity design an algorithm that identifies local maxima of validity (as a function of k) to provide a basis for a good choice of k. For a solution that applies to image segmentation, see the work by Ray and Turi.

REFERENCES

1. Birkhoff G. Lattice Theory. 3rd ed. Providence, RI: American Mathematical Society; 1967.
2. Berkhin P. A survey of clustering data mining techniques. In: Kogan J, Nicholas C, Teboulle M, editors, Grouping Multidimensional Data—Recent Advances in Clustering. Berlin: Springer-Verlag; 2006. p 25–72.
3. Berkhin P, Becher J. Learning simple relations: theory and applications. Proceedings of the 2nd SIAM International Conference on Data Mining; Arlington, VA; 2002.
4. http://www2.sims.berkeley.edu/research/projects/how-much-info/
5. Jain AK, Murty MN, Flynn PJ. Data clustering: a review. ACM Comput Surv 1999;31:264–323.
6. Jain AK, Dubes RC. Algorithms for Clustering Data. Englewood Cliffs: Prentice Hall; 1988.
7. Kaufman L, Rousseeuw PJ. Finding Groups in Data — An Introduction to Cluster Analysis. New York: Wiley-Interscience; 1990.
8. Kleinberg J. An impossibility theorem for clustering. Proceedings of the 16th Conference on Neural Information Processing Systems; 2002.

9. Kurita T. An efficient agglomerative clustering algorithm using a heap. Pattern Recogn 1991;24:205–209.

10. Ng RN, Han J. Efficient and effective clustering methods for spatial data mining. Proceedings of the 20th VLDB Conference; Santiago, Chile; 1994. p 144–155.

11. Rand WM. Objective criteria for the evaluation of clustering methods. J Am Stat Assoc 1971;61:846–850.

12. Ray S, Turi R. Determination of number of clusters in *k*-means clustering in colour image segmentation. Proceedings of the 4th International Conference on Advances in Pattern Recognition and Digital Technology; Calcutta, India. New Delhi, India: Narosa Publishing House. p 137–143.

13. Steinbach M, Karypis G, Kumar V. A comparison of document clustering techniques. KDD Workshop on Text Mining; 2000.

14. Tan PN, Steinbach M, Kumar V. Introduction to Data Mining. Reading, MA: Addison-Wesley; 2005.

CHAPTER 7

Data Mining Algorithms II: Frequent Item Sets

DAN A. SIMOVICI

7.1 INTRODUCTION

Association rules have received a lot of attention in data mining due to their many applications in marketing, advertising, inventory control, and many other areas. The area of data mining has been initiated in the seminal paper [5].

A typical supermarket may well have several thousand items on its shelves. Clearly, the number of subsets of the set of items is immense. Even though a purchase by a customer involves a small subset of this set of items, the number of such subsets is very large. In principle, there are $\sum_{i=1}^{5} \binom{10000}{i}$ subsets T having no more than 5 elements of a set that has 10,000 items and this is indeed a large number!

The supermarket is interested in identifying associations between item sets; for example, it may be interested to know how many of the customers who bought bread and cheese also bought milk. This knowledge is important because if it turns out that many of the customers who bought bread and cheese also bought milk, the supermarket will place milk physically close to bread and cheese in order to stimulate the sales of milk. Of course, such a piece of knowledge is especially interesting when there is a substantial number of customers who buy all three items and a large fraction of those individuals who buy bread and cheese also buy milk. Informally, if this is the case, we shall say that we have identified the association rule bread cheese $\rightarrow$ milk. Two numbers will play a role in evaluating such a rule: N_{bcm}/N and $N_{\text{bcm}}/N_{\text{bc}}$. Here, N is the total number of purchases, N_{bcm} denotes the number of transactions involving bread, cheese, and milk, and N_{bc} gives the number of transactions involving bread and cheese. The first number is known as the *support* of the association rule; the second is its *confidence* and approximates the probability that a customer who bought bread and cheese will buy milk.

Thus, identifying association rules requires the capability to identify item sets that occur in large sets of transactions; these are the *frequent item sets*. Identifying association rules amounts essentially to finding frequent item sets. If N_{bcm} is

Handbook of Applied Algorithms: Solving Scientific, Engineering and Practical Problems
Edited by Amiya Nayak and Ivan Stojmenović

large, then N_{bc} is larger still. We formalize this problem and explore its algorithmic aspects.

7.2 FREQUENT ITEM SETS

Suppose that I is a finite set; we refer to the elements of I as *items*.

Definition 1 A transaction data set over I is a function $T : \{1, \ldots, n\} \longrightarrow \mathcal{P}(I)$. The set $T(k)$ is the kth transaction of T. The numbers $1, \ldots, n$ are the transaction identifiers (tids).

An example of a transaction set is the set of items present in the shopping cart of a consumer who completed a purchase in a store.

Example 1 The table below describes a transaction data set over the set of over-the-counter medicines in a drugstore.

Transactions	Content
$T(1)$	{Aspirin, Vitamin C}
$T(2)$	{Aspirin, Sudafed}
$T(3)$	{Tylenol}
$T(4)$	{Aspirin, Vitamin C, Sudafed}
$T(5)$	{Tylenol, Cepacol}
$T(6)$	{Aspirin, Cepacol}
$T(7)$	{Aspirin, Vitamin C}

The same data set can be presented as a 0/1 table as follows:

	Aspirin	Vitamin C	Sudafed	Tylenol	Cepacol
$T(1)$	1	1	0	0	0
$T(2)$	1	0	1	0	0
$T(3)$	0	0	0	1	0
$T(4)$	1	1	1	0	0
$T(5)$	1	0	0	0	1
$T(6)$	1	0	0	0	1
$T(7)$	1	1	0	0	0

The entry in the row $T(k)$ and the column i_j is set to 1 if $i_j \in T(k)$; otherwise, it is set to 0. □

Example 1 shows that we have the option of two equivalent frameworks for studying frequent item sets: tables or transaction item sets.

Given a transaction data set T on the set I, we would like to determine those subsets of I that occur often enough as values of T.

Definition 2 Let $T : \{1, \ldots, n\} \longrightarrow \mathcal{P}(I)$ be a transaction data set over a set of items I. The support count of a subset K of the set of items I in T is the number $\text{suppcount}_T(K)$ given by

$$\text{suppcount}_T(K) = |\{k | 1 \leq k \leq n \text{ and } K \subseteq T(k)\}|.$$

The support of an item set K is the number

$$\text{supp}_T(K) = \frac{\text{suppcount}_T(K)}{n}.$$

Example 2 For the transaction data set T considered in Example 1 we have

$$\text{suppcount}_T(\{\text{Aspirin, VitaminC}\}) = 3,$$

because $\{$Aspirin, Vitamin C$\}$ is a subset of three of the sets $T(k)$. Therefore, $\text{supp}_T(\{\text{Aspirin, Vitamin C}\}) = \frac{3}{7}$. □

To simplify our notation we will denote item sets by the sequence of their elements. For instance, a set $\{a, b, c\}$ will be denoted from now on by abc.

Example 3 Let $I = \{i_1, i_2, i_3, i_4\}$ be a collection of items. Consider the transaction data set T given by

$$\begin{aligned} T(1) &= i_1 i_2, \\ T(2) &= i_1 i_3, \\ T(3) &= i_1 i_2 i_4, \\ T(4) &= i_1 i_3 i_4, \\ T(5) &= i_1 i_2, \\ T(6) &= i_3 i_4. \end{aligned}$$

Thus, the support count of the item set $i_1 i_2$ is 3; similarly, the support count of the item set $i_1 i_3$ is 2. Therefore, $\text{supp}_T(i_1 i_2) = \frac{1}{2}$ and $\text{supp}_T(i_1 i_3) = \frac{1}{3}$. □

The following rather straightforward statement is fundamental for the study of frequent item sets.

Theorem 1 *Let $T : \{1, \ldots, n\} \longrightarrow \mathcal{P}(I)$ be a transaction data set over a set of items I. If K and K' are two item sets, then $K' \subseteq K$ implies* $\text{supp}_T(K') \geq \text{supp}_T(K)$.

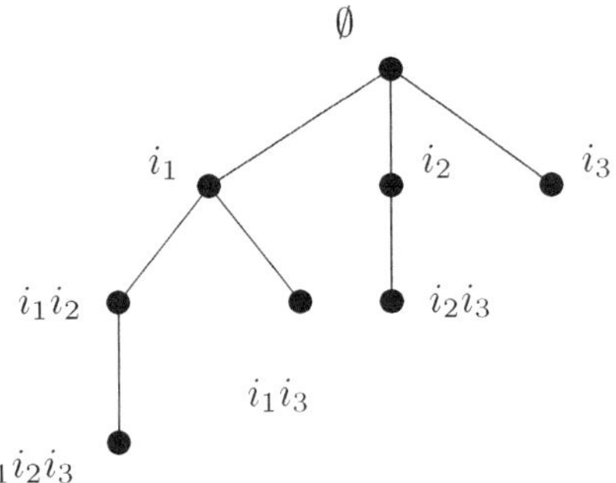

FIGURE 7.1 The Rymon tree of $\mathcal{P}(\{i_1, i_2, i_3\})$.

Proof. Note that every transaction that contains K also contains K'. The statement follows immediately. ■

If we seek those item sets that enjoy a minimum support level relative to a transaction data set T, then it is natural to start the process with the smallest nonempty item sets.

Definition 3 An item set K is μ-frequent relatively to the transaction data set T if $\text{supp}_T(K) \geq \mu$.

We denote by $\mathcal{F}_T^\mu$ the collection of all μ-frequent item sets relative to the transaction data set T, and by $\mathcal{F}_{T,r}^\mu$ the collection of μ-frequent item sets that contain r items for $r \geq 1$.

Note that

$$\mathcal{F}_T^\mu = \bigcup_{r \geq 1} \mathcal{F}_{T,r}^\mu.$$

If μ and T are clear from the context, then we may omit either or both adornments from this notation.

Let $I = \{i_1, \ldots, i_n\}$ be an item set that contains n elements. We use a graphical representation of $\mathcal{P}(I)$, the set of subsets of I, known as the *Rymon tree*.

The root of the tree is $\emptyset$. A vertex $K = i_{p_1} \cdots i_{p_k}$ with $i_{p_1} < i_{p_2} < \cdots < i_{p_k}$ has $n - i_{p_k}$ children $K \cup \{j\}$, where $i_{p_k} < j \leq n$. We shall denote this tree by $\mathcal{R}_I$.

Example 4 Let $I = \{i_1, i_2, i_3\}$. The Rymon tree $\mathcal{R}_I$ is shown in Figure 7.1. □

Let $\mathcal{S}_r$ be the collection of item sets that have r elements. The next theorem suggests a technique for generating $\mathcal{S}_{r+1}$ starting from $\mathcal{S}_r$.

Theorem 2 *Let $\mathcal{R}_I$ be the Rymon tree of the set of subsets of $I = \{i_1, \ldots, i_n\}$. If $W \in \mathcal{S}_{r+1}$, where $r \geq 2$, then there exists a unique pair of distinct sets $U, V \in \mathcal{S}_r$ that has a common immediate ancestor $T \in \mathcal{S}_{r-1}$ in $\mathcal{R}_I$ such that $U \cap V \in \mathcal{S}_{r-1}$ and $W = U \cup V$.*

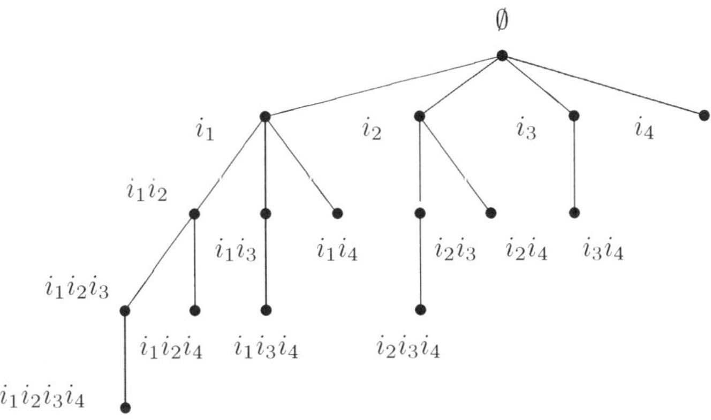

FIGURE 7.2 Rymon tree for $\mathcal{P}(\{i_1, i_2, i_3, i_4\})$.

Proof. Let u, v be the largest and the second largest subscript of an item that occurs in W, respectively. Consider the sets $U = W - \{u\}$ and $V = W - \{v\}$. Both sets belong to $\mathcal{S}_r$. Moreover, $Z = U \cap V$ belongs to $\mathcal{S}_{r-1}$ because it consists of the first $r - 1$ elements of W. Note that both U and V are descendants of Z and that $U \cup V = W$.

The pair (U, V) is unique. Indeed, suppose that W can be obtained in the same manner from another pair of distinct sets U', $V' \in \mathcal{S}_r$, such that U', V' are immediate descendants of a set $Z' \in \mathcal{S}_{r-1}$. The definition of the Rymon tree $\mathcal{R}_I$ implies that $U' = Z' \cup \{i_m\}$ and $V' = Z' \cup \{i_q\}$, where the letters in Z' are indexed by number smaller than $\min\{m, q\}$. Then, Z' consists of the first $r - 1$ symbols of W, so $Z' = Z$. If $m < q$, then m is the second highest index of a symbol in W and q is the highest index of a symbol in W, so $U' = U$ and $V' = V$. ■

Example 5 Consider the Rymon tree of the collection $\mathcal{P}(\{i_1, i_2, i_3, i_4)$ shown in Figure 7.2. The set $i_1i_3i_4$ is the union of the sets i_1i_3 and i_1i_4 that have the common ancestor i_1. □

Next we discuss an algorithm that allows us to compute the collection $\mathcal{F}_T^\mu$ of all μ-frequent item sets for a transaction data set T. The algorithm is known as the *Apriori algorithm*.

We begin with the procedure `apriori_gen` that starts with the collection $\mathcal{F}_{T,k}^\mu$ of frequent item sets for the transaction data set T that contain k elements and generates a collection $\mathcal{C}_{k+1}$ of sets of items that contains $\mathcal{F}_{T,k+1}^\mu$, the collection the frequent item sets that have $k + 1$ elements. The justification of this procedure is based on the next statement.

Theorem 3 *Let T be a transaction data set over a set of items I and let $k \in \mathbb{N}$ such that $k > 1$.*

If W is a μ-frequent item set and $|W| = k + 1$, then there exist a μ-frequent item set Z and two items i_m and i_q such that $|Z| = k - 1$, $Z \subseteq W$, $W = Z \cup \{i_m, i_q\}$ and both $Z \cup \{i_m\}$ and $Z \cup \{i_q\}$ are μ-frequent item sets.

Proof. If W is an item set such that $|W| = k + 1$, then we already know that W is the union of two subsets U, V of I such that $|U| = |V| = k$ and that $Z = U \cap V$ has $k - 1$ elements. Since W is a μ-frequent item set and Z, U, V are subsets of W, it follows that each of these sets is also a μ-frequent item set. ■

Note that the reciprocal statement of Theorem 3 is not true, as the next example shows.

Example 6 Let T be the transaction data set introduced in Example 3. Note that both i_1i_2 and i_1i_3 are $\frac{1}{3}$-frequent item sets; however,

$$\text{supp}_T(i_1i_2i_3) = 0,$$

so $i_1i_2i_3$ fails to be a $\frac{1}{3}$-frequent item set. □

The procedure `apriori_gen` mentioned above is introduced next. This procedure starts with the collection of item sets $\mathcal{F}_{T,k}$ and produces a collection of item sets $\mathcal{C}_{T,k+1}$ that includes the collection of item sets $\mathcal{F}_{T,k+1}$ of frequent item sets having $k + 1$ elements.

```
apriori_gen(μ, F^μ_{T,k}){
        C^μ_{T,k+1} = ∅;
        for each L, M ∈ F^μ_{T,k} such that
          L ≠ M and L ∩ M ∈ F^μ_{T,k−1} do
            begin
             add L ∪ M to C^μ_{T,k+1};
             remove all sets K in C^μ_{T,k+1} where
                 there is a subset of K containing k elements
                 that does not belong to F^μ_{T,k};
            end
}
```

Note that in `apriori_gen` no access to the transaction data set is needed.

The *Apriori* algorithm is introduced next. The algorithm operates on "levels." Each level k consists of a collection $\mathcal{C}^{\mu}_{T,k}$ of candidate item sets of μ-frequent item sets. To build the initial collection of candidate item sets $\mathcal{C}^{\mu}_{T,1}$, every single item set is considered for membership in $\mathcal{C}^{\mu}_{T,1}$. The initial set of frequent item set consists of those singletons that pass the minimal support test. The algorithm alternates between a candidate generation phase (accomplished by using ***apriori_gen***) and an evaluation phase, which involves a data set scan and is, therefore, the most expensive component of the algorithm.

```
Apriori(T, μ){
        C^μ_{T,1} = {{i}|i ∈ I};
        i = 1;
        while (C^μ_{T,i} ≠ ∅) do
        /* evaluation phase */
```

$\mathbb{F}^{\mu}_{T,i} = \{L \in \mathcal{C}^{\mu}_{T,i} | supp_T(L) \geq \mu\}$;
/* candidate generation */
$\mathbb{C}^{\mu}_{T,i+1} = \texttt{apriori\_gen}(\mathbb{F}^{\mu}_{T,i})$;
$i++$;
endwhile;
output $\mathcal{F}^{\mu}_T = \bigcup_{j<i} \mathcal{F}^{\mu}_{T,j}$;
}

Example 7 Let T be the data set given by

Transactions	i_1	i_2	i_3	i_4	i_5
$T(1)$	1	1	0	0	0
$T(2)$	0	1	1	0	0
$T(3)$	1	0	0	0	1
$T(4)$	1	0	0	0	1
$T(5)$	0	1	1	0	1
$T(6)$	1	1	1	1	1
$T(7)$	1	1	1	0	0
$T(8)$	0	1	1	1	1

The support counts of various subsets of $I = \{i_1, \ldots, i_5\}$ are given below:

i_1	i_2	i_3	i_4	i_5
5	6	5	2	5

i_1i_2	i_1i_3	i_1i_4	i_1i_5	i_2i_3	i_2i_4	i_2i_5	i_3i_4	i_3i_5	i_4i_5
3	2	1	3	5	2	3	2	3	2

$i_1i_2i_3$	$i_1i_2i_4$	$i_1i_2i_5$	$i_1i_3i_4$	$i_1i_3i_5$	$i_1i_4i_5$	$i_2i_3i_4$	$i_2i_3i_5$	$i_2i_4i_5$	$i_3i_4i_5$
2	1	1	1	1	1	2	3	2	2

$i_1i_2i_3i_4$	$i_1i_2i_3i_5$	$i_1i_2i_4i_5$	$i_1i_3i_4i_5$	$i_2i_3i_4i_5$
1	1	1	1	2

$i_1i_2i_3i_4i_5$
0

Starting with $\mu = 0.25$ and with $\mathcal{F}^{\mu}_{T,0} = \{\emptyset\}$, the *Apriori* algorithm computes the following sequence of sets:

$$\mathcal{C}^{\mu}_{T,1} = \{i_1, i_2, i_3, i_4, i_5\},$$

$$\mathcal{F}^{\mu}_{T,1} = \{i_1, i_2, i_3, i_4, i_5\},$$

$$\mathcal{C}^{\mu}_{T,2} = \{i_1i_2, i_1i_3, i_1i_4, i_1i_5, i_2i_3, i_2i_4, i_2i_5, i_3i_4, i_3i_5, i_4i_5\},$$

$$\mathcal{F}^{\mu}_{T,2} = \{i_1i_2, i_1i_3, i_1i_5, i_2i_3, i_2i_4, i_2i_5, i_3i_4, i_3i_5, i_4i_5\},$$

$$\mathcal{C}^{\mu}_{T,3} = \{i_1i_2i_3, i_1i_2i_5, i_1i_3i_5, i_2i_3i_4, i_2i_3i_5, i_2i_4i_5, i_3i_4i_5\},$$
$$\mathcal{F}^{\mu}_{T,3} = \{i_1i_2i_3, i_2i_3i_4, i_2i_3i_5, i_2i_4i_5, i_3i_4i_5\},$$
$$\mathcal{C}^{\mu}_{T,4} = \{i_2i_3i_4i_5\},$$
$$\mathcal{F}^{\mu}_{T,4} = \{i_2i_3i_4i_5\},$$
$$\mathcal{C}^{\mu}_{T,5} = \emptyset.$$

Thus, the algorithm will output the collection:

$$\begin{aligned}\mathcal{F}^{\mu}_{T} &= \bigcup_{i=1}^{4} \mathcal{F}^{\mu}_{T,i} \\ &= \{i_1, i_2, i_3, i_4, i_5, i_1i_2, i_1i_3, i_1i_5, i_2i_3, i_2i_4, i_2i_5, i_3i_4, i_3i_5, i_4i_5, \\ &\quad i_1i_2i_3, i_2i_3i_4, i_2i_3i_5, i_2i_4i_5, i_3i_4i_5, i_2i_3i_4i_5\}.\end{aligned}$$

□

Let I be a set of items and $T : \{1, \ldots, n\} \longrightarrow \mathcal{P}(I)$ be a transaction data set. Denote by D the set of transaction identifiers, $D = \{1, \ldots, n\}$. The functions $\text{items}_T : \mathcal{P}(D) \longrightarrow \mathcal{P}(I)$ and $\text{tids}_T : \mathcal{P}(I) \longrightarrow \mathcal{P}(D)$ are defined by

$$\text{items}_T(E) = \bigcap \{T(k) | k \in E\},$$
$$\text{tids}_T(H) = \{k \in D | H \subseteq T(k)\}$$

for every $E \in \mathcal{P}(D)$ and every $H \in \mathcal{P}(I)$.

Note that $\text{suppcount}_T(H) = |\text{tids}_T(H)|$ for every item set $H \in \mathcal{P}(I)$.

The next statement shows that the mappings items_T and tids_T form a Galois connection between the partial ordered sets $\mathcal{P}(D)$ and $\mathcal{P}(I)$ (see the works by Birkhoff [7] and Ganter and Wille [10] for this concept and related results). The use of Galois connections in data mining was initiated in the work by Pasquier et al. [15] and continued in the work by Zaki [19].

Theorem 4 *Let $T : \{1, \ldots, n\} \longrightarrow \mathcal{P}(I)$ be a transaction data set. We have*

1. *if* $E \subseteq E'$, *then* $\text{items}_T(E') \subseteq \text{items}_T(E)$,
2. *if* $H \subseteq H'$, *then* $\text{tids}_T(H') \subseteq \text{tids}_T(H)$,
3. $E \subseteq \text{tids}_T(\text{items}_T(E))$, *and*
4. $H \subseteq \text{items}_T(\text{tids}_T(H))$,

for every $E, E' \in \mathcal{P}(D)$ *and every* $H, H' \in \mathcal{P}(I)$.

Proof. The first two parts of the theorem follow immediately from the definitions of the functions items_T and tids_T.

To prove part (iii) let $k \in E$ be a transaction identifier. Then, the item set $T(e)$ includes $\text{items}_T(E)$, by the definition of $\text{items}_T(E)$. By part (ii), $\text{tids}_T(T(e)) \subseteq \text{tids}_T(\text{items}_T(E))$. Since $e \in \text{tids}_T(T(e))$ it follows that $e \in \text{tids}_T(\text{items}_T(E))$, so $E \subseteq \text{tids}_T(\text{items}_T(E))$.

The argument for part (iv) is similar. ■

Corollary 1 *Let $T : D \longrightarrow \mathcal{P}(I)$ be a transaction data set and let $\boldsymbol{I} : \mathcal{P}(I) \longrightarrow \mathcal{P}(I)$ and $\boldsymbol{D} : \mathcal{P}(D) \longrightarrow \mathcal{P}(D)$ be defined by $\boldsymbol{I}(H) = \text{items}_T(\text{tids}_T(H))$ for $H \in \mathcal{P}(I)$ and $\boldsymbol{D}(E) = \text{tids}_T(\text{items}_T(E))$ for $E \in \mathcal{P}(D)$. Then, $\boldsymbol{I}$ and $\boldsymbol{D}$ are closure operators on I and D, respectively.*

Proof. Let H, H' be two subsets of I such that $H \subseteq H'$. By part (ii) of Theorem 4 we have $\text{tids}_T(H') \subseteq \text{tids}_T(H)$; part (i) of the same theorem yields $\boldsymbol{I}(H) = \text{items}_T(\text{tids}_T(H)) \subseteq \text{items}_T(\text{tids}_T(H')) = \boldsymbol{I}(H')$, so $\boldsymbol{I}$ is monotonic. The proof of monotonicity for $\boldsymbol{D}$ is similar.

Since $E \subseteq \text{tids}_T(\text{items}_T(E))$, by part (i) of Theorem 4 we have

$$\text{items}_T(\text{tids}_T(\text{items}_T(E))) \subseteq \text{items}_T(E).$$

On the contrary, by the expansiveness of $\boldsymbol{I}$ we can write

$$\text{items}_T(E) \subseteq \text{items}_T(\text{tids}_T(\text{items}_T(E))),$$

which implies the equality

$$\text{items}_T(\text{tids}_T(\text{items}_T(E))) = \text{items}_T(E) \tag{7.1}$$

for every $E \in \mathcal{P}(D)$. This, in turn means that

$$\text{tids}_T(\text{items}_T(\text{tids}_T(\text{items}_T(E)))) = \text{tids}_T(\text{items}_T(E)),$$

which proves that $\boldsymbol{D}$ is idempotent. The proof for the idempotency of $\boldsymbol{I}$ makes use of the equality

$$\text{tids}_T(\text{items}_T(\text{tids}_T(H))) = \text{tids}_T(H) \tag{7.2}$$

and is similar; we omit it. ■

Closed sets of items, that is, sets of items H such that $H = \boldsymbol{I}(H)$, can be characterized as follows:

Theorem 5 *Let $T : \{1, \ldots, n\} \longrightarrow \mathcal{P}(I)$ be a transaction data set.*

A set of items H is closed if and only if for every set $L \in \mathcal{P}(I)$ such that $H \subset L$, we have $\text{supp}_T(H) > \text{supp}_T(L)$.

Proof. Suppose that for every superset L of H we have $\text{supp}_T(H) > \text{supp}_T(L)$ and that H is not a closed set of items. Therefore, the set $\boldsymbol{I}(H) = \text{items}_T(\text{tids}_T(H))$ is a superset of H, and consequently $\text{suppcount}_T(H) > \text{suppcount}_T(\text{items}_T(\text{tids}_T(H)))$. Since $\text{suppcount}_T(\text{items}_T(\text{tids}_T(H))) = |\text{tids}_T(\text{items}_T(\text{tids}_T(H)))| = |\text{tids}_T(H)|$, this leads to a contradiction. Thus, H must be closed.

Conversely, suppose that H is a closed set of items, that is

$$H = \boldsymbol{I}(H) = \text{items}_T(\text{tids}_T(H))$$

and let L be a strict superset of H. Suppose that $\text{supp}_T(L) = \text{supp}_T(H)$. This means that $|\text{tids}_T(L)| = |\text{tids}_T(H)|$.

Since $H = \text{items}_T(\text{tids}_T(H)) \subset L$, it follows that

$$\text{tids}_T(L) \subseteq \text{tids}_T(\text{items}_T(\text{tids}_T(H))) = \text{tids}_T(H),$$

which implies the equality $\text{tids}_T(L) = \text{tids}_T(\text{items}_T(\text{tids}_T(H)))$ because the sets $\text{tids}_T(L)$ and $\text{tids}_T(H)$ have the same number of elements. Thus, by equality (7.1), $\text{tids}_T(L) = \text{tids}_T(H)$. In turn, this yields

$$H = \text{items}_T(\text{tids}_T(H)) = \text{items}_T(\text{tids}_T(L)) \supseteq L,$$

which contradicts the initial assumption $H \subset L$. ■

Theorem 6 *For any transaction data set $T : \{1, \ldots, n\} \longrightarrow \mathcal{P}(I)$ and set of items L we have $\text{supp}_T(L) = \text{supp}_T(\boldsymbol{I}(L))$. In other words, the support of an item set in T equals the support of its closure.*

Proof. Equality (7.2) implies that

$$\text{tids}_T(\boldsymbol{I}(L)) = \text{tids}_T(\text{items}_T(\text{tids}_T(L))) = \text{tids}_T(L).$$

Since $\text{suppcount}_T(H) = |\text{tids}_T(H)|$ for every item set H, it follows that

$$\text{suppcount}_T(\boldsymbol{I}(L)) = \text{suppcount}_T(L). \blacksquare$$

A special class of subsets of closed sets is helpful for obtaining a concise representation of μ-frequent item sets.

Definition 4 A μ-maximal frequent item set is a μ-frequent item set that is closed.

Thus, once the μ-maximal frequent item sets have been identified, then all frequent item sets can be obtained as subsets of these sets.

Several improvements of the standard Apriori algorithm are very interesting to explore. Park et al. [14] hash tables used for substantially decreasing the sizes of the candidate sets. In a different direction, an algorithm that picks a random sample from a transaction data set, detects association rules satisfied in this sample, and verifies the results on the remaining transactions has been proposed in the work by Toivonen [18].

7.3 ASSOCIATION RULES

Definition 5 An association rule on an item set I is a pair of nonempty disjoint item sets (X, Y).

Note that if $|I| = n$, then there exist $3^n - 2^{n+1} + 1$ association rules on I. Indeed, suppose that the set X contains k elements; there are $\binom{n}{k}$ ways of choosing X. Once X is chosen, Y can be chosen among the remaining $2^{n-k} - 1$ nonempty subsets of $I - X$. In other words, the number of association rules is

$$\sum_{k=1}^{n} \binom{n}{k} (2^{n-k} - 1) = \sum_{k=1}^{n} \binom{n}{k} 2^{n-k} - \sum_{k=1}^{n} \binom{n}{k}.$$

By taking $x = 2$ in the equality

$$(1 + x)^n = \sum_{k=0}^{n} \binom{n}{k} x^{n-k},$$

we obtain

$$\sum_{k=1}^{n} \binom{n}{k} 2^{n-k} = 3^n - 2^n.$$

Since $\sum_{k=1}^{n} \binom{n}{k} = 2^n - 1$, we obtain immediately the desired equality. The number of association rules can be quite considerable even for small values of n. For example, for $n = 10$ we have $3^{10} - 2^{11} + 1 = 57,002$ association rules.

An association rule (X, Y) is denoted by $X \Rightarrow Y$. The *support* of $X \Rightarrow Y$ is the number $\text{supp}_T(XY)$. The *confidence of* $X \Rightarrow Y$ is the number

$$\text{conf}_T(X \Rightarrow Y) = \frac{\text{supp}_T(XY)}{\text{supp}_T(X)}.$$

Definition 6 An association rule holds in a transaction data set T with support μ and confidence c if $\text{supp}_T(XY) \geq \mu$ and $\text{conf}_T(X \Rightarrow Y) \geq c$.

Once a μ-frequent item set Z is identified, we need to examine the support levels of the subsets X of Z to ensure that an association rule of the form $X \Rightarrow Z - X$ has a sufficient level of confidence, $\text{conf}_T(X \Rightarrow Z - X) = \mu/\text{supp}_T(X)$. Observe that $\text{supp}_T(X) \geq \mu$ because X is a subset of Z. To obtain a high level of confidence for $X \Rightarrow Z - X$, the support of X must be as small as possible.

Clearly, if $X \Rightarrow Z - X$ does not meet the level of confidence, then it is pointless to look rules of the form $X' \Rightarrow Z - X'$ among the subsets X' of X.

Example 8 Let T be the transaction data set introduced in Example 7. We saw that the item set $L = i_2i_3i_4i_5$ has the support count equal to 2, and therefore, $\text{supp}_T(L) = 0.25$. This allows us to obtain the following association rules having three item sets in their antecedent, which are subsets of L.

Rule	$\text{suppcount}_T(X)$	$\text{conf}_T(X \Rightarrow Y)$
$i_2i_3i_4 \Rightarrow i_5$	2	1
$i_2i_3i_5 \Rightarrow i_4$	3	$\frac{2}{3}$
$i_2i_4i_5 \Rightarrow i_3$	2	1
$i_3i_4i_5 \Rightarrow i_2$	2	1

Note that $i_2i_3i_4 \Rightarrow i_5$, $i_2i_4i_5 \Rightarrow i_3$, and $i_3i_4i_5 \Rightarrow i_2$ have 100 percent confidence. We refer to such rules as *exact association rules*.

The rule $i_2i_3i_5 \Rightarrow i_4$ has confidence $(\frac{2}{3})$. It is clear that the confidence of rules of the form $U \Rightarrow V$ with $U \subseteq i_2i_3i_5$ and $UV = L$ will be lower than $(\frac{2}{3})$ since $\text{supp}_T(U)$ is at least 3. Indeed, the possible rules of this form are

Rule	$\text{suppcount}_T(X)$	$\text{conf}_T(X \Rightarrow Y)$
$i_2i_3 \Rightarrow i_4i_5$	5	$\frac{2}{5}$
$i_2i_5 \Rightarrow i_3i_4$	3	$\frac{2}{3}$
$i_3i_5 \Rightarrow i_2i_4$	3	$\frac{2}{3}$
$i_2 \Rightarrow i_3i_4i_5$	6	$\frac{2}{6}$
$i_3 \Rightarrow i_2i_4i_5$	5	$\frac{2}{5}$
$i_5 \Rightarrow i_2i_3i_4$	5	$\frac{2}{5}$

Obviously, if we seek association rules having a confidence larger than $\frac{2}{3}$ no such rule $U \Rightarrow V$ can be found such that U is a subset of $i_2i_3i_5$.

Suppose, for example, that we seek association rules $U \Rightarrow V$ that have a minimal confidence of 80 percent. We need to examine subsets U of the other sets: $i_2i_3i_4$, $i_2i_4i_5$, or $i_3i_4i_5$, which are not subsets of $i_2i_3i_5$ (since the subsets of $i_2i_3i_5$ cannot yield levels of confidence higher than $\frac{2}{3}$. There are five such sets.

Rule	$\text{suppcount}_T(X)$	$\text{conf}_T(X \Rightarrow Y)$
$i_2i_4 \Rightarrow i_3i_5$	2	1
$i_3i_4 \Rightarrow i_2i_5$	2	1
$i_4i_5 \Rightarrow i_2i_3$	2	1
$i_3i_4 \Rightarrow i_2i_5$	2	1
$i_4 \Rightarrow i_2i_3i_5$	2	1

Indeed, all these sets yield exact rules, that is, rules having 100 percent confidence. □

Many transaction data sets produce huge number of frequent item sets, and therefore, huge number of association rules particularly when the levels of support and confidence required are relatively low. Moreover, it is well known (see the work by Tan et al. [17]) that limiting the analysis of association rules to the support/confidence framework can lead to dubious conclusions. The data mining literature contains many references that attempt to derive interestingness measures for association rules in order to focus data analysis of those rules that may be more relevant (see, other works [4,6,8,11,12,16]).

7.4 LEVELWISE ALGORITHMS AND POSETS

The focus of this section is the levelwise algorithms, a powerful and elegant generalization of the Apriori algorithm that was introduced in the work by Mannila and Toivonen [13].

Let $(P, \leq)$ be a partially ordered set and let Q be a subset of P.

Definition 7 The border of Q is the set

$$\mathrm{BD}(Q) = \{p \in P | u < p \text{ implies } u \in Q \text{ and } p < v \text{ implies } v \notin Q\}.$$

The positive border of Q is the set

$$\mathrm{BD}^{+}(Q) = \mathrm{BD}(Q) \cap Q,$$

while the negative border of Q is

$$\mathrm{BD}^{-}(Q) = \mathrm{BD}(Q) - Q.$$

Clearly, we have $\mathrm{BD}(Q) = \mathrm{BD}^{+}(Q) \cup \mathrm{BD}^{-}(Q)$.

An alternative terminology exists that makes use of the terms *generalization* and *specialization*. If $r, p \in P$ and $r < p$, then we say that r is a *generalization* of p, or that p is a *specialization* of r. Thus, the border of a set Q consists of those elements p of P such that all their generalizations are in Q and none of their specializations is in Q.

Theorem 7 *Let $(P, \leq)$ be a partially ordered set. If Q, Q' are two disjoint subsets of P, then* $\mathrm{BD}(Q \cup Q') \subseteq \mathrm{BD}(Q) \cup \mathrm{BD}(Q')$.

Proof. Let $p \in \mathrm{BD}(Q \cup Q')$. Suppose that $u < p$, so $u \in Q \cup Q'$. Since Q and Q' are disjoint we have either $u \in Q$ or $u \in Q'$. On the contrary, if $p < v$, then $v \notin Q \cup Q'$, so $v \notin Q$ and $v \notin Q'$. Thus, we have $p \in \mathrm{BD}(Q) \cup \mathrm{BD}(Q')$. ■

The notion of a hereditary subset of a poset is an immediate generalization of the notion of hereditary family of sets.

Definition 8 A subset Q of a poset $(P, \leq)$ is said to be hereditary if $p \in Q$ and $r \leq p$ imply $r \in Q$.

Theorem 8 *If Q be a hereditary subset of a poset $(P, \leq)$, then the positive and the negative borders of Q are given by*

$$\mathrm{BD}^{+}(Q) = \{p \in Q | p < v \textit{ implies } v \notin Q\}$$

and

$$\mathrm{BD}^{-}(Q) = \{p \in P - Q | u < p \textit{ implies } u \in Q\},$$

respectively.

Proof. Let t be an element of the positive border $\mathrm{BD}^{+}(Q) = \mathrm{BD}(Q) \cap Q$. We have $t \in Q$ and $t < v$ implies $v \notin Q$, because $t \in \mathrm{BD}(Q)$.

Conversely, suppose that t is an element of Q such that $t < v$ implies $v \notin Q$. Since Q is hereditary, $u < t$ implies $u \in Q$, so $t \in \mathrm{BD}(Q)$. Therefore, $t \in \mathrm{BD}(Q) \cap Q = \mathrm{BD}^{+}(Q)$.

Let now s be an element of the negative border of Q, that is, $s \in \mathrm{BD}(Q) - Q$. We have immediately $s \in P - Q$. If $u < s$, then $u \in Q$, because Q is hereditary. Thus, $\mathrm{BD}^{-}(Q) \subseteq \{p \in P - Q | u < p$ implies $u \in Q\}$.

Conversely, suppose that $s \in P - Q$ and $u < s$ implies $u \in Q$. If $s < v$, then v cannot belong to Q because this would entail $s \in Q$ due to the hereditary property of Q. Consequently, $s \in \mathrm{BD}(Q)$, and so, $s \in \mathrm{BD}(Q) - Q = \mathrm{BD}^{-}(Q)$. ■

Theorem 8 can be paraphrased by saying that for a hereditary subset Q of P the positive border consists of the maximal elements of Q, while the negative border of Q consists of the minimal elements of $P - Q$.

Note that if Q, Q' are two hereditary subsets of P and $\mathrm{BD}^{+}(Q) = \mathrm{BD}^{+}(Q')$, then $Q = Q'$. Indeed, if $z \in P$, one of the following two cases may occur:

1. If z is not a maximal element of Q, then there is a maximal element w of Q such that $z < w$. Since $w \in \mathrm{BD}^{+}(Q) = \mathrm{BD}^{+}(Q')$, it follows that $w \in Q'$; hence $z \in Q'$, because Q' is hereditary.
2. If z is a maximal element of Q, then $z \in \mathrm{BD}^{+}(Q) = \mathrm{BD}^{+}(Q')$; hence $z \in Q'$.

In either case $z \in Q'$, so $Q \subseteq Q'$. The reverse inclusion can be proven in a similar way, so $Q = Q'$.

Similarly, we can show that for two hereditary collections Q, Q' of subsets of I, $\mathrm{BD}^{-}(Q) = \mathrm{BD}^{-}(Q')$ implies $Q = Q'$. Indeed, suppose that $z \in Q - Q'$. Since $z \notin Q'$, there exists a minimal element v such that $v \notin Q'$ and each of its lower bounds is in Q'. Since v belongs to the negative border $\mathrm{BD}^{-}(Q')$, it follows that

$v \in \mathrm{BD}^-(Q)$. This leads to a contradiction because $z \in Q$ and v (for which we have $v < z$) does not, thereby contradicting the fact that Q is a hereditary subset. Since no such z may exist, it follows that $Q \subseteq Q'$. The reverse inclusion can be shown in the same manner.

Definition 9 Let $\mathcal{D}$ be a relational database, $\mathcal{S}_{\mathcal{D}}$ be the set of states of $\mathcal{D}$, and let $(B, \leq, h)$ be a ranked poset, referred to as the ranked poset of objects.

A query is a function $q : \mathcal{S}_{\mathcal{D}} \times B \longrightarrow \{0, 1\}$ such that $D \in \mathcal{S}_{\mathcal{D}}$, $b \leq b'$, and $q(D, b') = 1$ imply $q(D, b) = 1$.

Definition 9 is meant to capture the framework of the *Apriori* algorithm for identification of frequent item sets. As shown in the work by Mannila and Toivonen [13], this framework can capture many other situations.

Example 9 Let $\mathcal{D}$ be a database that contains a tabular variable (T, H) and let $\theta = (T, H, \rho)$ be the table that is the current value of (T, H) contained by the current state D of $\mathcal{D}$.

The graded poset $(B, \leq, h)$ is $(\mathcal{P}(H), \subseteq, h)$, where $h(X) = |X|$. Given a number μ, the query is defined by

$$q(D, K) = \begin{cases} 1 \text{ if } \mathrm{supp}_T(K) \leq \mu, \\ 0 \text{ otherwise.} \end{cases}$$

Since $K \subseteq K'$ implies $\mathrm{supp}_T(K') \leq \mathrm{supp}_T(K)$, it follows that q satisfies the condition of Definition 9.

Example 10 As in Example 9, let $\mathcal{D}$ be a database that contains a tabular variable (T, H), and let $\theta = (T, H, \rho)$ be the table that is the current value of (T, H) contained by the current state D of $\mathcal{D}$. The graded poset $(\mathcal{P}(H), \supseteq, g)$ is the dual of the graded poset considered in Example 9, where $g(K) = |H| - |K|$. If L is a set of attributes the function q_L is defined by

$$q_L(D, K) = \begin{cases} 1 \text{ if } K \to L \text{ holds in } \theta, \\ 0 \text{ otherwise.} \end{cases}$$

Note that if $K' \subseteq K$ and D satisfies the functional dependency $K' \to L$, then D satisfies $K \to L$. Thus, q is a query in the sense of Definition 9. □

Definition 10 The set of interesting objects for the state D of the database and the query q is given by

$$\mathrm{INT}(D, q) = \{b \in B \mid q(D, b) = 1\}.$$

Note that the set of interesting objects is a hereditary set $(B, \leq)$. Indeed, if $b \in \mathrm{INT}(D, q)$ and $c \leq b$, then $c \in \mathrm{INT}(D, q)$, according to Definition 9. Thus,

$$BD^+(\mathrm{INT}(D,q)) = \{b \in \mathrm{INT}(D,q)\,|b < v \text{ implies } v \notin \mathrm{INT}(D,q)\},$$
$$BD^-(\mathrm{INT}(D,q)) = \{b \in B - \mathrm{INT}(D,q)\,|u < b \text{ implies } u \in \mathrm{INT}(D,q)\}.$$

In other words, $BD^+(\mathrm{INT}(D, q))$ is the set of maximal objects that are interesting, while $BD^-(\mathrm{INT}(D, q))$ is the set of minimal objects that are not interesting.

Next, we discuss a general algorithm that seeks to compute the set of interesting objects for a database state. The algorithm is known as the *levelwise algorithm* because it identifies these objects by scanning successively the levels of the graded poset of objects.

If $L_0, L_1, \ldots$ are the levels of the graded poset $(B, \leq, h)$, then the algorithm begins by examining all objects located on the initial level. The set of interesting objects located on the level L_i is denoted by $\mathcal{F}_i$; for each level L_i the computation of $\mathcal{F}_i$ is preceded by a computation of the set of potentially interesting objects $\mathcal{C}_i$ referred to as the set of *candidate objects*.

The first set of candidate objects $\mathcal{C}_1$ coincides with the level L_i. Only the interesting objects on this level are retained for the set $\mathcal{F}_1$.

The next set of candidate objects $\mathcal{C}_{i+1}$ is constructed by examining the level L_{i+1} and keeping those objects b having all their subobjects c in the interesting sets of the previous levels.

Generic_levelwise_algorithm$(D, (B, \leq, h), q)\{$

$\mathbb{C}_1 = L_1;$

$i = 1;$

while $(\mathbb{C}_i \neq \emptyset)$ ***do***

/* evaluation phase */

$\mathbb{F}_i = \{b \in \mathbb{C}_i | q(D, b) = 1\};$

/* candidate generation */

$\mathbb{C}_{i+1} = \{b \in L_{i+1} | c < b \text{ implies } c \in \bigcup_{j \leq i} \mathbb{F}_j\} - \bigcup_{j \leq i} \mathbb{C}_j$

$i++;$

endwhile;

output $\bigcup_{j<i} \mathbb{F}_j;$

$\}$

Example 11 For frequent item sets we can work in the framework described in Example 9. The algorithm, which is essentially the Apriori algorithm described in Section 7.2, goes through the ***while*** loop no more than $k + 1$ times, where

$$k = \max\{|X|\,|X \subseteq H, \mathrm{supp}_T(X) > \mu\}. \qquad \square$$

Example 12 In Example 10, we defined the grading query q_L as

$$q_L(D, K) = \begin{cases} 1 \text{ if } K \rightarrow L \text{ holds in } \theta, \\ 0 \text{ otherwise.} \end{cases}$$

for $K \in \mathcal{P}(H)$. The levelwise algorithm allows us to identify those subsets K such that a table $\theta = (T, H, \rho)$ satisfies the functional dependency $K \rightarrow L$. The first level consists of all subsets K of H that have $|H| - 1$ attributes. There are, of course, $|H| - 1$ such subsets and the set $\mathcal{F}_1$ will contain all these sets such that $K \rightarrow H$ is satisfied. Successive levels contain sets that have fewer and fewer attributes. Level L_i contains sets that have $|H| - i$ attributes.

The algorithm will go through the ***while*** loop at most $1 + |H - K|$, where K is the smallest set such that $K \rightarrow L$ holds. □

Observe that the computation of $\mathcal{C}_{i+1}$ in the generic levelwise algorithm,

$$\mathcal{C}_{i+1} = \left\{ b \in L_{i+1} \,|\, c < b \text{ implies } c \in \bigcup_{j \leq i} \mathcal{F}_j \right\} - \bigcup_{j \leq i} \mathcal{C}_j$$

can be written as

$$\mathcal{C}_{i+1} = \mathrm{BD}^- \left(\bigcup_{j \leq i} \mathcal{F}_j \right) - \bigcup_{j \leq i} \mathcal{C}_j.$$

This shows that the set of candidate objects at level L_{i+1} is the negative border of the interesting sets located on lower level excluding those objects that have been already evaluated.

The most expensive component of the levelwise algorithm is the evaluation of $q(D, b)$ since this requires a scan of the database state D. Clearly, we need to evaluate this function for each candidate element, so we will require $|\bigcup_{i=1}^{\ell} \mathcal{C}_i|$ evaluations, where ℓ is the number of levels that are scanned. Some of these evaluations will result in including the evaluated object b in the set $\mathcal{F}_i$. Objects that will not be included in $INT(D, q)$ are such that any of their generalizations are in $INT(D, q)$, even though they fail to belong to this set. They belong to $BD^-(INT(D, q))$. Thus, the levelwise algorithm performs $|INT(D, q)| + |BD^-(INT(D, q))|$ evaluations of $q(D, b)$.

Exercises 5–8 are reformulations of results obtained in the work by Mannila and Toivonen [13].

7.5 FURTHER READINGS

In addition to general data mining references [17], the reader should consult [1], a monograph dedicated to frequent item sets and association rules. Seminal work in this area, in addition to the original paper [5], has been done by Mannila and Toivonen [13] and by Zaki [19]; these references and others, such as [2] and [3], lead to an interesting and rewarding journey through the data mining literature. An alternative method for detecting frequent item sets based on a very interesting condensed representation of the data set was developed by Han et al. [9].

7.6 EXERCISES

1. Let $I = \{a, b, c, d\}$ be a set of items and let T be a transaction data set defined by

$$\begin{aligned} T(1) &= abc, \\ T(2) &= abd, \\ T(3) &= acd, \\ T(4) &= bcd, \\ T(5) &= ab. \end{aligned}$$

   (a) Find item sets whose support it at least 0.25.

   (b) Find association rules having support at least 0.25 and a confidence at least 0.75.

2. Let $I = i_1 i_2 i_3 i_4 i_5$ be a set of items. Find the 0.6-frequent item sets of the transaction data set T over I defined by

$$\begin{array}{ll} T(1) = i_1 & T(6) = i_1 i_2 i_4 \\ T(2) = i_1 i_2 & T(7) = i_1 i_2 i_5 \\ T(3) = i_1 i_2 i_3 & T(8) = i_2 i_3 i_4 \\ T(4) = i_2 i_3 & T(9) = i_2 i_3 i_5 \\ T(5) = i_2 i_3 i_4 & T(10) = i_3 i_4 i_5 \end{array}$$

   Also, determine all rules whose confidence is at least 0.75.

3. Let T be a transaction data set T over an item set I, $T : \{1, \ldots, n\} \longrightarrow \mathcal{P}(I)$. Define the bit sequence of an item set X as sequence $\boldsymbol{b}^X = (b_1, \ldots, b_n) \in \boldsymbol{Seq}_n(\{0, 1\})$, where

$$b_i = \begin{cases} 1 \text{ if } X \subseteq T(i), \\ 0 \text{ otherwise,} \end{cases}$$

   for $1 \le i \le n$.

   For $\boldsymbol{b} \in \boldsymbol{Seq}_n(\{0, 1\})$ the number $\sqrt{|\{i \mid 1 \le i \le n, b_i = 1\}|}$ is denoted by $\|\boldsymbol{b}\|$. The distance between the sequences $\boldsymbol{b}, \boldsymbol{c}$ is defined as $\|\boldsymbol{b} \oplus \boldsymbol{c}\|$. Prove that

(a) $\boldsymbol{b}^{X \cup Y} = \boldsymbol{b}^X \wedge \boldsymbol{b}^Y$ for every $X, Y \in \mathcal{P}(I)$;

(b) $\boldsymbol{b}^{K \oplus L} = \boldsymbol{b}^L \oplus \boldsymbol{b}^K$, where $K \oplus L$ is the symmetric difference of the item sets K and L;

(c) $|\sqrt{\text{supp}_T(K)} - \sqrt{\text{supp}_T(L)}| \leq d(\boldsymbol{b}^K, \boldsymbol{b}^L)/\sqrt{|T|}$.

4. For a transaction data set T over an item set $I = \{i_1, \ldots, i_n\}, T : \{1, \ldots, n\} \longrightarrow \mathcal{P}(I)$ and a number h, $1 \leq h \leq n$, define the number $\nu_T(h)$ by

$$\nu_T(h) = 2^{n-1} b_n + \cdots + 2b_2 + b_1,$$

where

$$b_k = \begin{cases} 1 & \text{if } i_k \in T(h), \\ 0 & \text{otherwise,} \end{cases}$$

for $1 \leq k \leq n$. Prove that $i_k \in T(h)$ if and only if the result of the integer division $\nu_T(h)/k$ is an odd number.

Suppose that the tabular variables of a database $\mathcal{D}$ are $(T_1, H_1), \ldots, (T_p, H_p)$. An *inclusion dependency* is an expression of the form $T_i[K] \subseteq T_j[L]$, where $K \subseteq H_i$ and $L \subseteq H_j$ for some i, j, where $1 \leq i, j \leq p$ are two sets of attributes having the same cardinality. Denote by $\text{ID}_{\mathcal{D}}$ the set of inclusion dependences of $\mathcal{D}$.

Let $D \in \mathcal{S}_{\mathcal{D}}$ be a state of the database $\mathcal{D}$, $\phi = T_i[K] \subseteq T_j[L]$ be an inclusion dependency and let $\theta_i = (T_i, H_i, \rho_i)$, $\theta_j = (T_j, H_j, \rho_j)$ be the tables that correspond to the tabular variables (T_i, H_i) and (T_j, H_j) in D. The inclusion dependency ϕ is satisfied in the state D of $\mathcal{D}$ if for every tuple $t \in \rho_i$ there is a tuple $s \in \rho_j$ such that $t[K] = s[L]$.

5. For $\phi = T_i[K] \subseteq T_j[L]$ and $\psi = T_d[K'] \subseteq T_e[L']$ define the relation $\phi \leq \psi$ if $d = i, e = j$, $K \subseteq K'$, and $H \subseteq H'$. Prove that "$\leq$" is a partial order on $\text{ID}_{\mathcal{D}}$.

6. Prove that the triple $(\text{ID}_{\mathcal{D}}, \leq, h)$ is a graded poset, where $h(T_i[K] \subseteq T_j[L]) = |K|$.

7. Prove that the function $q : \mathcal{S}_{\mathcal{D}} \times \text{ID}_{\mathcal{D}} \longrightarrow \{0, 1\}$ defined by

$$q(D, \phi) = \begin{cases} 1 & \text{if } \phi \text{ is satisfied in } D, \\ 0 & \text{otherwise} \end{cases}$$

is a query (as in Definition 9).

8. Specialize the generic levelwise algorithm to an algorithm that retrieves all inclusion dependences satisfied by a database state.

Let $T : \{1, \ldots, n\} \longrightarrow \mathcal{P}(D)$ be a transaction data set over an item set D. The contingency matrix of two item sets X, Y is the 2×2 matrix:

$$M_{XY} = \begin{pmatrix} m_{11} & m_{10} \\ m_{01} & m_{00} \end{pmatrix},$$

where

$$
\begin{aligned}
m_{11} &= |\{k | X \subseteq T(k) \text{ and } Y \subseteq T(k)\}|,\\
m_{10} &= |\{k | X \subseteq T(k) \text{ and } Y \not\subseteq T(k)\}|,\\
m_{01} &= |\{k | X \not\subseteq T(k) \text{ and } Y \subseteq T(k)\}|,\\
m_{00} &= |\{k | X \not\subseteq T(k) \text{ and } Y \not\subseteq T(k)\}|.
\end{aligned}
$$

Also, let $m_{1.} = m_{11} + m_{10}$ and $m_{.1} = m_{11} + m_{01}$.

9. Let $X \Rightarrow Y$ be an association rule. Prove that

$$\text{supp}_T(X \Rightarrow Y) = \frac{m_{11} + m_{10}}{n} \quad \text{and} \quad \text{conf}_T(X \Rightarrow Y) = \frac{m_{11}}{m_{11} + m_{10}}.$$

Which significance has the number m_{10} for $X \Rightarrow Y$?

10. Let $T : \{1, \ldots, n\} \longrightarrow \mathcal{P}(I)$ be a transaction data set over a set of items I and let π be a partition of the set $\{1, \ldots, n\}$ of transaction identifiers, $\pi = \{B_1, \ldots, B_p\}$. Let $n_i = |B_i|$ for $1 \leq i \leq p$.

    A *partitioning* of T is a sequence $T_1, \ldots, T_p$ of transaction data sets over I such that $T_i : \{1, \ldots, n_i\} \longrightarrow \mathcal{P}(I)$ is defined by $T_i(\ell) = T(k_\ell)$, where $B_i = \{k_1, \ldots, k_{n_i}\}$ for $1 \leq i \leq p$.

    Intuitively, this corresponds to splitting horizontally the table of T into p tables that contain $n_1, \ldots, n_p$ consecutive rows, respectively.

    Let K be an item set. Prove that if $\text{supp}_T(K) \geq \mu$, there exists j, $1 \leq j \leq p$, such that $\text{supp}_{T_j}(K) \geq \mu$. Give an example to show that the reverse implication does not hold; in other words, give an example of a transaction data set T, a partitioning $T_1, \ldots, T_p$ of T, and an item set K such that K is μ-frequent in some T_i but not in T.

11. Piatetsky-Shapiro [16] formulated three principles that a rule interestingness measure R should satisfy:

    (a) $R(X \Rightarrow Y) = 0$ if $m_{11} = m_1 m_1 / n$;

    (b) $R(X \rightarrow Y)$ increases with m_{11} when other parameters are fixed;

    (c) $R(X \rightarrow Y)$ decreases with $m_{.1}$ and with $m_{1.}$ when other parameters are fixed.

    The *lift* of a rule $X \Rightarrow Y$ is the number $lift(X \Rightarrow Y) = (nm_{11})/(m_1 m_1)$. The *PS* measure is $PS(X \rightarrow Y) = m_{11} - (m_1 m_1)/(n)$. Do *lift* and *PS* satisfy Piatetsky-Shapiro's principles? Give examples of interestingness measures that satisfy these principles.

REFERENCES

1. Adamo JM. Data Mining for Association Rules and Sequential Patterns. New York: Springer-Verlag; 2001.

2. Agarwal RC, Aggarwal CC, Prasad VVV. A tree projection algorithm for generation of frequent item sets. J Parallel Distrib Comput 2001;61(3):350–371.
3. Agarwal RC, Aggarwal CC, Prasad VVV. Depth first generation of long patterns. Proceedings of Knowledge Discovery and Data Mining; 2000. p 108–118.
4. Aggarwal CC, and Yu PS. Mining associations with the collective strength approach. IEEE Trans. Knowledge Data Eng 2001;13(6):863–873.
5. Agrawal R, Imielinski T, Swami A. Mining association rules between sets of items in very large databases. Proceedings of the ACM SIGMOD Conference on Management of Data; 1993. p 207–216.
6. Bayardo R, Agrawal R. Mining the most interesting rules. Proceedings of the 5th KDD. San Diego; 1999. p 145–153.
7. Birkhoff G. Lattice Theory. 3rd ed. Providence, RI: American Mathematical Society; 1967.
8. Brin S, Motwani R, Silverstein C. Beyond market baskets: generalizing association rules to correlations. Proceedings of ICMD; 1997. p 255–264.
9. Han J, Pei J, Yin Y. Mining frequent patterns without candidate generation. Proceedings of the ACM–SIGMOD International Conference on Management of Data; Dallas; 2000. p 1–12.
10. Ganter B, Wille R. Formal Concept Analysis. Berlin: Springer-Verlag; 1999.
11. Hilderman R, Hamilton H. Knowledge discovery and interestingness measures: a survey. Technical Report No. CS 99-04. Department of Computer Science, University of Regina; October 1999.
12. Jaroszewicz S, Simovici D. Interestingness of frequent item sets using Bayesian networks as background knowledge. Proceedings of the 10th KDD International Conference; Seattle; 2004. p 178–186.
13. Mannila H, Toivonen H. Levelwise search and borders of theories in knowledge discovery. TR C-1997-8. Helsinki, Finland: University of Helsinki; 1997.
14. Park JS, Chen MS, Yu PS. An Effective Hash based algorithm for mining association rules. Proceedings of the 1995 ACM SIGMOD International Conference on Management of Data; San Jose, CA; 1995. p 175–186.
15. Pasquier N, Bastide Y, Taouil R, Lakhal L. Discovering Frequent Closed Itemsets for Association Rules. Lecture Notes in Computer Science. Volume 1540. New York: Springer-Verlag; 1999. p 398–416.
16. Piatetsky-Shapiro G. Discovery, analysis and presentation of strong rules. In: Piatetsky-Shapiro G, Frawley W, editors. Knowledge Discovery in Databases. Cambridge, MA: MIT Press; 1991. p 229–248.
17. Tan PN, Steinbach M, Kumar V. Introduction to Data Mining. Reading, MA: Addison-Wesley; 2005.
18. Toivonen H. Sampling large databases for association rules. Proceedings of the 22nd VLDB Conference; Mumbai, India; 1996. p 134–145.
19. Zaki MJ. Mining non-redundant association rules. Data Mining Knowledge Discov 2004;9:223–248.

CHAPTER 8

Algorithms for Data Streams

CAMIL DEMETRESCU and IRENE FINOCCHI

8.1 INTRODUCTION

Efficient processing over massive data sets has taken an increased importance in the last few decades due to the growing availability of large volumes of data in a variety of applications in computational sciences. In particular, monitoring huge and rapidly changing streams of data that arrive online has emerged as an important data management problem: Relevant applications include analyzing network traffic, online auctions, transaction logs, telephone call records, automated bank machine operations, and atmospheric and astronomical events. For these reasons, the streaming model has recently received a lot of attention. This model differs from computation over traditional stored data sets since algorithms must process their input by making one or a small number of passes over it, using only a limited amount of working memory. The streaming model applies to settings where the size of the input far exceeds the size of the main memory available and the only feasible access to the data is by making one or more passes over it.

Typical streaming algorithms use space at most polylogarithmic in the length of the input stream and must have fast update and query times. Using sublinear space motivates the design for summary data structures with small memory footprints, also known as synopses [34]. Queries are answered using information provided by these synopses, and it may be impossible to produce an exact answer. The challenge is thus to produce high quality approximate answers, that is, answers with confidence bounds on the possible error: Accuracy guarantees are typically made in terms of a pair of user-specified parameters, ε and δ, meaning that the error in answering a query is within a factor of $1 + \varepsilon$ of the true answer with probability at least $1 - \delta$. The space and update time will depend on these parameters and the goal is to limit this dependence as much as possible.

Major progress has been achieved in the last 10 years in the design of streaming algorithms for several fundamental data sketching and statistics problems, for which several different synopses have been proposed. Examples include number of distinct

Handbook of Applied Algorithms: Solving Scientific, Engineering and Practical Problems
Edited by Amiya Nayak and Ivan Stojmenović

items, frequency moments, L_1 and L_2 norms of vectors, inner products, frequent items, heavy hitters, quantiles, histograms, and wavelets. Recently, progress has been achieved for other problem classes, including computational geometry (e.g., clustering and minimum spanning trees) and graphs (e.g., triangle counting and spanners). At the same time, there has been a flurry of activity in proving impossibility results, devising interesting lower bound techniques, and establishing important complementary results.

This chapter is intended as an overview of this rapidly evolving area. The chapter is not meant to be comprehensive, but rather aims at providing an outline of the main techniques used for designing algorithms or for proving lower bounds. We refer the interested reader to the works by Babcock et al. [7], Gibbons and Matias [34] and Muthukrishnan [57] for an extensive discussion of problems and results not mentioned here.

8.1.1 Applications

As observed before, the primary application of data stream algorithms is to monitor continuously huge and rapidly changing streams of data in order to support exploratory analyses and to detect correlations, rare events, fraud, intrusion, and unusual or anomalous activities. Such streams of data may be, for example, performance measurements in traffic management, all detail records in telecommunications, transactions in retail chains, ATM operations in banks, bids in online auctions, log records generated by Web Servers, or sensor network data. In all these cases, the volumes of data are huge (several terabytes or even petabytes), and records arrive at a rapid rate. Other relevant applications for data stream processing are related, for example, to processing massive files on secondary storage and to monitoring the contents of large databases or data warehouse environments. In this section, we highlight some typical needs that arise in these contexts.

8.1.1.1 Network Management Perhaps the most prominent application is related to network management. This involves monitoring and configuring network hardware and software to ensure smooth operations. Consider, for example, traffic analysis in the Internet. Here, as IP packets flow through the routers, we would like to monitor link bandwidth usage, to estimate traffic demands, to detect faults, congestion, and usage patterns. Typical queries that we would be able to answer are thus the following. How many IP addresses used a given link in a certain period of time? How many bytes were sent between a pair of IP addresses? Which are the top 100 IP addresses in terms of traffic? What is the average duration of an IP session? Which sessions transmitted more than 1000 bytes? Which IP addresses are involved in more than 1000 sessions? All these queries are heavily motivated by traffic analysis, fraud detection, and security.

To get a rough estimate of the amount of data that need to be analyzed to answer one such query, consider that each router can forward up to 1 billion packets per hour, and each Internet Service Provider may have many hundreds of routers: thus, many terabytes of data per hour need to be processed. These data arrive at a rapid rate, and we therefore need algorithms to mine patterns, process queries, and compute statistics on such data streams in almost real time.

8.1.1.2 *Database Monitoring* Many commercial database systems have a query optimizer used for estimating the cost of complex queries. Consider, for example, a large database that undergoes transactions (including updates). Upon the arrival of a complex query q, the optimizer may run some simple queries in order to decide an optimal query plan for q: In particular, a principled choice of an execution plan by the optimizer depends heavily on the availability of statistical summaries such as histograms, the number of distinct values in a column for the tables referenced in a query, or the number of items that satisfy a given predicate. The optimizer uses this information to decide between alternative query plans and to optimize the use of resources in multiprocessor environments. The accuracy of the statistical summaries greatly impacts the ability to generate good plans for complex SQL queries. The summaries, however, must be computed quickly: In particular, examining the entire database is typically regarded as prohibitive.

8.1.1.3 *Online Auctions* During the last few years, online implementations of auctions have become a reality, thanks to the Internet and to the wide use of computer-mediated communication technologies. In an online auction system, people register to the system, open auctions for individual items at any time, and then submit continuously items for auction and bids for items. Statistical estimation of auction data is thus very important for identifying items of interest to vendors and purchasers, and for analyzing economic trends.

Typical queries may require to convert the prices of incoming bids between different currencies, to select all bids of a specified set of items, to maintain a table of the currently open auctions, to select the items with the most bids in a specified time interval, to maintain the average selling price over the items sold by each seller, to return the highest bid in a given period of time, or to monitor the average closing price (i.e., the price of the maximum bid, or the starting price of the auction in case there were no bids) across items in each category.

8.1.1.4 *Sequential Disk Accesses* In modern computing platforms, the access times to main memory and disk vary by several orders of magnitude. Hence, when the data reside on disk, it is much more important to minimize the number of I/Os (i.e., the number of disk accesses) than the CPU computation time as it is done in traditional algorithms theory. Many *ad hoc* algorithmic techniques have been proposed in the external memory model for minimizing the number of I/Os during a computation (see, e.g., the work by Vitter [64]).

Due to the high sequential access rates of modern disks, streaming algorithms can also be effectively deployed for processing massive files on secondary storage, providing new insights into the solution of several computational problems in external memory. In many applications managing massive data sets, using secondary and tertiary storage devices is indeed a practical and economical way to store and move data: such large and slow external memories, however, are best optimized for sequential access, and thus naturally produce huge streams of data that need to be processed in a small number of sequential passes. Typical examples include data access to database systems [39] and analysis of Internet archives stored on tape [43]. The streaming algorithms designed with these applications in mind may have a greater flexibility: Indeed, the rate at which data are processed can be adjusted, data can be processed in chunks, and more powerful processing primitives (e.g., sorting) may be available.

8.1.2 Overview of the Literature

The problem of computing in a small number of passes over the data appears already in papers from the late 1970s. Morris, for instance, addressed the problem of keeping approximate counts of large numbers [55]. Munro and Paterson [56] studied the space required for selection when at most P passes over the data can be performed, giving almost matching upper and lower bounds as a function of P and of the input size. The paper by Alon et al. [5,6], awarded in 2005 with the Gödel Prize for outstanding papers in the area of theoretical computer science, provided the foundations of the field of streaming and sketching algorithms. This seminal work introduced the novel technique of designing small randomized linear projections that allow the approximation (to user specified precision) of the frequency moments of a data set and other quantities of interest. The computation of frequency moments is now fully understood, with almost matching (up to polylogarithmic factors) upper bounds [12,20,47] and lower bounds [9,14,46,62]. Namely, Indyk and Woodruff [47] presented the first algorithm for estimating the kth frequency moment using space $\tilde{O}(n^{1-2/k})$. A simpler one-pass algorithm is described in [12].

Since 1996, many fundamental data statistics problems have been efficiently solved in streaming models. For instance, the computation of frequent items is particularly relevant in network monitoring applications and has been addressed, for example, in many other works [1,16,22,23,51,54]. A plethora of other problems have been studied in the last few years, designing solutions that hinge upon many different and interesting techniques. Among them, we recall sampling, probabilistic counting, combinatorial group testing, core sets, dimensionality reduction, and tree-based methods. We will provide examples of application of some of these techniques in Section 8.3. An extensive bibliography can be found in the work by Muthukrishnan [57]. The development of advanced techniques made it possible to solve progressively more complex problems, including the computation of histograms, quantiles, norms, as well as geometric and graph problems.

Histograms capture the distribution of values in a data set by grouping values into buckets and maintaining suitable summary statistics for each bucket. Different kinds of histograms exist: for example, in an equidepth histogram the number of values falling into each bucket is uniform across all buckets. The problem of computing these histograms is strictly related to the problem of maintaining the quantiles for the data set: quantiles represent indeed the bucket boundaries. These problems have been addressed, for example, in many other works [18,36,37,40,41,56,58,59]. Wavelets are also widely used to provide summarized representations of data: works on computing wavelet coefficients in data stream models include [4,37,38,60].

A few fundamental works consider problems related to norm estimation, for example, dominance norms and L_p sums [21,44]. In particular, Indyk pioneered the design of sketches based on random variables drawn from stable distributions (which are known to exist) and applied this idea to the problem of estimating L_p sums [44].

Geometric problems have also been the subject of much recent research in the streaming model [31,32,45]. In particular, clustering problems received special attention: given a set of points with a distance function defined on them, the goal is to find a clustering solution (a partition into clusters) that optimizes a certain objective function. Classical objective functions include minimizing the sum of distances of points to their closest median (k-median) or minimizing the maximum distance of a point to its closest center (k-center). Streaming algorithms for such problem are presented, for example, in the works by Charikar [17] and Guha et al. [42].

Differently from most data statistics problems, where $O(1)$ passes and polylogarithmic working space have been proven to be enough to find approximate solutions, many classical graph problems seem to be far from being solved within similar bounds: for many classical graph problems, linear lower bounds on the space $\times$ passes product are indeed known [43]. A notable exception is related to counting triangles in graphs, as discussed in the works by Bar-Yossef et al. [10], Buriol et al. [13], and Jowhari and Ghodsi [49]. Some recent papers show that several graph problems can be solved with one or few passes in the semi-streaming model [26–28,53] where the working memory size is $O(n \cdot \text{polylog } n)$ for an input graph with n vertices: in other words, akin to semi-external memory models [2,64] there is enough space to store vertices, but not edges of the graph. Other works, such as [3,25,61], consider the design of streaming algorithms for graph problems when the model allows more powerful primitives for accessing stream data (e.g., use of intermediate temporary streams and sorting).

8.1.3 Chapter Outline

This chapter is organized as follows. In Section 8.2 we describe the most common data stream models: such models differ in the interpretation of the data on the stream (each item can either be a value itself or indicate an update to a value) and in the primitives available for accessing and processing stream items. In Section 8.3 we focus on techniques for proving upper bounds: we describe some mathematical and algorithmic tools that have proven to be useful in the construction of synopsis data structures (including randomization, sampling, hashing, and probabilistic counting) and we first show how these techniques can be applied to classical data statistics problems. We then move to consider graph problems as well as techniques useful in streaming models that provide more powerful primitives for accessing stream data in a nonlocal fashion (e.g., simulations of parallel algorithms). In Section 8.4 we address some lower bound techniques for streaming problems, using the computation of the number of distinct items in a data stream as a running example: we explore the use of reductions of problems in communication complexity to streaming problems, and we discuss the use of randomization and approximation in the design of efficient synopses. In Section 8.5 we summarize our contribution.

8.2 DATA STREAM MODELS

A variety of models exist for data stream processing: the differences depend on how stream data should be interpreted and which primitives are available for accessing stream items. In this section we overview the main features of the most commonly used models.

8.2.1 Classical Streaming

In *classical data streaming* [5,43,56,57], input data are accessed sequentially in the form of a data stream $\Sigma = x_1, ..., x_n$ and need to be processed using a working memory that is small compared to the length n of the stream. The main parameters of the model are the number p of sequential passes over the data, the size s of the working memory, and the per-item processing time. All of them should be kept small: typically, one strives for one pass and polylogarithmic space, but this is not a requirement of the model.

There exist at least three variants of classical streaming, dubbed (in increasing order of generality) *time series*, *cash register*, and *turnstile* [57]. Indeed, we can think of stream items $x_1, ..., x_n$ as describing an underlying signal A, that is, a one-dimensional function over the reals. In the time series model, each stream item x_i represents the ith value of the underlying signal, that is, $x_i = A[i]$. In the other models, each stream item x_i represents an update of the signal: namely, x_i can be thought of as a pair (j, U_i), meaning that the jth value of the underlying signal must be changed by the quantity U_i, that is, $A_i[j] = A_{i-1}[j] + U_i$. The partially dynamic scenario in which the signal can be only incremented, that is, $U_i \geq 0$, corresponds to the cash register model, while the fully dynamic case yields the turnstile model.

8.2.2 Semi-Streaming

Despite the heavy restrictions of classical data streaming, we will see in Section 8.3 that major success has been achieved for several data sketching and statistics problems, where $O(1)$ passes and polylogarithmic working space have been proven to be enough to find approximate solutions. On the contrary, there exist many natural problems (including most problems on graphs) for which linear lower bounds on $p \times s$ are known, even using randomization and approximation: these problems cannot be thus solved within similar polylogarithmic bounds. Some recent papers [27,28,53] have therefore relaxed the polylog space requirements considering a *semi-streaming* model, where the working memory size is $O(n \cdot \text{polylog } n)$ for an input graph with n vertices: in other words, akin to semi-external memory models [2,64], there is enough space to store vertices, but not edges of the graph. We will see in Section 8.3.3 that some complex graph problems can be solved in semi-streaming, including spanners, matching, and diameter estimation.

8.2.3 Streaming with a Sorting Primitive

Motivated by technological factors, some authors have recently started to investigate the computational power of even less restrictive streaming models. Today's computing platforms are equipped with large and inexpensive disks highly optimized for sequential read/write access to data, and among the primitives that can efficiently access data in a nonlocal fashion, sorting is perhaps the most optimized and well understood. These considerations have led to introduce the *stream-sort* model [3,61]. This model extends classical streaming in two ways: the ability to write intermediate temporary streams and the ability to reorder them at each pass for free. A stream-sort algorithm alternates streaming and sorting passes: a streaming pass, while reading data from the input stream and processing them in the working memory, produces items that are sequentially appended to an output stream; a sorting pass consists of reordering the input stream according to some (global) partial order and producing the sorted stream as output. Streams are pipelined in such a way that the output stream produced during pass i is used as input stream at pass $i + 1$. We will see in Section 8.3.4 that the combined use of intermediate temporary streams and of a sorting primitive yields enough power to solve efficiently (within polylogarithmic passes and memory) a variety of graph problems that cannot be solved in classical streaming. Even without sorting, the model is powerful enough for achieving space–passes trade-offs [25] for graph problems for which no sublinear memory algorithm is known in classical streaming.

8.3 ALGORITHM DESIGN TECHNIQUES

Since data streams are potentially unbounded in size, when the amount of computation memory is bounded it may be impossible to produce an exact answer. In this case, the challenge is to produce high quality approximate answers, that is, answers with confidence bounds on the possible error. The typical approach is to maintain a "lossy" summary of the data stream by building up a *synopsis data structure* with memory footprint substantially smaller than the length of the stream. In this section we describe some mathematical and algorithmic techniques that have proven to be useful in the construction of such synopsis data structures. Besides the ones considered in this chapter, many other interesting techniques have been proposed: the interested reader can find pointers to relevant works in Section 8.1.2. Rather than being comprehensive, our aim is to present a small amount of results in sufficient detail that the reader can get a feeling of some common techniques used in the field.

The most natural approach to designing streaming algorithms is perhaps to maintain a small *sample* of the data stream: if the sample captures well the essential characteristics of the entire data set with respect to a specific problem, evaluating a query over the sample may provide reliable approximation guarantees for that problem. In Section 8.3.1 we discuss how to maintain a bounded size sample of a (possibly unbounded) data stream and describe applications of sampling to the problem of finding frequent items in a data stream.

Useful randomized synopses can also be constructed hinging upon hashing techniques. In Section 8.3.2 we address the design of *hash-based sketches* for estimating the number of distinct items in a data stream. We also discuss the main ideas behind the design of randomized sketches for the more general problem of estimating the frequency moments of a data set: the seminal paper by Alon et al. [5] introduced the technique of designing small randomized linear projections that summarize large amounts of data and allow frequency moments and other quantities of interest to be approximated to user-specified precision. As quoted from the Gödel Award Prize ceremony, this paper "set the pattern for a rapidly growing body of work, both theoretical and applied, creating the now burgeoning fields of streaming and sketching algorithms."

Sections 8.3.3 and 8.3.4 are mainly devoted to the semi-streaming and stream-sort models. In Section 8.3.3 we focus on techniques that can be applied to solve complex graph problems in $O(1)$ passes and $\tilde{O}(n)$ space. In Section 8.3.4, finally, we analyze the use of more powerful primitives for accessing stream data, showing that sorting yields enough power to solve efficiently a variety of problems for which efficient solutions in classical streaming cannot be achieved.

8.3.1 Sampling

A small random sample S of the data often captures certain characteristics of the entire data set. If this is the case, the sample can be maintained in memory and queries can be answered over the sample. In order to use sampling techniques in a data stream context, we first need to address the problem of maintaining a sample of a specified size over a possibly unbounded stream of data that arrive online. Note that simple coin tossing is not possible in streaming applications, as the sample size would be unbounded. The standard solution is to use Vitter's *reservoir sampling* [63] that we describe in the following Sections.

8.3.1.1 Reservoir Sampling This technique dates back to the 1980s [63]. Given a stream Σ of n items that arrive online, at any instant of time reservoir sampling guarantees to maintain a uniform random sample S of fixed size m of the part of stream observed up to that time. Let us first consider the following natural sampling procedure.

> At the beginning, add to S the first m items of the stream. Upon seeing the stream item x_t at time t, add x_t to S with probability m/t. If x_t is added, evict a random item from S (other than x_t).

It is easy to see that at each time $|S| = m$ as desired. The next theorem proves that, at each time, S is actually a uniform random sample of the stream observed so far.

Theorem 1 [63] *Let S be a sample of size m maintained over a stream $\Sigma = x_1, \ldots, x_n$ by the above algorithm. Then, at any time t and for each $i \leq t$, the probability that $x_i \in S$ is m/t.*

Proof. We use induction on t. The base step is trivial. Let us thus assume that the claim is true up to time; t that is, by inductive hypothesis $\Pr[x_i \in S] = m/t$ for each $i \leq t$. We now examine how S can change at time $t+1$, when item x_{t+1} is considered for addition. Consider any item x_i with $i < t+1$. If x_{t+1} is not added to S (this happens with probability $1 - m/(t+1)$), then x_i has the same probability of being in S of the previous step (i.e., m/t). If x_{t+1} is added to S (this happens with probability $m/(t+1)$), then x_i has a probability of being in S equal to $(m/t)(1 - 1/m)$, since it must have been in S at the previous step and must not be evicted at the current step. Thus, for each $i \leq t$, at time $t+1$ we have

$$\Pr[x_i \in S] = \left(1 - \frac{m}{t+1}\right)\frac{m}{t} + \frac{m}{t+1}\left[\frac{m}{t}\left(1 - \frac{1}{m}\right)\right] = \frac{m}{t+1}.$$

The fact that x_{t+1} is added to S with probability $m/(t+1)$ concludes the proof. ■

Instead of flipping a coin for each element (that requires to generate n random values), the reservoir sampling algorithm randomly generates the number of elements to be skipped before the next element is added to S. Special care is taken to generate these skip numbers, so as to guarantee the same properties that we discussed in Theorem 1 for the naïve coin-tossing approach. The implementation based on skip numbers has the advantage that the number of random values to be generated is the same as the number of updates of the sample S. We refer to the work by Vitter [63] for the details and the analysis of this implementation.

We remark that reservoir sampling works well for insert and updates of the incoming data, but runs into difficulties if the data contain deletions. In many applications, however, the timeliness of data is important, since outdated items expire and should be no longer used when answering queries. Other sampling techniques have been proposed that address this issue: see, for example, [8,35,52] and the references therein. Another limitation of reservoir sampling derives from the fact that the stream may contain duplicates, and any value occurring frequently in the sample is a wasteful use of the available space: concise sampling overcomes this limitation representing elements in the sample by pairs (value, count). As described by Gibbons and Matias [33], this natural idea can be used to compress the samples and allows it to solve, for example, the top-k problem, where the k most frequent items need to be identified.

In the rest of this section, we provide a concrete example of how sampling can be effectively applied to certain nontrivial streaming problems. However, as we will see in Section 8.4, there also exist classes of problems for which sampling-based approaches are not effective, unless using a prohibitive (almost linear) amount of memory.

8.3.1.2 An Application of Sampling: Frequent Items Following an approach proposed by Manku and Motwani [51], we will now show how to use sampling to address the problem of identifying frequent items in a data stream, that is, items whose frequency exceeds a user-specified threshold. Intuitively, it should be possible to estimate frequent items by a good sample. The algorithm that we discuss, dubbed *sticky sampling* [51], supports this intuition. The algorithm accepts two user-specified thresholds: a frequency threshold $\varphi \in (0, 1)$, and an error parameter $\varepsilon \in (0, 1)$ such that $\varepsilon < \varphi$. Let Σ be a stream of n items $x_1, ..., x_n$. The goal is to report

- all the items whose frequency is at least φn (i.e., there must be no *false negatives*)
- no item with frequency smaller than $(\varphi - \varepsilon)n$.

We will denote by $f(x)$ the true frequency of an item x, and by $f_e(x)$ the frequency estimated by sticky sampling. The algorithm also guarantees small error in individual frequencies; that is, the estimated frequency is less than the true frequency by at most εn. The algorithm is randomized, and in order to meet the two goals with probability at least $1 - \delta$, for a user-specified probability of failure $\delta \in (0, 1)$, it maintains a sample with expected size $2\varepsilon^{-1} \log(\varphi^{-1}\delta^{-1}) = 2t$. Note that the space is independent of the stream length n.

The sample $\mathcal{S}$ is a set of pairs of the form $(x, f_e(x))$. In order to handle potentially unbounded streams, the sampling rate r is not fixed, but is adjusted so that the probability $1/r$ of sampling a stream item decreases as more and more items are considered. Initially, $\mathcal{S}$ is empty and $r = 1$. For each stream item x, if $x \in \mathcal{S}$, then $f_e(x)$ is increased by 1. Otherwise, x is sampled with rate r, that is, with probability $1/r$: if x is sampled, the pair $(x, 1)$ is added to $\mathcal{S}$, otherwise we ignore x and move to the next stream item.

After sampling with rate $r = 1$ the first $2t$ items, the sampling rate increases geometrically as follows: the next $2t$ items are sampled with rate $r = 2$, the next $4t$ items with rate $r = 4$, the next $8t$ items with rate $r = 8$, and so on. Whenever the sampling rate changes, the estimated frequencies of sample items are adjusted so as to keep them consistent with the new sampling rate: for each $(x, f_e(x)) \in \mathcal{S}$, we repeatedly toss an unbiased coin until the coin toss is successful, decreasing $f_e(x)$ by 1 for each unsuccessful toss. We evict $(x, f_e(x))$ from $\mathcal{S}$ if $f_e(x)$ becomes 0 during this process. Effectively, after each sampling rate doubling, $\mathcal{S}$ is transformed to exactly the state it would have been in, if the new rate had been used from the beginning.

Upon a frequency items query, the algorithm returns all sample items whose estimated frequency is at least $(\varphi - \varepsilon)n$.

The following technical lemma will be useful in the analysis of sticky sampling. Although pretty straightforward, we report the proof for the sake of completeness.

Lemma 1 *Let $r \geq 2$ and let n be the number of stream items considered when the sampling rate is r. Then $1/r \geq t/n$, where $t = \varepsilon^{-1} \log(\varphi^{-1}\delta^{-1})$.*

Proof. It can be easily proved by induction on r that $n = rt$ at the beginning of the phase in which sampling rate r is used. The base step, for $r = 2$, is trivial: at the beginning $\mathcal{S}$ contains exactly $2t$ elements by construction. During the phase with sampling rate r, as far as the algorithm works, rt new stream elements are considered; thus, when the sampling rate doubles at the end of the phase, we have $n = 2rt$, as needed to prove the induction step. This implies that during any phase it must be $n \geq rt$, which proves the claim. ■

We can now prove that sticky sampling meets the goals in the definition of the frequent items problem with probability at least $1-\delta$ using space independent of n.

Theorem 2 [51] *For any $\varepsilon, \varphi, \delta \in (0,1)$, with $\varepsilon < \varphi$, sticky sampling solves the frequent items problems with probability at least $1-\delta$ using a sample of expected size $(2/\varepsilon)\log(\varphi^{-1}\delta^{-1})$.*

Proof. We first note that the estimated frequency of a sample element x is an underestimate of the true frequency, that is, $f_e(x) \leq f(x)$. Thus, if the true frequency is smaller than $(\varphi-\varepsilon)n$, the algorithm will not return x, since it must also be $f_e(x) < (\varphi-\varepsilon)n$.

We now prove that there are no false negatives with probability $\geq 1-\delta$. Let k be the number of elements with frequency at least φ, and let $y_1, \ldots, y_k$ be those elements. Clearly, it must be $k \leq 1/\varphi$. There are no false negatives if and only if all the elements $y_1, \ldots, y_k$ are returned by the algorithm. We now study the probability of the complementary event, proving that it is upper bounded by δ.

$$\Pr[\exists \text{ false negative}] \leq \sum_{i=1}^{k} \Pr[y_i \text{ is not returned}] = \sum_{i=1}^{k} \Pr[f_e(y_i) < (\varphi-\varepsilon)n].$$

Since $f(y_i) \geq \varphi n$ by definition of y_i, we have $f_e(y_i) < (\varphi-\varepsilon)n$ if and only if the estimated frequency of y_i is underestimated by at least ϵn. Any error in the estimated frequency of an element corresponds to a sequence of unsuccessful coin tosses during the first occurrences of the element. The length of this sequence exceeds εn with probability

$$\left(1-\frac{1}{r}\right)^{\varepsilon n} \leq \left(1-\frac{t}{n}\right)^{\varepsilon n} \leq e^{-t\varepsilon},$$

where the first inequality follows from Lemma 1. Hence,

$$\Pr[\exists \text{ false negative}] < k\, e^{-t\varepsilon} < \frac{e^{-t\varepsilon}}{\varphi} = \delta$$

by definition of t. This proves that the algorithm is correct with probability $\geq 1-\delta$.

It remains to discuss the space usage. The number of stream elements considered at the end of the phase in which sampling rate r is used must be at most $2rt$ (see the proof of Lemma 1 for details). The algorithm behaves as if each element was sampled with probability $1/r$: the expected number of sampled elements is therefore $2t$. ■

Manku and Motwani also provide a deterministic algorithm for estimating frequent items: this algorithm guarantees no false negatives and returns no false positives with true frequency smaller than $(\varphi-\varepsilon)n$ [51]. However, the price paid for being deterministic is that the space usage increases to $O((1/\varepsilon)\log(\varepsilon n))$. Other works that describe different techniques for tracking frequent items are, for example, Refs. 1,16,22,23,54.

8.3.2 Sketches

In this section we exemplify the use of sketches as randomized estimators of the frequency moments of a data stream. Let $\Sigma = x_1, \ldots, x_n$ be a stream of n values taken from a universe U of size u, and let f_i, for $i \in U$, be the frequency (number of occurrences) of value i in Σ, that is, $f_i = |\{j : x_j = i\}|$. The kth frequency moment F_k of Σ is defined as

$$F_k = \sum_{i \in U} f_i^k.$$

Frequency moments represent useful statistical information on a data set and are widely used in database applications. In particular, F_0 and F_1 represent the number of distinct values in the data stream and the length of the stream, respectively. F_2, also known as Gini's index, provides valuable information about the skew of the data. F_∞, finally, is related to the maximum frequency element in the data stream, that is, $\max_{i \in U} f_i$.

8.3.2.1 Probabilistic Counting We begin our discussion from the estimation of F_0. The problem of counting the number of distinct values in a data set using small space has been studied since the early 1980s by Flajolet and Martin [29,30], who proposed a hash-based *probabilistic counter*. We first note that a naïve approach to compute the exact value of F_0 would use a counter $c(i)$ for each value i of the universe U, and would therefore require $O(1)$ processing time per item, but linear space. The probabilistic counter of Flajolet and Martin [29,30] relies on hash functions to find a good approximation of F_0 using only $O(\log u)$ bits of memory, where u is the size of the universe U.

The counter consists of an array C of $\log u$ bits. Each stream item is mapped to one of the $\log u$ bits by means of the combination of two functions h and t. The hash function $h : U \to [0, u-1]$ is drawn from a set of strongly 2-universal hash functions: it transforms values of the universe into integers sufficiently uniformly distributed over the set of binary strings of length $\log u$. The function t, for any integer i, gives the number $t(i)$ of trailing zeros in the binary representation of i. Updates and queries work as follows:

- *Counter update*: Upon seeing a stream value x, set $C[t(h(x))]$ to 1.
- *Distinct values query*: Let R be the position of the rightmost 1 in the counter C, with $1 \le R \le \log u$. Return 2^R.

Notice that all stream items by the same value will repeatedly set the same counter bit to 1. Intuitively, the fact that h distributes items uniformly over $[0, u-1]$ and the use of function t guarantee that counter bits are selected in accordance with a geometric distribution; that is, $1/2$ of the universe items will be mapped to the first counter bit, $1/4$ will be mapped to the second counter bit, and so on. Thus, it seems reasonable to expect that the first $\log F_0$ counter bits will be set to 1 when the stream contains

F_0 distinct items: this suggests that R, as defined above, yields a good approximation for F_0. We will now give a more formal analysis. We will denote by Z_j the number of distinct stream items that are mapped (by the composition of functions t and h) to a position $\geq j$. Thus, R is the maximum j such that $Z_j > 0$.

Lemma 2 *Let Z_j be the number of distinct stream items x for which $t(h(x)) \geq j$. Then, $E[Z_j] = F_0/2^j$ and* $\text{Var}[Z_j] < E[Z_j]$.

Proof. Let W_x be an indicator random variable whose value is 1 if and only if $t(h(x)) \geq j$. Then, by definition of Z_j,

$$Z_j = \sum_{x \in U \cap \Sigma} W_x. \tag{8.1}$$

Note that $|U \cap \Sigma| = F_0$. We now study the probability that $W_x = 1$. It is not difficult to see that the number of binary strings of length $\log u$ that have exactly j trailing zeros, for $0 \leq j < \log u$, is $2^{\log u - (j+1)}$. Thus, the number of strings that have at least j trailing zeros is $1 + \sum_{i=j}^{\log u - 1} 2^{\log u - (i+1)} = 2^{\log u - j}$. Since h distributes items uniformly over $[0, u-1]$, we have that

$$\Pr[W_x = 1] = \Pr[t(h(x)) \geq j] = \frac{2^{\log u - j}}{u} = 2^{-j}.$$

Hence, $E[W_x] = 2^{-j}$ and $\text{Var}[W_x] = E[W_x^2] - E[W_x]^2 = 2^{-j} - 2^{-2j} = 2^{-j}(1 - 2^{-j})$. We are now ready to compute $E[Z_j]$ and $\text{Var}[Z_j]$. By (8.1) and by linearity of expectation we have

$$E[Z_j] = F_0 \cdot \left(1 \cdot \frac{1}{2^j} + 0 \cdot \left(1 - \frac{1}{2^j}\right)\right) = \frac{F_0}{2^j}.$$

Due to pairwise independence (guaranteed by the choice of the hash function h) we have $\text{Var}[W_x + W_y] = \text{Var}[W_x] + \text{Var}[W_y]$ for any $x, y \in U \cap \Sigma$ and thus

$$\text{Var}[Z_j] = \sum_{x \in U \cap \Sigma} \text{Var}[W_x] = \frac{F_0}{2^j}\left(1 - \frac{1}{2^j}\right) < F_0 2^j = E[Z_j].$$

This concludes the proof. ■

Theorem 3 [5,29,30] *Let F_0 be the exact number of distinct values and let 2^R be the output of the probabilistic counter to a distinct values query. For any $c > 2$, the probability that 2^R is not between F_0/c and $c\,F_0$ is at most $2/c$.*

Proof. Let us first study the probability that the algorithm overestimates F_0 by a factor of c. We begin by noticing that Z_j takes only nonnegative values, and thus we can apply Markov's inequality to estimate the probability that $Z_j \geq 1$, obtaining

$$\Pr[Z_j \geq 1] \leq \frac{E[Z_j]}{1} = \frac{F_0}{2^j}, \tag{8.2}$$

where the equality is by Lemma 2. If the algorithm overestimates F_0 by a factor of c, then it must exist an index j such that $C[j] = 1$ and $2^j/F_0 > c$ (i.e., $j > \log_2(c\,F_0)$). By definition of Z_j, this implies $Z_{\log_2(c\,F_0)} \geq 1$. Thus,

$$\Pr[\exists j : C[j] = 1 \text{ and } 2^j/F_0 > c\,] \leq \Pr[\,Z_{\log_2(c\,F_0)} \geq 1\,] \leq \frac{F_0}{2^{\log_2(c\,F_0)}} = \frac{1}{c},$$

where the last inequality follows from (8.2). The probability that the algorithm overestimates F_0 by a factor of c is therefore at most $1/c$.

Let us now study the probability that the algorithm underestimates F_0 by a factor of $1/c$. Symmetrically to the previous case, we begin by estimating the probability that $Z_j = 0$. Since Z_j takes only nonnegative values, we have

$$\Pr[\,Z_j = 0\,] = \Pr[\,|Z_j - E[Z_j]| \geq E[Z_j]\,] \leq \frac{\mathrm{Var}[Z_j]}{E[Z_j]^2} < \frac{1}{E[Z_j]} = \frac{2^j}{F_0} \tag{8.3}$$

using Chebyshev inequality and Lemma 2. If the algorithm underestimates F_0 by a factor of $1/c$, then there must exist an index j such that $2^j < F_0/c$ (i.e., $j < \log_2(F_0/c)$) and $C[p] = 0$ for all positions $p \geq j$. By definition of Z_j, this implies $Z_{\log_2(F_0/c)} = 0$, and with reasonings similar to the previous case and by using (8.3), we obtain that the probability that the algorithm underestimates F_0 by a factor of $1/c$ is at most $2^{\log_2(F_0/c)}/F_0 = 1/c$.

The upper bounds on the probabilities of overestimates and underestimates imply that the probability that 2^R is not between F_0/c and $c\,F_0$ is at most $2/c$. ■

The probabilistic counter of Flajolet and Martin [29,30] assumes the existence of hash functions with some ideal random properties. This assumption has been more recently relaxed by Alon et al. [5], who adapted the algorithm so as to use simpler linear hash functions. We remark that streaming algorithms for computing a $(1+\varepsilon)$-approximation of the number of distinct items are presented, for example, in the work by Bar-Yossef et al. [11].

8.3.2.2 Randomized Linear Projections and AMS Sketches We now consider the more general problem of estimating the frequency moments F_k of a data set, for $k \geq 2$, focusing on the seminal work by Alon et al. [5].

In order to estimate F_2, Alon et al. introduced a fundamental technique based on the design of small randomized linear projections that summarize some essential properties of the data set. The basic idea of the sketch designed in the work by Alon et al. [5] for estimating F_2 is to define a random variable whose expected value is F_2, and whose variance is relatively small. We follow the description from the work Alon et al. [4].

The algorithm computes μ random variables $Y_1, ..., Y_\mu$ and outputs their median Y as the estimator for F_2. Each Y_i is in turn the average of α independent, identically distributed random variables X_{ij}, with $1 \le j \le \alpha$. The parameters μ and α need to be carefully chosen in order to obtain the desired bounds on space, approximation, and probability of error: such parameters will depend on the approximation guarantee λ and on the error probability δ.

Each X_{ij} is computed as follows. Select at random a hash function ξ mapping the items of the universe U to $\{-1, +1\}$: ξ is selected from a family of 4-wise independent hash functions. Informally, 4-wise independence means that for every four distinct values $u_1, ..., u_4 \in U$ and for every 4-tuple $\varepsilon_1, ..., \varepsilon_4 \in \{-1, +1\}$, exactly (1/16)-fraction of the hash functions in the family map u_i to ε_i, for $i = 1, ..., 4$. Given ξ, we define $Z_{ij} = \sum_{u \in U} f_u\, \xi(u)$ and $X_{ij} = Z_{ij}^2$. Notice that Z_{ij} can be considered as a random linear projection (i.e., an inner product) of the frequency vector of the values in U with the random vector associated with such values by the hash function ξ.

It can be proved that $E[Y] = F_2$ and that, thanks to averaging of the X_{ij}, each Y_i has small variance. Computing Y as the median of Y_i allows it to boost the confidence using standard Chernoff bounds. We refer the interested reader to the work by Alon et al. [5] for a detailed proof. We limit here to formalize the statement of the result proved in the work by Alon et al. [5].

Theorem 4 [5] *For every $k \ge 1$, $\lambda > 0$, and $\delta > 0$, there exists a randomized algorithm that computes a number Y that deviates from F_2 by more than λF_2 with probability at most δ. The algorithm uses only*

$$O\left(\frac{\log(1/\delta)}{\lambda^2}(\log u + \log n)\right)$$

memory bits and performs one pass over the data.

Let us now consider the case of F_k, for $k \ge 2$. The basic idea of the sketch designed in the work by Alon et al. [5] is similar to that described above, but each X_{ij} is now computed by sampling the stream Σ as follows: an index $p = p_{ij}$ is chosen uniformly at random in $[1, n]$ and the number r of occurrences of x_p in the stream following position p is computed by keeping a counter. X_{ij} is then defined as $n(r^k - (r - 1)^k)$. We refer the interested reader to the works by Alon et al. [4–6] for a detailed description of this sketch and for the extension to the case where the stream length n is not known. We limit here to formalize the statement of the result proved in the work by Alon et al. [5]:

Theorem 5 [5] *For every $k \ge 1$, $\lambda > 0$ and $\delta > 0$, there exists a randomized algorithm that computes a number Y such that Y deviates from F_k by more than λF_k with probability at most δ. The algorithm uses*

$$O\left(\frac{k\log(1/\delta)}{\lambda^2}u^{1-1/k}(\log u + \log n)\right)$$

memory bits and performs only one pass over the data.

Notice that Theorem 5 implies that F_2 can be estimated using $O((\log(1/\delta)/\lambda^2)\sqrt{u}$ $(\log u + \log n))$ memory bits: this is worse by a $\sqrt{u}$ factor than the bound obtained in Theorem 4.

8.3.3 Techniques for Graph Problems

In this section we focus on techniques that can be applied to solve graph problems in the classical streaming and semi-streaming models. In Section 8.3.4 we will consider results obtained in less restrictive models that provide more powerful primitives for accessing stream data in a nonlocal fashion (e.g., stream-sort). Graph problems appear indeed to be difficult in classical streaming, and only few interesting results have been obtained so far. This is in line with the linear lower bounds on the space $\times$ passes product proved in the work by Henzinger et al. [43], even using randomization and approximation.

One problem for which sketches could be successfully designed is counting the number of triangles: if the graphs have certain properties, the algorithm presented in the work by Bar-Yossef et al. [10] uses sublinear space. Recently, Cormode and Muthukrishnan [24] studied three fundamental problems on multigraph degree sequences: estimating frequency moments of degrees, finding the heavy hitter degrees, and computing range sums of degree values. In all cases, their algorithms have space bounds significantly smaller than storing complete information. Due to the lower bounds in the work by Henzinger et al. [43], most work has been done in the semi-streaming model, in which problems such as distances, spanners, matchings, girth, and diameter estimation have been addressed [27,28,53]. In order to exemplify the techniques used in these works, in the rest of this section we focus on one such result, related to computing maximum weight matchings.

8.3.3.1 Approximating Maximum Weight Matchings Given an edge weighted, undirected graph $G(V, E, w)$, the weighted matching problem is to find a matching M^* such that $w(M^*) = \sum_{e \in M^*} w(e)$ is maximized. We recall that edges in a matching are such that no two edges have a common end point. We now present a one-pass semi-streaming algorithm that solves the weighted matching problem with approximation ratio $1/6$; that is, the matching M returned by the algorithm is such that

$$w(M^*) \leq 6\, w(M).$$

The algorithm has been proposed in the work by Feigenbaum et al. [27] and is very simple to describe. Algorithms with better approximation guarantees are described in the work by McGregor [53].

As edges are streamed, a matching M is maintained in main memory. Upon arrival of an edge e, the algorithm considers the set $C \subseteq M$ of matching edges that share an end point with e. If $w(e) > 2w(C)$, then e is added to M while the edges in C are removed; otherwise ($w(e) \leq 2w(C)$) e is ignored.

Note that, by definition of matching, the set C of conflicting edges has cardinality at most 2. Furthermore, since any matching consists of at most $n/2$ edges, the space requirement in bits is clearly $O(n \log n)$.

In order to analyze the approximation ratio, we will use the following notion of replacement tree associated with a matching edge (see also Fig. 8.1). Let e be an edge that belongs to M at the end of the algorithm's execution: the nodes of its replacement tree T_e are edges of graph G, and e is the root of T_e. When e has been added to M, it may have replaced one or two other edges e_1 and e_2 that were previously in M: e_1 and e_2 are children of e in T_e, which can be fully constructed by applying the reasoning recursively. It is easy to upper bound the total weight of nodes of each replacement tree.

Lemma 3 *Let $R(e)$ be the set of nodes of the replacement tree T_e, except for the root e. Then, $w(R(e)) \leq w(e)$.*

Proof. The proof is by induction. When e is a leaf in T_e (base step), $R(e)$ is empty and $w(R(e)) = 0$. Let us now assume that e_1 and e_2 are the children of e in T_e (the case of a unique child is similar). By inductive hypothesis, $w(e_1) \geq w(R(e_1))$ and $w(e_2) \geq w(R(e_2))$. Since e replaced e_1 and e_2, it must have been $w(e) \geq 2\,(w(e_1) + w(e_2))$. Hence, $w(e) \geq w(e_1) + w(e_2) + w(R(e_1)) + w(R(e_2)) = w(R(e))$. ■

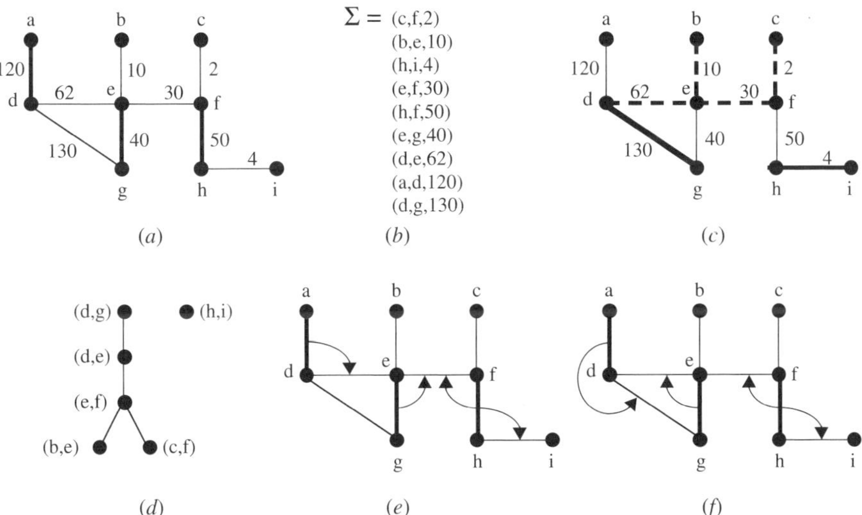

FIGURE 8.1 (*a*) A weighted graph and an optimal matching OPT (bold edges); (*b*) order in which edges are streamed; (*c*) matching M computed by the algorithm (bold solid edges) and edges in the history $H \setminus M$ (dashed edges); (*d*) replacement trees of edges in M; (*e*) initial charging of the weights of edges in OPT; (*f*) charging after the redistribution.

Theorem 6 [27] *In one pass and space $O(n \log n)$, the above algorithm constructs a $(1/6)$-approximate weighted matching M.*

Proof. Let $\text{OPT} = \{o_1, o_2, ...\}$ be the set of edges in a maximum weight matching and let $H = \bigcup_{e \in M}(R(e) \cup \{e\})$ be the set of edges that have been part of the matching at some point during the algorithm's execution (these are the nodes of the replacement trees).

We will show an accounting scheme that charges the weight of edges in OPT to edges in H. The charging strategy, for each edge $o \in$OPT, is the following:

- If $o \in H$, we charge $w(o)$ to o itself.
- If $o \notin H$, let us consider the time when o was examined for insertion in M, and let C be the set of edges that share an end point with o and were in M at that time. Since o was not inserted, it must have been $|C| \geq 1$ and $w(o) \leq 2\,w(C)$. If C contains only one edge, we charge $w(o)$ to that edge. If C contains two edges e_1 and e_2, we charge $w(o)w(e_1)/(w(e_1) + w(e_2)) \leq 2\,w(e_1)$ to e_1 and $w(o)w(e_2)/(w(e_1) + w(e_2)) \leq 2\,w(e_2)$ to e_2.

The following two properties hold: (a) the charge of o to any edge e is at most $2\,w(e)$; (b) any edge of H is charged by at most two edges of OPT, one per end point (see also Fig. 8.1).

We now redistribute some charges as follows: if an edge $o \in$ OPT charges an edge $e \in H$ and e gets replaced at some point by an edge $e' \in H$ that also shares an end point with o, we transfer the charge of o from e to e'. With this procedure, property (a) remains valid since $w(e') \geq w(e)$. Moreover, o will always charge an incident edge, and thus property (b) also remains true. In particular, each edge $e \in H \setminus M$ will be now charged by at most one edge in OPT: if at some point there are two edges charging e, the charge of one of them will be transferred to the edge of H that replaced e. Thus, only edges in M can be charged by two edges in OPT. By the above discussion we get

$$w(\text{OPT}) \leq \sum_{e \in H \setminus M} 2w(e) + \sum_{e \in M} 4w(e) = \sum_{e \in M} 2w(R(e)) + \sum_{e \in M} 4w(e)$$
$$\leq \sum_{e \in M} 6w(e) = 6w(M),$$

where the first equality is by definition of H and the last inequality is by Lemma 3. ■

8.3.4 Simulation of PRAM Algorithms

In this section we show that a variety of problems for which efficient solutions in classical streaming are not known or impossible to obtain can be solved very efficiently in the stream-sort model discussed in Section 8.2.3. In particular, we show that parallel algorithms designed in the PRAM model [48] can yield very efficient algorithms in the stream-sort model. This technique is very similar to previous methods developed in the context of external memory management for deriving I/O efficient algorithms (see, e.g., the work by Chiang et al. [19]). We recall that the PRAM is a popular model of parallel computation: it consists of a number of processors (each processor is a standard Random Access Machine) that communicate through a common, shared memory. The computation proceeds in synchronized steps: no processor will proceed with instruction $i + 1$ before all other processors complete the ith step.

Theorem 7 *Let A be a PRAM algorithm that uses N processors and runs in time T. Then, A can be simulated in stream-sort in $p = O(T)$ passes and space $s = O(\log N)$.*

Proof. Let $\Sigma = (1, \mathrm{val}_1)(2, \mathrm{val}_2) \cdots (M, \mathrm{val}_M)$ be the input stream that represents the memory image given as input to algorithm A, where val_j is the value contained at address j, and $M = O(N)$. At each step of algorithm A, processor p_i reads one memory cell at address in_i, updates its internal state st_i, and possibly writes one output cell at address out_i. In a preprocessing pass, we append to Σ the N tuples:

$$(p_1, \mathrm{in}_1, \mathrm{st}_1, \mathrm{out}_1) \cdots (p_N, \mathrm{in}_N, \mathrm{st}_N, \mathrm{out}_N),$$

where in_i and out_i are the cells read and written by p_i at the first step of algorithm A, respectively, and st_i is the initial state of p_i. Each step of A can be simulated by performing the following sorting and scanning passes:

1. We sort the stream so that each (j, val_j) is immediately followed by tuples $(p_i, \mathrm{in}_i, \mathrm{st}_i, \mathrm{out}_i)$ such that $\mathrm{in}_i = j$; that is, the stream has the form

   $$\begin{array}{l}(1, \mathrm{val}_1)(p_{i_{11}}, 1, \mathrm{st}_{i_{11}}, \mathrm{out}_{i_{11}})(p_{i_{12}}, 1, \mathrm{st}_{i_{12}}, \mathrm{out}_{i_{12}}) \cdots \\ (2, \mathrm{val}_2)(p_{i_{21}}, 2, \mathrm{st}_{i_{21}}, \mathrm{out}_{i_{21}})(p_{i_{22}}, 2, \mathrm{st}_{i_{22}}, \mathrm{out}_{i_{22}}) \cdots \\ \ldots \\ (M, \mathrm{val}_M)(p_{i_{M1}}, M, \mathrm{st}_{i_{M1}}, \mathrm{out}_{i_{M1}})(p_{i_{M2}}, M, \mathrm{st}_{i_{M2}}, \mathrm{out}_{i_{M2}}) \cdots\end{array}$$

   This can be done, for example, by using $2j$ as sorting key for tuples (j, val_j) and $2\mathrm{in}_i + 1$ as sorting key for tuples $(p_i, \mathrm{in}_i, \mathrm{st}_i, \mathrm{out}_i)$.
2. We scan the stream, performing the following operations:
   - If we read (j, val_j), we let currval $= \mathrm{val}_j$ and we write $(j, \mathrm{val}_j,$"old") to the output stream.
   - If we read $(p_i, \mathrm{in}_i, \mathrm{st}_i, \mathrm{out}_i)$, we simulate the task performed by processor p_i, observing that the value val_{in_i} that p_i would read from cell in_i is readily available in currval. Then we write to the output stream $(\mathrm{out}_i, \mathrm{res}_i,$"new"), where res_i is the value that p_i would write at address out_i, and we write tuple $(p_i, \mathrm{in}'_i, \mathrm{st}'_i, \mathrm{out}'_i)$, where in'_i and out'_i are the cells to be read and written at the next step of A, respectively, and st'_i is the new state of processor p_i.
3. Notice that at this point, for each j we have in the stream a triple of the form $(j, \mathrm{val}_j,$"old"), which contains the value of cell j before the parallel step, and possibly one or more triples $(j, \mathrm{res}_i,$"new"), which store the values written by processors to cell j during that step. If there is no "new" value for cell j, we simply drop the "old" tag from $(j, \mathrm{val}_j,$"old"). Otherwise, we keep for cell j

one of the new triples pruned of the "new" tag, and get rid of the other triples. This can be easily done with one sorting pass, which lets triples by the same j be consecutive, followed by one scanning pass, which removes tags and duplicates.

To conclude the proof, we observe that if A performs T steps, then our stream-sort simulation requires $p = O(T)$ passes. Furthermore, the number of bits of working memory required to perform each processor task simulation and to store curryal is $s = O(\log N)$. ■

Theorem 7 provides a systematic way of constructing streaming algorithms (in the stream-sort model) for several fundamental problems. Prominent examples are list ranking, Euler tour, graph connectivity, minimum spanning tree, biconnected components, and maximal independent set, among others: for these problems there exist parallel algorithms that use a polynomial number of processors and polylogarithmic time (see, e.g., the work by Jájá [48]). Hence, according to Theorem 7, these problems can be solved in the stream-sort model within polylogarithmic space and passes. Such bounds essentially match the results obtainable in more powerful computational models for massive data sets, such as the parallel disk model [64]. As observed by Aggarwal et al. [3], this suggests that using more powerful, harder to implement models may not always be justified.

8.4 LOWER BOUNDS

An important technique for proving streaming lower bounds is based on communication complexity lower bounds [43]. A crucial restriction in accessing a data stream is that items are revealed to the algorithm sequentially. Suppose that the solution of a computational problem needs to compare two items directly; one may argue that if the two items are far apart in the stream, one of them must be kept in main memory for long time by the algorithm until the other item is read from the stream. Intuitively, if we have limited space and many distant pairs of items to be compared, then we cannot hope to solve the problem unless we perform many passes over the data. We formalize this argument by showing reductions of communication problems to streaming problems. This allows us to prove lower bounds in streaming based on lower bounds in communication complexity. To illustrate this technique, we prove a lower bound for the element distinctness problem, which clearly implies a lower bound for the computation of the number of distinct items F_0 addressed in Section 8.3.2.

Theorem 8 *Any deterministic or randomized algorithm that decides whether a stream of n items contains any duplicates requires $p = \Omega(n/s)$ passes using s bits of working memory.*

Proof. The proof follows from a two-party communication complexity lower bound for the bit-vector-disjointness problem. In this problem, Alice has an n-bit-vector A and Bob has an n-bit-vector B. They want to know whether $A \cdot B > 0$, that is, whether there is at least one index $i \in \{1, \ldots, n\}$ such that $A[i] = B[i] = 1$. By a well-known communication complexity lower bound [50], Alice and Bob must communicate $\Omega(n)$ bits to solve the problem. This results holds also for randomized protocols: any algorithm that outputs the correct answer with high probability must communicate $\Omega(n)$ bits.

We now show that bit-vector-disjointness can be reduced to the element distinctness streaming problem. The reduction works as follows. Alice creates a stream of items S_A containing indices i such that $A[i] = 1$. Bob does the same for B, that is, he creates a stream of items S_B containing indices i such that $B[i] = 1$. Alice runs a streaming algorithm for element distinctness on S_A, then she sends the content of her working memory to Bob. Bob continues to run the same streaming algorithm starting from the memory image received from Alice, and reading items from the stream S_B. When the stream is over, Bob sends his memory image back to Alice, who starts a second pass on S_A, and so on. At each pass, they exchange $2s$ bits. At the end of the last pass, the streaming algorithm can answer whether the stream obtained by concatenating S_A and S_B contains any duplicates; since this stream contains duplicates if and only if $A \cdot B > 0$, this gives Alice and Bob a solution to the problem.

Assume by contradiction that the number of passes performed by Alice and Bob over the stream is $o(n/s)$. Since at each pass they communicate $2s$ bits, then the total number of bits sent between them over all passes is $o(n/s) \cdot 2s = o(n)$, which is a contradiction as they must communicate $\Omega(n)$ bits as noticed above. Thus, any algorithm for the element distinctness problem that uses s bits of working memory requires $p = \Omega(n/s)$ passes. ■

Lower bounds established in this way are information-theoretic, imposing no restrictions on the computational power of the algorithms. The general idea of reducing a communication complexity problem to a streaming problem is very powerful, and allows it to prove several streaming lower bounds. Those range from computing statistical summary information such as frequency moments [5] to graph problems such as vertex connectivity [43], and imply that for many fundamental problems there are no one-pass exact algorithms with a working memory significantly smaller than the input stream.

A natural question is whether approximation can make a significant difference for those problems, and whether randomization can play any relevant role. An interesting observation is that there are problems, such as the computation of frequency moments, for which neither randomization nor approximation is powerful enough for getting a solution in one pass and sublinear space, unless they are used together.

8.4.1 Randomization

As we have seen in the proof of Theorem 8, lower bounds based on the communication complexity of the bit-vector-disjointness problem hold also for randomized algorithms, which yields clear evidence that randomization without approximation may not help. The result of Theorem 8 can be generalized for all one-pass frequency moments. In particular, it is possible to prove that any randomized algorithm for computing the frequency moments that outputs the correct result with probability higher than $1/2$ in one pass must use $\Omega(n)$ bits of working memory. The theorem can be proven using communication complexity tools.

Theorem 9 [6] *For any nonnegative integer* $k \neq 1$*, any randomized algorithm that makes one pass over a sequence of at least* $2n$ *items drawn from the universe* $U = \{1, 2, \ldots, n\}$ *and computes* F_k *exactly with probability* $> 1/2$ *must use* $\Omega(n)$ *bits of working memory.*

8.4.2 Approximation

Conversely, we can show that any deterministic algorithm for computing the frequency moments that approximates the correct result within a constant factor in one pass must use $\Omega(n)$ bits of working memory. Differently from the lower bounds addressed earlier in this section, we give a direct proof of this result without resorting to communication complexity arguments.

Theorem 10 [6] *For any nonnegative integer* $k \neq 1$*, any deterministic algorithm that makes one pass over a sequence of at least* $n/2$ *items drawn from the universe* $U = \{1, 2, \ldots, n\}$ *and computes a number* Y *such that* $|Y - F_k| \leq F_k/10$ *must use* $\Omega(n)$ *bits of working memory.*

Proof. The idea of the proof is to show that if the working memory is not large enough, for any deterministic algorithm (which does not use random bits) there exist two subsets S_1 and S_2 in a suitable collection of subsets of U such that the memory image of the algorithm is the same after reading either S_1 or S_2; that is, S_1 and S_2 are indistinguishable. As a consequence, the algorithm has the same memory image after reading either $S_1 : S_1$ or $S_2 : S_1$, where $A : B$ denotes the stream of items that starts with the items of A and ends with the items of B. If S_1 and S_2 have a small intersection, then the two streams $S_1 : S_1$ and $S_2 : S_1$ must have rather different values of F_k, and the algorithm must necessarily make a large error on estimating F_k on at least one of them. We now give more details on the proof assuming that $k \geq 2$. The case $k = 0$ can be treated symmetrically.

Using a standard construction in coding theory, it is possible to build a family $\mathcal{F}$ of $2^{\Omega(n)}$ subsets of U of size $n/4$ each such that any two of them have at most $n/8$ common items. Notice that, for every set in $\mathcal{F}$, the frequency of any value of U in that set is either 0 or 1. Fix a deterministic algorithm and let $s < \log_2 \mathcal{F}$ be the size of its working memory. Since the memory can assume at most 2^s different configurations and we have $|\mathcal{F}| > 2^s$ possible distinct input sets in $\mathcal{F}$, then by the pigeonhole principle there must be two input sets $S_1, S_2 \in \mathcal{F}$ such that the memory image of the algorithm after reading either one of them is the same. Now, if we consider the two streams $S_1 : S_1$ and $S_2 : S_1$, the memory image of the algorithm after processing either one of them is the same. Since by construction of $\mathcal{F}$, S_1 and S_2 contain $n/4$ items each, and have at most $n/8$ items in common, then

- Each of the $n/4$ distinct items in $S_1 : S_1$ has frequency 2, thus

$$F_k^{S_1:S_1} = \sum_{i=1}^{n} f_i^k = 2^k \cdot \frac{n}{4}.$$

- If S_1 and S_2 have exactly $n/8$ items in common, then $S_2 : S_1$ contains exactly $n/8 + n/8 = n/4$ items with frequency 1 and $n/8$ items with frequency 2. Hence,

$$F_k^{S_2:S_1} = \sum_{i=1}^{n} f_i^k = \frac{n}{4} + 2^k \cdot \frac{n}{8}.$$

  Notice that, for $k \geq 2$, $F_k^{S_2:S_1}$ can only decrease as $|S_1 \cap S_2|$ decreases, and therefore we can conclude that

$$F_k^{S_2:S_1} \leq \frac{n}{4} + 2^k \cdot \frac{n}{8}.$$

To simplify the notation, let $A = F_k^{S_2:S_1}$ and $B = F_k^{S_1:S_1}$. The maximum relative error performed by the algorithm on either input $S_2 : S_1$ or input $S_1 : S_1$ is

$$\max\left\{\frac{|Y - A|}{A}, \frac{|Y - B|}{B}\right\}.$$

In order to prove that the maximum relative error is always $\geq 1/10$, it is sufficient to show that

$$\frac{|Y - B|}{B} < \frac{1}{10} \quad \Rightarrow \quad \frac{|Y - A|}{A} \geq \frac{1}{10}. \tag{8.4}$$

Let $C = n/4 + 2^k \cdot n/8$. For $k \geq 2$, it is easy to check that $A \leq C \leq B = 2^k \cdot n/4$. Moreover, the maximum relative error obtained for any $Y < A$ is larger than the maximum relative error obtained for $Y = A$ (similarly for $Y > B$): thus, the value of Y that minimizes the relative error is such that $A \leq Y \leq B$. Under this hypothesis, $|Y - B| = B - Y$ and $|Y - A| = Y - A$. With simple calculations, we can show that proving (8.4) is equivalent to proving that

$$Y > \frac{9}{10}B \quad \Rightarrow \quad Y \geq \frac{11}{10}A.$$

Notice that $C = n/4 + B/2$. Using this fact, it is not difficult to see that $9B \geq 11C$ for any $k \geq 2$, and therefore the above implication is always satisfied since $C \geq A$.

Since the maximum relative error performed by the algorithm on either input $S_1 : S_1$ or input $S_2 : S_1$ is at least $1/10$, we can conclude that if we use fewer than $\log_2 \mathcal{F} = \Omega(n)$ memory bits, there is an input on which the algorithm outputs a value Y such that $|Y - F_k| > F_k/10$, which proves the claim. ■

8.4.3 Randomization and Approximation

A natural approach that combines randomization and approximation would be to use random sampling to get an estimator of the solution. Unfortunately, this may not always work: as an example, Charikar et al. [15] have shown that estimators based on random sampling do not yield good results for F_0.

Theorem 11 [15] *Let E be a (possibly adaptive and randomized) estimator of F_0 that examines at most r items in a set of n items and let* $\mathrm{err} = \max\{E/F_0, F_0/E\}$ *be the error of the estimator. Then, for any $p > 1/e^r$, there is a choice of the set of items such that* $\mathrm{err} \geq \sqrt{((n-r)/2r)\ln(1/p)}$ *with probability at least p.*

The result of Theorem 11 states that no good estimator can be obtained if we only examine a fraction of the input. On the contrary, as we have seen in Section 8.3.2, hashing techniques that examine all items in the input allow it to estimate F_0 within an arbitrary fixed error bound with high probability using polylogarithmic working memory space for any given data set.

We notice that, while the ideal goal of a streaming algorithm is to solve a problem using a working memory of size polylogarithmic in the size of the input stream, for some problems this is impossible even using approximation and randomization, as shown in the following theorem from the work by Alon et al. [6].

Theorem 12 [6] *For any fixed integer $k > 5$, any randomized algorithm that makes one pass over a sequence of at least n items drawn from the universe $U = \{1, 2, \ldots, n\}$ and computes an approximate value Y such that $|Y - F_k| > F_k/10$ with probability $< 1/2$ requires at least $\Omega(n^{1-5/k})$ memory bits.*

Theorem 12 holds in a streaming scenario where items are revealed to the algorithm in an online manner and no assumptions are made on the input. We finally notice that in the same scenario there are problems for which approximation and randomization do not help at all. A prominent example is given by the computation of F_∞, the maximum frequency of any item in the stream.

Theorem 13 [6] *Any randomized algorithm that makes one pass over a sequence of at least $2n$ items drawn from the universe $U = \{1, 2, \ldots, n\}$ and computes an approximate value Y such that $|Y - F_\infty| \geq F_\infty/3$ with probability $< 1/2$ requires at least $\Omega(n)$ memory bits.*

8.5 SUMMARY

In this chapter we have addressed the emerging field of data stream algorithmics, providing an overview of the main results in the literature and discussing computational models, applications, lower bound techniques, and tools for designing efficient algorithms. Several important problems have been proven to be efficiently solvable despite the strong restrictions on the data access patterns and memory requirements of the algorithms that arise in streaming scenarios. One prominent example is the computation of statistical summaries such as frequency moments, histograms, and wavelet coefficient, which are of great importance in a variety of applications including network traffic analysis and database optimization. Other widely studied problems include norm estimation, geometric problems such as clustering and facility location, and graph problems such as connectivity, matching, and distances.

From a technical point of view, we have discussed a number of important tools for designing efficient streaming algorithms, including random sampling, probabilistic counting, hashing, and linear projections. We have also addressed techniques for graph problems and we have shown that extending the streaming paradigm with a sorting primitive yields enough power for solving a variety of problems in external memory, essentially matching the results obtainable in more powerful computational models for massive data sets.

Finally, we have discussed lower bound techniques, showing that tools from the field of communication complexity can be effectively deployed for proving strong streaming lower bounds. We have discussed the role of randomization and approximation, showing that for some problems neither one of them yields enough power, unless they are used together. We have also shown that other problems are intrinsically hard in a streaming setting even using approximation and randomization, and thus cannot be solved efficiently unless we consider less restrictive computational models.

ACKNOWLEDGMENTS

We are indebted to Alberto Marchetti-Spaccamela for his support and encouragement, and to Andrew McGregor for his very thorough reading of this survey. This work has been partially supported by the Sixth Framework Programme of the EU under Contract IST-FET 001907 ("DELIS: Dynamically Evolving Large Scale Information Systems") and by MIUR, the Italian Ministry of Education, University and Research, under Project ALGO-NEXT ("Algorithms for the Next Generation Internet and Web: Methodologies, Design and Experiments").

REFERENCES

1. Agrawal D, Metwally A, El Abbadi, A. Efficient computation of frequent and top-*k* elements in data stream. Proceedings of the 10th International Conference on Database Theory; 2005. p 398–412.
2. Abello J, Buchsbaum A, Westbrook JR. A functional approach to external graph algorithms. Algorithmica 2002;32(3):437–458.
3. Aggarwal G, Datar M, Rajagopalan S, Ruhl M. On the streaming model augmented with a sorting primitive. Proceedings of the 45th Annual IEEE Symposium on Foundations of Computer Science (FOCS'04); 2004.

4. Alon N, Gibbons P, Matias Y, Szegedy M. Tracking join and self-join sizes in limited storage. Proceedings of the 18th ACM Symposium on Principles of Database Systems (PODS'99); 1999. p 10–20.
5. Alon N, Matias Y, Szegedy M. The space complexity of approximating the frequency moments. Proceedings of the 28th Annual ACM Symposium on Theory of Computing (STOC'96). ACM Press: 1996. p 20–29.
6. Alon N, Matias Y, Szegedy M. The space complexity of approximating the frequency moments. J Comput Syst Sci 1999; 58(1):137–147.
7. Babcock B, Babu S, Datar M, Motwani R, Widom J. Models and issues in data stream systems. Proceedings of the 21st ACM Symposium on Principles of Database Systems (PODS'02); 2002. p 1–16.
8. Babcock B, Datar M, Motwani R. Sampling from a moving window over streaming data. Proceedings of the 13th Annual ACM-SIAM Symposium on Discrete Algorithms (SODA'02); 2002. p 633–634.
9. Bar-Yossef Z, Jayram T, Kumar R, Sivakumar D. Information statistics approach to data stream and communication complexity. Proceedings of the 43rd Annual IEEE Symposium on Foundations of Computer Science (FOCS'02); 2002.
10. Bar-Yossef Z, Kumar R, Sivakumar D. Reductions in streaming algorithms, with an application to counting triangles in graphs. Proceedings of the 13th Annual ACM-SIAM Symposium on Discrete Algorithms (SODA'02); 2002. p 623–632.
11. Bar-Yossef Z, Jayram T, Kumar R, Sivakumar D, Trevisan L. Counting distinct elements in a data stream. Proceedings of the 6th International Workshop on Randomization and Approximation Techniques in Computer Science; 2002. p 1–10.
12. Bhuvanagiri L, Ganguly S, Kesh D, Saha C. Simpler algorithm for estimating frequency moments of data streams. Proceedings of the 17th Annual ACM-SIAM Symposium on Discrete Algorithms (SODA'06); 2006. p 708–713.
13. Buriol L, Frahling G, Leonardi S, Marchetti-Spaccamela A, Sohler C. Counting triangles in data streams. Proceedings of the 25th ACM Symposium on Principles of Database Systems (PODS'06); 2006. p 253–262.
14. Chakrabarti A, Khot S, Sun X. Near-optimal lower bounds on the multi-party communication complexity of set disjointness. Proceedings of the IEEE Conference on Computational Complexity; 2003. p 107–117.
15. Charikar M, Chaudhuri S, Motwani R, Narasayya V. Towards estimation error guarantees for distinct values. Proceedings of the 19th ACM Symposium on Principles of Database Systems (PODS'00); 2000. p 268–279.
16. Charikar M, Chen K, Farach-Colton M. Finding frequent items in data streams. Proceedings of the 29th International Colloquium on Automata, Languages and Programming (ICALP'02); 2002. p 693–703.
17. Charikar M, O'Callaghan L, Panigrahy R. Better streaming algorithms for clustering problems. Proceedings of the 35th Annual ACM Symposium on Theory of Computing (STOC'03); 2003.
18. Chaudhuri S, Motwani R, Narasayya V. Random sampling for histogram construction: How much is enough? Proceedings of the ACM SIGMOD International Conference on Management of Data; 1998. p 436–447.

19. Chiang Y, Goodrich MT, Grove EF, Tamassia R, Vengroff DE, Vitter JS. External-memory graph algorithms. Proceedings of the 6th Annual ACM-SIAM Symposium on Discrete Algorithms (SODA'95); 1995. p 139–149.

20. Coppersmith D, Kumar R. An improved data stream algorithm for frequency moments. Proceedings of the 15th Annual ACM-SIAM Symposium on Discrete Algorithms (SODA'04); 2004. p 151–156.

21. Cormode G, Muthukrishnan S. Estimating dominance norms on multiple data streams. Proceedings of the 11th Annual European Symposium on Algorithms (ESA'03); 2003. p 148–160.

22. Cormode G, Muthukrishnan S. What is hot and what is not: Tracking most frequent items dynamically. Proceedings of the 22nd ACM Symposium on Principles of Database Systems (PODS'03); 2003.

23. Cormode G, Muthukrishnan S. An improved data stream summary: the count-min sketch and its applications. J Algorithms 2005;55(1):58–75.

24. Cormode G, Muthukrishnan S. Space efficient mining of multigraph streams. Proceedings of the 24th ACM Symposium on Principles of Database Systems (PODS'05); 2005.

25. Demetrescu C, Finocchi I, Ribichini A. Trading off space for passes in graph streaming problems. Proceedings of the 17th Annual ACM-SIAM Symposium on Discrete Algorithms (SODA'06); 2006. p 714–723.

26. Elkin M, Zhang J. Efficient algorithms for constructing $(1 + \epsilon, \beta)$-spanners in the distributed and streaming models. Proceedings of the 23rd Annual ACM Symposium on Principles of Distributed Computing (PODC'04); 2004. p 160–168.

27. Feigenbaum J, Kannan S, McGregor A, Suri S, Zhang J. On graph problems in a semi-streaming model. Proceedings of the 31st International Colloquium on Automata, Languages and Programming (ICALP'04); 2004.

28. Feigenbaum J, Kannan S, McGregor A, Suri S, Zhang J. Graph distances in the streaming model: the value of space. Proceedings of the 16th ACM/SIAM Symposium on Discrete Algorithms (SODA'05); 2005. p 745–754.

29. Flajolet P, Martin GN. Probabilistic counting. Proceedings of the 24th Annual Symposium on Foundations of Computer Science; 1983. p 76–82.

30. Flajolet P, Martin GN. Probabilistic counting algorithms for database applications. J Comput Syst Sci 1985;31(2):182–209.

31. Frahling G, Indyk P, Sohler C. Sampling in dynamic data streams and applications. Proceedings of the 21st ACM Symposium on Computational Geometry; 2005. p 79–88.

32. Frahling G, Sohler C. Coresets in dynamic geometric data streams. Proceedings of the 37th Annual ACM Symposium on Theory of Computing (STOC'05); 2005.

33. Gibbons PB, Matias Y. New sampling-based summary statistics for improving approximate query answers. Proceedings of the ACM SIGMOD International Conference on Management of Data; 1998.

34. Gibbons PB, Matias Y. Synopsis data structures for massive data sets. In: External Memory Algorithms. DIMACS Series in Discrete Mathematics and Theoretical Computer Science. Volume 50. American Mathematical Society; 1999. p 39–70.

35. Gibbons PB, Matias Y, Poosala V. Fast incremental maintenance of approximate histograms. Proceedings of 23rd International Conference on Very Large Data Bases (VLDB'97); 1997.

36. Gilbert A, Kotidis Y, Muthukrishnan S, Strauss M. How to summarize the universe: dynamic maintenance of quantiles. Proceedings of 28th International Conference on Very Large Data Bases (VLDB'02); 2002. p 454–465.
37. Gilbert AC, Guha S, Indyk P, Kotidis Y, Muthukrishnan S, Strauss M. Fast, small-space algorithms for approximate histogram maintenance. Proceedings of the 34th ACM Symposium on Theory of Computing (STOC'04); 2002. p 389–398.
38. Gilbert AC, Kotidis Y, Muthukrishnan S, Strauss M. Surfing wavelets on streams: one-pass summaries for approximate aggregate queries. Proceedings of 27th International Conference on Very Large Data Bases (VLDB'01); 2001. p 79–88.
39. Golab L, Ozsu MT. Data stream management issues—a survey. Technical Report No. TR CS-2003-08. School of Computer Science, University of Waterloo; 2003.
40. Guha S, Indyk P, Muthukrishnan S, Strauss M. Histogramming data streams with fast per-item processing. Proceedings of the 29th International Colloquium on Automata, Languages and Programming (ICALP'02); 2002. p 681–692.
41. Guha S, Koudas N, Shim K. Data streams and histograms. Proceedings of the 33rd Annual ACM Symposium on Theory of Computing (STOC'01); 2001. p 471–475.
42. Guha S, Mishra N, Motwani R, O'Callaghan L. Clustering data streams. Proceedings of the 41st Annual IEEE Symposium on Foundations of Computer Science (FOCS'00); 2000. p 359–366.
43. Henzinger M, Raghavan P, Rajagopalan S. Computing on data streams. In: External Memory Algorithms. DIMACS Series in Discrete Mathematics and Theoretical Computer Science. Volume 50. American Mathematical Society; 1999. 107–118.
44. Indyk P. Stable distributions, pseudorandom generators, embeddings and data stream computation. Proceedings of the 41st Annual IEEE Symposium on Foundations of Computer Science (FOCS'00); 2000. p 189–197.
45. Indyk P. Algorithms for dynamic geometric problems over data streams. Proceedings of the 36th Annual ACM Symposium on Theory of Computing (STOC'04); 2004. p 373–380.
46. Indyk P, Woodruff D. Tight lower bounds for the distinct elements problem. Proceedings of the 44th Annual IEEE Symposium on Foundations of Computer Science (FOCS'03); 2003.
47. Indyk P, Woodruff D. Optimal approximations of the frequency moments. Proceedings of the 37th Annual ACM Symposium on Theory of Computing (STOC'05); 2005.
48. Jájá J. An Introduction to Parallel Algorithms. Addison-Wesley; 1992.
49. Jowhari H, Ghodsi M. New streaming algorithms for counting triangles in graphs. Proceedings of the 11th Annual International Conference on Computing and Combinatorics (COCOON'05); 2005. p 710–716.
50. Kushilevitz E, Nisan N. Communication Complexity. Cambridge University Press; 1997.
51. Manku GS, Motwani R. Approximate frequency counts over data streams. Proceedings 28th International Conference on Very Large Data Bases (VLDB'02); 2002. p 346–357.
52. Matias Y, Vitter JS, Wang M. Dynamic maintenance of wavelet-based histograms. Proceedings of 26th International Conference on Very Large Data Bases (VLDB'00); 2000.
53. McGregor A. Finding matchings in the streaming model. Proceedings of the 8th International Workshop on Approximation Algorithms for Combinatorial Optimization Problems (APPROX'05), LNCS 3624; 2005. p 170–181.

54. Misra J, Gries D. Finding repeated elements. Sci Comput Program 1982;2:143–152.

55. Morris R. Counting large numbers of events in small registers. Commun ACM 1978;21(10):840–842.

56. Munro I, Paterson M. Selection and sorting with limited storage. Theor Comput Sci 1980; 12:315–323. A preliminary version appeared in IEEE FOCS'78.

57. Muthukrishnan S. Data streams: algorithms and applications. Technical report; 2003. Available at http://athos.rutgers.edu/~muthu/stream-1-1.ps.

58. Muthukrishnan S, Strauss M. Maintenance of multidimensional histograms. Proceedings of the FSTTCS; 2003. p 352–362.

59. Muthukrishnan S, Strauss M. Rangesum histograms. Proceedings of the 14th Annual ACM-SIAM Symposium on Discrete Algorithms (SODA'03); 2003.

60. Muthukrishnan S, Strauss M. Approximate histogram and wavelet summaries of streaming data. Technical report, DIMACS TR 2004-52; 2004.

61. Ruhl M. Efficient algorithms for new computational models. Ph.D. thesis. Department of Electrical Engineering and Computer Science, Massachusetts Institute of Technology; 2003.

62. Saks M, Sun X. Space lower bounds for distance approximation in the data stream model. Proceedings of the 34th Annual ACM Symposium on Theory of Computing (STOC'02); 2002. p 360–369.

63. Vitter JS. Random sampling with a reservoir. ACM Trans Math Software 1995;11(1):37–57.

64. Vitter JS. External memory algorithms and data structures: dealing with massive data. ACM Comput Surv 2001;33(2):209–271.

CHAPTER 9

Applying Evolutionary Algorithms to Solve the Automatic Frequency Planning Problem

FRANCISCO LUNA, ENRIQUE ALBA, ANTONIO J. NEBRO, PATRICK MAUROY, and SALVADOR PEDRAZA

9.1 INTRODUCTION

The global system for mobile communications (GSM) [14] is an open, digital cellular technology used for transmitting mobile voice and for data services. GSM is also referred to as 2G, because it represents the second generation of this technology, and it is certainly the most successful mobile communication system. Indeed, by mid-2006 GSM services are in use by more than 1.8 billion subscribers across 210 countries, representing approximately 77 percent of the world's cellular market. GSM differs from the first-generation wireless systems in that it uses digital technology and frequency division multiple access/time division multiple access (FDMA/TDMA) transmission methods. It is also widely accepted that the Universal Mobile Telecommunication system (UMTS) [15], the third-generation mobile telecommunication system, will coexist with the enhanced releases of the GSM standard (GPRS [9] and EDGE [7]) at least in the first phases. Therefore, GSM is expected to play an important role as a dominating technology for many years.

The success of this multiservice cellular radio system lies in efficiently using the scarcely available radio spectrum. GSM uses *frequency division multiplexing* and *time division multiplexing* schemes to maintain several communication links "in parallel." The available frequency band is slotted into channels (or frequencies) that have to be allocated to the elementary transceivers (TRXs) installed in the base stations of the network. This problem is known as the automatic frequency planning (AFP), frequency assignment problem (FAP), or channel assignment problem (CAP). Several different problem types are subsumed under these general terms and many mathematical models have been proposed since the late 1960s [1,6,12]. This chapter,

[1] http://www.wirelessintelligence.com/.

Handbook of Applied Algorithms: Solving Scientific, Engineering and Practical Problems
Edited by Amiya Nayak and Ivan Stojmenović

however, is focused on concepts and models that are relevant for current GSM frequency planning and not on simplified models of the abstract problem. In GSM, a network operator has usually a small number of frequencies (few dozens) available to satisfy the demand of several thousands of TRXs. A reuse of these frequencies is therefore unavoidable. However, reusing frequencies is limited by interferences that could lead the quality of service (QoS) for subscribers to be reduced down to unsatisfactory levels. The automatic generation of frequency plans in real GSM networks [5] is a very important task for present GSM operators not only in the initial deployment of the system, but also in the subsequent expansions or modifications of the network, solving unpredicted interference reports, and/or handling anticipated scenarios (e.g., an expected increase in the traffic demand in some areas).

This optimization problem is a generalization of the graph coloring problem, and thus it is an NP-hard problem [10]. As a consequence, using exact algorithms to solve real-sized instances of AFP problems is not practical, and therefore other approaches are required. Many different methods have been proposed in the literature [1], and among them, metaheuristic algorithms have proved to be particularly effective. Metaheuristics [3,8] are stochastic algorithms that sacrifice the guarantee of finding optimal solutions for the sake of (hopefully) getting accurate (also optimal) ones in a reasonable time. This fact is even more important in commercial tools, in which the GSM operator cannot wait very long times for a frequency plan (e.g., several weeks). Among the existing metaheuristic techniques, evolutionary algorithms (EAs) [2] have been widely used [6]. EAs work iteratively on a population of individuals. Every individual is the encoded version of a tentative solution to which a fitness value is assigned indicating its suitability to the problem. The canonical algorithm applies stochastic operators such as selection, crossover (merging two or more parents to yield one or more offsprings), and mutation (random alterations of the problem variables) on an initial population in order to compute a whole generation of new individuals. However, it has been reported in the literature that crossover operators do not work properly for this problem [4,17]. In this scenario, our algorithmic proposal is a fast and accurate $(1 + 10)$ EA (see the work by Schwefel [16] for details on this notation) in which recombination of individuals is not performed. The main contributions of this chapter are the following:

- We have developed and analyzed a new $(1 + 10)$ EA. Several seeding methods as well as several mutation operators have been proposed.
- The evaluation of the algorithm has been performed by using a real-world instance provided by Optimi Corp.™ This is a currently operating GSM network in which we are using real traffic data, accurate models for all the system components (signal propagation, TRX, locations, etc.), and actual technologies such as frequency hopping. This evaluation of the tentative frequency plans is carried out with a powerful commercial simulator that enables users to simulate and analyze those plans prior to implementation in a real environment.
- Results show that this simple algorithm is able to compute accurate frequency plans, which can be directly deployed in a real GSM network.

The chapter is structured as follows. In the next section, we provide the reader with some details on the frequency planning in GSM networks. Section 9.3 describes the algorithm proposed along with the different genetic operators used. The results of the experimentation are analyzed in Section 9.4. Finally, conclusions and future lines of research are discussed in the last section.

9.2 AUTOMATIC FREQUENCY PLANNING IN GSM

This section is devoted to presenting some details on the frequency planning task for a GSM network. We first provide the reader with a brief description of the GSM architecture. Next, we give the relevant concepts to the frequency planning problem that will be used along this chapter.

9.2.1 The GSM System

An outline of the GSM network architecture is shown in Figure 9.1. The solid lines connecting components carry both traffic information (voice or data) and the "in-band" signaling information. The dashed lines are signaling lines. The information exchanged over these lines is necessary for supporting user mobility, network features,

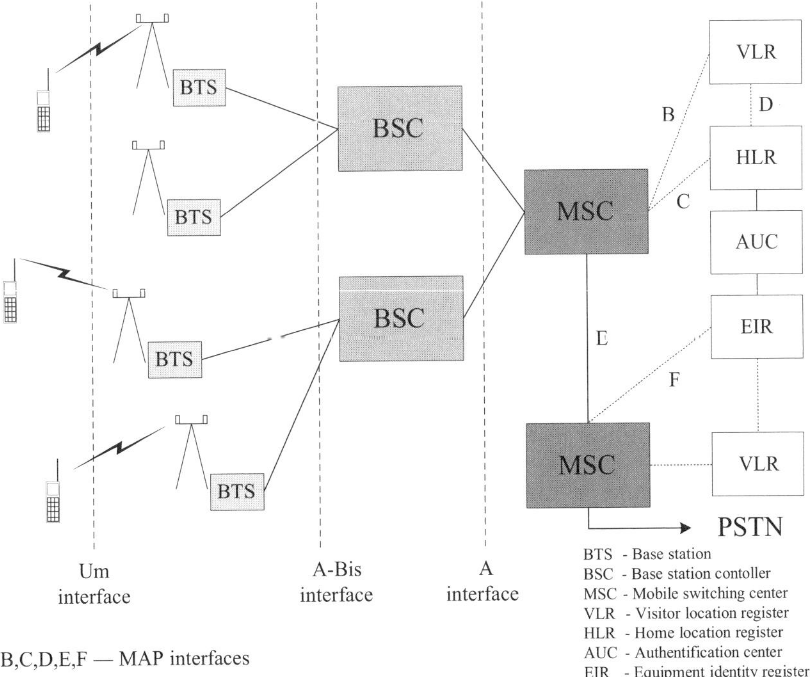

FIGURE 9.1 Outline of the GSM network architecture.

operation and maintenance, authentication, encryption, and many other functions necessary for the network's proper operation. Figure 9.1 shows the different network components and interfaces within a GSM network.

As it can be seen, GSM networks are built out of many different components. The most relevant ones to frequency planning are briefly described next.

9.2.1.1 Mobile Terminals

Mobile terminals are the (only) part of the system's equipment that the user is aware of. Usually, the mobile terminal is designed in the form of a phone. The GSM mobile phone is designed as a unity of two parts that are both functionally and physically separated:

1. Hardware and software specific to the GSM radio interface.
2. Subscriber identity module (SIM). The SIM is a removable part of the mobile terminal that stores a subscriber's unique identification information. The SIM allows the subscriber to access the network regardless of the particular mobile station being used.

9.2.1.2 Base Transceiver Station (BTS)

In essence, the BTS is a set of TRXs. In GSM, one TRX is shared by up to eight users in TDMA mode. The main role of a TRX is to provide conversion between the digital traffic data on the network side and radio communication between the mobile terminal and the GSM network. The site at which a BTS is installed is usually organized in sectors: one to three sectors are typical. Each sector defines a cell. A single GSM BTS can host up to 16 TRXs.

9.2.1.3 Base Station Controller (BSC)

The BSC plays a role of a small digital exchange station with some mobility-specific tasks and it has a substantial switching capability. It is responsible for intra-BTS functions (e.g., allocation and release of radio channels), as well as for most processing involving inter-BTS handovers.

9.2.1.4 Other Components

Every BSC is connected to one mobile service switching center (MSC), and the core network interconnects the MSC core network MSCs. Specially equipped gateway MSCs (GMSCs) interface with other telephony and data networks. The home location registers (HLRs) and the visitors location registers (VLRs) are database systems, which contain VLR subscriber data and facilitate mobility management. Each gateway MSC consults its home location register if an incoming call has to be routed to a mobile terminal. The HLR is also used in the authentication of the subscribers together with the authentication center (AuC).

9.2.2 Automatic Frequency Planning

The frequency planning is the last step in the layout of a GSM network. Prior to tackling this problem, the network designer has to address some other issues: where to install the BTSs, how to dimension signaling propagation parameters of the antennas (tilt, azimuth, etc.), how to connect BTSs to BSCs, or how to connect MSCs among each other and to the BSCs [13]. Once the sites for the BTSs are selected and the sector layout is decided, the number of TRXs to be installed per sector has to be fixed. This number depends on the traffic demand that the corresponding sector has to support. The result from this process is a quantity of TRXs per cell. A channel has to be allocated to every TRX and this is the main goal of the automatic frequency planning [5]. Essentially, three kinds of allocation exist: fixed channel allocation (FCA), dynamic channel allocation (DCA), and hybrid channel allocation. In FCA, the channels are permanently allocated to each TRX, while in DCA the channels are allocated dynamically upon request. Hybrid channel allocation (HCA) schemes combine FCA and DCA. Neither DCA nor HCA are supported in GSM, so we only consider here FCA.

We now explain the most important parameters to be taken into account in GSM frequency planning. Let us consider the example network shown in Figure 9.2, in which each site has three installed sectors (e.g., site A operates $A1$, $A2$, and $A3$). The first issue that we want to remark is the implicit topology that results from the previous steps in the network design. In this topology, each sector has an associated list of neighbors containing the possible handover candidates for the mobile residing

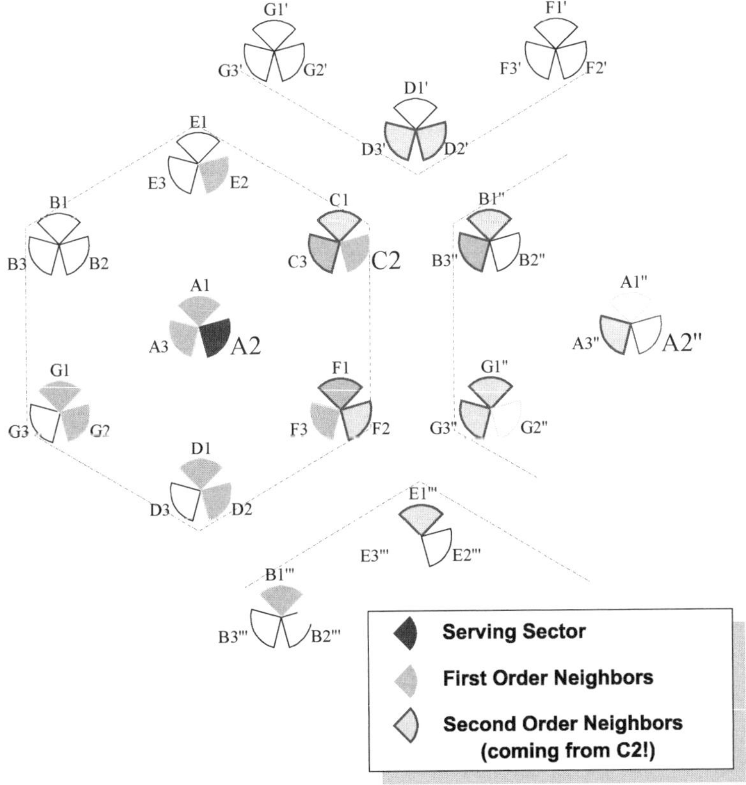

FIGURE 9.2 An example of GSM network.

in a specific cell. These neighbors are further distinguished into first-order (those that can potentially provoke strong interference to the serving sector) and second-order neighbors. In Figure 9.2, $A2$ is the serving sector and the first-order neighbors defined are $A1$, $A3$, $C2$, $D1$, $D2$, $E2$, $F3$, $G1$, $G2$, and $B1'''$, whereas the second-order neighbors coming from $C2$ are $F1$, $F2$, $C1$, $C3$, $D2'$, $D3'$, $A3''$, $B1''$, $B3''$, $G1''$, $G3''$, and $E1'''$.

As stated before, each sector in a site defines a cell; the number of TRXs installed in each cell depends on the traffic demand. A valid channel from the available spectrum has to be allocated to each TRX. Owing to technical and regulatory restrictions, some channels in the spectrum may not be available in every cell. Such channels are called locally blocked and they can be specified for each cell.

Each cell operates one broadcast control channel (BCCH), which broadcasts cell organization information. The TRX allocating the BCCH can also carry user data. When this channel does not meet the traffic demand, some additional TRXs have to be installed to which new dedicated channels are assigned for traffic data. These are called traffic channels (TCHs).

In GSM, significant interference may occur if the same or adjacent channels are used in neighboring cells. Correspondingly, they are named co-channel and adj-channel interference. Many different constraints are defined to avoid strong interference in the GSM network. These constraints are based on how close the channels assigned to a pair of TRXs may be. These are called separation constraints, and they seek to ensure that there is proper transmission and reception at each TRX and/or that the call handover between cells is supported. Several sources of constraint separation exist: co-site separation, when two or more TRXs are installed in the same site, or co-cell separation, when two TRXs serve the same cell (i.e., they are installed in the same sector).

This is intentionally an informal description of the automatic frequency problem in GSM networks. It is out of the scope of this chapter to propose a precise model of the problem, since we use a proprietary software that is aware of all these concepts, as well as the consideration of all the existing advanced techniques, such as frequency hopping, power control, discontinuous transmission, and so on [5], developed for efficiently using the scarce frequency spectrum available in GSM.

9.3 EAs FOR SOLVING THE AFP PROBLEM

EAs have been widely used for solving the many existing flavors of the frequency assignment problem [1,5,6,11]. However, it has been shown that well-known crossover operators such as single-point crossover do not perform well on this problem [4]. Indeed, it does not make sense for a frequency planning to randomly exchange two different, possibly nonrelated assignments. Our approach here is to use an $(1 + 10)$ EA, in which the recombination operator is not required. In the following, we first describe the generic $(\mu + \lambda)$ EA. The solution encoding used, the fitness function, and several proposals for generating the initial solutions and the perturbing individuals are discussed afterward.

```
P = new Population(μ);
PAux = new Population(μ + λ);
init(P);
evaluate(P);
PAux = addTo(PAux,P);
for iteration = 0 to NUMBER_OF_ITERATIONS do
  for i = 1 to λ do
    individual = select(P);
    perturbed = perturb(individual);
    evaluate(perturbed);
    PAux = addTo(PAux,perturbed);
  end for
  P = bestIndividuals(PAux,μ);
  PAux = P;
end for
```

FIGURE 9.3 Pseudocode of the $(\mu + \lambda)$ EA.

9.3.1 $(\mu + \lambda)$ Evolutionary Algorithm

This optimization technique first generates μ initial solutions. Next, the algorithm perturbs and evaluates these μ individuals at each iteration, from which λ new ones are obtained. Then, the best μ solutions taken from the $\mu + \lambda$ individuals are moved to the next iteration. An outline of the algorithm is shown in Figure 9.3. Other works using this algorithmic approach for the AFP problem can be found in works by Dorne and Hao [4] and Vidyarthi et al. [18].

As stated before, the configuration used in this chapter for μ and λ is 1 and 10, respectively. This means that 10 new solutions are generated from single initial random one, and the best from the 11 is selected as the current solution for the next iteration. With this configuration, the seeding procedure for generating the initial solution and the perturbation (mutation) operator are the core components defining the exploration capabilities of the $(1 + 10)$ EA. Several approaches for these two procedures are detailed in Sections 9.3.4 and 9.3.5.

9.3.2 Solution Encoding

A major issue in this kind of algorithms is how solutions are encoded, because it will determine the set of search operators that can be applied during the exploration of the search space.

Let T be the number of TRXs needed to meet the traffic demand of a given GSM network. Each TRX has to be assigned with a channel. Let $F_i \subset \mathbb{N}$ be the set of valid channels for transceiver i, $i = 1, 2, 3, \ldots, T$. A solution p (a frequency plan) is encoded

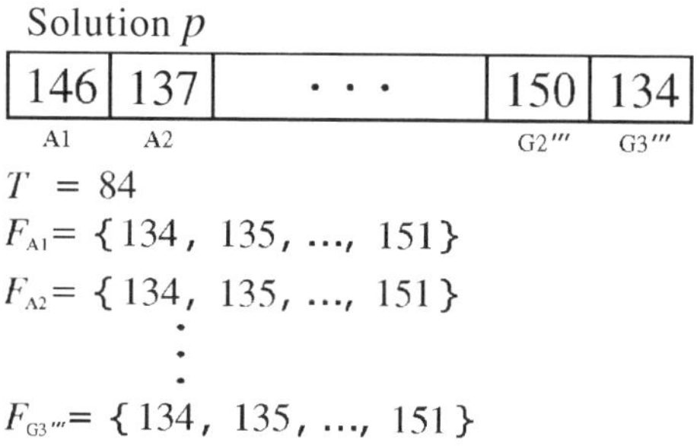

Solution p

146	137	· · ·	150	134
A1	A2		G2'''	G3'''

$T = 84$

$F_{A1} = \{134, 135, \ldots, 151\}$

$F_{A2} = \{134, 135, \ldots, 151\}$

$\vdots$

$F_{G3'''} = \{134, 135, \ldots, 151\}$

FIGURE 9.4 Solution encoding example.

as a T-length integer array $p = [f_1, f_2, f_3, \ldots, f_T]$, $p \in F_1 \times F_2 \times \cdots \times F_T$, where $f_i \in F_i$ is the channel assigned to TRX i. The fitness function (see the next section) is aware of adding problem-specific information to each transceiver, that is, whether it allocates a BCCH channel or a TCH channel, whether it is a frequency hopping TRX or not, and so on.

As an example, Figure 9.4 displays the representation of a frequency plan p for the GSM network shown in Figure 9.2. We have assumed that the traffic demand in the example network is fulfilled by one single TRX per sector (TRX $A1$, TRX $A2$, etc.).

9.3.3 Fitness Function

As it was stated before, we have used a proprietary application provided by Optimi Corp.™, which allows us to estimate the performance of the tentative frequency plans generated by the optimizer. Factors like frame erasure rate, block error rate, RxQual, and BER are evaluated. This commercial tool combines all aspects of network configuration (BCCHs, TCHs, frequency hopping, etc.) in a unique cost function, F, which measures the impact of proposed frequency plans on capacity, coverage, QoS objectives, and network expenditures. This function can be roughly defined as

$$F = \sum_{v} (\text{CostIM}(v) \times E(v) + \text{CostNeighbor}(v)), \tag{9.1}$$

that is, for each sector v that is a potential victim of interference, the associated cost is composed of two terms: a signaling cost computed with the interference matrix (CostIM (v)) that is scaled by the traffic allocated to v, $E(v)$, and a cost coming from the current frequency assignment in the neighbors of v. Of course, the lower the total cost, the better the frequency plan; that is, this is a minimization problem.

9.3.4 Initial Solution Generation

Two different initializations of individuals have been developed: *Random Init* and *Advanced Init*.

1. *Random Init*. This is the most usual seeding method used in the evolutionary field. Individuals are randomly generated: each TRX in the individual is assigned with a channel that is randomly chosen from the set of its valid channels.
2. *Advanced Init*. In this initialization method, individuals are not fully generated at random; instead, we have used a constructive method [3], which uses topological information of the GSM network. It first assigns a random channel to the first TRX of the individual; then, for the remainder of the TRXs, several attempts (as many as the number of valid channels of the considered TRXs) are tried with assignments that minimize interference as follows.

   Let t and F_t be the TRX to be allocated a new channel and its set of valid channels, respectively. A random valid channel $f \in F_t$ is generated. However, f is assigned to t if no co-channel or adj-channel interference occurs with any channel already assigned to a TRX installed in the same or any first-order neighboring sector of t. This procedure is repeated $|F_t|$ times. If no channel is allocated to t in this process, the *Random Init* strategy is used.

   If we continue the GSM network of Figure 9.2 (assuming a TRX per sector), generating an initial solution with the Advanced Init strategy might take first TRX $A1$. Let us suppose that the randomly chosen channel is 146 (Fig. 9.4). Next, a channel has to be allocated to TRX $A2$. In this case, channels 145, 146, and 147 are forbidden since $A2$ is a first-order neighbor of $A1$ (see Fig. 9.2) and this will provoke co-channel (channel 146) and adj-channel (channels 145 and 147) interference. Then, TRX $A2$ is assigned with channel number 137 after several possible attempts at randomly selecting a channel from its set of valid channels. Of course, the Random Init scheme will surely be used for many assignments in the last sectors of each first-order neighborhood.

9.3.5 Perturbation Operators

In $(\mu + \lambda)$ EAs, the perturbation (or mutation) operator largely determines the search capabilities of the algorithm. The mutation mechanisms proposed are based on modifying the channels allocated to a number of transceivers. Therefore, two steps must be performed:

1. *Selection of the transceivers*. The perturbation has first to determine the set of transceivers to be modified.
2. *Selection of channels*. Once a list of TRXs have been chosen, a new channel allocation must be performed.

9.3.5.1 *Strategies for Selecting Transceivers* This is the first decision to be made in the perturbation process. It is a major decision because it determines how explorative the perturbation is; that is, how different the resulting plan is from its original solution. Several strategies have been developed, which consist of reallocating channels on neighborhoods of TRXs. These neighborhoods are defined based on the topological information of the network:

TABLE 9.1 Weights Used in the *Interference-Based* Strategy

	Sector	First-order neighbor
Co-channel	16	8
Adj-channel	4	1

1. *OneHopNeighborhood.* Set of TRXs belonging to the first-order neighbors of a given transceiver.
2. *TwoHopNeighborhood.* The same, but using not only the first-order neighbors, but also the second-order ones. That is, a larger number of TRXs are reassigned.

We now need to specify the TRX from which the corresponding neighborhood is generated. In the experiments, the following selection schemes have been used:

1. *Random.* The TRX is randomly chosen from the set of all transceiver of the given problem instance.
2. *Interference-based.* This selection scheme uses a binary tournament. This method randomly chooses two TRXs of the network and returns the one with the higher interfering cost value. This cost value is based on counting the number of co-channel and adj-channel constraint violations provoked by these two TRXs in the current frequency planning. Since the closer the TRXs the stronger the interference, we further distinguish between co-channel and adj-channel within the same sector or within a first-order neighboring sector. Consequently, the cost value is computed as a weighted sum with four addends. The weights used are included in Table 9.1.

   Since we are looking for frequency plans with minimal interference, we have used this information for perturbing those TRXs with high values of this measurement in order to hopefully reach better assignments. Note that this interference-based value is only computed for two TRXs each time the perturbation method is invoked.

   Let us illustrate this with an example. Consider the GSM network shown in Figure 9.5, where the traffic demand is met with one single TRX per cell. This way, the number next to the name of each sector is the current channel allocated to the TRX. No intrasector interference can therefore occur. Let us now suppose that the two TRXs selected by the binary tournament are $B1$ and $D2$. Their corresponding first-order neighbors are the sets $\{B2, B3, E1, E3\}$ and $\{D1, D3, F3\}$, respectively (see the gray-colored sectors in Fig. 9.5). With the current assignment, the interference-based value of $B1$ is $8 \times 1 + 1 \times 1 = 9$, that is, a co-channel with $E1$ plus an adj-channel with $B2$. Concerning $D2$, this value is $8 \times 2 + 1 \times 1 = 17$, which corresponds to two co-channels with $D1$ and $F3$ plus an adj-channel with $D3$. So $D2$ would be the chosen sector to be perturbed in this case.

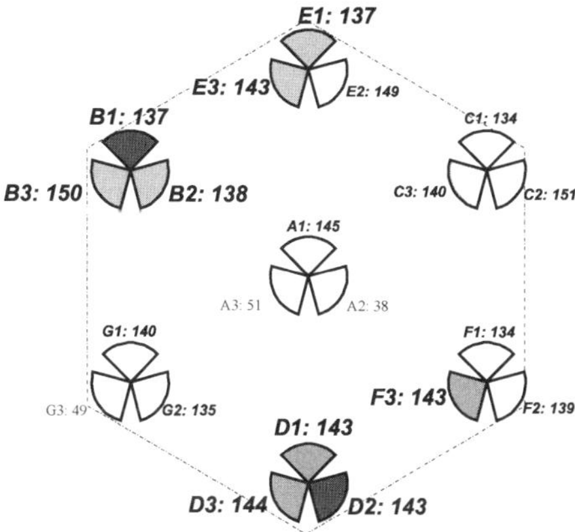

FIGURE 9.5 A tentative frequency planning for a GSM network composed of 21 sectors.

9.3.5.2 Frequency Selection At this point, the perturbation method has defined a set of TRXs whose channels can be modified. The modification is to determine which channel is allocated to each TRX. Again, two different schemes have been used:

1. *Random*. The channel allocated is randomly chosen from the set of valid channels of each TRX.
2. *Interference-based*. In this scheme, all the valid channels of a TRX are assigned sequentially and the interference-based cost value described previously is computed. The channel showing the lowest value for this interference-based cost is then allocated to the TRX.

For instance, let us continue with the example shown in Figure 9.5. Now, the TRX installed in sector $D2$, $F_{D2} = \{134, 143, 144, 145\}$, has to be assigned with a new channel. This strategy computes the cost value for all the valid channels of $D2$ (see Table 9.2), and the one with the lowest value is chosen (channel 134).

TABLE 9.2 Interference-Based Cost Values for the Single TRX Installed in Sector *D2* from the GSM Network in Figure 9.5

	Interference-based cost		
Channel	Co-channel	Adj-channel	Value
134	0	0	0
143	2	1	17
144	1	2	10
145	0	1	1

TABLE 9.3 Configurations of the (1 + 10) EA That Have Been Tested

Config name	Init	TRXs	Selection scheme for TRXs	Selection scheme for channels
Rand&Rand-1	Random	OneHopNeighborhood	Random	Random
Rand&Rand-2	Advanced	OneHopNeighborhood	Random	Random
Rand&Rand-3	Random	TwoHopNeighborhood	Random	Random
Rand&Rand-4	Advanced	TwoHopNeighborhood	Random	Random
Interf&Rand-1	Random	OneHopNeighborhood	Interference-based	Random
Interf&Rand-2	Advanced	OneHopNeighborhood	Interference-based	Random
Interf&Rand-3	Random	TwoHopNeighborhood	Interference-based	Random
Interf&Rand-4	Advanced	TwoHopNeighborhood	Interference-based	Random
Rand&Interf-1	Random	OneHopNeighborhood	Random	Interference-based
Rand&Interf-2	Advanced	OneHopNeighborhood	Random	Interference-based
Rand&Interf-3	Random	TwoHopNeighborhood	Random	Interference-based
Rand&Interf-4	Advanced	TwoHopNeighborhood	Random	Interference-based
Interf&Interf-1	Random	OneHopNeighborhood	Interference-based	Interference-based
Interf&Interf-2	Advanced	OneHopNeighborhood	Interference-based	Interference-based
Interf&Interf-3	Random	TwoHopNeighborhood	Interference-based	Interference-based
Interf&Interf-4	Advanced	TwoHopNeighborhood	Interference-based	Interference-based

9.4 EXPERIMENTS

In this section we now turn to present the experiments conducted to evaluate the (1 + 10) EAs proposed when solving a real-world instance of the AFP problem. We first detail the parameterization of the algorithms and the different configurations used in the EA. A discussion of the results is carried out afterward.

9.4.1 Parameterization

Several seeding and mutation operators for the (1 + 10) EA have been defined in the previous section. Table 9.3 summarizes all the combinations that have been studied. The number of iterations that are allowed to run is 2 000 in all the cases.

We also want to provide the reader with some details about the AFP instance that is being solved. The GSM network used has 711 sectors with 2 612 TRXs installed. That is, the length of the individuals in the EA is 2 132. Each TRX has 18 available channels (from 134 to 151). Additional topological information indicates that, on average, each TRX has 25.08 first-order neighbors and 96.60 second-order neighbors, thus showing the high complexity of this AFP instance, in which the available spectrum is much smaller that the average number of neighbors. Indeed, only 18 channels can be allocated to TRXs with 25.08 potential first-order neighbors. We also want to remark that this real network operates with advanced technologies, such as frequency hopping, and it employs accurate interference information that has been actually measured at a cell-to-cell level (neither predictions nor distance-driven estimations are used).

TABLE 9.4 Initial Cost Reached with the Two Initialization Methods

Initialization method	AFP cost	
	$\bar{x}$	σ_n
Random Init	180,631,987	15,438,987
Advanced Init	113,789,997	11,837,857

9.4.2 Discussion of the Results

All the values included in Table 9.4 are the average, $\bar{x}$, and the standard deviation, σ_n, of five independent runs. Although it is commonly accepted that 30 independent runs should be performed at least, we were only able to run five because of the very high complexity of such a large problem instance (2 612 TRXs) and the many different configurations used.

Let us start showing the performance of the two initialization methods. We present in Table 9.4 the AFP costs of the frequency plannings that result from both Random Init and Advanced Init. As expected, the latter reaches more accurate frequency assignments since it prevents the network from initially incurring in many interferences.

For each configuration of the EAs, the AFP costs of these final solutions are included in Table 9.5. If we analyze these results as a whole, it can be noticed that the configuration Rand&Rand-1 gets the lowest AFP cost on average, thus indicating that the computed frequency plannings achieve the smaller interference and therefore the better QoS for subscribers. Similar high quality frequency assignments are computed by the Rand&Interf-1, Rand&Interf-1, and Interf&Interf-3, where the cost values are around 20,000 units. We also want to remark two additional facts here. The first one was already mentioned before and it lies in the huge reduction of the AFP costs that

TABLE 9.5 Resulting AFP Costs (Average Over Five Executions)

Config	AFP cost		
	$\bar{x}$	σ_n	Best run
Rand&Rand-1	18,808	12,589	9,966
Rand&Rand-2	31,506	10,088	13,638
Rand&Rand-3	34,819	24,756	13,075
Rand&Rand-4	76,115	81,103	13,683
Interf&Rand-1	56,191	87,562	14,224
Interf&Rand-2	63,028	96,670	11,606
Interf&Rand-3	108,146	99,839	18,908
Interf&Rand-4	72,043	83,198	15,525
Rand&Interf-1	21,279	11,990	9,936
Rand&Interf-2	19,754	7,753	11,608
Rand&Interf-3	34,292	16,178	12,291
Rand&Interf-4	28,422	20,473	11,493
Interf&Interf-1	147,062	273,132	14,011
Interf&Interf-2	26,346	10,086	15,304
Interf&Interf-3	20,087	10,468	13,235
Interf&Interf-4	32,982	19,814	16,818

EAs can achieve starting from randomly generated solution (from more than 110 million to several thousand cost units). This means that the strongest interference in the network has been avoided. The second fact concerns the best solutions found so far by the solvers, which are included in the column "best" of Table 9.5. They point out that all the configurations of the (1 + 10) EA are able to compute very accurate frequency assignments. As a consequence, we can conclude that these algorithms are very suitable for solving this optimization problem.

We now turn to further analyze how the different strategies proposed for initializing and perturbing work within the (1 + 10) EA framework. With this goal in mind, Figure 9.6 displays the average costs of the configurations using

1. The Random Init strategy *versus* those using the Advanced Init method
2. OneHopNeighborhood *versus* TwoHopNeighborhood strategies for determining the number of TRXs to be reallocated a channel
3. The random scheme *versus* interference-based one for selecting the TRXs
4. The random *versus* interference-based channel selection strategies.

Concerning the initialization method, Figure 9.6 shows that the (1 + 10) EAs using the Advanced Init scheme reach, on average, better frequency assignments than the configurations with Random Init. It is clear from these results that our proposed EAs can profit from good initial plannings that guide the search toward promising regions of the search space.

If we compare the different strategies used in the perturbation method, several conclusions can be drawn. First of all, configurations of the (1 + 10) EA that reallocate the channel to a smaller number of TRXs, that is, OneHopNeighborhood strategy, against using the TwoHopNeighborhood scheme report a small improvement in the AFP cost. However, it is clear that randomly choosing the TRX (and its corresponding

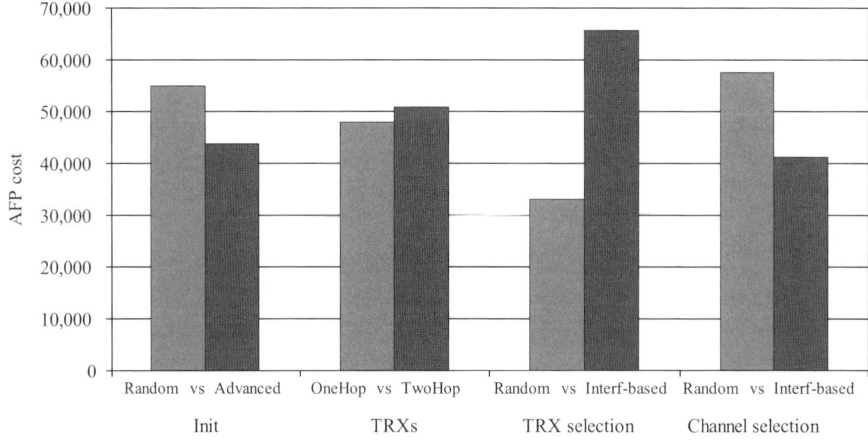

FIGURE 9.6 Performance of the initialization and perturbation methods in the (1 + 10) EA.

neighborhood) comes up with a large reduction in the AFP costs of the configurations using this selection strategy (see Fig. 9.6). Indeed, the interference-based scheme leads the (1 + 10) EA to converge prematurely to a local minimum because of an excessive intensification of the search. This also means that the many existing works advising sophisticated local searches work only on easy conceptualizations of low dimensionality of this problem, which is an important fact [1,5].

Even though this interference-based selection strategy does not work properly for selecting the TRXs to be perturbed, the EA configurations using this strategy for choosing channels show better performance (lower AFP costs) than those applying the random one (see the last columns in Fig. 9.6). That is, perturbations using this scheme allow the (1 + 10) EA to reach accurate frequency plans, which means that interference information is very useful at the channel selection stage of the perturbation, whereas random selection is preferred when the TRXs have to be chosen.

9.5 CONCLUSIONS AND FUTURE WORK

This chapter describes the utilization of (1 + 10) EAs to solve the automatic frequency planning in a real-world GSM network composed of 2132 transceivers. Instead of using a mathematical formulation of this optimization problem, we have used a commercial application that allows the target frequency plannings to be evaluated in a real scenario where current technologies are in use (e.g., frequency hopping, discontinuous transmission, etc.).

Two different methods for generating initial solutions along with several perturbation methods have been proposed. We have analyzed all the possible configurations of an (1 + 10) EA using these operators. The results show that the configuration called Rand&Rand-1 gets the lowest cost values for the final frequency planning computed, thus reaching an assignment that avoids major interference in the network. We have then compared the different seeding and perturbation methods among them to provide insight into their search capabilities within the (1 + 10) EA framework. Concerning the seeding methods, the configurations using the Advanced Init scheme outperforms those endowed with Random Init. In the perturbation operator, OneHopNeighborhood and TwoHopNeighborhood strategies for selecting how many TRXs have to be reallocated a channel are very similar. However, significant reductions in the cost values are reached when using the random scheme to choose which TRX (and its corresponding neighboring sectors) will be perturbed, instead of the interference-based approach. We want to remark that this is contraintuitive and brings into discredit simplified works of k-coloring and small instances of 200/300 TRXs like those included in COST, CELAR, or OR Library, for example. Conversely, the interference-based strategy performs the best when a channel has to be chosen to be allocated a TRX.

As future work, we plan to develop new search operators and new metaheuristic algorithms to solve this problem. Their evaluation with the current instance and other real-world GSM networks is also an ongoing research line. The formulation of the AFP problem as a multiobjective optimization problem will be investigated as well.

ACKNOWLEDGMENTS

This work has been partially funded by the Ministry of Science and Technology and FEDER under contract TIN2005-08818-C04-01 (the OPLINK project).

REFERENCES

1. Aardal KI, van Hoesen SPM, Koster AMCA, Mannino C, Sassano A. Models and solution techniques for frequency assignment problems. 4OR 2003;1(4):261–317.
2. Bäck T. Evolutionary Algorithms: Theory and Practice. New York: Oxford University Press; 1996.
3. Blum C, Roli A. Metaheuristics in combinatorial optimization: overview and conceptual comparison. ACM Comput Surv 2003;35(3):268–308.
4. Dorne R, Hao J-K. An evolutionary approach for frequency assignment in cellular radio networks. Proceedings of the IEEE International Conference on Evolutionary Computation; 1995. p 539–544.
5. Eisenblätter A. Frequency assignment in GSM networks: models, heuristics, and lower bounds. Ph.D. thesis. Institut für Mathematik, Technische Universität Berlin; 2001.
6. FAP Web. http://fap.zib.de/
7. Furuskar A, Naslund J, Olofsson H. EDGE—enhanced data rates for GSM and TDMA/136 evolution. Ericsson Rev 1999;72(1):28–37.
8. Glover FW, Kochenberger GA. Handbook of Metaheuristics. International Series in Operations Research and Management Science. Norwell, MA: Kluwer 2003.
9. Granbohm H, Wiklund J. GPRS—general packet radio service. Ericsson Rev 1999;76(2):82–88.
10. Hale WK. Frequency assignment: theory and applications. Proc the IEEE 1980;68(12):1497–1514.
11. Kampstra P, van der Mei RD, Eiben AE. Evolutionary Computing in Telecommunication Network Design: A Survey. Forthcoming.
12. Kotrotsos S, Kotsakis G, Demestichas P, Tzifa E, Demesticha V, Anagnostou M. Formulation and computationally efficient algorithms for an interference-oriented version of the frequency assignment problem. Wireless Personal Commun 2001;18:289–317.
13. Mishra AR. Radio network planning and optimisation. Fundamentals of cellular network planning and optimisation: 2G/2.5G/3G... Evolution to 4G. Wiley; 2004. p 21–54.
14. Mouly M, Paulet MB. The GSM System for Mobile Communications. Palaiseau: Mouly et Paulet; 1992.
15. Rapeli J. UMTS: targets, system concept, and standardization in a global framework. IEEE Personal Commun 1995;2(1):30–37.
16. Schwefel H-P. Numerical Optimization of Computer Models. Wiley; 1981.
17. Smith DH, Allen SM, Hurley S. Characteristics of good meta-heuristics algorithms for the frequency assignment problem. Ann Oper Res 2001;107:285–301.
18. Vidyarthi G, Ngom A, Stojmenović I. A hybrid channel assignment approach using an efficient evolutionary strategy in wireless mobile networks. IEEE Trans Vehicular Technol 2005;54(5):1887–1895.

CHAPTER 10

Algorithmic Game Theory and Applications

MARIOS MAVRONICOLAS, VICKY PAPADOPOULOU, and PAUL SPIRAKIS

10.1 INTRODUCTION

Most of the existing and foreseen complex networks, such as the Internet, are operated and built by thousands of large and small entities (autonomous *agents*), which collaborate to process and deliver end-to-end *flows* originating from and terminating at any of them. The distributed nature of the Internet implies a lack of coordination among its users. Instead, each user attempts to obtain maximum performance according to his own parameters and objectives.

Methods from game theory and mathematical economics have been proven to be a powerful modeling tool, which can be applied to understand, control, and efficiently design such dynamic, complex networks. Game theory provides a good starting point for computer scientists in their endeavor to understand selfish rational behavior in complex networks with many *agents* (*players*). Such scenarios are readily modeled using techniques from game theory, where players with potentially conflicting goals participate in a common setting with well-prescribed interactions.

Nash equilibrium [73,74] distinguishes itself as the predominant concept of rationality in noncooperative settings. So, game theory and its various concepts of equilibria provide a rich framework for modeling the behavior of selfish agents in these kinds of distributed or networked environments; they offer mechanisms to achieve efficient and desirable global outcomes in spite of the selfish behavior.

Mechanism design, a subfield of game theory, asks how one can design systems so that agents' selfish behavior results to desired systemwide goals. *Algorithmic mechanism design* additionally considers computational tractability to the set of concerns of mechanism design. Work on algorithmic mechanism design has focused on the complexity of centralized implementations of game-theoretic mechanisms for distributed optimization problems. Moreover, in such huge and heterogeneous networks, each agent does not have access to (and may not process) complete information.

Handbook of Applied Algorithms: Solving Scientific, Engineering and Practical Problems
Edited by Amiya Nayak and Ivan Stojmenović

The notion of *bounded rationality* for agents and the design of corresponding incomplete-information *distributed* algorithms have been successfully utilized to capture the aspect of lack of global knowledge in information networks.

In this chapter, we review some of the most thrilling algorithmic problems and solutions, and corresponding advances, achieved on the account of game theory. The areas addressed are the following.

Congestion Games A central problem arising in the management of large-scale communication networks is that of routing traffic through the network. However, due to the large size of these networks, it is often impossible to employ a centralized traffic management. A natural assumption to make in the absence of central regulation is that network users behave selfishly and aim at optimizing their own individual welfare. One way to address this problem is to model this scenario as a noncooperative multiplayer game and formalize it using *congestion* game. Congestion games (either *unweighted* or *weighted*) offer a very natural framework for resource allocation in large networks like the Internet. In a nutshell, the main feature of congestion games is that they model *congestion* on a resource as a function of the number (or total weight) of all agents sharing the resource.

Price of Anarchy We survey precise and approximate estimations for the *price of anarchy*; this is the cost of selfish behavior in dynamic, large-scale networks compared to hypothetical centralized solutions. We consider the price of anarchy for some of the most important network problems that are modeled by noncooperative games; for example, we consider *routing* and *security* problems. A natural variant of the price of anarchy is the *price of stability* [5], which is the *best-case* cost of selfish behavior in complex networks, compared to a hypothetical centralized solution. The best-case assumption in the formulation of the price of stability implies that this cost can be enforced to the agents since they are interested in paying as low cost as possible.

Selfish Routing with Incomplete Information The impact of bounded rationality in networks with incomplete information can be addressed in two successful ways: either by *Bayesian games* or by congestion games with *player-specific payoff functions*. We will survey methods and tools for approximating network equilibria and network flows for a selfish system comprised of agents with bounded rationality.

Mechanism Design Mechanism design is a subfield of game theory and microeconomics, which deals with the design of protocols for rational agents. Generally, a mechanism design problem can be described as the task of selecting, out of a collection of feasible games, one that will yield desirable results for the designer. So, mechanism design can be thought of as the "inverse problem" in game theory, where the input is a game's outcome and the output is a game guaranteeing the desired outcome. The study of mechanism design from the algorithmic point of view starts with the seminal paper of Nisan and Ronen [76].

The routing problem in large-scale networks, where users are instinctively selfish, can be modeled by a noncooperative game. Such a game could impose strategies that might induce an equilibrium close to the overall optimum. These strategies can be enforced through *pricing mechanisms* [28], *algorithmic mechanisms* [76], and *network design* [57,87].

Stackelberg Games We will examine network routing games from the *network designer*'s point of view. In particular, the network *administrator* or *designer* can define prices and rules, or even construct the network, in a way that induces near-optimal performance when the users act selfishly inside the system. Particularly interesting is the approach where the network manager takes part in the noncooperative game. The manager has the ability to control centrally a part of the system resources, while the rest resources are managed by the selfish users. This approach has been implemented through *Stackelberg* or *leader–follower* games [16,58].

The apparent advantage of this approach is that it might be easier to be deployed in large-scale networks. This is so since there is no need to add extra components to the network, or to exchange information between the users of the network.

In a typical Stackelberg game, one player acts as a *leader* (here, the centralized authority interested in optimizing system performance) and the rest act as *followers* (here, the selfish users). The problem is then to compute a strategy for the leader (a *Stackelberg strategy*) that induces the followers to react in a way that (at least approximately) minimizes the total latency in the system.

Selfish routing games can be modeled as a Stackelberg game. We will survey issues related to how the manager should assign the flow under his control into the system so as to induce optimal cost incurred by the selfish users. In particular, we will be interested in the complexity of designing *optimal* Stackelberg strategies.

Pricing Mechanisms Pricing mechanisms for resource allocation problems aim at allocating resources in such a way that those users who derive greater utility from the network are not denied access due to other users placing a lower value on it. In other words, pricing mechanisms are designed to guarantee *economic* efficiency. We will survey *cost-sharing* mechanisms for pricing the competitive usage of a collection of resources by a collection of selfish agents, each coming with an individual *demand*.

Network Security Games We will also consider security problems in dynamic, large-scale, distributed networks. Such problems can be modeled as concise, noncooperative multiplayer games played on a graph. We will investigate the associated Nash equilibria for such network security games. In the literature, there have been studied at least two such interesting network security games.

Complexity of Computing Equilibria The investigation of the computational complexity of finding a Nash equilibrium in a general strategic game is definitely a fundamental task for the development of algorithmic game theory. Answers to such questions are expected to have great practical impact on both the analysis of the performance of antagonistic networks and the development and implementation of policies for the network designers themselves.

Finding a Nash equilibrium in a game with two players could potentially be easier (than for many players) for several reasons.

- First, the *zero-sum* version of the game can be solved in polynomial time by linear programming. This grooms hopes for the polynomial solvability of the general (nonconstant sum) version of the problem.
- Second, the two-player version of the game admits a polynomial size rational number solution, while there are games with three or more players that may only have solutions in irrational numbers.

This reasoning justified the identification of the problem of finding Nash equilibria for a two-player game as one of the most important open questions in the field of algorithmic game theory. The complexity of this problem was very recently settled in a perhaps surprising way in a series of breakthrough papers. In this chapter, we will later survey some of the worldwide literature related to this problem and the recent progress to it.

In this chapter, we only assume a basic familiarity of the reader with some central concepts of game theory such as strategic games and Nash equilibria; for more details, we refer the interested reader to the leading textbooks by Osborne [77] and Osborne and Rubinstein [78]. We also assume some acquaintance of the reader with the basic facts of the theory of computational complexity, as laid out, for example, in the leading textbook of Papadimitriou [80]. For readers interested in recalling the fundamental of algorithms design and analysis, we refer the reader to the prominent textbook of Kleinberg and Tardos [53]. For overwhelming motivation to delving into the secrets of algorithmic game theory, we cheerfully refer the reader to the inspirational and prophetic survey of Papadimitriou in STOC 2001 [81].

10.2 CONGESTION GAMES

10.2.1 The General Framework

10.2.1.1 Congestion Games Rosenthal [84] introduced a special class of strategic games, now widely known as *congestion* games and currently under intense investigation by researchers in algorithmic game theory. Here, the strategy set of each player is a subset of the power set of a set of *resources*; so, it is a set of sets of resources. Each player has an objective function, defined as the sum (over their chosen resources) of functions in the number of players sharing this resource. In his seminal work, Rosenthal showed with the help of a *potential function* that congestion games (in sharp contrast to *general* strategic games) always admit at least one pure Nash equilibrium.

An extension to congestion games are *weighted congestion games*, in which the players have *weights*, and thus exert different influences on the congestion of the resources. In (weighted) *network congestion games*, the strategy sets of the players correspond to paths in a network.

10.2.1.2 Price of Anarchy In order to measure the degradation of social welfare due to the selfish behavior of the players, Koutsoupias and Papadimitriou [60] introduced in their seminal work a global objective function, usually coined as *social cost*. It is quite remarkable that no notion similar in either spirit or structure to social cost had been studied in the game theory literature before. They defined the *price of anarchy*, also called *coordination ratio* and denoted as PoA, as the worst-case ratio between the value of social cost at a Nash equilibrium and that of some social optimum. The social optimum is the *best-case* social cost; so it is the least value of social cost achievable through cooperation. Thus, the coordination ratio measures the extent to which noncooperation approximates cooperation.

As a starting point for analyzing the price of anarchy, Koutsoupias and Papadimitriou considered a very simple weighted network congestion game, now known as the *KP model*. Here, the network consists of a single *source* and a single *destination* (in other words, it is a *single-commodity* network) that are connected together by parallel *links*. The *load* on a link is the total weight of players assigned to this link. Associated with each link is a *capacity* (or *speed*) representing the rate at which the link processes load. Each of the players selfishly routes from the source to the destination by using a probability distribution over the links. The private objective function of a player is its expected latency. The social cost is the expected maximum latency on a link, where the expectation is taken over all random choices of the players.

Fotakis et al. [34] have proved that computing social cost (in the form of expected maximum) is a $\#P$-complete problem. The stem of this negative result is the nature of exponential enumeration explicit in the definition of social cost (as an exponential-size expectation sum). An essentially identical $\#P$-hardness result has been proven recently by Daskalakis et al. [19]. This is one of the very few hard enumeration problems known in algorithmic game theory as of today. Determining more remains a great challenge.

Mavronicolas and Spirakis [69] introduced *fully mixed Nash equilibria* for the particular case of the KP model, in which each player chooses every link with positive probability. Gairing et al. [38,39] explicitly conjectured that, in case the fully mixed Nash equilibrium exists, it is the worst-case Nash equilibrium with respect to social cost. This so-called *fully mixed Nash equilibrium conjecture* is simultaneously intuitive and significant.

- It is intuitive because the fully mixed Nash equilibrium favors an increased number of collisions between different players, since each player assigns its load with positive probability to every link. This increased probability of collisions should favor an increase to social cost.
- The conjecture is also significant since it identifies the worst-case Nash equilibrium over all instances. The fully mixed Nash equilibrium conjecture has been studied very intensively in the last few years over a variety of settings and models relative to the KP model.

The KP model was recently extended to *restricted strategy sets* [9,35], where the strategy set of each player is a subset of the links. Furthermore, the KP model was extended to general latency functions and studied with respect to different definitions of social cost [36,37,63].

Inspired by the arisen interest in the price of anarchy, the much older Wardrop model was reinvestigated in the work by Roughgarden and Tordos[88] (see also references therein). In this weighted network congestion game, weights can be split into arbitrary pieces. The social welfare of the system is defined as the sum of the edge latencies (*sum* or *total* social cost). An equilibrium in the Wardrop model can be interpreted as a Nash equilibrium in a game with infinitely many players, each carrying an infinitesimal amount of weight. There has been a tremendous amount of work following the work by Roughgarden and Tordos[88] on the reinvestigation of the Wardrop model. For an exposition, see the book by Roughgarden [86], which gives an account of the earliest results.

Koutsoupias and Papadimitriou [60] initiated a systematic investigation of the social objective of (expected) maximum latency (also called *maximum* social cost) for a weighted congestion game on uniformly related parallel links. The price of anarchy for this game has been shown to be $\Theta\,(\log m/\log\log m)$ if either the users or the links are identical [18,59], and $\Theta\,(\log m/\log\log\log m)$ for weighted users and uniformly related links [18]. On the contrary, Czumaj et al. [17] showed that the price of anarchy is far worse and can be even unbounded for *arbitrary* latency functions. For uniformly related parallel links, identical users, and the objective of total latency, the price of anarchy is $1 - o(1)$ for the general case of mixed equilibria and $4/3$ for pure equilibria [63]. For identical users and *polynomial* latency functions of degree d, the price of anarchy is $d^{\Theta(d)}$ [8,15].

Christodoulou and Koutsoupias [15] consider the price of anarchy of pure Nash equilibria in congestion games with linear latency functions. They showed that for general (asymmetric) games, the price of anarchy for maximum social cost is $\Theta(\sqrt{n})$, where n is the number of players. For all other cases of symmetric or asymmetric games, and for both maximum and average social cost, the price of anarchy is shown to be $5/2$. Similar results were simultaneously obtained by Awerbuch et al. [15]

10.2.2 Pearls

A comprehensive survey of some of the most important recent advances in the literature on atomic congestion games is provided by Kontogiannis and Spirakis [55]. That work is an overview of the extensive expertise on (mainly, network) congestion games and the closely related *potential games* [71], which has been developed in various disciplines (e.g., economics, computer science and operations research) under a common formalization and modeling. In particular, the survey goes deep into the details of some of the most characteristic results in the area in order to compile a useful toolbox that game theory provides in order to study antagonistic behavior due to congestion phenomena in computer science settings.

10.2.2.1 Selfish Unsplittable Flows Fotakis et al. study congestion games where selfish users with varying service demands on the system resources may request a joint service from an arbitrary subset of resources [32]. Each user's demand has to be served *unsplittably* from a specific subset of resources. In that work, it is proved that the weighted congestion games are no longer isomorphic to the well-known potential games, although this was true for the case of users with identical service demands. The authors also demonstrate the power of the network structure in the case of users with varying demands. For very simple networks, they show that there may not exist a pure Nash equilibria, which is not true for the case of parallel links network or for the case of infinitely splittable service demands. Furthermore, the authors propose a family of networks (called *layered networks*) for which they show the existence of at least one pure Nash equilibrium when each resource charges its users with a delay equal to its load. Finally, the same work considers the price of anarchy for the family of layered networks in the same case. It is shown that the price of anarchy for this case is $\Theta(\log m/\log\log m)$. That is, within constant factors, the worst-case network is the simplest one (the *parallel links* network). This implies that, for this family of networks, the network structure does not affect the quality of the outcome of the congestion games played on the network in an essential way.

Panagopoulou and Spirakis [79] consider selfish routing in single-commodity networks, where selfish users select paths to route their loads (represented by arbitrary integer weights). They consider identical delay functions for the links of the network. That work focuses also on an algorithm suggested in the work by Fotakis et al. [32]; this is a potential-based algorithm for finding pure Nash equilibria in such networks. The analysis of this algorithm from the work by Fotakis et al. [32] has given an upper bound on its running time, which is polynomial in n (the number of users) and the sum W of their weights. This bound can be exponential in n when some weights are superpolynomial. Therefore, the algorithm is only known to be *pseudopolynomial.* The work of Panagopoulou and Spirakis [79] provides strong experimental evidence that this algorithm actually converges to a pure Nash equilibria in polynomial time in n (and, therefore, independent of the weights values).

In addition, Panagopoulou and Spirakis [79] propose an initial allocation of users to paths that dramatically accelerates this algorithm, as opposed to an arbitrary initial allocation. A by-product of that work is the discovery of a weighted potential function when link loads are exponential to their loads. This guarantees the existence of pure Nash equilibria for these delay functions, while it extends the results of Fotakis et al. [32].

10.2.2.2 *Worst-Case Equilibria* Fischer and Vöcking [30] reexamined the question of worst-case Nash equilibria for the selfish routing game associated with the KP model [60], where n weighted jobs are allocated to m identical machines. Recall that Gairing et al. [38,39] had conjectured that the fully mixed Nash equilibrium is the worst Nash equilibrium for this game (with respect to the expected maximum load over all machines). The known algorithms for approximating the price of anarchy relied on proven cases of that conjecture. Fischer and Vöcking [30], interestingly present a counterexample to the conjecture showing that fully mixed Nash equilibria cannot be generally used to approximate the price of anarchy within reasonable factors. In addition, they present an algorithm that constructs the so-called *concentrated Nash* *equilibria*, which approximate the worst-case Nash equilibrium within constant factors.

Although the work of Fischer and Vöcking [30] has disproved the fully mixed Nash equilibrium conjecture for the case of weighted users and identical links, the possibility that the conjecture holds for the case of identical users and arbitrary links is still open.

10.2.2.3 Symmetric Congestion Games Fotakis et al. [33] continued the work and studied computational and coordination issues of Nash equilibria in symmetric network congestion games. A game is symmetric if all users have the same strategy set and users costs are given by identical symmetric functions of other users' strategies. (Symmetric games were already considered in the original work of Nash [73,74].) In unweighted congestions games, users are identical, so that a common strategy set implies symmetry.

This work proposed a simple and natural greedy method (which is called the *Greedy Best Response—GBR*), to compute a pure Nash equilibria. In this algorithm, each user plays only once and allocates his traffic to a path selected via a shortest path computation. It is shown that this algorithm works for three special cases: (1) *series-parallel networks*, (2) users are identical, and (3) users are of varying demands but they have the same best response strategy for any initial network traffic (this is called the *Common Best Response* property).

The authors also give constructions where the algorithm fails if either the latter condition is violated (even for a series-parallel network) or the network is not series-parallel (even for the case of identical users). Thus, these results essentially indicate the limits of the applicability of this greedy approach.

The same work [33] also studies the price of anarchy for the objective of (expected) maximum latency. It is proved that for any network of m uniformly related links and for identical users, the price of anarchy is $\Theta(\log m/\log\log m)$. This result is complementary (and somewhat orthogonal) to a similar result proved in the work by Fotakis et al. [32] for the case of weighted users to be routed in a layered network.

10.2.2.4 Exact Price of Anarchy Obtaining exact *bounds* on price of anarchy is, of course, the ultimate wish providing a happy end to the story. Unfortunately, the cases where such exact bounds are known are truly rare as of today. We describe here a particularly interesting example of a success story for one of these rare cases.

Exact bounds on the price of anarchy for both unweighted and weighted congestion games with polynomial latency functions are provided in the work by Aland et al. [3]. The authors use the total latency as the social cost measure. The result in the work by Aland et al. [3] vastly improve on results by Awerbuch et al. [8] and Christodoulou and Koutsoupias [15], where nonmatching upper and lower bounds were given. (We will later discuss the precise relation of the newer result to the older results.)

For the case of *unweighted congestion games*, it is shown in the work by Aland et al. [3] that the price of anarchy is exactly

$$\mathrm{PoA} = \frac{(k+1)^{2d+1} - k^{d+1}(k+2)^d}{(k+1)^{d+1} - (k+2)^d + (k+1)^d - k^{d+1}},$$

where $k = \lfloor \Phi_d \rfloor$ and Φ_d is a natural generalization of the *golden ratio* to larger dimensions such that Φ_d is the solution to the equation $(\Phi_d + 1)^d = \Phi_d^{d+1}$. The best known upper and lower bounds had before been shown to be of the form $d^{d(1-o(1))}$ [15]. However, the term $o(1)$ was still hiding a significant gap between the upper and the lower bound.

For *weighted congestion games*, the authors show that the price of anarchy is exactly

$$\mathsf{PoA} = \Phi_d^{d+1}.$$

This result closes the gap between the so far best upper and lower bounds of $O(2^d d^{d+1})$ and $\Omega(d^{d/2})$ from the work by Awarbuch et al. [8].

Aland et al. [3] show that the above values on the price of anarchy also hold for the subclasses of unweighted and weighted network congestion games. For the upper bounds, the authors use a similar analysis as in the work by Christodoulou et al. [15]. The core of their analysis is to simultaneously determine parameters c_1 and c_2 such that

$$yf(x+1) \leq c_1 x f(x) + c_2 y f(y)$$

for all polynomial latency functions of maximum degree d and for all reals $x, y \geq 0$. For the case of unweighted users, it suffices to show the inequality for all pairs of integers x and y. (In order to prove their upper bound, Christodoulou and Koutsoupias [15] looked at the inequality with $c_1 = 1/2$ and gave an asymptotic estimate for c_2.) In the analysis presented in the work by Aland et al. [3], both parameters c_1 and c_2 are optimized. This optimization process required new mathematical ideas and is highly nontrivial. This optimization was successfully applied by Dumrauf and Gairing [24] to the so-called *polynomial Wardrop games*, where it yielded almost exact bounds on price of stability.

10.3 SELFISH ROUTING WITH INCOMPLETE INFORMATION

In his seminal work, Harsanyi [46] introduced an elegant approach to study noncooperative games with *incomplete information*, where the players are uncertain about some parameters of the game. To model such games, he introduced the *Harsanyi transformation*, which converts a game with incomplete information to a strategic game where players may have different *types*. In the resulting *Bayesian game*, the players' uncertainty about each other's types is described by a probability distribution over all possible *type profiles*. It was only recently that Bayesian games were investigated from the point of view of algorithmic game theory. Naturally, researchers were interested in formulating Bayesian versions of already studied routing games, as we described below.

In more detail, the problem of selfish routing with incomplete information has recently been faced via the introduction of new suitable models and the development of new methodologies that help to analyze such network settings. In particular, there were introduced new selfish routing games with incomplete information, called *Bayesian routing games* [40].

In a different piece of work, the same problem has been viewed as a congestion game where latency functions are *player-specific* [41], or a congestion game under the restriction that the link for each user must be chosen from a certain set of allowed links for the user [9,26].

10.3.1 Bayesian Routing Games

Gairing et al. [40] introduced a particular selfish routing game with incomplete information, called *Bayesian routing game*. Here, n selfish *users* wish to assign their *traffics* to one of m parallel *links*. Users do not know each other's traffic. Following Harsanyi's approach, the authors introduce for each user a set of *types*. Each type represents a possible traffic; so, the set of types captures the set of all possibilities for each user. Unfortunately, users know the set of all possibilities for each other, but not the actual traffic itself.

Gairing et al. [40] proved, with the help of a potential function, that every Bayesian routing game has a pure Bayesian Nash equilibrium. This result has also been generalized to a larger class of games, called *weighted Bayesian congestion games*. For the case of identical links and *independent* type distributions, it is shown that a pure Bayesian Nash equilibrium can be computed in polynomial time. (A probability distribution over all possible type profiles is *independent* if it can be expressed as the product of independent probability distributions, one for each type.)

In the same work, Gairing et al. study structural properties of *Bayesian fully mixed Nash equilibria* for the case of identical links; they show that those maximize individual cost. This implies, in particular, that Bayesian fully mixed Nash equilibria maximize social cost as sum of individual costs.

In general, there may exist more than one fully mixed Bayesian Nash equilibrium. Gairing et al. [40] provide a characterization of the class of fully mixed Bayesian Nash equilibria for the case of independent type distribution; the characterization determines, in turn, the *dimension* of Bayesian fully mixed Nash equilibria. (The *dimension* of Bayesian fully mixed Nash equilibria is the dimension of the smallest Euclidean space into which all Bayesian fully mixed Nash equilibria can be mapped.)

Finally, Gairing et al. [40] consider the price of anarchy for the case of identical links and for three different social cost measures; that is, they consider social cost as expected maximum congestion, as sum of individual costs, and as maximum individual cost. For the latter two measures, (asymptotic) tight bounds were provided using the proven structural properties of fully mixed Bayesian Nash equilibria.

10.3.2 Player-Specific Latency Functions

Gairing et al. [41] address the impact of incomplete knowledge in (weighted) network congestion games with either splittable or unsplittable flow. In this perspective, the proposed models generalize the two famous models of selfish routing, namely weighted (network) congestion games and Wardrop games, to accommodate player-specific latency functions. Latency functions may be arbitrary, nondecreasing functions; however, many of the shown results in the work by Gairing et al. [41] assume that the latency function for player i on resource j is a *linear* function $f_{ij}(x) = a_{ij}x + b_{ij}$, where $a_{ij} \geq 0$ and $b_{ij} \geq 0$. Gairing et al. use the term *player-specific capacities* to denote a game where $b_{ij} = 0$ in all (linear) latency functions.

Gairing et al. [41] derive several interesting results on the existence and computational complexity of (pure) Nash equilibria and on the price of anarchy. For routing games on parallel links with player-specific capacities, they introduce two new *potential* functions, one for unsplittable traffic and the other for splittable traffic. The first potential function is used to prove that games with unweighted players possess the *finite improvement property* in the case of unsplittable traffics. It is also shown in the work by Gairing et al. [41] that games with weighted players do not possess the finite improvement property in general, even if there are only three users. The second potential function is a convex function tailored to the case of splittable traffics. This convex function is minimized if and only if the corresponding assignment is a Nash equilibrium. Since such minimization of a convex latency function can be carried out in polynomial time, the established equivalence between minimizes of the potential function and Nash equilibria implies that a Nash equilibrium can be computed in polynomial time.

The same work [41] proves upper and lower bounds on the price of anarchy under a certain restriction on the linear latency functions. For the case of unsplittable traffics, the upper and lower bounds are asymptotically tight. All bounds on the price of anarchy translate to corresponding bounds for general congestion games.

10.3.3 Network Uncertainty in Selfish Routing

The problem of selfish routing in the presence of incomplete network information has also been studied by Georgiou et al. [43]. This work proposes an interesting new model for selfish routing in the presence of incomplete network information. The model proposed by Georgiou et al. captures situations where the users have incomplete information regarding the link capacities. Such uncertainty may be caused if the network links actually represent complex paths created by *routers*, which are constructed differently on separate occasions and sometimes according to the presence of congestion or link failures.

The new, extremely interesting model presented in the work by Georgiou et al. [43] consists of a number of users who wish to route their traffic on a network of m parallel links with the objective of minimizing their latency. In order to capture the lack of precise knowledge about the capacity of the network links, Georgiou et al. [43] assumed that links may present a number of different capacities. Each user's uncertainty about the capacity of each link is modeled via a probability distribution over all possibilities. Furthermore, it is assumed that users may have different sources of information regarding the network; therefore, Georgiou et al. assume the probability distributions of the various users to be (possibly) distinct from each other. This gives rise to a very interesting model with user-specific payoff functions, where each user uses its distinct probability distribution to take decisions as to how to route its traffic.

The authors propose simple polynomial-time algorithms to compute pure Nash equilibria in some special cases of the problem and demonstrate that a counterexample presented in the work by Milchtaich et al. [70], showing that pure Nash equilibria may not exist in the general case, does not apply to their model. Thus, Georgiou et al. identify an interesting open problem in this area, that of the existence of pure Nash equilibria in the general case of their model. Also, two different expressions for the social cost and the associated price of anarchy are identified and employed in the work by Georgiou et al. [43]. For the latter, Georgiou et al. obtain upper bounds for the general case and some better upper bounds for several special cases of their model.

In the same work, Georgiou et al. show how to compute the fully mixed Nash equilibrium in polynomial time; they also show that when it exists, it is unique. Also, Georgiou et al. prove that for certain instances of the game, fully mixed Nash equilibria assign all links to all users equiprobably. Finally, the work by Georgiou et al. [43] verifies the fully mixed Nash equilibrium conjecture, namely that the fully mixed Nash equilibrium maximizes social cost.

10.3.4 Restricted Selfish Scheduling

Elsässer et al. [26] further consider selfish routing problems in networks under the restriction that the link for each user must be chosen from a certain set of allowed links for the user. It is particularly assumed that each user has access (that is, finite cost) to only *two* machines; its cost on other machines is infinitely large, giving it no incentive to switch there. Interaction with just a few neighbors is a basic design principle to guarantee efficient use of resources in a distributed system. Restricting the number of interacting neighbors to just two is then a natural starting point for the theoretical study of the impact of selfish behavior in a distributed system with local interactions. In the model of Elsässer et al., the (expected) cost of a user is the (expected) load on the machine it chooses.

The particular way of modeling local interaction in the work by Elsässer et al. [26] has given rise to a simple, graph-theoretic model for selfish *scheduling* among m noncooperative *users* over a collection of n *machines* with local interaction. In their graph-theoretic model, Elsässer et al. [26] address these bounded interactions by using an *interaction graph,* whose vertices and edges are the machines and the users, respectively. Elsässer et al. [26] have been interested in the impact of their modeling on the properties of the induced Nash equilibria.

The main result of Elsässer et al. [26] is that the *parallel links* graph is the *best-case* interaction graph—the one that minimizes expected *makespan* of the *standard fully mixed Nash equilibrium*—among all 3-*regular* interaction graphs. (In the standard fully mixed Nash equilibria each user chooses each of its two admissible machines with probability $\frac{1}{2}$.) The proof employs a graph-theoretic lemma about *orientations* in 3-regular graphs, which may be of independent interest. This is a particularly pleasing case where algorithmic game theory rewards graph theory with a wealth of new interesting problems about orientations in regular graphs.

A lower bound on price of anarchy is also provided in the work of Elsässer et al. [26]. In particular, it is proved that there is an interaction graph incurring price of anarchy $\Omega(\log n/\log\log n)$. This bound relies on a proof employing pure Nash equilibria. Finally, the authors present counterexample interaction graphs to prove that a *fully mixed Nash equilibrium* may sometimes not exist at all. (A characterization of interaction graphs admitting fully mixed Nash equilibria is still missing.) Moreover, they prove existence and uniqueness properties of the fully mixed Nash equilibrium for *complete bipartite* graphs and *hypercube* graphs.

The problems left open in the work by Elsässer et al. [26] invite graph theory to a pleasing excursion into algorithmic game theory.

10.3.5 Adaptive Routing with Stale Information

Fischer and Vöcking [29] consider the problem of adaptive routing in networks by selfish users that lack central control. The main focus of this work is on simple adaption policies, or *dynamics*, that make possible use of stale information. The analysis provided in the work by Fischer and Vöcking [29] covers a wide class of dynamics encompassing the well-known *replicator dynamics* and other dynamics from *evolutionary game theory*; the basic milestone is the well-known fact that choosing the best option on the basis of out-of-date information can lead to undesirable oscillation effects and poor overall performance.

Fischer and Vöcking [29] show that it is possible to cope with this problem, and guarantee efficient convergence toward an equilibrium state, for all of this broad class of dynamics, if the function describing the cost of an edge depending on its load is not too steep. As it turns out, guaranteeing convergence depends solely on the size of a single parameter describing the greediness of the agents!

While the best response dynamics, which corresponds to always choosing the best option, performs well if information is always up-to-date, it is interestingly clear from the results in the work by Fischer and Vöcking [29] that this policy fails when information is stale. More interestingly, Fischer and Vöcking [29] present a dynamics that approaches the global optimal solution in networks of parallel links with linear latency functions as fast as the best response dynamics does, but which does not suffer from poor performance when information is out-of-date.

10.4 ALGORITHMIC MECHANISM DESIGN

Mechanism design is a subfield of game theory and microeconomics, which, generally speaking, deals with the design of protocols for rational agents. In most simple words, a mechanism design problem can be described as the task of selecting from a collection of (feasible) games, a game that will yield desirable results for the designer. Specifically, the theory of mechanism design has focused on problems where the goal is to satisfactorily aggregate privately known preferences of several agents toward a *social choice*. Intuitively, a mechanism design problem has two components:

- The usual algorithmic output specification.
- Descriptions of what the participating agents want, formally given as *utility functions* over the set of possible *outputs* (outcomes).

The origin of algorithmic mechanism design is marked with the seminal paper of Nisan and Romen [76].

A mechanism solves a given problem by assuring that the required outcome occurs, under the assumption that agents choose their strategies as to maximize their own selfish utilities. A mechanism needs thus to ensure that players' utilities (which it can influence by handing out *payments*) are compatible with the algorithm.

Recall that the routing problem in large-scale networks where users are instinctively selfish can be modeled as a noncooperative game. Such a game is expected to impose strategies that would induce an equilibrium as close to the overall optimum as possible. Two possible approach to formulate such strategies are through *pricing mechanisms* [28] and *network design* [57,87].

In the first approach, the network administrator defines *prices* (or *rules*) in a way that induces near optimal performance when the users act selfishly. This approach has been considered in the works by Caragiannis et al. [10] and Cole et al. [16] (see also references therein). In the second approach, the network manager takes part in the noncooperative game. The manager has the ability to control centrally a part of the system resources, while the rest of the resources are to be shared by the selfish users. This approach has been studied through *Stackelberg* or *leader–follower* games [50,85] (see also references therein). We here overview some issues related to how should the manager assign the flow he controls into the system, with the objective to induce optimal cost in spite of the behavior of the selfish users.

10.4.1 Stackelberg Games

Roughgarden [85], studied the problem of optimizing the performance of a system shared by selfish, noncooperative users assigned to shared machines with load-dependent latency functions. Roughgarden measured system performance by the total latency of the system. (This measure is different from that used in the KP model.) Assigning jobs according to the selfish interests of individual users typically results in suboptimal system performance. However, in many systems of this type, there is a mixture of "selfishly controlled" and "centrally controlled" jobs; as the assignment of centrally controlled jobs will influence the subsequent actions by selfish users, the degradation in system performance due to selfish behavior can be reduced by scheduling the centrally controlled jobs in the best possible way. Stackelberg games provide a framework that fits this situation in an excellent way.

A *Stackelberg game* is a special game where there are two kinds of entities: a number of selfish entities, called *players*, that are interested in optimizing their own utilities, and a distinguished *leader* controlling a number of non-self-interested entities called *followers*; the leader aims at improving the social welfare and decides on the strategies of the followers so that the resulting situation will induce suitable decisions for the players that will optimize social welfare (as much as possible).

Roughgarden [85] formulated this particular goal for such a selfish routing system as an optimization problem via Stackelberg games. The problem is then to compute a strategy for the leader (a Stackelberg strategy) that induces the followers to react in a way that (at least approximately) minimizes the total latency in the system. Roughgarden [85] proved that, perhaps not surprisingly, it is $\mathcal{NP}$-hard to compute the *optimal* Stackelberg strategy; he also presented simple strategies with provable performance guarantees.

More precisely, Roughgarden [85] gave a simple algorithm to compute a strategy inducing a job assignment with total latency no more than a small constant times that of the optimal assignment for all jobs; in the absence of centrally controlled jobs and a Stackelberg strategy, no result of this type is possible. Roughgarden also proved stronger performance guarantees in the special case where every latency function is linear in the load.

10.4.1.1 The Price of Optimum Kaporis and Spirakis [50] continued the study of the Stackelberg games from the work by Roughgarden [85]. They considered a system of parallel machines, each with a strictly increasing and differentiable load-dependent latency function. The users of such a system are of infinite number and act selfishly, routing their infinitesimally small portion of the total flow they control to machines of currently minimum delay. In that work, such a system is modeled as a Stackelberg or leader–follower game motivated by the work by Roughgarden and Tardos [88].

Roughgarden [85] had presented the `LLF` Stackelberg strategy for a *leader* in a Stackelberg game with an infinite number of *followers*, each routing its infinitesimal flow through machines of currently minimum delay (this is called the *flow model* in the work by Roughgarden [85]). An important question posed there was the computation of the *least* portion β_{M} that a leader must control in order to enforce the overall optimum cost on the system. An algorithm that computes β_{M} was presented and its optimality was also shown [50]. Most importantly, it was proved that the algorithm presented is *optimal* for *any* class of latency functions for which Nash and optimum assignments can be efficiently computed. This is one of a very few known cases where the computation of optimal Stackelberg strategies is reduced to the computation of (pure) Nash equilibria and optimal assignments.

10.4.2 Cost Sharing Mechanisms

In its most general form, a *cost sharing mechanism* specifies how costs originating from resource consumption in a selfish system should be shared among the users of the system. Apparently, not all sharing ways are good. Intuitively, a cost sharing mechanism is good if it can induce equilibria optimizing social welfare as much as possible. This point of view was adopted in a recent work by Mavronicolas et al. [65].

In more detail, a simple and intuitive *cost mechanism* that assigns *costs* for the competitive usage of *m resources* by *n* selfish *agents* was proposed by Mavronicolas et al. [65]. Each agent has an individual *demand*; demands are drawn according to some (unknown) probability distribution coming from a (known) class of probability distributions. The cost paid by an agent for a resource he chooses is the total demand put on the resource divided by the number of agents who chose that same resource. So, resources charge costs in an equitable, fair way, while each resource makes no *profit* out of the agents. This simple model was called *fair pricing* in the work by Mavronicolas et al. [65]. [1]

Mavronicolas et al. [65] analyzed the *Nash equilibria* (both *pure* and *mixed*) for the induced game; in particular, they consider the *fully mixed Nash equilibrium*, where each agent selects each resource with nonzero probability. While offering (in addition) an advantage with respect to convenience in handling, the fully mixed Nash equilibrium is suitable for that economic framework under the very natural assumption that each resource offers usage to all agents without imposing any access restrictions.

The most significant contribution of the work by Mavronicolas [65] was the introduction of the *diffuse price of anarchy* for the analysis of Nash equilibria in the induced game. Roughly speaking, the diffuse price of anarchy is an extension to the price of anarchy that takes into account the probability distribution of the demands. Roughly speaking, the diffuse price of anarchy is the *worst case*, over all allowed probability distributions, of the expectation (according to each specific probability distribution) of the ratio of social cost over optimum in the *worst-case* Nash equilibrium. The diffuse price of anarchy is meant to alleviate the sometimes overly pessimistic Price of Anarchy due to Koutsoupias and Papadimitriou [60] (which is a *worst-case* measure) by introducing and analyzing stochastic assumptions on the system inputs.

Mavronicolas et al. [65] proved that pure Nash equilibria may not exist unless all chosen demands are identical; in contrast, a fully mixed Nash equilibrium exists for all possible choices of the demands. Further on, it was proved that the fully mixed Nash equilibrium is the *unique* Nash equilibrium in case there are only two agents. It was also shown that, in the *worst-case* choice of demands, the price of anarchy is $\Theta(n)$; for the special case of two agents, the price of anarchy is less than $2 - 1/m$.

A plausible assumption is that demands are drawn from a *bounded, independent probability distribution*, where all demands are *identically distributed* and each is at most a (*universal* for the class) constant times its expectation. Under this very general assumption, it is proved in the work by Mavronicolas et al. [65] that the diffuse price of anarchy is at most that same universal constant; the constant is just 2 when each demand is distributed symmetrically around its expectation.

10.4.3 Tax Mechanisms

How much can *taxes* improve the performance of a selfish system? This is a very general question since it leaves three important dimensions of it completely unspecified: the precise way of modeling taxes, the selfish system itself, and the measure of performance. Making specific choices for these three dimensions gives rise to specific interesting questions about taxes. There is already a sizeable amount of lit-

[1] One could argue that this pricing scheme is *unfair* in the sense that players with smaller demands can be forced to support those players with larger demands that share the same resource. However, the model can also be coined as fair on account of the fact that it treats all players sharing the same resource equally, and players are not overcharged beyond the actual cost of the resource they choose.

erature addressing such questions and variants of them (see, e.g., the works by Caragiannis et al. [10], Cole et al. [16], and Fleischer et al. [31] and references therein). In this section, we briefly describe the work of Caragiannis et al. [10], and we refer the reader to the work by Cole et al. [16] and Fleischer et al. [16,31] for additional related results.

Caragiannis et al. [10] consider the (by now familiar) class of congestion games due to Rosenthal [84] as their selfish system; they consider several measures for social welfare, including total latency and a new interesting measure they introduce, called *total disutility*, which is the sum of latencies plus taxes incurred to players. Caragiannis et al. [10] focus on the well-studied case of linear latency functions, and they provide many (both positive and negative) interesting results.

Their most interesting positive result is (in our opinion) the fact that there is a way to assign taxes that can improve the performance of congestion games by forcing players to follow strategies by which the total latency is within a factor of two of the least possible; Caragiannis et al. prove that, most interestingly, this is the *best* possible way of assigning taxes. Furthermore, Caraginannis et al. [10] consider cases where the system performance may be very poor in the absence of taxes; they prove that, fortunately, in such cases the total disutility *cannot* be much larger than the *optimal* total latency. Another interesting result emanating from the work of Caragiannis et al. [10] is that there is a polynomial-time algorithm (based on solving convex quadratic programs) to compute good taxes; this represents the *first* result on the efficiency of taxes for linear congestion games.

10.5 NETWORK SECURITY GAMES

It is an undeniable fact that the huge growth of the Internet has significantly extended the importance of *network security* [90]. Unfortunately, as it is well known, many widely used Internet systems and components are prone to security risks (see, e.g., the work by Cheswick and Bellovin [14]); some of these risks have even led to successful and well-publicized attacks [89]. Typically, an *attack* exploits the discovery of loopholes in the security mechanisms of the Internet. Attacks and *defenses* are currently attracting a lot of interest in major forums of communication research. A current challenge for algorithmic game theory is to invent and analyze appropriate theoretical models of security attacks and defenses for emerging networks like the Internet.

Two independent research teams, one consisting of Aspnes et al. [6] and another consisting of Mavronicolas et al. [67,68], initiated recently the introduction of strategic games on graphs (and the study of their associated Nash equilibria) as a means of studying security problems in networks with selfish entities. The nontrivial results achieved by these two teams exhibit a novel interaction of ideas, arguments, and techniques from two seemingly diverse fields, namely *game theory* and *graph theory*. This research line invites a simultaneously game-theoretic and graph-theoretic analysis of network security problems, where not only threats seek to maximize their caused damage to the network, but also the network seeks to protect itself as much as possible.

The two graph-theoretic models of Internet security can be cast as particular cases of the so-called *interdependent security* games studied earlier by Kearns and Ortiz [52]. There, a large number of players must make individual decisions related to security. The ultimate safety of each player may depend in a complex way on the actions of the entire population.

10.5.1 A Virus Inoculation Game

Aspnes et al. [6] consider an interesting graph-theoretic game with an interesting security flavor, modeling containment of the spread of *viruses* on a network with installable *antivirus* software. In this game, the antivirus software may be installed at individual nodes; a virus damages a node if it can reach the node starting at a random initial node and proceeding to it without crossing a node with installed antivirus software. Aspnes et al. [6] prove several algorithmic properties for their graph-theoretic game and establish connections to a certain graph-theoretic problem called *sum-of-squares partition*.

Moscibroda et al. [72] initiate the study of *Byzantine game theory* in the context of the specific virus inoculation game introduced by Aspnes et al. [6]. In their extension, they allow some players to be malicious or *Byzantine* rather than selfish. They ask the very natural question of what the impact of Byzantine players on the performance of the system compared to either the purely selfish setting (where all players are self-interested and there are no Byzantine players) or to the social optimum is.

To address such questions, they introduce the very interesting notion of the *price of malice* that captures the efficiency degradation due to the presence of Byzantine players (on top of selfish players). Moscibroda et al. [72] use the price of malice to quantify how much the presence of Byzantine players can deteriorate the social welfare of the distributed system corresponding to the virus inoculation game of Aspnes et al. [6]. Most interestingly, Moscibroda et al. [72] demonstrate that in case the selfish players are highly *risk-averse*, the social welfare of the system can improve as a result of taking Byzantine players into account!

We expect that Byzantine game theory will further develop in the upcoming years and be applied successfully to evaluate the impact of Byzantine players on the performance of selfish computer systems.

10.5.2 A Network Security Game

The work of Mavronicolas et al. [67,68] considers a security problem on a distributed network modeled as a multiplayer noncooperative game with *attackers* (e.g., viruses) and a *defender* (e.g., a security software) entities. More specifically, there are two classes of confronting randomized players on a graph: ν *attackers*, each choosing vertices and wishing to minimize the probability of being caught, and a single *defender*, who chooses edges and gains the expected number of attackers it catches. The authors exploit both game-theoretic and graph-theoretic tools for analyzing the associated Nash equilibria.

In a subsequent work, Mavronicolas et al. [64] introduced the *price of defense* in order to evaluate the loss in the provided security guarantees due to the selfish nature of attacks and defenses. The work address the question of whether there are Nash equilibria that both are computationally tractable and offer good price of defense. An extensive collection of trade-offs between price of defense and the computational complexity of Nash equilibria is provided in the work of Mavronicolas et al. [64]. Most interestingly, the work of Mavronicolas et al. [64,66–68] introduce certain natural classes of Nash equilibria for their network security game on graphs, including *matching Nash equilibria* [67,68] and *perfect matching Nash equilibria* [64]; they prove that deciding the existence of equilibria from such classes is precisely equivalent to the recognition problem for *König–Egervary* graphs [25,54]. So, this establishes a very interesting (and perhaps unexpected) link to some classical pearls in graph theory.

10.6 COMPLEXITY OF COMPUTING EQUILIBRIA

By Nash's celebrating result [73,74] every strategic game has at least one Nash equilibrium (and an odd number of them). What is the complexity of computing one? Note that this question is meaningful exactly when the payoff table is given in some implicit way that allows for a succinct representation. The celebrated algorithm of Lemke and Howson [61] shows that for bimatrix games this complexity is no more than exponential.

10.6.1 Pure Nash Equilibria

A core question in the study of Nash equilibria is which games have pure Nash equilibria. Also, under what circumstances can we find one (assuming that there is one) in polynomial time?

Recall that congestion games make a class of games that are guaranteed to have pure Nash equilibria. In a classical paper [84], Rosenthal proves that, in any such game, the *Nash dynamics* converges; equivalently, the directed graph with action combinations as nodes and payoff-improving deviations by individual players as edges is acyclic. Hence, the game has pure Nash equilibria that are the *sinks* of this graph. The proof is based on a simple potential function. This existence theorem, however, again left open the question of whether there is a polynomial-time algorithm for finding pure Nash equilibria in congestion games.

Fabrikant et al. [27] prove that the answer to this general question is positive when all players have the same origin and destination (the so-called *symmetric* case); a pure Nash equilibrium is found by computing the optimum of Rosenthal's potential function through a reduction to *min-cost flow*. However, it is shown that computing a pure Nash equilibrium in the general network case is $\mathcal{PLS}$-complete [49]. Intuitively, this means that it is as hard to compute as any object whose existence is guaranteed by a potential function. (The precise definition of the complexity class $\mathcal{PLS}$ is beyond the scope of this chapter.) The proof of Fabrikant et al. [27] has the interesting consequence: the existence of examples with exponentially long shortest paths, as well as the $\mathcal{PSPACE}$-completeness for the problem of computing a Nash equilibrium reachable from a specified state.

The completeness proof requires reworking the reduction to the problem of finding local optimal of weighted MAX2SAT instances. Ackermann et al. [1] present a significantly simpler proof based on a $\mathcal{PLS}$-reduction from MAX-CUT showing that finding Nash equilibria in network congestion games is $\mathcal{PLS}$-complete even for the case of linear latency functions. Additional results about the complexity of pure Nash equilibria in congestion games appear in the works of Ackermann et al. [1,2].

Gottlob et al. [45] provide a comprehensive study of complexity issues related to pure Nash equilibria. They consider restrictions of strategic games intended to capture certain aspects of bounded rationality. For example, they show that even in the settings where each player's payoff function depends on the strategies of at most three other players, and where each player is allowed to choose one out of at most three strategies, the problem of determining whether a game has a pure Nash equilibrium is $\mathcal{NP}$-complete. On the positive side, they also identified tractable classes of games.

10.6.2 Mixed Nash Equilibria

Daskalakis et al. [20] consider the complexity of Nash equilibria in a game with four or more players. They show that this problem is complete for the complexity class $\mathcal{PPAD}$. Intuitively, this means that a polynomial-time algorithm would imply a similar algorithm, for example, for computing *Brouwer fixpoints*; note that this is a problem for which quite strong lower bounds for large classes of algorithms are known [48]. (A precise definition of the complexity class $\mathcal{PPAD}$ is beyond the scope of this chapter.)

Nash [73,74] had shown his celebrated result on the existence of Nash equilibria by reducing the existence of Nash equilibria to the existence of Brouwer fixpoints. Given any strategic game, Nash constructs a Brouwer function whose fixpoints are precisely the equilibria of the game. In Nash's reduction, as well as in subsequent simplified ones [42], the constructed Brouwer function is quite specialized; this has led to the speculation that the fixpoints of such functions (thus, Nash equilibria) are easier to find than for *general* Brouwer functions. This question is answered in the negative by presenting a very interesting reduction in the opposite direction [20]: Any (computationally presented) Brouwer function can be simulated by a suitable game, so that Nash equilibria correspond to fixpoints.

It is proved that computing a Nash equilibrium in a three-player game is also $\mathcal{PPAD}$-complete [23]. The proof is based on a variant of an *arithmetical gadget* from [44], Independently, Chen and Deng [11] have also come up with a quite different proof of the same result.

In a very recent paper [12], Chen and Deng settle the complexity of Nash equilibria for two-player strategic games with a $\mathcal{PPAD}$-completeness proof. Their proof derived a direct reduction from a search problem called the *three-dimensional Brouwer* problem, which is known to be $\mathcal{PPAD}$-complete [20] to the objective problem. The completeness proof of the work by Chen and Deng[12] utilizes new gadgets for various arithmetic and logic operations.

10.6.3 Approximate Nash Equilibria

As it is always the case, an established intractability invites an understanding of the limits of approximation. Since it was established that computing a Nash equilibrium is $\mathcal{PPAD}$-complete [20], even for two-player strategic games [12], the question of computing approximate Nash equilibria has emerged as the central remaining open problem in the area of computing Nash equilibria.

Assume from this point on that all utilities have been normalized to be between 0 and 1. (Clearly, this assumption is without any loss of generality.) Say that a set of mixed strategies is an *ε-approximate Nash equilibrium*, where $\varepsilon > 0$, if for each player all strategies have expected payoff that is at most ε more that the expected payoff for its strategy in the given set. (So, ε is an additive approximation term.)

Lipton et al. [62] proved that an ε-approximate Nash equilibrium can be computed in time $O(n^{\varepsilon^2/\log n})$ (that is, in strictly subexponential time) by examining all supports of size $\log n/\epsilon^2$. It had been earlier pointed out [4] that no algorithm examining supports smaller than about $\log n$ can achieve an approximation better than $\frac{1}{4}$, even for zero-sum games. In addition, it is easy to see that a $\frac{3}{4}$-approximation Nash equilibrium can be found (in polynomial time) by examining all supports of size 2.

Two research teams, one consisting of Daskalakis et al. [21] and the other of Kontogiannis et al. [56], investigated very recently the approximability of Nash equilibria in two-player games, and established essentially identical, strong results. Most remarkably, there is a simple, linear-time algorithm in the work by Daskalakis et al. [21], which builds heavily on a corresponding algorithm from the work by Kontogiannis et al. [56]; it examines just two strategies per player and results in a $\frac{1}{2}$-approximate Nash equilibrium for any two-player game. Daskalakis et al. [21] also looked at the more demanding notion of *well-supported approximate Nash equilibria* introduced in the work by Daskalakis et al. [20] and present an interesting reduction (of the same problem) to *win–lose* games (that is, games with all utilities equal to 0 and 1). For this more demanding notion, Daskalakis et al. showed that an approximation of $\frac{5}{6}$ is possible contingent upon a graph-theoretic conjecture.

Chen et al. [13] establish strong inapproximability results for approximate Nash equilibria. Their results imply that it is unlikely to obtain a fully polynomial-time approximation scheme for Nash equilibria (unless $\mathcal{PPAD} \subseteq \mathcal{P}$).

10.6.4 Correlated Equilibria

Nash equilibrium [73,74] is widely accepted as the standard notion of rationality in game theory. However, there are several other competing formulations of rationality; chief among them is the *correlated equilibrium*, proposed by Aumann [7]. Observe that the mixed Nash equilibrium is a distribution on the strategy space that is *uncorrelated* or *independent*; that is, it is the product of independent probability distributions, one for each player. In sharp contrast, a *correlated equilibrium* is a *general* distribution over strategy profiles. It must, however, possess an equilibrium property: If a strategy profile is drawn according to this distribution, and each player is told separately his suggested strategy (that is, his own component in the profile), then no player has an incentive to switch to a different strategy (assuming that all other players also obey), because the suggested strategy is the best in expectation. Correlated equilibria enjoy a very nice combinatorial structure: The set of correlated equilibria of a multiplayer, noncooperative game is a convex polytope, and all Nash equilibria are not only included in this polytope but they all lie on the boundary of the polytope. (See the work by Nau et al. [75] for an elegant elementary proof of this latter result.)

As noted in the own words of Papadimitriou [82], the correlated equilibrium has several important advantages: It is a perfectly reasonable, simple, and plausible concept; it is guaranteed to always exist (simply because the Nash equilibrium is a particular case of a correlated equilibrium); and it can be found in polynomial time for any number of players and strategies by linear programming, since the inequalities specifying the satisfaction of all players are linear. In fact, it turns out that the correlated equilibrium that optimizes any linear function of the players' utilities (e.g., their sum) can be computed in polynomial time.

Succinct Games Equilibria in games, of which the correlated equilibrium is a prominent example, are objects worth of studying from the algorithmic point of view. *Multiplayer games* are the most compelling specimens in this regard. But, to be of algorithmic interest, they must be *represented succinctly*. Succinct representation is required since otherwise a typical (multiplayer) game would need an exponential size of bits in order to be described. Some well-known games that admit a succinct representation include

- *Symmetric* games, where all players are identical and indistinguishable.
- *Graphical* games [51], where the players are the vertices of a graph, and the payoff for each player only depends on its own strategy and those of its neighbors.
- *Congestion* games, where the payoff of each player only depends on its strategy and those choosing the same strategy as him.

Papadimitriou and Roughgarden [83] initiated the systematic study of algorithmic issues involved in finding equilibria (both Nash and correlated) in games with a large number of players, which are succinctly represented. The authors develop a general framework for obtaining polynomial-time algorithms for optimizing over correlated equilibria in such settings. They show how such algorithms can be applied successfully to symmetric games, graphical games, and congestion games, among others. They also present complexity results, implying that such algorithms are not in sight for certain other similar games. Finally, a polynomial-time algorithm, based on *quantifier elimination*, for finding a Nash equilibrium in symmetric games (when the number of strategies is relatively small) was presented.

Daskalakis and Papadimitriou [22] studied from the complexity point of view the problem of finding equilibria in games played on highly regular graphs with extremely succinct representation, such as the *d-dimensional grid*. There, it is argued that such games are of interest in modeling large systems of interacting agents. It has been shown by Daskalakis and Papadimitriou [22] that the problem of determining whether such a game on the d-dimensional grid has a pure Nash equilibrium depends on d, and the dichotomy is remarkably sharp: It is polynomial time solvable when $d = 1$, but $\mathcal{NEXP}$-complete for $d \geq 2$. In contrast, it was also proved that mixed Nash equilibria can be found in deterministic exponential time for any fixed d by quantifier elimination.

Recently, Papadimitriou [82] considered, and largely settled, the question of the existence of polynomial-time algorithms for computing correlated equilibria in succinctly representable multiplayer games. Papadimitriou developed a polynomial-time algorithm for finding correlated equilibria in a broad class of succinctly representable multiplayer games, encompassing essentially all kinds of such games we mentioned before.

The algorithm presented by Papadimitriou [82] was based on a careful mimicking of the existence proof due to Hart and Schmeidler [47], combined with an argument based on linear programming duality and the ellipsoid algorithm, Markov chain steady state computations, as well as application-specific methods for computing multivariate expectations.

10.7 DISCUSSION

In this chapter, we attempted a glimpse at the fascinating field of *algorithmic game theory*. This is a field that is currently undergoing a very intense investigation by the community of the *theory of computing*. Although some fundamental theoretical questions have been resolved (e.g., the complexity of computing Nash equilibria for two-player games), there are still a lot of challenges ahead of us. Among those, most important are, in our opinion, the further complexity classification of algorithmic problems in game theory, and the further application of systematic techniques from game theory to modeling and evaluating modern computer systems with selfish entities.

ACKNOWLEDGMENT

This work was partially supported by the IST Program of the European Union under contract number IST-2004-001907 (DELIS).

REFERENCES

1. Ackermann H, Röglin H, Vöcking B. On the impact of combinatorial structure on congestion games. Proceedings of the 47th Annual IEEE Symposium on Foundations of Computer Science (FOCS 2006). IEEE Press; 2006. p 613–622.
2. Ackermann H, Röglin H, Vöcking B. Pure Nash equilibria in player-specific and weighted congestion games. Proceedings of the 2nd International Workshop on Internet and

Network Economics (WINE 2006). Lecture Notes in Computer Science. Volume 4286. Springer; 2006. p 50–61.

3. Aland S, Dumrauf D, Gairing M, Monien B, Schoppmann F. Exact price of anarchy for polynomial congestion games. Proceedings of the 23rd International Symposium on Theoretical Aspects of Computer Science (STACS 2006). Lecture Notes in Computer Science. Volume 3884. Springer; 2006; p 218–229.
4. Althofer I. On sparse approximations to randomized strategies and convex combinations. Linear Algebra Appl 1994;199:339–355.
5. Anshelevich E, Dasgupta A, Kleinberg J, Tardos E, Wexler T, Roughgarden T. The price of stability for network design with fair cost allocation. Proceedings of the 45th Annual IEEE Symposium on Foundations of Computer Science (FOCS 2004). IEEE Press; 2004. p 295–304.
6. Aspnes J, Chang K, Yampolskiy A. Inoculation strategies for victims of viruses and the sum-of-squares partition problem. Proceedings of the 16th Annual ACM-SIAM Symposium on Discrete Algorithms (SODA 2005). Society for Industrial and Applied Mathematics; 2005. p 43–52.
7. Aumann RJ. Subjectivity and correlation in randomized strategies. J Math Econ 1974;1: 67–96.
8. Awerbuch B, Azar Y, Epstein A. The price of routing unsplittable flow. Proceedings of the 37th Annual ACM Symposium on Theory of Computing (STOC 2005). ACM Press; 2005. p 57–66.
9. Awerbuch B, Azar Y, Richter Y, Tsur D. Tradeoffs in worst-case equilibria. Theor Comput Sci 2006;361(2–3):200–209.
10. Caragiannis I, Kaklamanis C, Kanellopoulos P. Taxes for linear atomic congestion games. Proceedings of the 13th Annual European Symposium on Algorithms (ESA 2006). Volume 4168. 2006. p 184–195.
11. Chen X, Deng X. 3-Nash is $\mathcal{PPAD}$-complete. Technical Report No. TR05-134. Electronic Colloquium in Computational Complexity (ECCC); 2005.
12. Chen X, Deng X. Settling the complexity of 2-player Nash-equilibrium. Proceedings of the 47th Annual IEEE Symposium on Foundations of Computer Science. IEEE Press; 2006. p 261–272.
13. Chen X, Deng X, Teng S. Computing Nash equilibria: approximation and smoothed complexity. Proceedings of the 47th Annual IEEE Symposium on Foundations of Computer. IEEE Press; 2006. p 603–612.
14. Cheswick ER, Bellovin SM. Firewalls and Internet Security. Addison-Wesley; 1994.
15. Christodoulou G, Koutsoupias E. The price of anarchy of finite congestion games. Proceedings of the 37th Annual ACM Symposium on Theory of Computing (STOC 2005). ACM Press; 2005. p 67–73.
16. Cole R, Dodis Y, Roughgarden T. Pricing network edges for heterogeneous selfish users. Proceedings of the 35th Annual ACM Symposium on Theory of Computing (STOC 2003). ACM Press; 2003. p 521–530.
17. Czumaj A, Krysta P, Vöcking B. Selfish traffic allocation for server farms. Proceedings of the 34th Annual ACM Symposium on Theory of Computing (STOC 2002). ACM Press; 2002. p 287–296.

18. Czumaj A, Vöcking B. Tight bounds for worst-case equilibria. Proceedings of the 13th Annual ACM-SIAM Symposium on discrete Algorithms (SODA 2002). Society for Industrial and Applied Mathematics; 2002. p 413–420.

19. Daskalakis C, Fabrikant A, Papadimitriou CH. The game world is flat: the complexity of Nash equilibria in succinct games. Proceedings of the 33rd International Colloquium on Automata, Languages and Programming (ICALP 2006). Lecture Notes in Computer Science. Volume 4051. Springer; 2006. p 513–524.

20. Daskalakis C, Goldberg PW, and Papadimitriou CH. The complexity of computing a Nash equilibrium. Proceedings of the 38th Annual ACM Symposium on Theory of Computing (STOC 2006). ACM Press; 2006. p 71–78.

21. Daskalakis C, Mehta A, Papadimitriou C. A note on approximate Nash equilibria. Proceedings of the 2nd International Workshop on Internet and Network Economics (WINE 2006). Lecture Notes in Computer Science. Volume 4286. Springer; 2006. p 297–306.

22. Daskalakis C, Papadimitriou CH. The complexity of equilibria in highly regular graph games. Proceedings of the 13th Annual European Symposium on Algorithms (ESA 2005). Lecture Notes in Computer Science. Volume 3669. Springer; 2005. p 71–82.

23. Daskalakis C, Papadimitriou CH. Three-player games are hard. Technical report TR05-139. Electronic Colloquium in Computational Complexity (ECCC); 2005.

24. Dumrauf D, Gairing M. Price of anarchy for polynomial wardrop games. Proceedings of the 2nd International Workshop on Internet and Network Economics (WINE 2006). Lecture Notes in Computer Science. Volume 4286. Springer; 2006. p 319–330.

25. Egerváry J. Matrixok kombinatorius tulajdonságairól. Matematikai és Fizikai Lapok 1931;38:16–28.

26. Elsässer R, Gairing M, Lücking T, Mavronicolas M, Monien B. A simple graph-theoretic model for selfish restricted scheduling. Proceedings of the 1st International Workshop on Internet and Network Economics (WINE 2005). Lecture Notes in Computer Science. Volume 3828. Springer; 2005. p 195–209.

27. Fabrikant A, Papadimitriou CH, Talwar K. The complexity of pure Nash equilibria. Proceedings of the 36th Annual ACM Symposium on Theory of Computing (STOC 2004). ACM Press; 2004. p 604–612.

28. Feigenbaum J, Papadimitriou CH, Shenker S. Sharing the cost of muliticast transmissions. J Comput Sys Sci 2001;63:21–41.

29. Fischer S, Vöcking B. Adaptive routing with stale information. Proceedings of the 24th Annual ACM Symposium on Principles of Distributed Computing (PODC 2005). ACM Press; 2005. p 276–283.

30. Fischer S, Vöcking B. On the structure and complexity of worst-case equilibria. Proceedings of the 1st Workshop on Internet and Network Economics (WINE 2005). Lecture Notes in Computer Science. Volume 3828. Springer Verlag; 2005. p 151–160.

31. Fleischer L, Jain K, Mahdian M. Tolls for heterogeneous selfish users in multicommodity networks and generalized congestion games. Proceedings of the 45th Annual IEEE Symposium on Foundations of Computer Science (FOCS 2004). IEEE Press; 2004. p 277–285.

32. Fotakis D, Kontogiannis S, Spirakis P. Selfish unsplittable flows. Theor Comp Sci 2005;348(2–3):226–239.

33. Fotakis D, Kontogiannis S, Spirakis P. Symmetry in network congestion games: pure equilibria and anarchy cost. Proceedings of the 3rd International Workshop on Approximation and Online Algorithms (WAOA 2005). Lecture Notes in Computer Science. Volume 3879. Springer; 2006. p 161–175.

34. Fotakis D, Kontogiannis SC, Koutsoupias E, Mavronicolas M, Spirakis PG, The structure and complexity of Nash equilibria for a selfish routing game. Proceedings of the 29th International Colloquium on Automata, Languages and Programming (ICALP 2002). Lecture Notes in Computer Science. Volume 2380. Springer; 2002. p 123–134.

35. Gairing M, Lücking T, Mavronicolas M, Monien B. Computing Nash equilibria for scheduling on restricted parallel links. Proceedings of the 36th Annual ACM Symposium on Theory of Computing (STOC 2004). ACM Press; 2004. p 613–622.

36. Gairing M, Lücking T, Mavronicolas M, Monien B. The price of anarchy for polynomial social cost. Proceedings of the 29th International Symposium on Mathematical Foundations of Computer Science (MFCS 2004). Lecture Notes in Computer Science. Volume 3153. Springer; 2004. p 574–585.

37. Gairing M, Lücking T, Mavronicolas M, Monien B, Rode M. Nash equilibria in discrete routing games with convex latency functions. Proceedings of the 31st International Colloquium on Automata, Languages and Programming (ICALP 2004). Lecture Notes in Computer Science. Volume 3142. Springer; 2004. p 645–657.

38. Gairing M, Lücking T, Mavronicolas M, Monien B, Spirakis PG. Extreme Nash equilibria. Proceedings of the 8th Italian Conference of Theoretical Computer Science (ICTCS 2003). Lecture Notes in Computer Science. Volume 2841. Springer; 2003. p 1–20.

39. Gairing M, Lücking T, Mavronicolas M, Monien B, Spirakis PG. Structure and complexity of extreme Nash equilibria. Theor Comput Sci 2005;343(1–2):133–157. (Special issue titled Game Theory Meets Theoretical Computer Science, M. Mavronicolas and S. Abramsky, guest editors).

40. Gairing M, Monien B, Tiemann K. Selfish routing with incomplete information. Proceedings of the 17th Annual ACM Symposium on Parallelism in Algorithms and Architectures (SPAA 2005), ACM Press; 2005. p 203–212. Extended version accepted to Theory of Computing Systems, Special Issue with selected papers from the 17th Annual ACM Symposium on Parallelism in Algorithms and Architectures (SPAA 2005).

41. Gairing M, Monien B, Tiemann K. Routing (un-)splittable flow in games with player-specific linear latency functions. Proceedings of the 33rd International Colloquium on Automata, Languages and Programming (ICALP 2006). Lecture Notes in Computer Science. Volume 4051. Springer; 2006. p 501–512.

42. Geanakoplos J. Nash and Walras equilibrium via Brouwer. Econ Theor 2003;2(2–3): 585–603.

43. Georgiou C, Pavlides T, Philippou A. Network uncertainty in selfish routing. CD-ROM Proceedings of the 20th IEEE International Parallel and Distributed Processing Symposium (IPDPS 2006); 2006.

44. Goldberg PW, Papadimitriou CH. Reducibility among equilibrium problems. Proceedings of the 38th Annual ACM Symposium on Theory of Computing (STOC 2006). ACM Press; 2006. p 61–70.

45. Gottlob G, Greco G, Scarcello F. Pure Nash equilibria: hard and easy games. J Artif Intell Res 2005;24:357–406.

46. Harsanyi JC. Games with incomplete information played by Bayesian players, I, II, III. Manage Sci 1967;14:159–182, 320–332, 468–502.
47. Hart S, Schmeidler D. Existence of correlated equilibria. Math Oper Res 1989;14(1): 18–25.
48. Hirsch M, Papadimitriou CH, Vavasis S. Exponential lower bounds for finding brouwer fixpoints. J Complexity 1989;5:379–41.
49. Johnson DS, Papadimitriou CH, Yannakakis M. How easy is local search? J Comp Sys Sci 1988;17(1):79–100.
50. Kaporis A, Spirakis P. The price of optimum in stackelberg games. Proceedings of the 18th Annual ACM Symposium on Parallelism in Algorithms and Architectures (SPAA 2006); 2006. p 19–28.
51. Kearns M, Littman M, Singh S. Graphical models for game theory. Proceedings of the 17th Conference on Uncertainty in Artificial Intelligence; 2001. p 253–260.
52. Kearns M, Ortiz L. Algorithms for interdependent security games. Proceedings of the 16th Annual Conference on Neural Information Processing Systems (NIPS 2004). MIT Press; 2004. p 288–297.
53. Kleinberg J, Tardos É. Algorithm Design. Addison-Wesley; 2005.
54. König D. Graphok és Matrixok. Matematikai és Fizikai Lapok 1931;38:116–119.
55. Kontogiannis S, Spirakis P. Atomic selfish routing in networks: a survey. Proceedings of the 1st International Workshop on Internet and Network Economics (WINE 2005). Lecture Notes in Computer Science. Volume 3828. Springer; 2005. p 989–1002.
56. Kontogiannis SC, Panagopoulou PN, Spirakis PG. Polynomial algorithms for approximating Nash equilibria of bimatrix games. Proceedings of the 2nd International Workshop on Internet and Network Economics (WINE 2006); Lecture Notes in Computer Science. Volume 4286. Springer; 2006. p 286–296.
57. Korilis YA, Lazar A, Orda A. The designer's perspective to noncooperative networks. Proceedings of the 14th Annual Joint Conference of the IEEE Computer and Communications Societies (IEEE INFOCOM 1995). Volume 2; 1995. p 562–570.
58. Korilis YA, Lazar A, Orda A. Achieving network optima using Stackelberg routing strategies. IEEE/ACM T Netw 1997;5(1):161–173.
59. Koutsoupias E, Mavronicolas M, Spirakis PG. Approximate equilibria and ball fusion. Theor Comput Syst 2003;36(6):683–693.
60. Koutsoupias E, Papadimitriou CH. Worst-case equilibria. Proceedings of the 16th International Symposium on Theoretical Aspects of Computer Science (STACS 1999). Lecture Notes in Computer Science. Volume 1563. Springer; 1999. p 404–413.
61. Lemke CE, Howson JT, Jr. Equilibrium points of bimatrix games. J Soc Ind Appl Math 1964;12:413–423.
62. Lipton RJ, Markakis E, Mehta A. Playing large games using simple strategies. Proceedings 4th ACM Conference on Electronic Commerce (EC-2003). ACM Press; 2003. p 36–41.
63. Lücking T, Mavronicolas M, Monien B, Rode M. A new model for selfish routing. Proceedings of the 21st International Symposium on Theoretical Aspects of Computer Science (STACS 2004). Lecture Notes in Computer Science. Volume 2996. Springer; 2004. p 547–558.
64. Mavronicolas M, Michael L, Papadopoulou VG, Philippou A, Spirakis PG. The price of defense. Proceedings of the 31st International Symposium on Mathematical Foundations

of Computer Science (MFCS 2006). Lecture Notes in Computer Science. Volume 4162. Springer, 2006. p 717–728.

65. Mavronicolas M, Panagopoulou P, Spirakis P. A cost mechanism for fair pricing of resource usage. Proceedings of the 1st International Workshop on Internet and Network Economics (WINE 2005). Lecture Notes in Computer Science. Volume 3828. Springer; 2005. p 210–224.

66. Mavronicolas M, Papadopoulou VG, Persiano G, Philippou A, Spirakis P. The price of defense and fractional matchings. Proceedings of the 8th International Conference on Distributed Computing and Networking (ICDCN 2006). Lecture Notes in Computer Science. Volume 4308. Springer; 2006. p 115–126.

67. Mavronicolas M, Papadopoulou VG, Philippou A, Spirakis PG. A graph-theoretic network security game. Proceedings of the 1st International Workshop on Internet and Network Economics (WINE 2005). Lecture Notes in Computer Science. Volume 3828. Springer; 2005. p 969–978.

68. Mavronicolas M, Papadopoulou VG, Philippou A, Spirakis PG. A network game with attacker and protector entities. Proceedings of the 16th Annual International Symposium on Algorithms and Computation (ISAAC 2005). Lecture Notes in Computer Science. Volume 3827. Springer; 2005. p 288–297.

69. Mavronicolas M, Spirakis P. The price of selfish routing. Proceedings of the 33th Annual ACM Symposium on Theory of Computing (STOC 2001). ACM Press; 2001. p 510–519. Full version accepted to Algorithmica.

70. Milchtaich I. Congestion games with player-specific payoff functions. Games Econ Behav 1996;13(1):111–124.

71. Monderer D, Shapley LS. Potential games. Games Econ Behav 1996;14(1):124–143.

72. Moscibroda T, Schmid S, Wattenhofer R. When selfish meets evil: byzantine players in a virus inoculation game. Proceedings of the 25th Annual ACM Symposium on Principles of Distributed Computing (PODC 2006). ACM Press; 2006.

73. Nash JF. Equilibrium points in *N*-person games. Proc Natl Acad Sci USA 1950;36: 48–49.

74. Nash JF. Non-cooperative games. Ann Math 1951;54(2):286–295.

75. Nau R, Canovas SG, Hansen P. On the geometry of Nash equilibria and correlated equilibria. Int J Game Theor 2003;32(4):443–453.

76. Nisan N, Ronen A. Algorithmic mechanism design. Games Econ Behav 2001;35(1-2): 166–196.

77. Osborne M. An Introduction to Game Theory. Oxford University Press; 2003.

78. Osborne M, Rubinstein A. A Course in Game Theory. MIT Press; 1994.

79. Panagopoulou P, Spirakis P. Efficient convergence to pure Nash equilibria in weighted network congestion games. Proceedings of the 4th International Workshop on Efficient and Experimental Algorithms (WEA 2005). Lecture Notes in Computer Science. Volume 3503. Springer; 2005. p 203–215.

80. Papadimitriou CH. Computational Complexity. Addison-Wesley; 1994.

81. Papadimitriou CH. Algorithms, games, and the Internet. Proceedings of the 33th Annual ACM Symposium on Theory of Computing (STOC 2001). ACM Press; 2001. p 749–753.

82. Papadimitriou CH. Computing correlated equilibria in multi-player games. Proceedings of the 37th Annual ACM Symposium on Theory of Computing (STOC 2005). ACM Press; 2005. p 49–56.

83. Papadimitriou CH, Roughgarden T. Computing equilibria in multi-player games. Proceedings of the 16th Annual ACM–SIAM Symposium on Discrete Algorithms (SODA 2005). Society for Industrial and Applied Mathematics; 2005. p 82–91.

84. Rosenthal RW. A class of games possessing pure-strategy Nash equilibria. Int J Game Theor 1973;2:65–67.

85. Roughgarden T. Stackelberg scheduling strategies. SIAM J Comput 2003;33(2):332–350.

86. Roughgarden T. Selfish Routing and the Price of Anarchy. MIT Press; 2005.

87. Roughgarden T. On the severity of Braess's paradox: designing networks for selfish users is hard. J Comput Syst Sci 2006. p 922–953.

88. Roughgarden T, Tardos É. How bad is selfish routing? J ACM 2002;49(2):236–259.

89. Spafford EH. The Internet worm: crisis and aftermath. Commun ACM 1989;6(2–3): 678–687.

90. Stallings W. Cryptography and Network Security: Principles and Practice. 3rd ed. Prentice-Hall; 2003.

CHAPTER 11

Algorithms for Real-Time Object Detection in Images

MILOS STOJMENOVIC

11.1 INTRODUCTION

11.1.1 Overview of Computer Vision Applications

The field of Computer Vision (CV) is still in its infancy. It has many real-world applications, and many breakthroughs are yet to be made. Most of the companies in existence today that have products based on CV can be divided into three main categories: auto manufacturing, computer circuit manufacturing, and face recognition. There are other smaller categories of this field that are beginning to be developed in industry such as pharmaceutical manufacturing applications and traffic control. Auto manufacturing employs CV through the use of robots that put the cars together. Computer circuit manufacturers use CV to visually check circuits in a production line against a working template of that circuit. CV is used as quality control in this case. The third most common application of CV is in face recognition. This field has become popular in the last few years with the advent of more sophisticated and accurate methods of facial recognition. Applications of this technology are used in security situations like checking for hooligans at sporting events and identifying known thieves and cheats in casinos. There is also the related field of biometrics where retinal scans, fingerprint analysis, and other identification methods are conducted using CV methods.

Traffic control is also of interest because CV software systems can be applied to already existing hardware in this field. By traffic control, we mean the regulation or overview of motor traffic by means of the already existing and functioning array of police monitoring equipment. Cameras are already present at busy intersections, highways, and other junctions for the purposes of regulating traffic, spotting problems, and enforcing laws such as running red lights. CV could be used to make all of these tasks automatic.

Handbook of Applied Algorithms: Solving Scientific, Engineering and Practical Problems
Edited by Amiya Nayak and Ivan Stojmenović

11.2 MACHINE LEARNING IN IMAGE PROCESSING

AdaBoost and support vector machines (SVMs) are, among others, two very popular and conceptually similar machine learning tools for image processing. They are both based on finding a set of hyperplanes to separate the sets of positive and negative examples. Current image processing culture involving machine learning for real-time performance almost exclusively uses AdaBoost instead of SVMs. AdaBoost is easier to program and has proven itself to work well. There are very few papers that deal with real-time detection using SVM principles. This makes the AdaBoost approach a better choice for real-time applications. A number of recent papers, using both AdaBoost and SVMs, confirm the same, and even apply a two-phase process. Most windows are processed in the first phase by AdaBoost, and in the second phase, an SVM is used on difficult cases that could not be easily eliminated by AdaBoost. This way, the real-time constraint remains intact.

Le and Satoh [16] maintain that "The pure SVM has constant running time of 554 windows per second (WPS) regardless of complexity of the input image, the pure AdaBoost (cascaded with 37 layers—5924 features) has running time of 640, 515 WPS." If a pure SVM approach was applied to our test set, it would take $17,500,000/554 \approx 9$ h of pure run time to test the 106 images. It would take roughly 2 min to process an image of size 320×240. Thus, Lee and Satoh [16] claim that cascaded AdaBoost is 1000 times faster than SVMs. A regular AdaBoost with 30 features was presented in the works by Stojmenovic [24,25]. A cascaded design cannot speed up the described version by more than 30 times. Thus, the program in the works by Stojmenovic [24,25] is faster than SVM by over $1000/30 > 30$ times.

Bartlett et al. [3] used both AdaBoost and SVMs for their face detection and facial expression recognition system. Although they state that "AdaBoost is significantly slower to train than SVMs," they only use AdaBoost for face detection, and it is based on Viola and Jones' approach [27]. For the second phase, facial expression recognition on detected faces, they use three approaches: AdaBoost, SVMs, and a combined one (all applied on Gabor representation), and reported differences within 3 percent of each other. They gave a simple explanation for choosing AdaBoost in the face detection phase, "The average number of features that need to be evaluated for each window is very small, making the overall system very fast" [3]. Moreover, each of these features is evaluated in constant time, because of integral image preprocessing. That performance is hard to beat, and no other approach in image processing literature for real-time detection is seriously considered now.

AdaBoost was proposed by Freund and Schapire [8]. The connection between AdaBoost and SVMs was also discussed by them [9]. They even described two very similar expressions for both of them, where the difference was that the Euclidean norm was used by SVMs while the boosting process used Manhattan (city block) and maximum difference norms. However, they also list several important differences. Different norms may result in very different margins. A different approach is used to efficiently search in high dimensional spaces. The computation requirements are different. The computation involved in maximizing the margin is mathematical programming, that is, maximizing a mathematical expression given a set of inequalities. The difference between the two methods in this regard is that SVM corresponds to *quadratic programming*, while AdaBoost corresponds only to *linear programming* [9]. Quadratic programming is more computationally demanding than linear programming [9].

AdaBoost is one of the approaches where a “weak” learning algorithm, which performs just slightly better than random guessing, is “boosted” into an arbitrarily accurate “strong” learning algorithm. If each weak hypothesis is slightly better than random, then the training error drops exponentially fast [9]. Compared to other similar learning algorithms, AdaBoost is adaptive to the error rates of the individual weak hypotheses, while other approaches required that all weak hypotheses need to have accuracies over a parameter threshold. It is proven [9] that AdaBoost is indeed a boosting algorithm in the sense that it can efficiently convert a weak learning algorithm into a strong learning algorithm (which can generate a hypothesis with an arbitrarily low error rate, given sufficient data).

Freund and Schapire [8] state “Practically, AdaBoost has many advantages. It is fast, simple, and easy to program. It has no parameters to tune (except for the number of rounds). It requires no prior knowledge about the weak learner and so can be flexibly combined with *any* method for finding weak hypotheses. Finally, it comes with a set of theoretical guarantees given sufficient data and a weak learner that can reliably provide only moderately accurate weak hypotheses. This is a shift in mind set for the learning-system designer: instead of trying to design a learning algorithm that is accurate over the entire space, we can instead focus on finding weak learning algorithms that only need to be better than random. On the other hand, some caveats are certainly in order. The actual performance of boosting on a particular problem is clearly dependent on the data and the weak learner. Consistent with theory, boosting can fail to perform well given insufficient data, overly complex weak hypotheses, or weak hypotheses that are too weak. Boosting seems to be especially susceptible to noise.”

Schapire and Singer [23] described several improvements to Freund and Schapire’s [8] original AdaBoost algorithm, particularly in a setting in which hypotheses may assign confidences to each of their predictions. More precisely, weak hypotheses can have a range over all real numbers rather than the restricted range $[-1, +1]$ assumed by Freund and Schapire [8]. While essentially proposing a general fuzzy AdaBoost training and testing procedure, Howe and coworkers [11] do not describe any specific variant, with concrete fuzzy classification decisions. We propose in this chapter a specific variant of fuzzy AdaBoost. Whereas Freund and Schapire [8] prescribe a specific choice of weights for each classifier, Schapire and Singer [23] leave this choice unspecified, with various tunings. Extensions to multiclass classifications problems are also discussed.

In practice, the domain of successful applications of AdaBoost in image processing is any set of objects that are typically seen from the same angle and have a constant orientation. AdaBoost can successfully be trained to identify any object if this object is viewed from an angle similar to that in the training set. Practical real-world examples that have been considered so far include faces, buildings, pedestrians, some animals, and cars. The backbone of this research comes from the face detector work done by Viola et al. [27]. All subsequent papers that use and improve upon AdaBoost are inspired by it.

11.3 VIOLA AND JONES' FACE DETECTOR

The face detector proposed by Viola and Jones [27] was the inspiration for all other AdaBoost applications thereafter. It involves different stages of operation. The training of the AdaBoost machine is the first part and the actual use of this machine is the second part. Viola and Jones' contributions come in the training and assembly of the AdaBoost machine. They had three major contributions: integral images, combining features to find faces in the detection process, and use of a cascaded decision process when searching for faces in images. This machine for finding faces is called cascaded AdaBoost by Viola and Jones [27]. Cascaded AdaBoost is a series of smaller AdaBoost machines that together provide the same function as one large AdaBoost machine, yet evaluate each subwindow more quickly, which results in real-time performance. To understand cascaded AdaBoost, regular AdaBoost will have to be explained first. The following sections will describe Viola and Jones' face detector in detail.

Viola and Jones' machine takes in a square region of size equal to or greater than 24×24 pixels as input and determines whether the region is a face or is not a face. This is the smallest size of window that can be declared a face according to Viola and Jones. We use such a machine to analyze the entire image, as illustrated in Figure 11.1. We pass every subwindow of every scale through this machine to find all subwindows that contain faces. A sliding window technique is therefore used. The window is shifted 1 pixel after every analysis of a subwindow. The subwindow grows in size 10 percent every time all of the subwindows of the previous size were exhaustively searched. This means that the window size grows exponentially at a rate of $(1.1)^p$,

FIGURE 11.1 Subwindows of an image.

where p is the number of scales. In this fashion, more than 90 percent of faces of all sizes can be found in each image.

As with any other machine learning approach, the machine must be trained using positive and negative examples. Viola and Jones used 5000 positive examples of randomly found upright, forward-facing faces and 10,000 negative examples of any other nonface objects as their training data. The machine was developed by trying to find combinations of common attributes, or features of the positive training set that are not present in the negative training set.

The library of positive object (head) representatives contains face pictures that are concrete examples. That is, faces are cropped from larger images, and positive examples are basically closeup portraits only. Moreover, positive images should be of the same size (that is, when cut out of larger images, they need to be scaled so that all positive images are of the same size). Furthermore, all images are frontal upright faces. The method is not likely to work properly if the faces change orientation.

11.3.1 Features

An image *feature* is a function that maps an image into a number or a vector (array). Viola and Jones [27] used only features that map images into numbers. Moreover, they used some specific types of features, obtained by selecting several rectangles within the training set, finding the sum of pixel intensities in each rectangle, assigning a positive or negative sign and/or weight to each sum, and then summing them. The pixel measurements used by Viola and Jones were the actual grayscale intensities of pixels. If the areas of the dark (positive sign) and light (negative sign) regions are not equal, the weight of the lesser region is raised. For example, feature 2.1 in Figure 11.2 has a twice greater light area than a dark one. The area of the dark rectangle in this case would be multiplied by 2 to normalize the feature. The main problem is to find which of these features, among the thousands available, would best distinguish positive and negative examples, and how to combine them into a learning machine.

Figure 11.2 shows the set of basic shapes used by Viola and Jones [27]. Adding features to the feature set can increase the accuracy of the AdaBoost machine at the cost of additional training time. Each of the shapes seen in Figure 11.2 is scaled and translated anywhere in the test images, consequently forming features. Therefore, each feature includes a basic shape (as seen in Fig. 11.2), its translated position in the image, and its scaling factors (height and width scaling). These features define the separating ability between positive and negative sets. This phenomenon is illustrated in Figure 11.3. Both of the features seen in Figure 11.3 (each defined by its position and scaling factors) are derived from the basic shapes in Figure 11.2.

1. Edge features 2. Line features 3. Point features 4. Diagonal feature

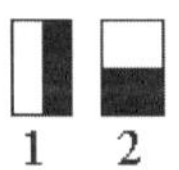

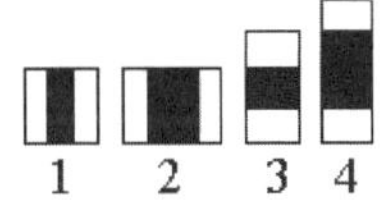

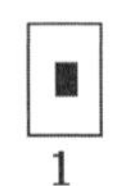

FIGURE 11.2 Basic shapes that generate features by translation and scaling.

FIGURE 11.3 First and second features in Viola and Jones face detection.

Figure 11.3 shows the first and second features selected by the program [27]. Why are they selected? The first feature shows the difference in pixel measurements for the eye area and area immediately below it. The "black" rectangle covering the eyes is filled with predominantly darker pixels, whereas the area immediately beneath the eyes is covered with lighter pixels. The second feature also concentrates on the eyes, showing the contrast between two rectangles containing eyes and the area between them. This feature corresponds to feature 2.1 in Figure 11.2 where the light and dark areas are inverted. This is not a separate feature; it was drawn this way in Figure 11.3 to better depict the relatively constant number obtained by this feature when it is evaluated in this region on each face.

11.3.2 Weak Classifiers (WCs)

A WC is a function of the form $h(x, f, s, \theta)$, where x is the tested subimage, f is the feature used, s is the sign (+ or −), and θ is the threshold. The sign s defines on what side of the threshold the positive examples are located. Threshold θ is used to establish whether a given image passes a classifier test in the following fashion: when feature f is evaluated on image x, the resulting number is compared to threshold θ to determine how this image is categorized by the given feature. The equation is given as $sf(x) < s\theta$. If the equation evaluates true, the image is classified as positive. The function $h(x, f, s, \theta)$ is then defined as follows: $h(x, f, s, \theta) = 1$ if $sf(x) < s\theta$ and 0 otherwise. This is expected to correspond to positive and negative examples, respectively. There are a few ways to determine the threshold θ. In the following example, the green numbers are considered to be the positive set, and the red letters are considered to be the negative set. The threshold is set to be the black vertical line after the "7" since at this location overall classification error is minimal. All of the positions are tried, and the one with minimal error is selected. The error function that is used is the number of misclassifications divided by the total number of examples. The array of evaluated feature values is sorted by the values of $f(x)$, and it shows positive examples as 1, 2, 3, . . . in green and negatives as A, B, C, D, . . . in red. The error of the threshold selected below is $3/17 \approx 0.17$.

1 6 8 4 I 3 5 7 | B E A 9 C 2 F D G H

In general, the threshold is found to be the value θ that best separates the positive and negative sets. When a feature f is selected as a "good" distinguisher of images between positive and negative sets, its value would be similar for images in the positive set and different for all other images. When this feature is applied to an individual image, a number $f(x)$ is generated. It is expected that values $f(x)$ for positive and negative images can be separated by a threshold value of θ.

It is worthy to note that a single WC needs only to produce results that are slightly better than chance to be useful. A combination of WCs is assembled to produce a strong classifier as seen in the following text.

11.3.3 Strong Classifiers

A strong classifier is obtained by running the AdaBoost machine. It is a linear combination of WCs. We assume that there are T WCs in a strong classifier, labelled $h_1, h_2, \ldots, h_T$, and each of these comes with its own weight labeled $\alpha_1, \alpha_2, \ldots, \alpha_T$. Tested image x is passed through the succession of WCs $h_1(x), h_2(x), \ldots, h_T(x)$, and each WC assesses if the image passed its test. The assessments are discrete values: $h_i(x) = 1$ for a pass and $h_i(x) = 0$ for a fail. $\alpha_i(x)$ are in the range $[0, +\infty]$. Note that $h_i(x) = h_i(x, f_i, s_i, \theta_i)$ is abbreviated here for convenience. The decision that classifies an image as being positive or negative is made by the following inequality:

$$\alpha_1 h_1(x) + \alpha_2 h_2(x) + \ldots + \alpha_T h_T(x) > \alpha/2 \quad \text{where } \alpha = \sum_{i=1}^{T} \alpha_i .$$

From this equation, we see that images that pass a weighted average of half of the WC tests are cataloged as positive. It is therefore a weighted voting of selected WCs.

11.3.4 AdaBoost: Meta Algorithm

In this section we explain the general principles of the AdaBoost (an abbreviation of Adaptive Boosting) learning strategy [8]. First, a huge (possibly hundreds of thousands) "panel" of experts is identified. Each expert, or WC, is a simple threshold-based decision maker, which has a certain accuracy. The AdaBoost algorithm will select a small panel of these experts, consisting of possibly hundreds of WCs, each with a weight that corresponds to its contribution in the final decision. The expertise of each WC is combined in a classifier so that more accurate experts carry more weight.

The selection of WCs for a classifier is performed iteratively. First, the best WC is selected, and its weight corresponds to its overall accuracy. Iteratively, the algorithm identifies those records in the training data that the classifier built so far was unable to capture. The weights of the misclassified records increase since it becomes more important to correctly classify them. Each WC might be adjusted by changing its threshold to better reflect the new weights in the training set. Then a single WC is selected, whose addition to the already selected WCs will make the greatest contribution to improving the classifier's accuracy. This process continues iteratively until a satisfactory accuracy is achieved, or the limit for the number of selected WCs is reached. The details of this process may differ in particular applications, or in particular variants of the AdaBoost algorithm.

There exist several AdaBoost implementations that are freely available in Weka (Java-based package http://www.cs.waikato.ac.nz/ml) and in R (http://www.r-project.org). Commercial data mining toolkits that implement AdaBoost include TreeNet, Statistica, and Virtual Predict. We did not use any of these packages for two main reasons. First, our goal was to achieve real-time performance, which restricted the choice of programming languages. Next, we have modified the general algorithm to better suit our needs, which required us to code it from scratch.

AdaBoost is a general scheme adaptable to many classifying tasks. Little is assumed about the learners (WCs) used. They should merely perform only a little better than random guesses in terms of error rates. If each WC is always better than a chance, then AdaBoost can be proven to converge to a perfectly accurate classifier (no training error). Boosting can fail to perform if there is insufficient data or if WCs are overly complex. It is also susceptible to noise. Even when the same problem is being solved by different people applying AdaBoost, the performance greatly depends on the training set being selected and the choice of WCs (that is, features).

In the next subsection, the details of the AdaBoost training algorithm, as used by Viola and Jones [27], will be given. In this approach, positive and negative training sets are separated by a cascade of classifiers, each constructed by AdaBoost. Real time performance is achieved by selecting features that can be computed in constant time. The training time of the face detector appears to be slow, even taking months according to some reports. Viola and Jones' face finding system has been modified in literature in a number of articles. The AdaBoost machine itself was modified in literature in several ways.

11.3.5 AdaBoost Training Algorithm

We now show how to create a classifier with the AdaBoost machine. It follows the algorithm given in the work by Viola and Jones [27]. The machine is given images $(x_1, y_1), \ldots, (x_q, y_q)$ as input, where $y_i = 1$ or *0* for positive and negative examples, respectively. In iteration t, the ith image is assigned the weight $w(t, i)$, which corresponds to the importance of that image for a good classification. The initial weights are $w(1, i) = 1/(2p), 1/(2n)$, for $y_i = 0$ or 1, respectively, where n and p are the numbers of negatives and positives, respectively, $q = p + n$. That is, all positive images have equal weight, totaling $\frac{1}{2}$, and similarly for all negative images. The algorithm will select, in step t, the tth feature f, its threshold value θ, and its direction of inequality $s (s = 1 \, or - 1)$. The classification function is $h(x, f, s, \theta) = 1$ (declared positive) if $sf(x) < s\theta$, and 0 otherwise (declared negative).

The expression $|h(x_i, f, s, \theta) - y_i|$ indicates whether or not $h(x, f, s, \theta)$ correctly classified image x_i. Its value is 0 for correct classification, and 1 for incorrect classification. The sum $\sum_{i=1}^{N} w(t, i) \times |h(x_i, f, s, \theta) - y_i|$ then represents the weighted misclassification error when using $h(x, f, s, \theta)$ as the feature-based classifier. The goal is to minimize that sum when selecting the next WC.

We revisit the classification of numbers and letters example to illustrate the assignment of weights in the training procedure. We assume that feature 1 classifies the example set in the order seen below. The threshold is chosen to be just after the "7" since this position minimizes the classification error. We will call the combination of feature 1 with its threshold WC 1. We notice that "*I*", "9," and "2" were incorrectly classified. The number of incorrect classifications determines the weight α_1 of this classifier. The fewer errors that it makes, the heavier the weight it is awarded.

1 6 8 4 I 3 5 7 | B E A 9 C 2 F D G H

The weights of the incorrectly classified examples (*I*, 9, and 2) are increased before finding the next feature in an attempt to find a feature that can better classify cases that are not easily sorted by previous features. We assume that feature two orders the example set as seen below.

E 6 9 I 1 3 5 7 2 | B A C G 8 F H D 4

Setting the threshold just after the "2" minimizes the error in classification. We notice that this classifier makes more mistakes in classification than its predecessor. This means that its weight, α_2, will be less that α_1. The weights for elements "*E*", "*I*," "8," and "4" are increased. These are the elements that were incorrectly classified by WC 2. The actual training algorithm will be described in pseudocode below.

For t=1 to T do:

Normalize the weights $w(t, i)$, by dividing each of them with their sum (so that the new sum of all weights becomes 1);
swp ← sum of weights of all positive images
swn ← sum of weights of all negative images
(* note that $swp + swn = 1$ *)

FOR each candidate feature *f*, find $f(x_i)$ and $w(t, i)^* f(x_i)$, $i = 1, \ldots, q$.

- Consider records $(f(x_i), y_i, w(t, i))$. Sort these records by the $f(x_i)$ field with mergesort, in increasing order. Let the obtained array of the $f(x_i)$ field be $g_1, g_2, \ldots, g_q$. The corresponding records are $(g_j, status(j), w'(j)) = (f(x_i), y_i, w(t, i))$, where $g_j = f(x_i)$. That is, if the *j*th element g_j is equal to *i*th element from the original array $f(x_i)$ then $status(j) = y_i$ and $w'(j) = w(t, i)$.

(*Scan through the sorted list, looking for threshold θ and direction *s* that minimizes the error $e(f, s, \theta)$*)

$sp \leftarrow 0$; $sn \leftarrow 0$; (*weight sums for positives/negatives below a considered threshold *)
emin ← minimal total weighted classification error
If $swn < swp$ **then** $\{emin \leftarrow swn; smin \leftarrow 1; \theta min \leftarrow g_n + 1$ (*all declared positive*)

else { $emin \leftarrow swp; smin \leftarrow 1; \theta min \leftarrow g_1 - 1$ } (*all declared negative *)

For $j \leftarrow 1$ **to** q-1 **do** {

If $status(j) = 1$ **then** $sp \leftarrow sp + w'(j)$ **else** $sn \leftarrow sn + w'(j)$

$\theta \leftarrow (g_j + g_{j+1})/2$

If $sp + swn - sn < emin$ **then** $\{emin \leftarrow sp + swn - sp; smin \leftarrow -1; \theta min \leftarrow \theta\}$

If $sn + swp - sp < emin$ **then** $\{emin \leftarrow sn + swp - sp; smin \leftarrow 1; \theta min \leftarrow \theta\}\}$

EndFOR

Set $s_t \leftarrow smin$; set $\theta_t \leftarrow \theta min$(*s and θ of current stage are determined*)

$\beta_t \leftarrow emin/(1 - emin)$;

$\alpha_T \leftarrow -\log(\beta_t)$ (* α_T is the output of AdaBoost for the second part*)

Update the weights for the next weak classifier, if needed:

$w(t+1, i) \leftarrow w(t, i)\beta_t^{1-e}$, where $e = \left\{\begin{array}{l} 0 \text{ if } x_i \text{is correctly classified bycurrent } h_t \\ 1\text{otherwise} \end{array}\right\}$

EndFor;

AdaBoost therefore assigns large weights with each good classification and small weights with each poor function. The selection of the next feature depends on selections made for previous features.

11.3.6 Cascaded AdaBoost

Viola and Jones [27] also described the option of designing a cascaded AdaBoost. For example, instead of one AdaBoost machine with 100 classifiers, one could design 10 such machines with 10 classifiers in each. In terms of precision, there will not be much difference, but testing for most images will be faster [27]. One particular image is first tested on the first classifier. If declared as nonsimilar, it is not tested further. If it cannot be rejected, then it is tested with the second machine. This process continues until either one machine rejects an image, or all machines "approve" it, and similarity is confirmed. Figure 11.4 illustrates this process. Each classifier seen in Figure 11.4 comprises one or more features. The features that define a classifier are chosen so that their combination eliminates as much as possible all negative images that are

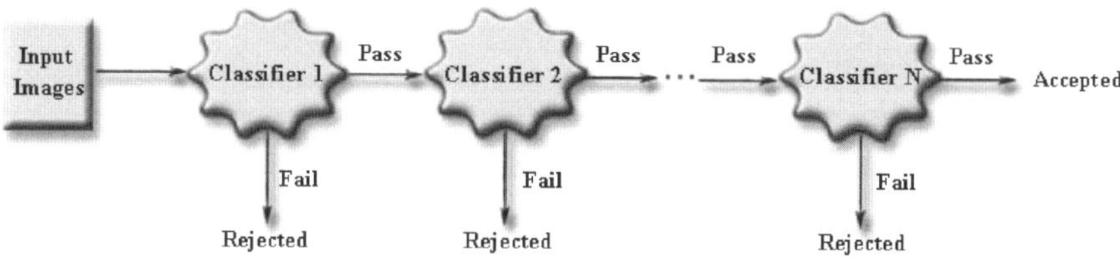

FIGURE 11.4 Cascaded decision process.

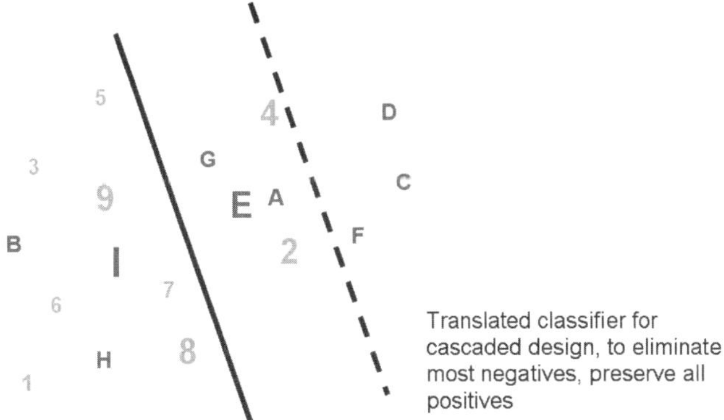

FIGURE 11.5 Concept of a classifier.

passed through this classifier, while at the same time accepting nearly 100 percent of the positives. It is desirable that each classifier eliminates at least 50 percent of the remaining negatives in the test set. A geometric progression of elimination is created until a desired threshold of classification is attained. The number of features in each classifier varies. It typically increases with the number of classifiers added. In Viola and Jones' face finder cascade, the first classifiers had 2, 10, 25, 25, and 50 features, respectively. The number of features grew very rapidly afterward. Typical numbers of features per classifier ranged in the hundreds. The total number of features used was roughly 6000 in Viola and Jones' application.

Figure 11.5 will help explain the design procedure of the cascaded design process. We revisit the letters and numbers example in our efforts to show the development of a strong classifier in the cascaded design. At the stage seen in Figure 11.5, we assume to have two WCs with weights α_1 and α_2. Together these two WCs make a conceptual hyperplane depicted by the solid dark blue line. In actuality, this line is not a hyperplane (in this case a line in two-dimensional space), but a series of orthonormal dividers. It is, however, conceptually easier to explain the design of a strong classifier in a cascade if we assume that WCs form hyperplanes.

So far in Figure 11.5, we have two WCs where the decision inequality would be of the form $\alpha_1 h_1(x) + \alpha_2 h_2(x) > \alpha/2$, where $\alpha = \alpha_1 + \alpha_2$. At this stage, the combination of the two WCs would be checked against the training set to see if they have a 99 percent detection rate (this 99 percent is a design parameter). If the detection rate is below the desired level, the threshold $\alpha/2$ is replaced with another threshold γ such that the detection rate increases to the desired level. This has the conceptual effect of translating the dark blue hyperplane in Figure 11.5 to the dotted line. This also has a residual effect of increasing the false positive rate. At the same time, once we are happy with the detection rate, we check the false positive rate of the shifted threshold detector. If this rate is satisfactory, for example, below 50 percent (also a design parameter), then the construction of the classifier is completed. The negative examples that were correctly identified by this classifier are ignored from further consideration by future classifiers. There is no need to consider them if they are already successfully eliminated by a previous classifier. In Figure 11.5, "D", "C," and "F" would be eliminated from future consideration if the classifier construction were completed at this point.

11.3.7 Integral Images

One of the key contributions in the work by Viola and Jones [27] (which is used and/or modified by Levi and Weiss [17], Luo et al. [19], etc.) is the introduction of a new image representation called the "integral image," which allows the features used by their detector to be computed very quickly.

In the preprocessing step, Viola and Jones [27] find the sums $ii(a, b)$ of pixel intensities $i(a', b')$ for all pixels (a', b') such that $a' \leq a, b' \leq b$. This can be done in one pass over the original image using the following recurrences:

$$s(a, b) = s(a, b - 1) + i(a, b),$$

$$ii(a, b) = ii(a - 1, b) + s(a, b),$$

where $s(a, b)$ is the cumulative row sum, $s(a, -1) = 0$, and $ii(-1, b) = 0$. In prefix sum notation, the expression for calculating the integral image values is

$$ii(a, b) = \sum_{a' \leq a, b' \leq b} i(a', b').$$

Figure 11.6 shows an example of how the "area" for rectangle "D" can be calculated using only four operations. Let the area mean the sum of pixel intensities of a rectangular region. The preprocessing step would have found the values of corners 1, 2, 3, and 4, which are in effect the areas of rectangles A, $A + B$, $A + C$, and $A + B + C + D$, respectively. Then the area of rectangle D is $= (A + B + C + D) +$

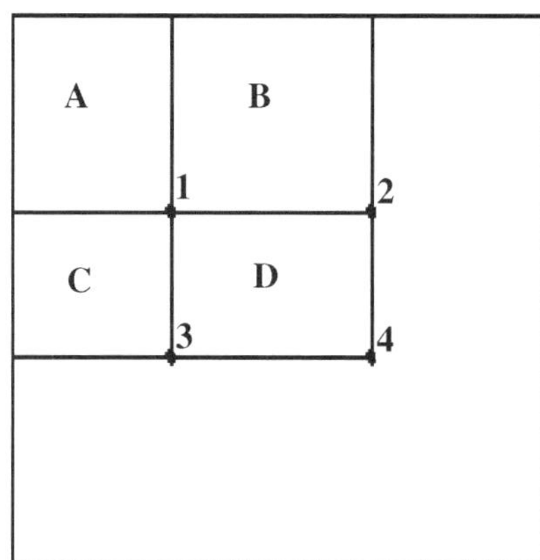

FIGURE 11.6 Integral image.

$(A) - (A + B) - (A + C) =$ "4" + "1" − "2" − "3". Jones and Viola [12] built one face detector for each view of the face. A decision tree is then trained to determine the viewpoint class (such as right profile or rotated 60 degrees) for a given window of the image being examined. The appropriate detector for that viewpoint can then be run instead of running all of the detectors on all windows.

11.4 CAR DETECTION

The most popular example of object detection is the detection of faces. The fundamental application that gave credibility to AdaBoost was Viola and Jones' real-time face finding system [27]. AdaBoost is the concrete machine learning method that was used by Viola and Jones to implement the system. The car detection application was inspired by the work of Viola and Jones. It is based on the same AdaBoost principles, but a variety of things, both in testing and in training, were adapted and enhanced to suit the needs of the CV system described in the works by Stojmenovic [24,25]. The goal of this chapter is to analyze the capability of current machine learning techniques of solving similar image retrieval problems. The "capability" of the system includes real-time performance, a high detection rate, low false positive rate, and learning with a small training set. Of particular interest are cases where the training set is not easily available, and most of it needs to be manually created.

As a particular case study, we will see the application of machine learning to the detection of rears of cars in images [24,25]. Specifically, the system is able to recognize cars of a certain type such as a Honda Accord 2004. While Hondas have been used as an instance, the same program, by just replacing the training sets, could be used to recognize other types of cars. Therefore, the input should be an arbitrary image, and the output should be that same image with a rectangle around any occurrence of the car we are searching for (see Fig. 11.7). The system will work by directly searching for an occurrence of the positive in the image, while treating all subwindows of the image the same way. It will not first search for a general vehicle class and then specify the model of the vehicle. This is a different and much more complicated task that is not easily solvable by machine learning. Any occurrence of a rectangle around a part of the image that is not a rear of a Honda Accord 2004 is considered a negative detection.

The image size in the testing set is arbitrary, while the image sizes in both the negative and positive training sets are the same. Positive training examples are the rears of Hondas. The data set was collected by taking pictures of Hondas (about

FIGURE 11.7 Input and output of the testing procedure.

300 of them) and other cars. The training set was actually manually produced by cropping and scaling positives from images to a standard size. Negative examples in the training set include any picture, of the same fixed size, that cannot be considered as a rear of a Honda. This includes other types of cars, as close negatives, for improving the classifier's accuracy. Thus, a single picture of a larger size contains thousands of negatives. When a given rectangle around a rear of a Honda is slightly translated and scaled, one may still obtain a positive example, visually and even by the classifier. That is, a classifier typically draws several rectangles at the back of each Honda. This is handled by a separate procedure that is outside the machine learning framework.

In addition to precision of detection, the second major main goal of the system was real-time performance. The program should quickly find all the cars of the given type and position in an image, in the same way that Viola and Jones finds all the heads. The definition of "real time" depends on the application, but generally speaking the system delivers an answer for testing an image within a second. The response time depends on the size of the tested image, thus what appears to be real-time for smaller images may not be so for larger ones.

Finally, this object detection system is interesting since it is based on a small number of training examples. Such criteria are important in cases where training examples are not easily available. For instance, in the works by Stojmenovic [24,25], photos of back views of a few hundred Honda Accords and other cars were taken manually to create training sets, since virtually no positive images were found on the Internet. In such cases, it is difficult to expect that one can have tens of thousands of images readily available, which was the case for the face detection problem. The additional benefit of a small training set is that the training time is reduced. This enabled us to perform a number of training attempts, adjust the set of examples, adjust the set of features, test various sets of WCs, and otherwise analyze the process by observing the behavior of the generated classifiers.

11.4.1 Limitations and Generalizations of Car Detection

Machine learning methods were applied in the work by Stojmenovic [24] in an attempt to solve the problem of detecting rears of a particular car type since they appear to be appropriate given the setting of the problem. Machine learning in similar image retrieval has proven to be reliable in situations where the target object does not change orientation. As in the work of Viola and Jones [27], cars are typically found in the same orientation with respect to the road. The situation Stojmenovic [24] is interested in is the rear view of cars. This situation is typically used in monitoring traffic since license plates are universally found at the rears of vehicles.

The positive images were taken such that all of the Hondas have the same general orthogonal orientation with respect to the camera. Some deviation occurred in the pitch, yaw, and roll of these images, which might be why the resulting detector has such a wide range of effectiveness. The machine that was built is effective for the following deviations in angles: pitch -15°; yaw -30° to 30°; and roll -15° to 15°. This means that pictures of Hondas taken from angles that are off by the stated amounts are still detected by the program. Yaw, pitch, and roll are common jargon in aviation describing the three degrees of freedom the pilot has to maneuver an aircraft.

Machine learning concepts in the CV field that deal with retrieving similar objects within images are generally faced with the same limitations and constraints. All successful real-time applications in this field have been limited to successfully finding objects from only one view and one orientation that generally does not vary much. There have been attempts to combine several strong classifiers into one machine, but discussing only individual strong classifiers, we conclude that they are all sensitive to variations in viewing angle. This limits their effective range of real-world applications to things that are generally seen in the same orientation. Typical applications include faces, cars, paintings, posters, chairs, some animals, and so on. The generalization of such techniques to problems that deal with widely varying orientations is possible only if the real-time performance constraint is lifted. Another problem that current approaches are faced with is the size of the training sets. It is difficult to construct a sufficiently large training database for rare objects.

11.4.2 Fast AdaBoost Based on a Small Training Set for Car Detection

This section describes the contributions and system [24] for detecting cars in real time. Stojmenovic [24] has revised the AdaBoost-based learning environment, for use in their object recognition problem. It is based on some of the ideas from literature, and some new ideas, all combined into a new machine.

The feature set used in the work Stojmenovic [24,25] initially included most of the feature types used by Viola and Jones [27] and Lienhart [14]. The set did not include rotated features [14], since the report on their usefulness was not convincing. Edge orientation histogram (EOH)-based features [17] were considered a valuable addition and were included in the set. New features that resemble the object being searched for, that is, custom-made features, were also added.

Viola and Jones [27] and most followers used weight-based AdaBoost, where the training examples receive weights based on their importance for selecting the next WC, and all WCs are consequently retrained in order to choose the next best one. Stojmenovic [24,25] states that it is better to rely on the Fast AdaBoost variant [30], where all of the WCs are trained exactly once, at the beginning. Instead of the weighted error calculation, Stojmenovic [24] believes that it is better to select the next WC to be added as the one that, when added, will make the best contribution (measured as the number of corrections made) to the already selected WCs. Each selected WC will still have an associated weight that depends on its accuracy. The reason for selecting the Fast AdaBoost variant is to achieve an $O(\log q)$ time speed-up in the training process, believing that the lack of weights for training examples can be compensated for by other "tricks" that were applied to the system.

Stojmenovic [24,25] has also considered a change in the AdaBoost logic itself. In existing logic, each WC returns a binary decision (0 or 1) and can therefore be referred to as the *binary WC*. In the machine proposed by Schapire and Singer [23], each WC will return a number in the range $[-1, 1]$ instead of returning a binary decision (0 or 1), after evaluating the corresponding example. Such a WC will be referred to as a *fuzzy*

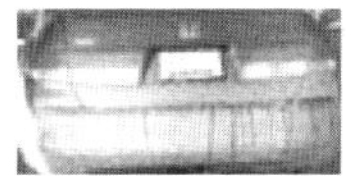    

FIGURE 11.8 Positive training examples.

WC. Evaluation of critical cases is often done by a small margin of difference from the threshold. Although the binary WC may not be quite certain about evaluating a particular feature against the adopted threshold (which itself is also determined heuristically, therefore is not fully accurate), the current AdaBoost machine assigns the full weight to the decision on the corresponding WC. Stojmenovic [24,25] therefore described an AdaBoost machine based on a fuzzy WC. More precisely, the described system proposes a specific function for making decisions, while Schapire [23] left this choice unspecified. The system produces a "doubly weighted" decision. Each WC receives a corresponding weight α, then each decision is made in the interval $[-1, 1]$. The WC then returns the product of the two numbers, that is, a number in the interval $[-\alpha, \alpha]$ as its "recommendation." The sum of all recommendations is then considered. If positive, the majority opinion is that the example is a positive one. Otherwise, the example is a negative one.

11.4.3 Generating the Training Set

All positives in the training set were fixed to be 100×50 pixels in size. The entire rear view of the car is captured in this window. Examples of positives are seen in Figure 11.8. The width of a Honda Accord 2004 is 1814 mm. Therefore, each pixel in each training image represents roughly $1814/100 = 18.14$ mm of the car.

A window of this size was chosen due to the fact that a typical Honda is unrecognizable to the human eye at lower resolutions; therefore, a computer would find it impossible to identify accurately. Viola and Jones used similar logic in determining their training example dimensions. All positives in the training set were photographed at a distance of a few meters from the camera. Detected false positives were added in the negative training set (bootstrapping), in addition to a set of manually selected examples, which included backs of other car models. The negative set of examples perhaps has an even bigger impact on the training procedure than the positive set. All of the positive examples look similar to the human eye. It is therefore not important to overfill the positive set since all of the examples there should look rather similar. The negative set should ideally combine a large variety of different images. The negative images should vary with respect to their colors, shapes, and edge quantities and orientations.

11.4.4 Reducing Training Time by Selecting a Subset of Features

Viola and Jones' faces were 24×24 pixels each. Car training examples are 100×50 pixels each. The implications of having such large training examples are immense from a memory consumption point of view. Each basic feature can be scaled in both height and width, and can be translated around each image. There are seven basic features used by Viola and Jones. They generated a total of 180,000 WCs [27]. Stojmenovic [24,25] also used seven basic features (as described below), and they generate a total of approximately 6.5 million WCs! Each feature is shifted to each position in the image and for every vertical and horizontal scale. By shifting our features by 2 pixels in each direction (instead of 1) and making scale increments of 2 during the training procedure, we were able to cut this number down to approximately 530,000, since every second position and scale of feature was used. In the initial training of the WCs, each WC is evaluated based on its cumulative error of classification (CE). The cumulative error of a classifier is CE = (false positives + number of missed examples)/total number of examples. WCs that had a CE that was greater than a predetermined threshold were automatically eliminated from further consideration. Details are given in the works by Stojmenovic [24,25].

11.4.5 Features Used in Training for Car Detection

Levi and Weiss [17] stress the importance of using the right features to decrease the sizes of the training sets, and increase the efficiency of training. A good feature is the one that separates the positive and negative training sets well. The same ideology is applied here in hopes of saving time in the training process. Initially, all of Viola and Jones' features were used in combination with the dominant edge orientation features proposed by Levi and Weiss [17] and the redness features proposed by Luo et al. [19]. It was determined that the training procedure never selected any of Viola and Jones' grayscale features to be in the strong classifier at the end of training. This is a direct consequence of the selected positive set. Hondas come in a variety of colors and these colors are habitually in the same relative locations in each positive case. The most obvious example is the characteristic red tail lights of the Honda accord. The redness features were included specifically to be able to use the redness of the tail lights as a WC. The training algorithm immediately exploited this distinguishing feature and chose the red rectangle around one of the tail lights as one of the first WCs in the strong classifier. The fact that the body of the Honda accord comes in its own subset of colors presented problems to the grayscale set of Viola and Jones' features. When these body colors are converted to a grayscale space, they basically cover the entire space. No adequate threshold can be chosen to beneficially separate positives from negatives. Subsequently, all of Viola and Jones' features were removed due to their inefficiency.

The redness features we refer to are taken from the work of Luo et al. [19]. More details are given in the works by Stojmenovic [24,25]. Several dominant edge orientation features were used in the training algorithm. To get a clearer idea of what edge orientation features are, we will first describe how they are made. Just as their name suggests, they arise from the orientation of the edges of an image. A Sobel gradient mask is a matrix used in determining the location of edges in an image. A typical mask of this sort is of size 3×3 pixels. It has two configurations, one for finding edges in the x-direction and the other for finding edges in the y-direction of source images ([7], p. 165). These two matrices, $\mathbf{h_x}$ and $\mathbf{h_y}$ (shown in Figs. 11.9 and 11.10), are known as the Sobel kernels.

$$\begin{bmatrix} -1 & 0 & 1 \\ -2 & 0 & 2 \\ -1 & 0 & 1 \end{bmatrix}$$

FIGURE 11.9 Kernel $\mathbf{h}_y$.

$$\begin{bmatrix} -1 & -2 & -1 \\ 0 & 0 & 0 \\ 1 & 2 & 1 \end{bmatrix}$$

FIGURE 11.10 Kernel $\mathbf{h}_x$.

Figure 11.9 shows the typical Sobel kernel for determining vertical edges (y-direction), and Figure 11.10 shows the kernel used for determining horizontal edges (x-direction). Each of these kernels is placed over every pixel in the image. Let P be the grayscale version of the input image. Grayscale images are determined from RGB color images by taking a weighted sampling of the red, green, and blue color spaces. The value of each pixel in a grayscale image was found by considering its corresponding color input intensities, and applying the following formula: $0.212671 \times R + 0.715160 \times G + 0.072169 \times B$, which is a built in function in OpenCV, which was used in the implementation.

Let $P(x, y)$ represent the value of the pixel at point (x, y) and $I(x, y)$ is a 3×3 matrix of pixels centered at (x, y). Let X and Y represent output edge orientation images in the x and y directions, respectively. X and Y are computed as follows:

$$\begin{aligned} X(i, j) = \mathbf{h}_x \cdot I(i, j) &= -P(i-1, j-1) + P(i+1, j-1) - 2P(i-1, j) \\ &\quad +2P(i+1, j) - P(i-1, j+1) + P(i+1, j+1), \end{aligned}$$

$$\begin{aligned} Y(i, j) = \mathbf{h}_y \cdot I(i, j) &= -P(i-1, j-1) - 2P(i, j-1) - P(i+1, j-1) \\ &\quad +P(i-1, j+1) + 2P(i, j+1) + P(i+1, j+1) \end{aligned}$$

A Sobel gradient mask was applied to each image to find the edges of that image. Actually, a Sobel gradient mask was applied both in the x-dimension, called $X(i, j)$, and in the y-dimension, called $Y(i, j)$. A third image, called $R(i, j)$, of the same dimensions as X, Y, and the original image, was generated such that $R(i, j) = \sqrt{X(i, j)^2 + Y(i, j)^2}$. The result of this operation is another grayscale image with a black background and varying shades of white around the edges of the objects in the image. The image $R(i, j)$ is called a Laplacian image in image processing literature, and values $R(i, j)$ are called Laplacian intensities. One more detail of our implementation is the threshold that was placed on the intensities of the Laplacian values. We used a threshold of 80 to eliminate the faint edges that are not useful. A similar threshold was employed in the work by Levi and Weiss [17].

The orientations of each pixel are calculated from the $X(i, j)$ and $Y(i, j)$ images. The orientation of each pixel $R(i, j)$ in the Laplacian image is found as

$$\text{orientation}(i,j) = \arctan(\mathrm{Y}(i,j), \mathrm{X}(i,j)) \times 180/\pi.$$

This formula gives the orientation of each pixel in degrees. The orientations are divided into six bins so that similar orientations can be grouped together. The whole circle is divided into six bins. Bin shifting (rotation of all bins by 15°) is applied to better capture horizontal and vertical edges. Details are given in the work by Stojmenovic [24].

11.5 NEW FEATURES AND APPLICATIONS

11.5.1 Rotated Features and Postoptimization

Lienhart and Maydt [14] add a set of classifiers (Haar wavelets) to those already proposed by Viola and Jones. Their new classifiers are the same as those proposed by Viola and Jones, but they are all rotated 45°. They claim to gain a 10 percent improvement in the false detection rate at any given hit rate when detecting faces. The features used by Lienhart were basically Viola and Jones' entire set rotated 45° counterclockwise. He added two new features that resembled the ones used by Viola and Jones, but they too failed to produce notable gains.

However, there is a postoptimization stage involved with the training process. This postoptimization stage is credited with over 90 percent of the improvements claimed by this paper. Therefore, the manipulation of features did not impact the results all that much; rather the manipulation of the weights assigned to the neural network at the end of each stage of training is the source of gains. OpenCV supports the integral image function on 45° rotated images since Lienhart was on the development team for OpenCV.

11.5.2 Detecting Pedestrians

Viola et al. [29] propose a system that finds pedestrians in motion and still images. Their system is based on the AdaBoost framework. It considers both motion information and appearance information. In the motion video pedestrian finding system, they train AdaBoost on pairs of successive frames of people walking. The intensity differences between pairs of successive images are taken as positive examples. They find the direction of motion between two successive frames, and also try to establish if the moving object can be a person. If single images are analyzed for pedestrians, no motion information is available, and just the regular implementation of AdaBoost seen for faces is applied to pedestrians. Individual pedestrians are taken as positive training examples. It does not work as well as the system that considers motion information since the pedestrians are relatively small in the still pictures, and also relatively low resolution (not easily distinguishable, even by humans). AdaBoost is easily confused in such situations. Their results suggest that the motion analysis system works better than the still image recognizer. Still, both systems are relatively inaccurate and have high false positive rates.

11.5.3 Detecting Penguins

Burghardt et al. [5] apply the AdaBoost machine to the detection of African penguins. These penguins have a unique chest pattern that AdaBoost can be trained on. They were able to identify not only penguins in images, but distinguish between individual penguins as well. Their database of penguins was small and taken from the local zoo. Lienhart's [14] adaptation of AdaBoost was used with the addition of an extra feature: the empty kernel. The empty kernel is not a combination of light and dark areas, but rather only a light area so that AdaBoost may be trained on "pure luminance information." AdaBoost was used to find the chests of penguins, and other methods were used to distinguish between different penguins. Their technique did not work very well for all penguins. They gave no statistics concerning how well their approach works. This is another example of how the applications of AdaBoost are limited to very specialized problems.

11.5.4 Redeye Detection: Color-Based Feature Calculation

Luo et al. [19] introduce an automatic redeye detection and correction algorithm that uses machine learning in the detection of red eyes. They use an adaptation of AdaBoost in the detection phase of redeye instances. Several novelties are introduced in the machine learning process. The authors used, in combination with existing features, color information along with aspect ratios (width to height) of regions of interest as trainable features in their AdaBoost implementation.

Viola and Jones [27] used only grayscale intensities, although their solution to face detection could have used color information. Finding red eyes in photos means literally finding red oval regions, which absolutely requires the recognition of color. Another unique addition in their work is a set of new features similar to those proposed by Viola and Jones [27], yet designed specifically to easily recognize circular areas. We see these feature templates in Figure 11.11. It is noticeable that the feature templates presented in this figure have three distinct colors: white, black, and gray. The gray and black regions are taken into consideration when feature values are calculated. Each of the shapes seen in Figure 11.11 is rotated around itself or reflected creating eight different positions. The feature value of each of the eight positions is calculated, and the minimum and maximum of these results are taken as output from the feature calculation.

The actual calculations are performed based on the RGB color space. The pixel values are transformed into a one-dimensional space before the feature values are calculated in the following way: Redness $= 4R - 3G + B$. This color space is biased toward the red spectrum (which is where red eyes occur). This redness feature was used in the car detection system [24].

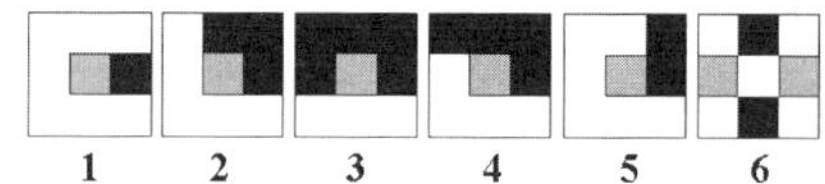

FIGURE 11.11 Features for redeye detection.

11.5.5 EOH-Based Features

Levi and Weiss [17] add a new perspective on the training features proposed by Viola and Jones [27]. They also detect upright, forward-facing faces. Among other contributions in their work [17], their most striking revelation was adding an edge orientation feature that the machine can be trained on. They also experimented with mean intensity features, which means taking the average pixel intensity in a rectangular area. These features did not produce good results in their experiments and were not used in their system. In addition to the features used by Viola and Jones [27], which considered sums of pixel intensities, Levi and Weiss [17] create features based on the most prevalent orientation of edges in rectangular areas. There are obviously many orientations available for each pixel but they are reduced to eight possible rotations for ease of comparison and generalization. For any rectangle, many possible features are extracted. One set of features is the ratio of any two pairs of the eight EOHs [17]. There are therefore 8 choose $2 = 28$ possibilities for such features. Another feature that is calculated is the ratio of the most dominant EOH in a rectangle to the sum of all other EOHs. Levi and Weiss [17] claim that using EOHs, they are able to achieve higher detection rates at all training database sizes.

Their goal was to achieve similar or better performance of the system to Viola and Jones' work while substantially reducing training time. They primarily achieve this because EOH gives good results with a much smaller training set. Using these orientation features, symmetry features are created and used. Every time a WC was added to their machine, its vertically symmetric version was added to a parallel yet independent cascade. Using this parallel machine architecture, the authors were able to increase the accuracy of their system by 2 percent when both machines were run simultaneously on the test data. The authors also mention detecting profile faces. Their results are comparable to those of other proposed systems but their system works in real-time and uses a much smaller training set.

11.5.6 Fast AdaBoost

Wu et al. [30] propose a training time performance increase over Viola and Jones' training method. They change the training algorithm in such a way that all of the features are tested on the training set only once (per each classifier). The ith classifier ($1 \leq i \leq N$) is given as input the desired minimum detection rate d_i and the maximum false positive rate fp_i. These rates are difficult to predetermine because the performance of the system varies greatly. The authors start with optimistic rates and gradually decrease expectations after including over 200 features until the criterion is met. Each feature is trained so that it has minimal false positive rate fp_i. The obtained WCs h_j are sorted according to their detection rates. The strong classifier is created by incrementally adding the feature that either increases the detection rate (if it is $<d_i$) or minimizes false positives until desired levels are achieved in both categories. Since the features are tested independently, the weights of the positive and negative training examples that are incorrectly classified are not changed. The decision of the

ensemble classifier is formed by a majority vote of the WCs (that is, each WC has equal weight in the work by wu et al. [30]). The authors state that using their model of training, the desired detection rate was more difficult to achieve than the desired false positive rate. To improve this defect, they introduce asymmetric feature selection. They incorporated a weighting scheme into the selection of the next feature. They chose weights of 1 for false positive costs and λ for false negative costs. λ is the cost ratio between false negatives and false positives. This setup allows the system to add features that increase the detection rate early on in the creation of the strong classifier.

Wu et al. [30] state that their method works almost as well as that of Viola and Jones when applied to the detection of upright, forward-facing faces. They however achieve a training time that is two orders of magnitude faster than that of Viola and Jones. This is achieved in part by using a training set that was much smaller than Viola and Jones' [27], yet generated similar results.

We will now explain the time complexity of both Viola and Jones' [27] and Wu's [30] training methods. There are three factors to consider when finding the time complexity of each training procedure: the number of features F, the number of WCs in a classifier T, and the number of examples in the training set q. One feature in one example takes $O(1)$ time because of integral images. One feature on q examples takes $O(q)$ time to evaluate, and $O(q \log(q))$ to sort and find the best WC. Finding the best feature takes $O(Fq \log(q))$ time. Therefore, the construction of the classifier takes $O(TFq \log q)$. Wu's [30] method takes $O(Fq \log q)$ time to train all of the classifiers in the initial stage. Testing each new WC while assuming that the summary votes of all classifiers are previously stored would take $O(q)$ time. It would then take $O(Fq)$ time to select the best WC. Therefore, it takes $O(TqF)$ time to chose T WC. We deduce that it would take $O(Fq \log q + TqF)$ time to complete the training using the methods described by Wu et al. [30]. The dominant term in the time complexity of Wu's [30] algorithm is $O(TqF)$. This is order $O(\log q)$ times faster than the training time for Viola and Jones' method [27]. For a training set of size $q = 10,000$, $\log_2 q \approx 13$. For the same size training sets, Wu's [30] algorithm would be 13 times faster to train, not a 100 times as claimed by the authors. The authors compared training times to achieve a predetermined accuracy rate, which requires fewer training items than Viola and Jones' method [27]. Froba et al. [13] elaborate on a face verification system. The goal of this system is to be able to recognize a particular person based on his/her face. The first step in face verification is face detection. The second is to analyze the detected sample and see if it matches one of the training examples in the database. The mouths of input faces into the system are cropped because the authors claim that this part of the face varies the most and produces unstable results. They however include the forehead since it helps with system accuracy. The authors use the same training algorithm for face detection as Viola and Jones [27], but include a few new features. They use AdaBoost to do the training, but the training set is cropped, which means that the machine is trained on slightly different input than Viola and Jones [27]. The authors mention that a face is detectable and verifiable with roughly 200 features that are determined by AdaBoost during the training phase. The actual verification or recognition step of individual people based on these images is done using information obtained in the detection step. Each face that is detected is made up of a vector of 200 numbers that are the evaluations of the different features that made up that face. These numbers more or less uniquely represent each face and are used as a basis of comparison of two faces. The sum of the weighted differences in the feature values between the detected face and the faces of the individual people in the database is found and compared against a threshold as the verification step. This is a sort of nearest-neighbor comparison that is used in many other applications.

11.5.7 Downhill Feature Search

McCane and Novins [20] described two improvements over the Viola and Jones' [27] training scheme for face detection. The first one is a 300-fold speed improvement over the training method, with an approximately three times slower execution time for the search. Instead of testing all features at each stag (exhaustive search), McCane and Novins [20] propose an optimization search, by applying a "downhill search" approach. Starting from a feature, a certain number of neighboring features are tested next. The best one is selected as the next feature, and the procedure is repeated until no improvement is possible. The authors propose to use same size adjacent features (e.g., rectangles "below" and "above" a given one, in each of the dimensions that share one common edge) as neighbors. They observe that the work by Viola and Jones [27] applies AdaBoost in each stage to optimize the overall error rate, and then, in a postprocessing step, adjust the threshold to achieve the desired detection rate on a set of training data. This does not exactly achieve the desired optimization for each cascade step, which needs to optimize the false positive rate subject to the constraint that the required detection rate is achieved. As such, sometimes adding a level in an AdaBoost classifier actually increases the false positive rate. Further, adding new stages to an AdaBoost classifier will eventually have no effect when the classifier improves to its limit based on the training data. The proposed optimization search allows it to add more features (because of the increased speed), and to add more parameters to the existing features, such as allowing some of the subsquares in a feature to be translated. The second improvement in the work by McCane and Novins [20] is a principled method for determining a cascaded classifier of optimal speed. However, no useful information is reported, except the guideline that the false positive rate for the first cascade stage should be between 0.5 and 0.6. It is suggested that exhaustive search [27] could be performed at earlier stages in the cascade, and replaced by optimized search [20] in later stages.

11.5.8 Bootstrapping

Sung and Poggio [22] applied the following "bootstrap" strategy to constrain the number of nonface examples in their face detection system. They incrementally select only those nonface patterns with high utility value. Starting with a small set of non-face examples, they train their classifier with current database examples and run the face detector on a sequence of random images (we call this set of images a "semitesting" set). All nonface examples that are wrongly classified by the current system as faces are collected and added to the training database as new negative examples. They notice that the same bootstrap technique can be applied to enlarge the set of positive examples. In the work by Bartlett et al. [3], a similar bootstrapping technique was applied. False alarms are collected and used as nonfaces for training the subsequent strong classifier in the sequence, when building a cascade of classifiers.

Li et al. [18] observe that the classification performance of AdaBoost is often poor when the size of the training sample set is small. In certain situations, there may be unlabeled samples available and labeling them is costly and time consuming. They propose an active learning approach, to select the next unlabeled sample that is at the minimum distance from the optimal AdaBoost hyperplane derived from the current set of labeled samples. The sample is then labeled and entered into the training set. Abramson and Freund [1] employ a selective sampling technique, based on boosting, which dramatically reduces the amount of human labor required for labeling images. They apply it to the problem of detecting pedestrians from a video camera mounted on a moving car. During the boosting process, the system shows subwindows with close classification scores, which are then labeled and entered into positive and negative examples. In addition to features from the work by Viola and Jones [27], authors also use features with "control points" from the work by Burghardt and Calic [2].

Zhang et al. [31] empirically observe that in the later stages of the boosting process, the nonface examples collected by bootstrapping become very similar to the face examples, and the classification error of Haar-like feature based WC is thus very close to 50 percent. As a result, the performance of a face detection method cannot be further improved. Zhang et al. [31] propose to use global features, derived from Principal component analysis (PCA), in later stages of boosting, when local features do not provide any further benefit. They show that WCs learned from PCA coefficients are better boosted, although computationally more demanding. In each round of boosting, one PCA coefficient is selected by AdaBoost. The selection is based on the ability to discriminate faces and nonfaces, not based on the size of coefficient.

11.5.9 Other AdaBoost Based Object Detection Systems

Treptow et al. [26] described a real-time soccer ball tracking system, using the described AdaBoost based algorithm [27]. The same features were used as in the work by Viola and Jones [27]. They add a procedure for predicting ball movement.

Cristinacce and Cootes [6] extend the global AdaBoost-based face detector by adding four more AdaBoost based algorithms that detect the left eye, right eye, left mouth corner, and right mouth corner within the face. Their placement within the face is probabilistically estimated. Training face images are centered at the nose and some flexibility in position of other facial parts with a certain degree of rotation is allowed in the main AdaBoost face detector, because of the help provided by the four additional machines.

FloatBoost [31,32] differs from AdaBoost in a step where the removal of previously selected WCs is possible. After a new WC is selected, if any of the previously added classifiers contributes to error reduction less than the latest addition, this classifier is removed. This results in a smaller feature set with similar classification accuracy. FloatBoost requires about a five times longer training time than AdaBoost. Because of the reduced set of selected WCs, Zhang et al. [31,32] built several face recognition learning machines (about 20), one for each of face orientation (from upfront to profiles). They also modified the set of features. The authors conclude that the method does not have the highest accuracy.

Howe [11] looks at boosting for image retrieval and classification, with comparative evaluation of several algorithms. Boosting is shown to perform significantly better than the nearest-neighbor approach. Two boosting techniques that are compared are based on feature- and vector-based boosting. Feature-based boosting is the one used in the work by Viola and Jones [27]. Vector-based boosting works differently. First, two vectors, toward positive and negative examples, are determined, both as weighted sums (thus corresponding to a kind of average value). A hyperplane bisecting the angle between them is used for classification. The dot product of the tested example that is orthogonal to that hyperplane is used to make a decision. Comparisons are made on five training sets containing suns, churches, cars, tigers, and wolves. The features used are color histograms, correlograms (probabilities that a pixel B at distance x from pixel A has the same color as A), stairs (patches of color and texture found in different image locations), and Viola and Jones' features. Vector boosting is shown to be much faster than feature boosting for large dimensions. Feature-based boosting gave better results than vector based when the number of dimensions in the image representation is small.

Le and Satoh [15] observe AdaBoost advantages and drawbacks, and propose to use it in the first two stages of the classification process. The first stage is a cascaded classifier with subwindows of size 36×36, the second stage is a cascaded classifier with subwindows of size 24×24. The third stage is an SVM classifier for greater precision. Silapachote et al. [21] use histograms of Gabor and Gaussian derivative responses as features for training and apply them for face expression recognition with AdaBoost and SVM. Both approaches show similar results and AdaBoost offers important feature selections that can be visualized.

Barreto et al. [4] described a framework that enables a robot (equipped with a camera) to keep interacting with the same person. There are three main parts of the framework: face detection, face recognition, and hand detection. For detection, they use Viola and Jones's features [27] improved by Lienhart and Maydt [14]. The eigenvalues and PCA are used in the face recognition stage of the system. For hand detection, they apply the same techniques used for face detection. They claim that the system recognizes hands in a variety of positions. This is contrary to the claims made by Kolsch et al. [13] who built one cascaded AdaBoost machine for every typical hand position and even rotation.

Kolsch and Turk [16,17] describe and analyze a hand detection system. They create a training set for each of the six posture/view combinations from different people's right hands. Then both training and validation sets were rotated and a classifier was trained for each angle. In contrast to the case of the face detector, they found poor accuracy with rotated test images for as little as a 4° rotation. They then added rotated example images to the same training set, showing that up to 15° of rotation can be efficiently detected with one detector.

11.5.10 Binary and Fuzzy Weak Classifiers

Most AdaBoost implementations that we found in literature use binary WCs, where the decision of a WC is either accept or reject, which will be valued at +1 and −1, respectively (and described in Chapter 2). We also consider *fuzzy WCs* [23] as follows. Instead of making binary decisions, fuzzy WCs make a 'weighted' decision, as a real number in the interval [−1, 1]. Fuzzy WCs can then simply replace binary WCs as basic ingredients in the training and testing programs, without affecting the code or structure of the other procedures.

A *fuzzy WC* is a function of the form $h(x, f, s, \theta, \theta_{mn}, \theta_{mx})$ where x is the tested sub image, f is the feature used, s is the sign (+ or −), θ is the threshold, and θ_{mn} and θ_{mx} are the adopted extreme values for positive and negative images. The sign s defines on what side the threshold the positive examples are located. Threshold θ is used to establish whether a given image passes a classifier test in the following fashion: when feature f is applied to image x, the resulting number is compared to threshold θ to determine how this image is categorized by the given feature. The equation is given below

$$sf(x) < s\theta.$$

If the equation evaluates true, the image is classified as positive. The function $h(x, f, s, \theta, \theta_{mn}, \theta_{mx})$ is then defined as follows. If the image is classified as positive ($sf(x) < s\theta$) then $h(x, f, s, \theta, \theta_{mn}, \theta_{mx}) = \min(1, |(f(x) - \theta)/(\theta_{mn} - \theta)|)$. Otherwise $h(x, f, s, \theta, \theta_{mn}, \theta_{mx}) = \max(-1, -|(f(x) - \theta)/(\theta_{mx} - \theta)|)$. This definition is illustrated in the following example.

1 6 8 4 I 3 5 7 | B E A 9 C 2 F D G H

θ_{mn} $f(x)$ θ θ_{mx}

Let $s = 1$, thus the test is $f(x) < \theta$. One way to determine θ_{mn} and θ_{mx} (used in our implementation) is to find the minimal feature value of the positive examples (example "1" seen here), and maximal negative value (example "H" seen here) and assign them to θ_{mn} and θ_{mx}, respectively. If $s = -1$, then the definitions are modified accordingly. Suppose that an image is evaluated to be around the letter "I" in the example (it could be exactly the letter "I" in the training process or a tested image at runtime). Since $f(x) < \theta$, the image is estimated as positive. The degree of confidence in the estimation is $|(f(x) - \theta)/(\theta_{mn} - \theta)|$, which is about 0.5 in the example. If the ratio is > 1, then it is replaced by 1. The result of the evaluation is then $h(x, f, s, \theta, \theta_{mn}, \theta_{mx}) = 0.5$, which is returned as the recommendation.

11.5.11 Strong Classifiers

A strong classifier is obtained by running the AdaBoost machine. It is a linear combination of WCs. We assume that there are T WCs in a strong classifier, labeled

$h_1, h_2, \ldots, h_T$, and each of these comes with its own weight labeled $\alpha_1, \alpha_2, \ldots, \alpha_T$. The tested image x is passed through the succession of WCs $h_1(x), h_2(x), \ldots, h_T(x)$, and each WC assesses if the image passed its test. In case of binary WCs, the recommendations are either -1 or 1. In case of using fuzzy WCs, the assessments are values ρ in the interval $[-1, 1]$. Values ρ from interval $(0, 1]$ correspond to a pass (with confidence ρ) and in the interval $[0, -1]$ a fail. Note that $h_i(x) = h_i(x, f_i, s_i, \theta_i, \theta_{mn}, \theta_{mx})$ is abbreviated here for convenience (parameters θ_{mn} and θ_{mx} are needed only for fuzzy WCs). The decision that classifies an image as being positive or negative is made by the following inequality:

$$\alpha = \alpha_1 h_1(x) + \alpha_2 h_2(x) + \cdots + \alpha_T h_T(x) > \delta.$$

From this equation, we see that images that pass (binary or weighted) weighted recommendations of the WC tests are cataloged as positive. It is therefore a (simple or weighted) voting of selected WCs. The value α also represents the confidence of overall voting. The error is expected to be minimal when $\delta = 0$, and this value is used in our algorithm. The α values are determined once at the beginning of the training procedure for each WC, and are not subsequently changed. Each $\alpha_i = -\log(e_i/(1 - e_i))$. Each e_i is equal to the cumulative error of the WC.

11.6 CONCLUSIONS AND FUTURE WORK

It is not so trivial to apply any AdaBoost approach to the recognition of a new vision problem. Pictures of the new object may not be readily available (such as those for faces). A positive training set numbering in the thousands is easily acquired with a few days spent on the internet hunting for faces. It took roughly a month to collect the data set required for the training and testing of the detection of the Honda Accord [24]. Even if a training set of considerable size could be assembled, how long would it take to train? Certainly, it would take in the order of months. It is therefore not possible to easily adapt Viola and Jones' standard framework to any vision problem. This is the driving force behind the large quantity of research that is being done in this field. Many authors still try to build upon the AdaBoost framework developed by Viola and Jones, which only credits this work further. The ideal object detection system in CV would be the one that can easily adapt to finding different objects in different settings while being autonomous from human input. Such a system is yet to be developed.

It is easy to see that there is room for improvement in the detection procedures seen here. The answer does not lie in arbitrarily increasing the number of training examples and WCs. The approach of increasing the number of training examples is brute force, and is costly when it comes to training time. Increasing the number of WCs would result in slower testing times. We propose to do further research in designing a cascaded classifier that will still work with a limited number of training examples, but can detect a wide range of objects. This new cascaded training procedure must also work in very limited time; in the order of hours, not days or months as proposed by predecessors.

The design of fuzzy WCs and the corresponding fuzzy training procedure may be worth further investigation. We have perhaps only seen applications that were solvable efficiently with standard binary WCs. There are perhaps some more difficult problems, with finer boundaries between positive and negative examples, where fuzzy WCs would produce better results. Since the change that is involved is quite small, affecting only a few lines of code, it is worth trying this method in future object detection cases.

All of the systems that were discussed here were mainly custom made to suit the purpose of detecting one object (or one class of objects). Research should be driven to find a flexible solution with a universal set of features that is capable of solving many detection problems quickly and efficiently.

An interesting open problem is to also investigate constructive learning of good features for object detection. This is different from applying an automatic feature triviality test on existing large set of features, proposed in the works by Stojmenovic [24,25]. The problem is to design a machine that will have the ability to build new features that will have good performance on a new object detection task. This appears to be an interesting ultimate challenge for the machine learning community.

REFERENCES

1. Abramson Y, Freund Y. Active Learning for Visual Object Recognition. Forthcoming.
2. Burghardt T, Calic J. Analysing animal behaviour in wildlife videos using face detection and tracking. IEE Proc Vision, Image Signal Proces. Special issue on the Integration of Knowledge, Semantics and Digital Media Technology; March 2005.
3. Bartlett MS, Littlewort G, Fasel I, Movellan JR. Real-time face detection and expression recognition: development and application to human-computer interaction. CVPR Workshop on Computer Vision and Pattern Recognition for Human–Computer Interaction, IEEE CVPR; Madison, Wi; 2003 June 17.
4. Barreto J, Menezes P, Dias J. Human–robot interaction based on Haar-like features and eigenfaces. Proceedings of the New Orleans International Conference on Robotics and Automation; 2004. p 1888–1893.
5. Burghardt T, Thomas B, Barham P, Calic J. Automated visual recognition of individual african penguins. Proceedings of the Fifth International Penguin Conference; Ushuaia, Tierra del Fuego, Argentina; September 2004.
6. Cristinacce D, Cootes T. Facial feature detection using AdaBoost with shape constraints. Proceedings of 14th BMVA British Machine Vision Conference; Volume 1; Norwich, UK; September 2003. p 231–240,
7. Efford N. Digital Image Processing: A Practical Introduction Using Java. Addison Wesley; 2000.
8. Freund Y, Schapire RE. A decision-theoretic generalization on on-line learning and an application to boosting. Proceedings of the 2nd European Conference on Computational Learning Theory (Eurocolt95); Barcelona, Spain; 1995. p 23–37. J Comput Syst Sci 1997;55(1):119–139.
9. Freund Y, Schapire RE. A short introduction to boosting. J J Soc Artif Intell 1999;14(5): 771–780.

10. Fröba B, Stecher S, Küblbeck C. Boosting a Haar-like feature set for face verification. Lecture Notes in Computer Science; 2003. p 617–624.
11. Howe NR. A closer look at boosted image retrieval. Proceedings of the International Conference on Image and Video Retrieval; July 2003. p 61–70.
12. Jones M, Viola P. Fast multi-view face detection. Mitsubishi Electric Research Laboratories, TR2003-96 July 2003, http://www.merl.com; shown as demo at IEEE Conference on Computer Vision and Pattern Recognition (CVPR); June 2003.
13. Kolsch M, Turk M. Robust hand detection. Proceedings of the IEEE Interanational Conference on Automatic Face and Gesture Recognition; May 2004. p 614–619.
14. Lienhart R, Maydt J. An extended set of haar-like features for rapid object detection. Proceedings of the IEEE International Conference Image Processing; Volume 1; 2002. p 900–903.
15. Le DD, Satoh S. Feature selection by AdaBoost for SVM-based face detection. Information Technology Letters, The Third Forum on Information Technology (FIT2004); 2004.
16. Le D, Satoh S. Fusion of local and global features for efficient object detection. IS & T/SPIE Symposium on Electronic Imaging; 2005.
17. Levi K, Weiss Y. Learning object detection from a small number of examples: the importance of good features. Proceedings of the International Conference on Computer Vision and Pattern Recognition (CVPR); Volume 2; 2004. p 53–60.
18. Li X, Wang L, Sung E. Improving AdaBoost for classification on small training sample sets with active learning. Proceedings of the Sixth Asian Conference on Computer Vision (ACCV); Korea; 2004.
19. Luo H, Yen J, Tretter D. An efficient automatic redeye detection and correction algorithm. Proceedings of the 17th IEEE International Conference on Pattern Recognition (ICPR'04); Volume 2; Aug 23–26, 2004; Cambridge, UK. p 883–886.
20. McCane B, Novins K. On training cascade face detectors. Image and Vision Computing. Palmerston North, New Zealand; 2003. p 239–244.
21. Silapachote P, Karuppiah DR, Hanson AR. Feature selection using AdaBoost for face expression recognition. Proceedings of the 4th IASTED International Conference on Visualization, Imaging, and Image Processing, VIIP 2004; Marbella, Spain; September 2004. p 452–273.
22. Sung K, Poggio T. Example based learning for view-based human face detection. IEEE Trans Pattern Anal Mach Intell 1998;20:39–51.
23. Schapire R, Singer Y. Improved boosting algorithms using confidence-rated predictions. Mach Learn 1999;37(3):297–336.
24. Stojmenovic M. Real time machine learning based car detection in images with fast training. Mach Vis Appl 2006;17(3):163–172.
25. Stojmenovic M. Real time object detection in images based on an AdaBoost machine learning approach and a small training set. Master thesis, Carleton University; June 2005.
26. Treptow A, Masselli A, Zell A. Real-time object tracking for soccer-robots without color information. Proceedings of the European Conference on Mobile Robotics ECMR; 2003.
27. Viola P, Jones M. Robust real-time face detection. Int J Comput Vis 2004; 57(2):137–154.
28. Viola P, Jones M. Fast and robust classification using asymmetric AdaBoost. Neural Inform Processing Syst 2002;14.

29. Viola P, Jones M, Snow D. Detecting pedestrians using patterns of motion and appearance. Proceedings of 9th International Conference on Computer Vision ICCV. Volume 2; 2003. p 734–741.

30. Wu J, Regh J, Mullin M. Learning a rare event detection cascade by direct feature selection. Proceedings of the Advances in Neural Information Processing Systems 16 (NIPS*2003). MIT Press; 2004.

31. Zhang D, Li S, Gatica-Perez D. Real-time face detection using boosting learning in hierarchical feature spaces. Proceedings of the International Conference on Pattern Recognition (ICPR); Cambridge, August. 2004. p 411–414.

32. Li SZ, Zhang Z. FloatBoost learning and statistical face detection. IEEE Trans Pattern Anal Machine Intell 2004;26(9):1112–1123.

CHAPTER 12

2D Shape Measures for Computer Vision

PAUL L. ROSIN and JOVIŠA ŽUNIĆ

12.1 INTRODUCTION

Shape is a critical element of computer vision systems. Its potential value is made more evident by considering how its effectiveness has been demonstrated in biological visual perception. For instance, in psychophysical experiments it was shown that for the task of object recognition, the outline of the shape was generally sufficient, rendering unnecessary the additional internal detail, texture, shading, and so on available in the control photographs [1,22]. A second example is the so-called shape bias. When children are asked to name new objects, generalizing from a set of previously viewed artificial objects, it was found that they tend to generalize on the basis of shape, rather than material, color, or texture [28,56].

There are many components in computer vision systems that can use shape information, for example, classification [43], shape partitioning [50], contour grouping [24], removing spurious regions [54], image registration [62], shape from contour [6], snakes [11], image segmentation [31], data mining [64], and content-based image retrieval [13], to name just a few.

Over the years, many ways have been reported in the literature for describing shape. Sometimes they provide a unified approach that can be applied to determine a variety of shape measures [35], but more often they are specific to a single aspect of shape. This material is covered in several reviews [26,32,53,67], and a comparison of some different shape representations has been carried out as part of the Core Experiment CE-Shape-1 for MPEG-7 [2,29,61].

Many shape representations (e.g., moments, Fourier, tangent angle) are capable of reconstructing the original data, possibly up to a transformation (e.g., modulo translation, rotation, scaling, etc.). However, for this chapter the completeness of the shape representations is not an issue. A simpler and more compact class of representation in common use is the one-dimensional signature (e.g., the histogram of tangent angles). This chapter does not cover such schemes either, but is focused on shape measures that compute single scalar values from a shape. Their advantage is that not only are

Handbook of Applied Algorithms: Solving Scientific, Engineering and Practical Problems
Edited by Amiya Nayak and Ivan Stojmenović

these measures extremely concise (benefiting storage and matching) but they tend to be designed to be invariant to rotations, translations, and uniform scalings, and often have an intuitive meaning (e.g., circularity) since they describe a single aspect of the shape. The latter point can be helpful for users of computer vision systems to understand their reasoning. The shapes themselves we assume to be extracted from images and are presented either in the form of a set of boundary or interior pixels, or as polygons.

The majority of the measures described have been normalized so that their values lie in the range $[0, 1]$ or $(0, 1]$. Nevertheless, even when measuring the same attribute (e.g., there are many measures of convexity) the values of the measures are not directly comparable since they have not been developed in a common framework (e.g., a probabilistic interpretation).

The chapter is organized as follows: Section 12.2 describes several shape descriptors that are derived by the use of minimum bounding rectangles. The considered shape descriptors are rectangularity, convexity, rectilinearity, and orientability. Section 12.3 extends the discussion to the shape descriptors that can be derived from other bounding shapes (different from rectangles). Fitting a shape model to the data is a general approach to the measurement of shape; an overview of this is given in Section 12.4. Geometric moments are widely used in computer vision, and their application to shape analysis is described in Section 12.5. The powerful framework of Fourier analysis has also been applied, and Fourier descriptors are a standard means of representing shape, as discussed in Section 12.6.

12.2 MINIMUM BOUNDING RECTANGLES

As we will see in the next section, using a bounding shape is a common method for generating shape measures, but here we will concentrate on a single shape, optimal bounding rectangles, and outline a variety of its applications to shape analysis.

Let $\mathbf{R}(S, \alpha)$ be the minimal area rectangle with edges parallel to the coordinate axes, which includes polygon S rotated by an angle α around the origin. Briefly, $\mathbf{R}(S)$ means $\mathbf{R}(S, \alpha = 0)$. Let $\mathbf{R}_{\min}(S)$ be the rectangle that minimizes $\mathbf{area}(\mathbf{R}(S, \alpha))$. This can be calculated in linear time with respect to the number of vertices of S by first computing the convex hull followed by Toussaint's [59] "rotating orthogonal calipers" method.

12.2.1 Measuring Rectangularity

There are a few shape descriptors that can be estimated from $\mathbf{R}_{\min}(S)$. For example, a standard approach to measure the rectangularity of a polygonal shape S is to compare S and $\mathbf{R}_{\min}(S)$. Of course, the shape S is said to be a perfectly rectangular shape (i.e., S is a rectangle) if and only if $S = \mathbf{R}_{\min}(S)$. Such a trivial observation suggests that rectangularity can be estimated by

$$\frac{\text{area}(S)}{\text{area}(\mathbf{R}_{\min}(S))}.$$

Also, the orientation of S can be defined by the orientation of $\mathbf{R}_{\min}(S)$, or more precisely, the orientation of S can be defined by the orientation of the longer edge of $\mathbf{R}_{\min}(S)$. Finally, the elongation of S can be derived from $\mathbf{R}_{\min}(S)$, where the elongation of S is estimated by the ratio of the lengths of the orthogonal edges of $\mathbf{R}_{\min}(S)$.

Analogous measures can be constructed using the minimum perimeter bounding rectangle instead of the minimum area bounding rectangle. Of course, in both cases where the bounding rectangles are used, a high sensitivity to boundary defects is expected.

12.2.2 Measuring Convexity

Curiously, the minimum area bounding rectangle can also be used to measure convexity [70]. Indeed, a trivial observation is that the total sum of projections of all the edges of a given shape S onto the coordinate axes is equal to the Euclidean perimeter of $\mathbf{R}(S)$, which will be denoted by $\mathcal{P}_2(\mathbf{R}(S))$. The sum of projections of all the edges of S onto coordinate axes can be written as $\mathcal{P}_1(S)$, where $\mathcal{P}_1(S)$ means the perimeter of S in the sense of l_1 distance (sometimes called the "city block distance"), and so we have

$$\mathcal{P}_1(S, \alpha) = \mathcal{P}_2(\mathbf{R}(S, \alpha)) \tag{12.1}$$

for every convex polygon S and all $\alpha \in [0, 2\pi)$ ($\mathcal{P}_1(S, \alpha)$ denotes the l_1 perimeter of S after rotation of an angle α).

The equality (12.1) could be satisfied for some nonconvex polygons as well (see Fig. 12.1), but a deeper observation (see the work by Žunić and Rosin [70]) shows that for any nonconvex polygonal shape S there is an angle α such that the strict inequality

$$\mathcal{P}_1(S, \alpha) > \mathcal{P}_2(\mathbf{R}(S, \alpha)) \tag{12.2}$$

holds.

Combining (12.1) and (12.2) the following theorem that gives a useful characterization of convex polygons can be derived.

Theorem 1 ([70]) *A polygon S is convex if and only if*

$$\mathcal{P}_1(S, \alpha) = \mathcal{P}_2(\mathbf{R}(S, \alpha))$$

holds for all $\alpha \in [0, 2\pi)$.

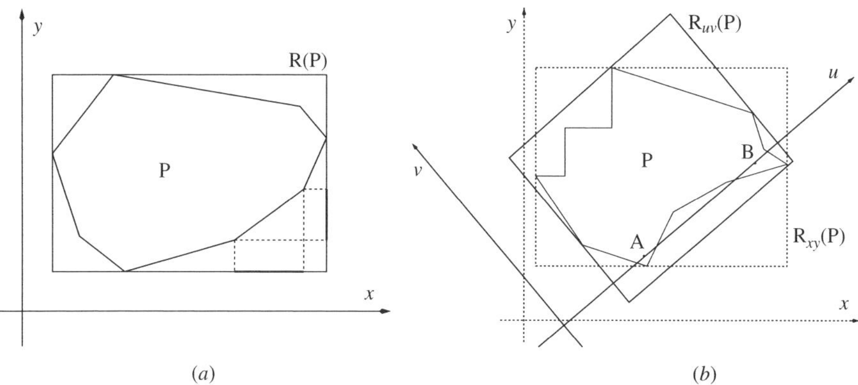

FIGURE 12.1 (*a*) Since S is convex, then $\mathcal{P}_1(S) = \mathcal{P}_2(\mathbf{R}(S))$. (*b*) If x and y are chosen to be the coordinate axes, then $\mathcal{P}_2(\mathbf{R}(S)) = \mathcal{P}_1(S)$. Since S is not convex, there is another choice of the coordinate axes, say u and v, such that the strict inequality $\mathcal{P}_2(\mathbf{R}(S)) < \mathcal{P}_1(S)$ holds.

Taking into account the previous discussion, inequality (12.2), and Theorem 1, the following convexity measure $\mathcal{C}(S)$ for a given polygonal shape S is very reasonable:

$$\mathcal{C}(S) = \min_{\alpha \in [0, 2\pi]} \frac{\mathcal{P}_2(\mathbf{R}(S, \alpha))}{\mathcal{P}_1(S, \alpha)}. \tag{12.3}$$

The convexity measure defined as above has several desirable properties:

- The estimated convexity is always a number from (0, 1].
- The estimated convexity is 1 if and only if the measured shape is convex.
- There are shapes whose estimated convexity is arbitrary close to 0.
- The new convexity measure is invariant under similarity transformations.

The minimum of the function $\mathcal{P}_2(\mathbf{R}(S, \alpha))/\mathcal{P}_1(S, \alpha)$ that is used to estimate the convexity of a given polygonal shape S cannot be given in a "closed" form. Also, it is obvious that the computation of $\mathcal{P}_2(\mathbf{R}(S, \alpha))/\mathcal{P}_1(S, \alpha)$ for a big enough number of uniformly distributed different values of $\alpha \in [0, 2\pi]$ would lead to an estimate of $\mathcal{C}(S)$ within an arbitrary required precision. But a result from the work by Žunić and Rosin [70] shows that there is a deterministic, very efficient algorithm that enables the exact computation of $\mathcal{C}(S)$. That is an advantage of the method. It turned out that it is enough to compute $\mathcal{P}_2(\mathbf{R}(S, \alpha))/\mathcal{P}_1(S, \alpha)$ for a number of $\mathcal{O}(n)$ different, precisely defined, values of α and take the minimum from the computed values (n denotes the number of vertices of S).

$\mathcal{C}(S)$ is a boundary-based convexity measure that implies a high sensitivity to the boundary defects. In the majority of computer vision tasks robustness (rather than sensitivity) is a preferred property, but in high precision tasks the priority has to be given to the sensitivity.

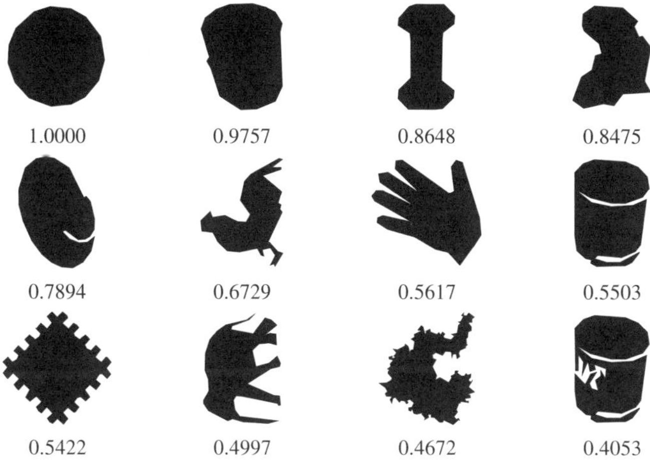

FIGURE 12.2 Shapes ranked by the $\mathcal{C}$ convexity measure.

Several shapes with their measured convexity values (the convexity measure $\mathcal{C}$ is used) are presented in Figure 12.2. Each shape S is rotated such that the function $\mathcal{P}_2(\mathbf{R}(S, \alpha))/\mathcal{P}_1(S, \alpha)$ reaches the minimum. The first shape (the first shape in the first row) is convex leading to the measured convexity equal to 1. Since the used measure $\mathcal{C}$ is boundary based, boundary defects are strongly penalized. For example, the first shape in the second row, the last shape in the second row, and the last shape in the third row all have measured convexity values that strongly depend on the intrusions. Also note that there are a variety of different shape convexity measures (e.g., [5,42,58]) including both boundary- and area-based ones.

12.2.3 Measuring Rectilinearity

In addition to the above, we give a sketch of two recently introduced shape descriptors with their measures that also use optimal (in a different sense) bounding rectangles. We start with rectilinearity. This shape measure has many possible applications such as shape partitioning, shape from contour, shape retrieval, object classification, image segmentation, skew correction, deprojection of aerial photographs, and scale selection (see the works by Rosin and Žunić [55,69]. Another application is the detection of buildings from satellite images. The assumption that extracted polygonal areas whose interior angles belong to $\{\pi/2, 3\pi/2\}$ very likely correspond to building footprints on satellite images seems to be reasonable. Consequently, a shape descriptor that would detect how much an extracted region differs from a polygonal area with interior angles belonging to $\{\pi/2, 3\pi/2\}$ could be helpful in detecting buildings on satellite images (see Fig. 12.3).

Thus, a shape with interior angles belonging to $\{\pi/2, 3\pi/2\}$ is named a "rectilinear shape," while a shape descriptor that measures the degree to which shape can be described as a rectilinear one is named "shape rectilinearity." It has turned out that

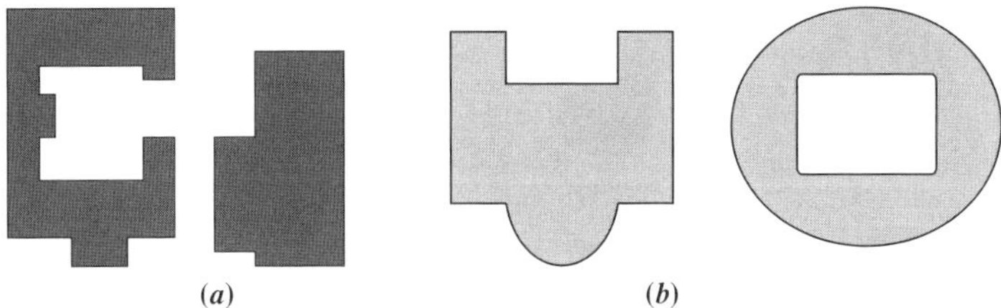

(a) (b)

FIGURE 12.3 *(a)* The presented rectilinear polygons correspond to building footprints. *(b)* The presented (nonpolygonal) shapes correspond to building footprints but they are not rectilinear polygons.

the following two quantities

$$\mathcal{R}_1(S) = \frac{4}{4-\pi} \cdot \left(\max_{\alpha \in [0,2\pi)} \frac{\mathcal{P}_2(S)}{\mathcal{P}_1(S,\alpha)} - \frac{\pi}{4} \right) \tag{12.4}$$

$$\mathcal{R}_2(S) = \frac{\pi}{\pi - 2\sqrt{2}} \cdot \left(\max_{\alpha \in [0,2\pi)} \frac{\mathcal{P}_1(S,\alpha)}{\sqrt{2}\mathcal{P}_2(S,\alpha)} - \frac{2\sqrt{2}}{\pi} \right) \tag{12.5}$$

are appropriate to be used as rectilinearity measures. For a detailed explanation see the work by Žunić and Rosin [69]. It is obvious that both $\mathcal{R}_1$ and $\mathcal{R}_2$ are boundary-based shape descriptors. An area-based rectilinear descriptor is not defined yet. A reasonably good area-based rectilinearity measure would be very useful as a building detection tool when working with low quality images.

The following desirable properties of rectilinearity measures $\mathcal{R}_1$ and $\mathcal{R}_2$ hold (for details see the works by Rosin and Žunić [55,69]):

- Measured rectilinearity values are numbers from $(0, 1]$.
- A polygon S has a measured rectilinearity equal to 1 if and only if S is rectilinear.
- For each $\varepsilon > 0$, there is a polygon whose measured rectilinearity belongs to $(0, \varepsilon)$.
- Measured rectilinearities are invariant under similarity transformations.

Although $\mathcal{R}_1$ and $\mathcal{R}_2$ are derived from the same source and give similar results, they are indeed different and they could lead to different shape ranking (with respect to the measured rectilinearity). For an illustration see Figure 12.4; shapes presented in Figure 12.4a are ranked with respect to $\mathcal{R}_1$ while the shapes presented in Figure 12.4b are ranked with respect to $\mathcal{R}_2$.

12.2.4 Measuring Orientability

To close this section on related shape measures based on bounding rectangles, we discuss "shape orientability" as a shape descriptor that should indicate the degree to

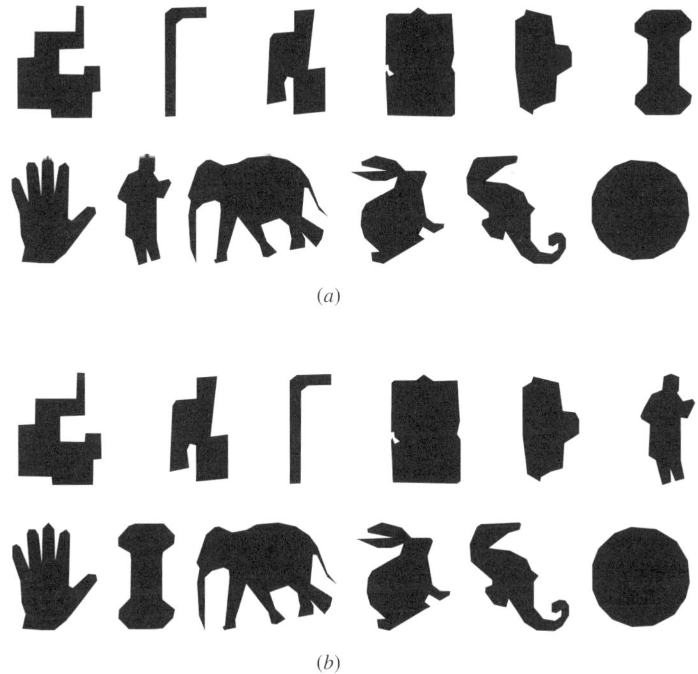

FIGURE 12.4 Shapes ranked by rectilinearity measures (*a*) $\mathcal{R}_1$ and (*b*) $\mathcal{R}_2$.

which a shape has a distinct (but not necessarily unique) orientation. This topic was recently investigated by the authors [71]. The definition of the orientability measure uses two optimal bounding rectangles. One of them is the minimum area rectangle $\mathbf{R}_{\min}(S)$ that inscribes the measured shape S while another is the rectangle $\mathbf{R}_{\max}(S)$ that maximizes area($\mathbf{R}(S, \alpha)$). A modification of Toussaint's [59] rotating orthogonal calipers method can be used for an efficient computation of $\mathbf{R}_{\max}(S)$. The orientability $\mathcal{D}(S)$ of a given shape S is defined as

$$\mathcal{D}(S) = 1 - \frac{\mathbf{R}_{\min}(S)}{\mathbf{R}_{\max}(S)}. \tag{12.6}$$

Defined as above, the shape orientability has the following desirable properties:

- $\mathcal{D}(S) \in [0, 1)$ for any shape S.
- A circle has measured orientability equal to 0.
- No polygonal shape has measured orientability equal to 0.
- The measured orientability is invariant with respect to similarity transformations.

Since both $\mathbf{R}_{\min}(S)$ and $\mathbf{R}_{\max}(S)$ are easily computable, it follows that the shape orientability of a given polygonal shape S is also easy to compute. For more details we refer to the work by Žunić et al. [71].

FIGURE 12.5 Trademarks ranked by orientability using $\mathcal{D}(S)$. The bounding rectangles $\mathbf{R}_{\min}(S)$ and $\mathbf{R}_{\max}(S)$ are displayed for each measured shape S.

Note that a trivial approach could be to measure shape orientability by the degree of elongation of the considered shape. Indeed, it seems reasonable to expect that the more elongated a shape, the more distinct its orientation. But if such an approach is used then problems arise with many-fold symmetric shapes, as described later in Sections 12.5.1 and 12.5.2. However, measuring shape orientability by the new measure $\mathcal{D}(S)$ is possible in the case of such many-fold symmetric shapes, as demonstrated in Figure 12.5. This figure gives several trademark examples whose orientability is computed by $\mathcal{D}(S)$. As expected, elongated shapes are considered to be the most orientable. Note, however, that the measure $\mathcal{D}(S)$ is also capable of distinguishing different degrees of orientability for several symmetric shapes that have similar compactness, such as the first and last examples in the top row.

12.3 FURTHER BOUNDING SHAPES

The approach taken to measure rectangularity (Section 12.2.1) can readily also be applied to other shape measures, as long as the bounding geometric primitive can be computed reasonably efficiently. However, in some cases it is not appropriate; for instance, sigmoidality (see Section 12.4) is determined more by the shape of its medial axis than its outline, while other measures such as complexity [40] or elongation (see Section 12.5.2) are not defined with respect to any geometric primitive.

A simple and common use of such a method is to measure convexity. If we denote the convex hull of polygon S by $\mathbf{CH}(S)$, then the standard convexity measure is defined as

$$\mathcal{C}_1(S) = \frac{\text{area}(S)}{\text{area}(\mathbf{CH}(S))}.$$

The computation time of the convex hull of a simple polygon is linear in the number of its vertices [36] and so the overall computational complexity of the measure is linear.

A perimeter-based version can be used in place of the area-based measure:

$$\mathcal{C}_2 = \frac{\mathcal{P}_2(\mathbf{CH}(S))}{\mathcal{P}_2(S)}.$$

It was straightforward to apply the same approach to compute triangularity [51]. Moreover, since linear time (w.r.t. number of polygon vertices) algorithms are available to determine the minimum area bounding triangle [37,39], this measure could be computed efficiently. Many other similar measures are possible, and we note that there are also linear time algorithms available to find bounding circles [18] and bounding ellipses [19] that can be used for estimating circularity and ellipticity.

A more rigorous test of shape is, given a realization of an ideal shape, to consider fluctuations in both directions, that is, intrusions and protrusions. Thus, in the field of metrology there is an ANSII standard for measuring roundness, which requires finding the minimum width annulus to the data. This involves determining the inscribing and circumscribing circles that have a common center and minimize the difference in their radii. Although the exact solution is computationally expensive, Chan [8] presented an $O(n + \epsilon^{-2})$ algorithm to find an approximate solution that is within a $(1 + \epsilon)$-factor of optimality, where the polygon contains n vertices and $\epsilon > 0$ is an input parameter. We note that, in general, inscribed shapes are more computationally expensive to compute than their equivalent circumscribing versions (even when the two are fitted independently). For instance, the best current algorithm for determining the maximum area empty (i.e., inscribed) rectangle takes $O(n^3)$ time [10] compared to the linear time algorithm for the minimum area bounding rectangle. Even more extreme is the convex skull algorithm; the optimal algorithm runs in $O(n^7)$ time [9] compared again to a linear time algorithm for the convex hull.

12.4 FITTING SHAPES

An obvious scheme for a general class of shape measures is to fit a shape model to the data and use the goodness of fit as the desired shape measure. There is of course great scope in terms of which fitting procedure is performed, which error measure is used, and the choice of the normalization of the error of fit.

12.4.1 Ellipse Fitting

For instance, to fit ellipses, Rosin [48] used the least median of squares (LMedS) approach that is robust to outliers and enables the ellipse to be fitted reliably even in the presence of outliers. The LMedS enables outliers to be rejected, and then a more accurate (and ellipse-specific) least squares fit to the inliers was found [15]. Calculating the shortest distance from each data point to the ellipse requires solving a quartic equation, and so the distances were approximated using the orthogonal conic distance approximation method [47]. The average approximated error over the data E was combined with the region's area A to make the ellipticity measure scale invariant [51]:

$$\left(1 + \frac{E}{\sqrt{A}}\right)^{-1}.$$

12.4.2 Triangle Fitting

For fitting triangles, a different approach was taken. The optimal three-line polygonal approximation that minimized the total absolute error to the polygon was found using dynamic programming. The average error was then normalized as above to give a triangularity measure [51].

12.4.3 Rectangle Fitting

An alternative approach to measure rectangularity [51] from the one introduced in Section 12.2 is to iteratively fit a rectangle R to S by maximizing the functional

$$1 - \frac{\text{area}(R \setminus S) + \text{area}(S \setminus R)}{\text{area}(S \cap R)} \tag{12.7}$$

based on the two set differences between R and S normalized by the union of R and S. This provides a trade-off between forcing the rectangle to contain most of the data while keeping the rectangle as small as possible, as demonstrated in Figure 12.6. Each iteration can be performed in $O(n \log n)$ time [12], where n is the number of vertices.

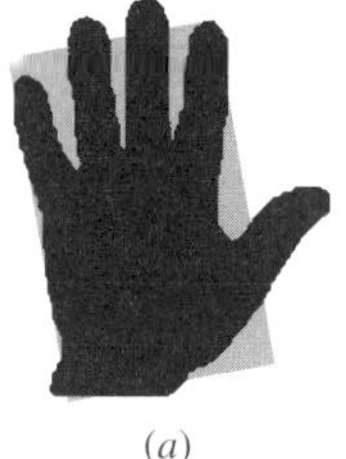

(*a*)

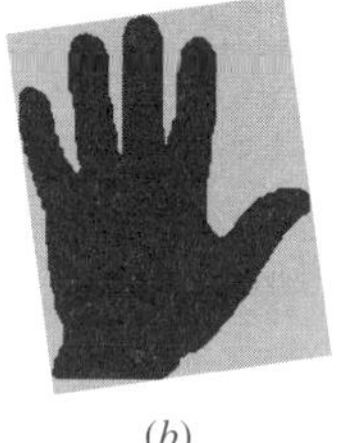

(*b*)

FIGURE 12.6 The rectangle shown in (*a*) was fitted according to (12.7) as compared to the minimum bounding rectangle shown in (*b*).

12.4.4 Sigmoid Fitting

To measure sigmoidality (i.e., how much a region is S-shaped), several methods were developed that analyze a single centerline curve that was extracted from the region by smoothing the region until the skeleton (obtained by any thinning algorithm) is nonbranching. The centerline is then rotated so that its principal axis lies along the x-axis. Fischer and Bunke [14] fitted a cubic polynomial $y = ax^3 + bx^2 + cx + d$ and classified the shape into linear, C-shaped, and sigmoid classes based on the coefficient values. A modified version specifically designed to produce only a sigmoidality measure [52] fitted the symmetric curve given by $y = ax^3 + bx + c$. The correlation coefficient ρ was used to measure the quality of fit between the data and the sampled model. Inverse correlation was not expected, and so the value was truncated at zero.

Rather than fit models directly to the coordinates, other derived data can be used instead. The following approach to compute sigmoidality used the tangent angle that was modeled by a generalized Gaussian distribution [52] (see Fig. 12.7). The probability density function is given by

$$p(x) = \frac{v\eta(v,\sigma)}{2\Gamma(1/v)} e^{-[\eta(v,\sigma)|x|]^v},$$

where $\Gamma(x)$ is the gamma function, σ is the standard deviation, v is a shape parameter controlling the peakiness of the distribution (values $v = 1$ and $v = 2$ correspond to Laplacian and Gaussian densities), and the following is a scaling function:

$$\eta(v,\sigma) = \frac{1}{\sigma}\sqrt{\frac{\Gamma(3/v)}{\Gamma(1/v)}}.$$

Mallat's method [34] for estimating the parameters was employed. First, the mean absolute value and variance of the data x_i are matched to the generalized Gaussian.

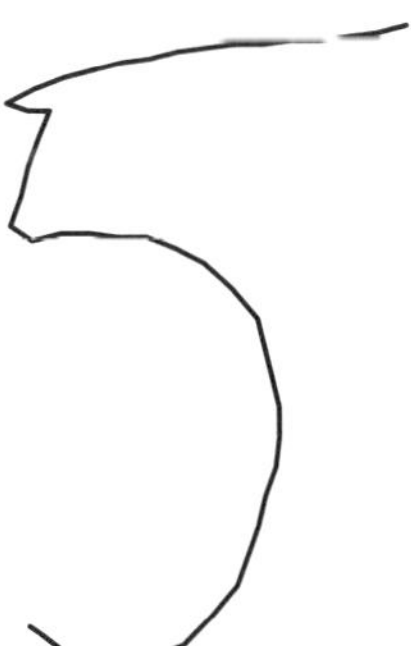

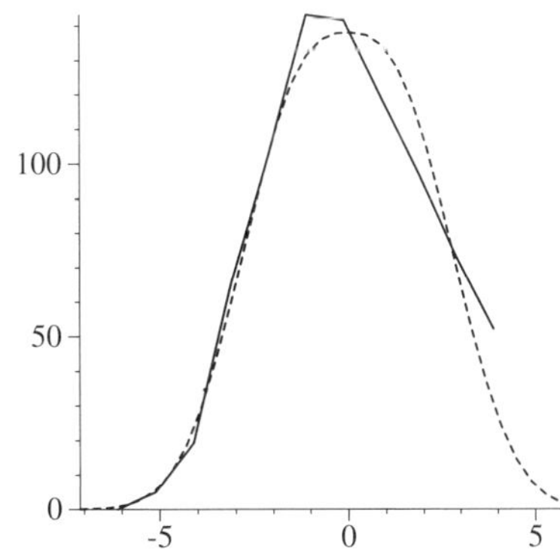

FIGURE 12.7 The tangent angle of the handwritten digit "5" is overlaid with the best fit generalized Gaussian (dashed) — the good fit yields a high sigmoidality measure.

If $m_1 = (1/n)\sum_{i=1}^{n} |x_i|$ and $m_2 = (1/n)\sum_{i=1}^{n} x_i^2$, then

$$v = F^{-1}\left(\frac{m_1}{\sqrt{m_2}}\right),$$

where

$$F(\alpha) = \frac{\Gamma(2/\alpha)}{\sqrt{\Gamma(1/\alpha)\Gamma(3/\alpha)}}.$$

In practice, the values of $F(\alpha)$ are precomputed, and the inverse function is determined by a lookup table with linear interpolation. Finally, the tangent angle is scaled so that the area under the curve sums to 1. It was found that rather than calculating the measure as the correlation coefficient, better results were obtained by taking the area of intersection A of the curves as an indication of the error of fit. An approximate normalization was found by experimentation as $\max(2A - 1, 0)$.

12.4.5 Using Circle and Ellipse Fits

Koprnicky et al. [27] fitted two model shapes M (a circle and ellipse) to the data S and for each considered four different error measures: the outer difference

$$\frac{\text{area}(S \cap \overline{M})}{\text{area}(S)},$$

the inner difference

$$\frac{\text{area}(\overline{S} \cap M)}{\text{area}(S)},$$

as well as the sum and difference of the above. This provided four different measures, from which the first three can be considered as circularity and ellipticity measures, focusing on the different aspects of the errors.

12.5 MOMENTS

Moments are widely used in shape analysis tasks. Shape normalization, shape encoding (characterization), shape matching, and shape identification are just some examples where moments techniques are successfully applied. To be precise, by "shape moments" we mean geometric moments. The geometric moment $m_{p,q}(S)$ of a given planar shape S is defined by

$$m_{p,q}(S) = \iint_S x^p y^q \, dx \, dy.$$

In real image processing applications, we are working with discrete data resulting from a particular digitization process applied to real data. In the most typical situation, real objects are replaced with a set of pixels whose centers belong to the considered shape. In such a case, the exact computation of geometric moments is not possible and each used moment $m_{p,q}(S)$ is usually replaced with its discrete analog $\mu_{p,q}(S)$, which is defined as

$$\mu_{p,q}(S) = \sum_{(i,j)\in S\cap \mathbf{Z}^2} i^p j^q,$$

where **Z** means the set of integers. The order of $m_{p,q}(S)$ is said to be $p+q$. Note that the zeroth-order moment $m_{0,0}(S)$ of a shape S coincides with the area of S.

12.5.1 Shape Normalization: Gravity Center and Orientation

Shape normalization is usually an initial step in image analysis tasks or a part of data preprocessing. It is important to provide an efficient normalization because a significant error in this early stage of image analysis would lead to a large cumulative error at the end of processing.

Shape normalization starts with the computation of image position. A common approach is that the shape position is determined by its gravity center (i.e., center of mass or, simply, centroid) of a given shape. Formally, for a given planar shape S its gravity center $(x_c(S), y_c(S))$ is defined as a function of the shape area (i.e., the zeroth-order moment of the shape) and the first-order moments

$$(x_c(S), y_c(S)) = \left(\frac{m_{1,0}(S)}{m_{0,0}(S)}, \frac{m_{0,1}(S)}{m_{0,0}(S)}\right). \tag{12.8}$$

Computation of shape orientation is another step in the shape normalization procedure, which is computed using moments. The orientation seems to be a very natural feature for many shapes, although obviously there are some shapes that do not have a distinct orientation. Many rotationally symmetric shapes are shapes that do not have a unique orientation while the circular disk is a shape that does not have any specific orientation at all. The standard approach defines the shape orientation by a line that minimizes the integral of the squared distances of points (belonging to the shape) to this line. Such a line is also known as the "axis of the least second moment of inertia." If $r(x, y, \delta, \rho)$ denotes the perpendicular distance from the point (x, y) to the line given in the form

$$x\cos\delta - y\sin\delta = \rho,$$

then the integral that should be minimized is

$$I(\delta, \rho, S) = \iint_S r^2(x, z, \delta, \rho)\, dx\, dy.$$

Elementary mathematics shows that the line that minimizes $I(\delta, \rho, S)$ passes through the centroid $(x_c(S), y_c(S))$ of S and consequently we can set $\rho = 0$. Thus, the problem of orientation of a given shape S is transformed to the problem of computing the angle δ for which the integral

$$I(\delta, S) = \iint_S (-x \sin\delta + y \cos\delta)^2 dx\, dy \tag{12.9}$$

reaches the minimum. Finally, if we introduce central geometric moments $\overline{m}_{p,q}(S)$ defined as usual

$$\overline{m}_{p,q}(S) \iint_S (x - x_c(S))^2 (y - y_c(S))^2 dx\, dy,$$

then the function $I(\delta, S)$ can be written as

$$I(\delta, S) = \overline{m}_{2,0}(S)(\sin\delta)^2 - 2\overline{m}_{1,1}(S)\sin\delta\cos\delta + \overline{m}_{0,2}(S)(\cos\delta)^2, \tag{12.10}$$

that is, as a polynomial in $\cos\delta$ and $\sin\delta$ whose coefficients are the second-order moments of S. The angle δ for which $I(\delta, S)$ reaches its maximum defines the orientation of S. Such an angle δ is easy to compute and it can be derived that the required δ satisfies the equation

$$\frac{\sin(2\delta)}{\cos(2\delta)} = \frac{2\overline{m}_{1,1}(S)}{\overline{m}_{2,0}(S) - \overline{m}_{0,2}(S)}. \tag{12.11}$$

It is worth mentioning that if working in discrete space, that is, if continuous shapes are replaced with their digitizations, then real moments have to be replaced with their discrete analogs. For example, the orientation of discrete shape that is the result of digitization of S is defined as a solution of the following optimization problem:

$$\min_{\delta\in[0,2\pi)} \left\{ \sum_{(i,j)\in S\cap\mathbf{Z}^2} (i \sin\delta - j\cos\delta)^2 \right\}.$$

The angle δ that is a solution of the above problem satisfies the equation

$$\frac{\sin(2\delta)}{\cos(2\delta)} = \frac{2\overline{\mu}_{1,1}(S)}{\overline{\mu}_{2,0}(S) - \overline{\mu}_{0,2}(S)},$$

which is an analog to (12.11).

So, the shape orientation defined by the axis of the least second moment of inertia is well motivated and easy to compute in both continuous and discrete versions. As expected, there are some situations when the method does not give any answer as to what the shape orientation should be. Such situations, where the standard method cannot be applied, are characterized by

$$I(\delta, S) = \text{constant}. \tag{12.12}$$

There are many regular and irregular shapes that satisfy (12.12). The result from the work by Tsai and Chou [60] says that (12.12) holds for all N-fold rotationally symmetric shapes with $N > 2$, where N-fold rotationally symmetric shapes are such shapes that are identical to themselves after being rotated through any multiple of $2\pi/N$.

In order to expand the class of shapes with a computable orientation, Tsai and Chou [60] suggested a use of the so-called Nth order central moments $I_N(\delta, S)$. For a discrete shape S those moments are defined by

$$I_N(\delta, S) = \sum_{(x,y)\in S} (-x \sin \delta + y \cos \delta)^N \tag{12.13}$$

assuming that the centroid of S is coincident with the origin.

Now, the shape orientation is defined by the angle δ for which $I_N(\delta, S)$ reaches the minimum. For $N = 2$, we have the standard method. Note that $I_N(\delta, S)$ is a polynomial in $\cos \delta$ and $\sin \delta$ while polynomial coefficients are central moments of S having the order less than or equal to N.

A benefit from this redefined shape orientation is that the method can be applied to a wider class of shapes. For example, since a square is a fourfold rotationally symmetric shape, the standard method does not work. If $I_4(\delta, S)$ is used, then the square can be oriented. A disadvantage is that there is not a closed formula (as (12.11)) that gives δ for which $I_N(\delta, S)$ reaches the minimum for an arbitrary shape S. Thus, a numerical computation has to be applied in order to compute shape orientation in the modified sense.

Figure 12.8 displays some shapes whose orientation is computed by applying the standard method ($N = 2$) and by applying the modified method with $N = 4$ and $N = 8$. Shapes **(1)**, **(2)**, and **(3)** are not symmetric, but they have a very distinct orientation. Because of that all three measured orientations are almost identical. Shapes **(4)**, **(5)**, and **(6)** have exactly one axis of symmetry and consequently their orientation is well determined. That is the reason why all three computed orientations coincide. The small variation in the case of the bull sketch (shape **(5)**) is caused by the fact that the sketch contains a relatively small number of (black) pixels, and consequently the digitization error has a large influence. Shapes **(7)**, **(8)**, **(9)**, and **(10)** do not have a distinct orientation. That explains the variation in the computed orientations. For shapes **(11)** and **(12)**, the standard method does not work. The presented regular triangle is a threefold rotationally symmetric shape and its orientation cannot be computed for $N = 4$, as well. For $N = 8$, the computed orientation is 150°, which is very reasonable. This is the direction of one of the symmetry axes. Of course, the modified method (in the case of $N = 8$) gives the angles $\delta = 270°$ and $\delta = 30°$ as the minimum of the function $I_8(\delta, S)$ and those angles can also be taken as the orientation of the

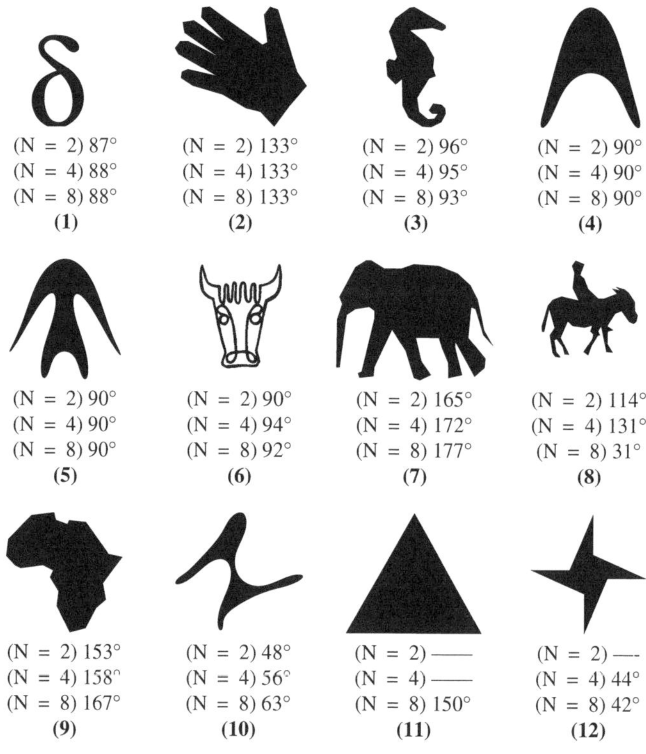

FIGURE 12.8 Computed orientation of the presented shapes for $N = 2$, $N = 4$, and $N = 8$ are given (in degrees).

presented triangle. The last shape is a fourfold rotationally symmetric shape whose orientation cannot be computed by the standard method.

12.5.2 Shape Elongation

Shape elongation is another shape descriptor with a clear intuitive meaning. A commonly used measure of elongatedness uses the central moments and is computed as the ratio of the maximum of $I(\delta, S)$ and the minimum of $I(\delta, S)$; that is, shape elongation is measured as [38]

$$\frac{\overline{\mu}_{20}(S) + \overline{\mu}_{02}(S) + \sqrt{(\overline{\mu}_{20}(S) - \overline{\mu}_{02}(S))^2 + 4\overline{\mu}_{11}(S)^2}}{\overline{\mu}_{20}(S) + \overline{\mu}_{02}(S) - \sqrt{(\overline{\mu}_{20}(S) - \overline{\mu}_{02}(S))^2 + 4\overline{\mu}_{11}(S)^2}}, \tag{12.14}$$

which can be simplified and reformulated as

$$\frac{\sqrt{(\overline{\mu}_{20}(S) - \overline{\mu}_{02}(S))^2 + 4\overline{\mu}_{11}(S)^2}}{\overline{\mu}_{20}(S) + \overline{\mu}_{02}(S)}$$

to provide a measure in the range [0, 1].

Similarly as in the previous subsection some problems arise when working with shapes satisfying $I(\delta, S) = \text{constant}$. All such shapes have the same measured elongation equal to 1. Rather, it is more reasonable that all the regular $2n$-gons have the same measured elongation. It seems natural that the elongation of regular $2n$-gons decreases if n increases. Partially, the problem can be avoided if higher order moments of inertia are used. A possibility (see the work by Žunić et al. [68]) is to define the elongation of a given shape S as

$$\frac{\max\{I_N(\delta, S) \mid \delta \in [0, 2\pi)\}}{\min\{I_N(\delta, S) \mid \delta \in [0, 2\pi)\}}. \tag{12.15}$$

Again, an advantage of the modified definition of the shape orientation is that a smaller class of shapes would have the measured elongation equal to 1. Such a minimum possible measured elongation should be reserved for the circular disk only. On the contrary, for $N > 2$ there is no closed formula (like (12.14)) that can be used for immediate computation of the shape elongation. More expensive numerical algorithms have to be applied. For more details about elongation of many-fold rotationally symmetric shapes see the work by Žunić et al. [68].

12.5.3 Other Shape Measures

A simple scheme for measuring rectangularity [49] considers the moments of a rectangle (dimensions $a \times b$) centered at the origin and aligned with the axes. The moments are $m_{00} = ab$ and $m_{22} = a^3b^3/144$, and so thc quantity

$$R = 144 \times \frac{m_{22}}{m_{00}^3}$$

is invariant for rectangles of variable aspect ratio and scaling, and can be normalized as

$$R_M = \begin{cases} R & \text{if } R \leq 1, \\ \dfrac{1}{R} & \text{otherwise.} \end{cases}$$

To add invariance to rotation and translation, the data are first normalized in the standard way by moving its centroid to the origin and orienting its principal axis to lie along the X-axis.

A straightforward scheme to measure similarity to shapes such as triangles and ellipses that do not change their category of shape under affine transformations is to use affine moment invariants [51]. The simplest version is to characterize shape using just the first, lowest order affine moment invariant [16]

$$I_1 = \frac{\overline{m}_{20}\overline{m}_{02} - \overline{m}_{11}^2}{\overline{m}_{00}^4}.$$

This has the advantage that it is less sensitive to noise than the higher order moments. The moments for the unit radius circle are

$$\mu_{pq} = \int_{-1}^{1} \int_{-\sqrt{r^2-x^2}}^{\sqrt{r^2-x^2}} x^p y^q \, dy \, dx$$

leading to the value of its invariant as $I_1 = 1/16\pi^2$. When normalized appropriately, this then provides a measure of ellipticity

$$E_I = \begin{cases} 16\pi^2 I_1 & \text{if} \quad I_1 \leq \dfrac{1}{16\pi^2}, \\ \dfrac{1}{16\pi^2 I_1} & \text{otherwise,} \end{cases}$$

which ranges over [0, 1], peaking at 1 for a perfect ellipse. The same approach was applied to triangles, all of which have the value $I_1 = 1/108$, and the triangularity measure is

$$T_I = \begin{cases} 108 I_1 & \text{if} \quad I_1 \leq \dfrac{1}{108}, \\ \dfrac{1}{108 I_1} & \text{otherwise.} \end{cases}.$$

Of course, using a single moment invariant is not very specific, and so the above two measures will sometimes incorrectly assign high ellipticity or triangularity values to some other nonelliptical or triangular shapes. This can be remedied using more moment values, either in the above framework, or as described next.

Voss and Süße describe a method for fitting geometric primitives by the method of moments [63]. The data are normalized into a (if possible unique) canonical frame, which is generally defined as the simplest instance of each primitive type, by applying an affine transformation. Applying the inverse transformation to the primitive produces the fitted primitive. For example, for an ellipse they take the unit circle as the canonical form, and the circle in the canonical frame is transformed back to an ellipse, thereby providing an ellipse fit to the data. For the purposes of generating shape measures, the inverse transformation is not necessary as the measures can be calculated in the canonical frame. This is done by computing the differences between the normalized moments of the data (m'_{ij}) and the moments of the canonical primitive (m_{ij}) where only the moments not used to determine the normalization are included:

$$\left(1 + \sum_{i+j\leq 4} (m'_{ij} - m_{ij})^2\right)^{-1}.$$

The above approach method was applied in this manner by Rosin [51] to generate measures of ellipticity and triangularity. Measuring rectangularity can be done in the same way, except that for fitting rectangles the procedure is modified to apply a similarity transformation rather than an affine transformation. After this transformation the rectangle's aspect ratio remains to be determined, and this is done by a one-dimensional optimization using the higher order moments (up to fourth order).

We note that the above methods can all compute the moments either from the polygon boundary directly (line moments) [57] or else can operate on the rasterized set of pixels inside the polygon (region) [33].

12.6 FOURIER DESCRIPTORS

Like moments, Fourier descriptors are a standard means of representing shape. This involves taking a Fourier expansion of the boundary function, which itself may be described in a variety of ways. If the boundary of the region is given by the points (x_j, y_j), $j = 1, \ldots, N$, then one approach is to represent the coordinates by complex numbers $z_j = x_j + iy_j$ [21]. Other possibilities are to represent the boundary by real 1D functions versus arc length such as tangent angle [66] or radius from the centroid.

Taking the representation $z_j = x_j + iy_j$ and applying the discrete Fourier transform leads to the complex coefficients that make up the descriptors

$$F_k = a_k + ib_k = \frac{1}{N} \sum_{m=0}^{N-1} z_m \exp(-i2\pi mk/N).$$

Often just the magnitude is used $r_k = \sqrt{a_k^2 + b_k^2}$, and since r_1 indicates the size of the region it can be used to make the descriptors scale invariant: $w_k = r_k / r_1$.

For a study of sand particles, Bowman et al. [4] used individual Fourier descriptors to describe specific aspects of shape, for example, w_{-3}, w_{-2}, w_{-1}, and w_{+1} to measure, respectively, squareness, triangularity, elongation, and asymmetry. However, this approach is rather crude. A modification [53] to make the measure more specific includes the relevant harmonics and also takes into account the remaining harmonics that do not contribute to squareness:

$$(w_{-3} + w_{-7} + w_{-11} + \cdots) / \sum_{\forall i \notin \{-1,0,1\}} w_i.$$

Kakarala [25] uses the same boundary representation and derives the following expression for the Fourier expansion of the boundary curvature:

$$K_n = \frac{1}{2} \sum_{m=-N}^{N} m \left[(m+n)^2 \bar{F}_m F_{m+n} + (m-n)^2 F_m \bar{F}_{m-n} \right],$$

where $\overline{F}$ is the complex conjugate of F.

He shows that for a convex contour

$$K_0 \geq 2 \sum_{n=1}^{2N} |K_n|$$

from which the following convexity shape measure is derived:

$$\frac{K_0 - 2\sum_{n=1}^{2N} |K_n|}{\sum_{n=-2N}^{2N} |K_n|}.$$

Another measure based on curvature is "bending energy," which considers the analog of the amount of energy required to deform a physical rod [65]. If a circle (which has minimum bending energy) is considered to be the simplest possible shape, then bending energy can be interpreted as a measure of shape complexity or deviation from circularity.

The normalized energy is the summed squared curvature values along the boundary, which can be expressed in the Fourier domain as

$$\sum_{m=-N}^{N} \left(\frac{2\pi m}{N}\right)^4 \left(|a_m|^2 + |b_m|^2\right)$$

although in practice the authors performed the computation in the spatial domain.

When the boundary is represented instead by the radius function, a "roughness coefficient" can be defined as

$$\sqrt{\frac{1}{2} \sum_{n=1}^{[(N+1)/2]-1} \left(a_n^2 + b_n^2\right)}.$$

This shape measure is effectively the mean squared deviation of the radius function from a circle of equal area [26].

12.7 CONCLUSIONS

This chapter has described several approaches for computing shape measures and has showed how each of these can then be applied to generate a variety of specific shape measures such as convexity, rectangularity, and so on. Figure 12.9 illustrates some of the geometric primitives that have been inscribed, circumscribed, or otherwise fitted to example data, and which are then used to generate shape measures.

Our survey is not complete, as there exist some methodologies in the literature that we have not covered. Here, for instance, *Information Theory* has been used to measure convexity [41] and complexity [17,40,44]. *Projections* are a common tool in image processing, and in the context of the Radon transform have also been used

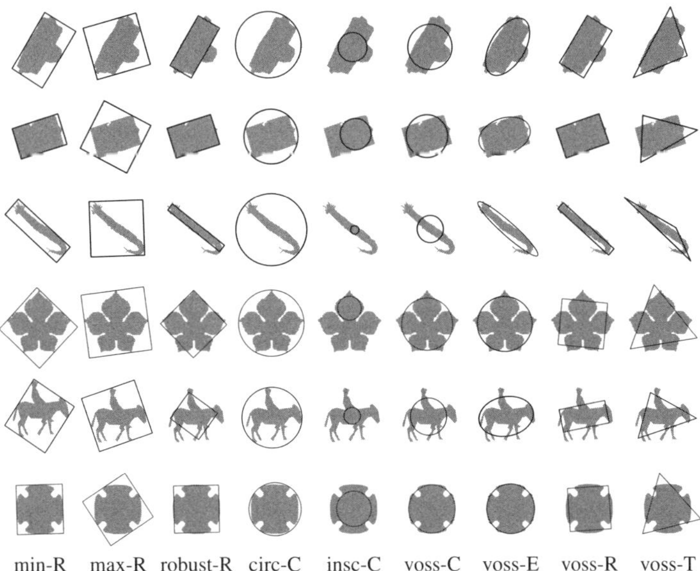

FIGURE 12.9 Geometric primitives fitted to shapes. min-R: minimum area rectangle; max-R: maximum area rectangle; robust-R: best fit rectangle — equation (12.7); circ-C: circumscribing circle; insc-C: inscribed circle; voss-C, voss-E, voss-R, voss-T: circle, ellipse, rectangle, and triangle fitted by Voss and Süße's moment-based method [63]. These primitives are used to generate some of the shape measures described in this chapter.

to compute convexity, elongation, and angularity shape measures [30]; a measure of triangularity was also based on projections [51]. Only a brief mention has been made to the issues of digitization, but it is important to note that this can have a significant effect. For instance, the popular compactness measure $\mathcal{P}_2(S)^2/\text{area}(S)$ in the continuous domain is minimized by a circle but this is not true when working with digital data [45]. Therefore, some measures explicitly consider the digitization process, for example, for convexity [46], digital compactness [3,7], and other shape measures [20].

Given these methodologies, it should be reasonably straightforward for the reader to construct new shape measures as necessary. For instance, consider an application requiring a "pentagonality" measure, that is, the similarity of a polygon to a regular pentagon. Considering the various methods discussed in this chapter, several seem to be readily applicable:

- A measure could be generated from the polygon's bounding pentagon; see Section 12.3.
- Once a pentagon is fitted to the polygon's coordinates, various shape measures can be produced; see Section 12.4.

- Rather than directly processing the polygon's coordinates the histogram of boundary tangents could be used instead, and it would be straightforward to fit five regular peaks and then compute a shape measure from the error of fit; see again Section 12.4.
- The two methods for generating shape measures from moments by Voss and Süße [63] and Rosin [51] could readily be applied; see Section 12.5.3.
- The Fourier descriptor method for calculating triangularity in Section 12.6 could also be readily adapted to computing pentagonality.

The natural question is, which is the best shape measure? While measures can be rated in terms of their computational efficiency, sensitivity to noise, invariance to transformations, and robustness to occlusion, ultimately their effectiveness depends on their application. For example, whereas for one application reliability in the presence of noise may be vital, for another sensitivity to subtle variations in shape may be more important. It should also be noted that, while there are many possible shape measures already available in the literature, and many more that can be designed, they are not all independent. Some analysis on this topic was carried out by Hentschel and Page [23] who computed the correlations between many similar measures as well as determined the most effective one for the specific task of powder particle analysis.

REFERENCES

1. Biederman I, Ju G. Surface versus edge-based determinants of visual recognition. Cogn Psychol 1988;20:38–64.
2. Bober M. MPEG-7 visual shape descriptors. IEEE Trans Circuits Syst Video Technol 2001;11(6):716–719.
3. Bogaert J, Rousseau R, Van Hecke P, Impens I. Alternative area–perimeter ratios for measurement of 2D-shape compactness of habitats. Appl Math Comput 2000;111:71–85.
4. Bowman ET, Soga K, Drummond T. Particle shape characterization using Fourier analysis. Geotechnique 2001;51(6):545–554.
5. Boxer L. Computing deviations from convexity in polygons. Pattern Recog Lett 1993;14:163–167.
6. Brady M, Yuille AL. An extremum principle for shape from contour. IEEE Trans Pattern Anal Mach Intell 1984;6(3):288–301.
7. Bribiesca E. Measuring 2D shape compactness usng the contacct perimeter. Pattern Recog 1997;33(11):1–9.
8. Chan TM. Approximating the diameter, width, smallest enclosing cylinder, and minimum-width annulus. Int J Comput Geom Appl 2002;12(1–2):67–85.
9. Chang JS, Yap CK. A polynomial solution for the potato-peeling problem. Discrete Comput Geom 1986;1:155–182.
10. Chaudhuri J, Nandy SC, Das S. Largest empty rectangle among a point set. J Algorithms 2003;46(1):54–78.

11. Cremers D, Tischhäuser F, Weickert J, Schnörr C. Diffusion snakes: introducing statistical shape knowledge into the Mumford–Shah functional. Int J Comput Vision 2002;50(3):295–313.

12. de Berg M, van Kreveld M, Overmars M, Schwarzkopf O. Computational Geometry: Algorithms and Applications. 2nd ed. Springer-Verlag; 2000.

13. Flickner M, Sawhney H, Niblack W, Ashley J, Huang Q, Dom B, Gorkani M, Hafner J, Lee D, Petkovic D, Steele D, Yanker P. Image and video content: the QBIC system. IEEE Comput 1995;28(9):23–32.

14. Fischer S, Bunke H. Identification using classical and new features in combination with decision tree ensembles. In: du Buf JMH, Bayer MM. editors. Automatic Diatom Identification. World Scientific; 2002. p 109–140.

15. Fitzgibbon AW, Pilu M, Fisher RB. Direct least square fitting of ellipses. IEEE Trans Pattern Anal Mach Intell 1999;21(5):476–480.

16. Flusser J, Suk T. Pattern recognition by affine moment invariants. Pattern Recog 1993;26:167–174.

17. Franco P, Ogier J.-M, Loonis P, Mullot R. A topological measure for image object recognition. Graphics recognition. Lecture Notes in Computer Science. Volume 3088. 2004. p 279–290.

18. Gärtner B. Fast and robust smallest enclosing balls. Algorithms—ESA. LNCS. Volume 1643. 1999. p 325–338.

19. Gärtner B, Schönherr S. Exact primitives for smallest enclosing ellipses. Inform Process Lett 1998;68(1):33–38.

20. Ghali A, Daemi MF, Mansour M. Image structural information assessment. Pattern Recog Lett 1998;19(5–6):447–453.

21. Granlund GH. Fourier preprocessing for hand print character recognition. IEEE Trans Comput 1972;21:195–201.

22. Hayward WG. Effects of outline shape in object recognition. J Exp Psychol: Hum Percept Perform 1998;24:427–440.

23. Hentschel ML, Page NW. Selection of descriptors for particle shape characterization. Part Part Syst Charact 2003;20:25–38.

24. Jacobs DW. Robust and efficient detection of salient convex groups. IEEE Trans Pattern Anal Mach Intell 1996;18(1):23–37.

25. Kakarala R. Testing for convexity with Fourier descriptors. Electron Lett 1998;34(14):1392–1393.

26. Kindratenko VV. On using functions to describe the shape. J Math Imaging Vis 2003;18(3):225–245.

27. Koprnicky M, Ahmed M, Kamel M. Contour description through set operations on dynamic reference shapes. International Conference on Image Analysis and Recognition. Volume 1. 2004. p 400–407.

28. Landau B, Smith LB, Jones S. Object shape, object function, and object name. J Memory Language 1998;38:1–27.

29. Latecki LJ, Lakämper R, Eckhardt U. Shape descriptors for non-rigid shapes with a single closed contour. Proceedings of the Conference on Computer Vision Pattern Recognition; 2000. p 1424–1429.

30. Leavers VF. Use of the two-dimensional radon transform to generate a taxonomy of shape for the characterization of abrasive powder particles. IEEE Trans Pattern Anal Mach Intell 2000;22(12):1411–1423.

31. Liu L, Sclaroff S. Deformable model-guided region split and merge of image regions. Image Vision Comput 2004;22(4):343–354.

32. Loncaric S. A survey of shape analysis techniques. Pattern Recog 1998;31(8):983–1001.

33. Maitra S. Moment invariants. Proc IEEE 1979;67:697–699.

34. Mallat SG. A theory for multiresolution signal decomposition: the wavelet representation. IEEE Trans Pattern Anal Mach Intell 1989;11(7):674–693.

35. Martin RR, Rosin PL. Turning shape decision problems into measures. Int J Shape Model 2004;10(1):83–113.

36. McCallum D, Avis D. A linear algorithm for finding the convex hull of a simple polygon. Inform Process Lett 1979;9:201–206.

37. Medvedeva A, Mukhopadhyay A. An implementation of a linear time algorithm for computing the minimum perimeter triangle enclosing a convex polygon. Canadian Conference on Computational Geometry; 2003. p 25–28.

38. Mukundan R, Ramakrishnan KR. Moment Functions in Image Analysis—Theory and Applications. World Scientific; 1998.

39. O'Rourke J, Aggarwal A, Maddila S, Baldwin M. An optimal algorithm for finding minimal enclosing triangles. J Algorithms 1986;7:258–269.

40. Page DL, Koschan A, Sukumar SR, Roui-Abidi B, Abidi MA. Shape analysis algorithm based on information theory. International Conference on Image Processing; Volume 1; 2003. p 229–232.

41. Pao HK, Geiger D. A continuous shape descriptor by orientation diffusion. Proceedings of the Workshop on Energy Minimization Methods in Computer Vision and Pattern Recognition. LNCS. Volume 2134. 2001. p 544–559.

42. Rahtu E, Salo M, Heikkila J. Convexity recognition using multi-scale autoconvolution. International Conference on Pattern Recognition; 2004. p 692–695.

43. Rangayyan RM, Elfaramawy NM, Desautels JEL, Alim OA. Measures of acutance and shape for classification of breast-tumors. IEEE Trans Med Imaging 1997;16(6):799–810.

44. Rigau J, Feixas M, Sbert M. Shape complexity based on mutual information. International Conference on Shape Modeling and Applications; 2005. p 357–362.

45. Rosenfeld A. Compact figures in digital pictures. IEEE Trans Syst Man Cybernet 1974;4:221–223.

46. Rosenfeld A. Measuring the sizes of concavities. Pattern Recog Lett 1985;3:71–75.

47. Rosin PL. Ellipse fitting using orthogonal hyperbolae and Stirling's oval. CVGIP: Graph Models Image Process 1998;60(3):209–213.

48. Rosin PL. Further five-point fit ellipse fitting. CVGIP: Graph Models Image Process 1999;61(5):245–259.

49. Rosin PL. Measuring rectangularity. Mach Vis Appl 1999;11:191–196.

50. Rosin PL. Shape partitioning by convexity. IEEE Trans Syst Man Cybernet A, 2000;30(2):202–210.

51. Rosin PL. Measuring shape: ellipticity, rectangularity, and triangularity. Mach Vis Appl 2003;14(3):172–184.

52. Rosin PL. Measuring sigmoidality. Pattern Recog 2004;37(8):1735–1744.

53. Rosin PL. Computing global shape measures. In: Chen CH, Wang PS-P, editors. Handbook of Pattern Recognition and Computer Vision. 3rd ed. World Scientific; 2005. p 177–196.
54. Rosin PL, Hervás J. Remote sensing image thresholding for determining landslide activity. Int J Remote Sensing 2005;26(6):1075–1092.
55. Rosin PL, Žunić J. Measuring rectilinearity. Comput Vis Image Understand 2005;99(2):175–188.
56. Samuelson LK, Smith LB. They call it like they see it: spontaneous naming and attention to shape. Dev Sci 2005;8(2):182–198.
57. Singer MH. A general approach to moment calculation for polygons and line segments. Pattern Recog 1993;26(7):1019–1028.
58. Stern HI. Polygonal entropy: a convexity measure. Pattern Recog Lett 1989;10:229–235.
59. Toussaint GT. Solving geometric problems with the rotating calipers. Proceedings of IEEE MELECON'83;1983. p A10.02/1–A10.02/4.
60. Tsai WH, Chou SL. Detection of generalized principal axes in rotationally symetric shapes. Pattern Recog 1991;24(1):95–104.
61. Veltkamp RC, Latecki LJ. Properties and performances of shape similarity measures. Conference on Data Science and Classification; 2006.
62. Ventura AD, Rampini A, Schettini R. Image registration by recognition of corresponding structures. IEEE Trans Geosci Remote Sensing 1990;28(3):305–314.
63. Voss K, Süße H. Invariant fitting of planar objects by primitives. IEEE Trans Pattern Anal Mach Intell 1997;19(1):80–84.
64. Wei L, Keogh E, Xi X. SAXually explicit images: finding unusual shapes. International Conference on Data Mining; 2006.
65. Young IT, Walker JE, Bowie JE. An analysis technique for biological shape. I. Inform Control 1974;25(4):357–370.
66. Zahn CT, Roskies RZ, Fourier descriptors for plane closed curves. IEEE Trans Comput 1972;C-21:269–281.
67. Zhang D, Lu G. Review of shape representation and description techniques. Pattern Recog 2004;37(1):1–19.
68. Žunić J, Kopanja L, Fieldsend JE. Notes on shape orientation where the standard method does not work. Pattern Recog 2006;39(2):856–865.
69. Žunić J, Rosin PL. Rectilinearity measurements for polygons. IEEE Trans Pattern Anal Mach Intell 2003;25(9):1193–1200.
70. Žunić J, Rosin PL. A new convexity measurement for polygons. IEEE Trans Pattern Anal Mach Intell 2004;26(7):923–934.
71. Žunić J, Rosin PL, Kopanja L. On the orientability of shapes. IEEE Trans Image Process 2006;15(11):3478–3487.

CHAPTER 13

Cryptographic Algorithms

BIMAL ROY and AMIYA NAYAK

13.1 INTRODUCTION TO CRYPTOGRAPHY

Cryptography is as old as writing itself and has been used for thousands of years to safeguard military and diplomatic communications. It has a long fascinating history. Kahn's *The Codebreakers* [23] is the most complete nontechnical account of the subject. This book traces cryptography from its initial and limited use by Egyptians some 4000 years ago, to the twentieth century where it played a critical role in the outcome of both the world wars. The name cryptography comes from the Greek words "kruptos" (means hidden) and "graphia" (means writing).

For electronic communications, cryptography plays an important role and that is why cryptography is quickly becoming a crucial part of the world economy. Organizations in both the public and private sectors have become increasingly dependent on electronic data processing. Vast amount of digital data are now gathered and stored in large computer databases and transmitted between computers and terminal devices linked together in complex communication networks. Without appropriate safeguards, these data are susceptible to interception (i.e., via wiretaps) during transmission, or they may be physically removed or copied while in storage. This could result in unwanted exposures of data and potential invasions of privacy. Before the 1980s, cryptography was used primarily for military and diplomatic communications, and in fairly limited contexts. But now cryptography is the only known practical method for protecting information transmitted through communications networks that use land lines, communications satellites, and microwave facilities. In some instances, it can be the most economical way to protect stored data.

A cryptosystem or cipher system is a method of disguising messages so that only certain people can see through the disguise. Cryptography, the art of creating and using cryptosystems, is one of the two divisions of the field called cryptology. The other division of cryptology is cryptanalysis, which is the art of breaking cryptosystems, seeing through the disguise even when you are not supposed to be able to. Thus, cryptology is the study of both cryptography and cryptanalysis. In cryptology, the

Handbook of Applied Algorithms: Solving Scientific, Engineering and Practical Problems
Edited by Amiya Nayak and Ivan Stojmenović

original message is called a plaintext. The disguised message is called a ciphertext, and the encryption means any procedure to convert plaintext into ciphertext, whereas decryption means any procedure to convert cipher text into plaintext.

The fundamental objective of cryptography is to enable two people, say A and B, to communicate over an insecure channel in such a way that an opponent O, cannot understand what is being said. Suppose A encrypts the plaintext using the predetermined key and sends the resulting ciphertext over the channel. O (opponent) on seeing the ciphertext in the channel by intercepting (i.e., wire tapping), cannot determine what the plaintext was; but B, who knows the key for encryption, can decrypt the ciphertext and reconstruct the plaintext. The plaintext message M that the sender wants to transmit will be considered to be a sequence of characters from a set of fixed characters called alphabet. M is encrypted to produce another sequence of characters from the set alphabet called the cipher C. In practice, we use the binary digits (bits) as alphabet. The encryption function $\mathcal{E}_{k_\mathrm{e}}$ operates on M to produce C, and the decryption function $\mathcal{D}_{k_\mathrm{d}}$ operates on C to recover original plaintext M. Both the encryption function $\mathcal{E}_{k_\mathrm{e}}$ and the decryption function $\mathcal{D}_{k_\mathrm{d}}$ are parameterized by the keys k_e and k_d, respectively, which are chosen from a very large set of possible keys called *keyspace*. The sender encrypts the plaintext by computing $C = \mathcal{E}_{k_\mathrm{e}}(M)$ and sends C to the receiver. Those functions have properties that receiver recovers the original text by computing $\mathcal{D}_{k_\mathrm{d}}(C) = \mathcal{D}_{k_\mathrm{d}}(\mathcal{E}_{k_\mathrm{e}}(M)) = M$ (see Fig. 13.1).

Two types of cryptographic schemes are typically used in cryptography. They are *private key* (symmetric key) cryptography and *public key* (asymmetric key) cryptography. Public key cryptography is a relatively new field. It was invented by Diffie and Hellman [11] in 1976. The idea behind a public key cryptosystem is that it might be possible to find a cryptosystem where it is computationally infeasible to determine the decryption rule given the encryption rule. Moreover, in public key cryptography, the encryption and the decryption are performed with different keys, whereas in private key cryptography both parties possesses the same key. Private key cryptography is again subdivided into *block cipher* and *stream cipher*. The stream ciphers operate with a time-varying transformation on smaller units of plane text, usually bits, whereas the block ciphers operate with a fixed transformation on larger blocks of data. Symmetric and asymmetric systems have their own strengths and weaknesses. In particular, asymmetric systems are vulnerable in different ways, such as through impersonation, and are much slower in execution than symmetric systems. However, they have particular benefits and, importantly, can work together with symmetric

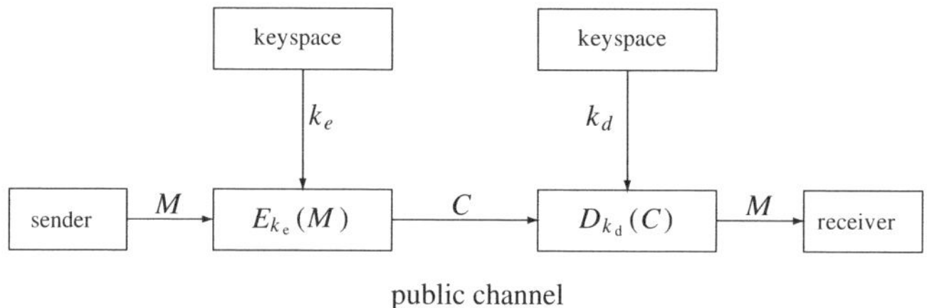

FIGURE 13.1 Basic cryptosystem.

systems to create cryptographic mechanisms that are elegant and efficient and can give an extremely high level of security. In this chapter, we will deal with both stream and block ciphers. Let us first talk about stream ciphers. In the following section, we will define and explain some of the important terms regarding stream ciphers.

13.2 STREAM CIPHERS

In stream ciphers, plaintext P is a binary string; keystream, K, is a pseudo-random binary string; ciphertext, C, is a bit-wise XOR (addition modulo 2) of plaintext and keystream. Decryption is bit-wise XOR of ciphertext and keystream. Let us consider the following example.

$$P : 1000110101011110110 11$$
$$K : 0100101011010011011 01$$
$$C : 1100011110001101101 10$$

In this example, one can observe that $C = P \oplus K$. Also, $P = C \oplus K$.

In 1949, Claude Shannon published a paper "Communication Theory of Secrecy Systems" [34] that had a great impact on the scientific study of cryptography. In the following subsection, we will discuss about Shannon's notion of perfect secrecy.

13.2.1 Shannon's Notion of Perfect Secrecy

Let $\mathcal{P}$, $\mathcal{K}$, and $\mathcal{C}$ denote the finite set of possible plaintexts, keys, and ciphertexts, respectively, for a given cryptosystem. We assume that a particular key $k \in \mathcal{K}$ is used for only one encryption. Let us suppose that there are probability distributions on both $\mathcal{P}$ and $\mathcal{K}$. Thus, two probability distributions on $\mathcal{P}$ and $\mathcal{K}$ induce a probability distribution on $\mathcal{C}$. Then, the cryptosystem has a perfect security, if

$$\Pr(x \mid y) = \Pr(x) \quad \text{for all } x \in \mathcal{P} \text{ and for all } y \in \mathcal{C}.$$

This basically means that the ciphertext has no information about the plaintext. The basic strength of stream- cipher lies in how "random" the keystream is. Random keystream will satisfy Shannon's notion [34]. Let us consider the following illustration.

Illustration Let us consider one bit encryption; $C = P \oplus K$. Here, K random means $\Pr(K = 0) = \Pr(K = 1) = \frac{1}{2}$. Let $\Pr(P = 0) = 0.6,\ \Pr(P = 1) = 0.4$. Then

$$\Pr(P = 0 \mid C = 1) = \frac{\Pr(P = 0,\ C = 1)}{\Pr(C = 1)}$$

$$= \frac{\Pr(P = 0,\ C = 1)}{\Pr(P = 0,\ C = 1) + \Pr(P = 1,\ C = 1)}$$

$$= \frac{\Pr(C=1 \mid P=0)\cdot\Pr(P=0)}{\Pr(C=1 \mid P=0)\cdot\Pr(P=0)+\Pr(C=1 \mid P=1)\cdot\Pr(P=1)}$$
$$= \frac{\Pr(K=1)\cdot\Pr(P=0)}{\Pr(K=1)\cdot\Pr(P=0)+\Pr(K=0)\cdot\Pr(P=1)}$$
$$= \frac{\frac{1}{2}\times 0.6}{\frac{1}{2}\times 0.6+\frac{1}{2}\times 0.4 = 0.6 = \Pr(P=0)}$$
$$= 0{\cdot}6 = \Pr(P=0).$$

Likewise, $\Pr(P=0 \mid C=0)=\Pr(P=0)$, $\Pr(P=1 \mid C=1)=\Pr(P=1)$, and $\Pr(P=1 \mid C=0)=\Pr(P=1)$.

The main objective of a stream cipher construction is to get K as much random as possible. So the measurement of randomness plays an important role in cryptography. In the following subsection, we will discuss about the randomness measurements.

13.2.2 Randomness Measurements

Randomness of a sequence is the unpredictability property of sequence. The aim is to measure randomness of the sequence generated by a deterministic method called a generator. The test is performed by taking a sample output sequence and subjecting it to various statistical tests to determine whether the sequence possesses certain kinds of attributes, a truly random sequence would be likely to exhibit. This is the reason the sequence is called pseudo-random sequence instead of random sequence and the generator is called pseudo-random sequence generator (PSG) in literature.

The sequence $s = s_0, s_1, s_2, \ldots$ is said to be *periodic* if there is some positive integer N such that $s_{i+N} = s_i$ and smallest N is called the *period* of sequence.

Golomb's Randomness Postulates is one of the initial attempts to establish some necessary conditions for a periodic pseudo-random sequence to look random.

13.2.2.1 Golomb's Randomness Postulates

R-1: In every period, the number of 1's differ from the number of 0's by at most 1. Thus, $|\sum_{i=0}^{N-1}(-1)^{s_i}| \leq 1$.

R-2: In every period, half the runs have length 1, one fourth have length 2, one-eighth have length 3, and so on, as long as the number of runs so indicated exceeds 1. Moreover, for each of these lengths, there are (almost) equally many runs of 0's and of 1's.

R-3: The autocorrelation function $C(\tau) = \sum_{i=0}^{N-1}(-1)^{s_i+s_{i+\tau}}$ is two valued. Explicitly

$$C(\tau) = \begin{cases} N \text{ if } \tau \equiv 0 (\bmod N) \\ T \text{ if } \tau \not\equiv 0 (\bmod N), \end{cases}$$

where T is a constant.

As an example, let us consider the periodic sequence s of period 15 with cycle

$$s^{15} = 011001000111101.$$

One can observe that

R-1: There are seven 0's and eight 1's.

R-2: Total runs are 8. Four runs of length 1 (2 for each 0's and 1's), two runs of length 2 (one for each 0's and 1's), one run of 0's of length 3, and one run of 1's of length 4.

R-3: The function $C(\tau)$ takes only two values: $C(0) = 15$ and $C(\tau) = -1$ for $1 \leq \tau \leq 14$.

13.2.3 Five Basic Tests

1. *Frequency test* (monobit test): To test whether the number of 0's and 1's in sequence s is approximately the same, as would be expected for a random sequence.
2. *Serial test* (two-bit test): To determine whether the number of 00, 01, 10, and 11 as subsequences of s are approximately the same, as would be expected for a random sequence.
3. *Poker test:* Let m be a positive integer. Divide the sequence into n/m nonoverlapping parts of length m. To test whether the number of each sequence of length m is approximately the same, as would be expected for a random sequence.
4. *Runs test:* To determine whether the number of runs of various lengths in the sequence satisfy the R-2, as expected for a random sequence.
5. *Autocorrelation test:* To check whether correlation between the sequence and its sifted version is approximately 0 when the number of shifts is not divisible by the period as expected for a random sequence. Here, autocorrelation is taken as $C(\tau)/N$, $C(\tau)$ is as defined in R-3.

For details on randomness measurement one can see the work by Gong [20].

In the next subsection, we will discuss about an efficient method of producing keystream in hardware using linear feedback shift register (LFSR).

13.2.4 LFSR

One of the basic constituents in many stream ciphers is a LFSR. An LFSR of length L consists of L stages numbered $0, 1, \ldots, L-1$, each storing one bit and having one input and one output; together with a clock that controls the movement of data. During each unit of time, the following operations are performed:

(i) The content of stage 0 is the output and forms part of the output sequence.

(ii) The content of stage i is moved to stage $i-1$.

(iii) The new content of stage $L-1$ is the feedback bit that is calculated by adding together modulo 2 the previous contents of a fixed subset of stages $0, 1, \ldots, L-1$.

The position of these previous contents may be thought of having a correspondence with a polynomial. A polynomial $\sum_{i=0}^{k} a_i X^i$ induces the recurrence on the output

$$\{D_n : n \geq 1\} \text{ as } D_n = \sum_{i=1}^{k} a_{k-i} D_{n-i}.$$

Let us consider the following example.

Example Consider an LFSR $\langle 4, 1 + X^3 + X^4 \rangle$. It induces the recurrence $D_n = D_{n-1} + D_{n-4}$.

t	D_3	D_2	D_1	D_0
0	0	1	1	0
1	0	0	1	1
2	1	0	0	1
3	0	1	0	0
4	0	0	1	0
5	0	0	0	1
6	1	0	0	0
7	1	1	0	0
8	1	1	1	0
9	1	1	1	1
10	0	1	1	1
11	1	0	1	1
12	0	1	0	1
13	1	0	1	0
14	1	1	0	1
15	0	1	1	0

Output: $s = 0, 1, 1, 0, 0, 1, 0, 0, 0, 1, 1, 1, 1, 0, 1, \ldots$.

For cryptographic use, the LFSR should have period as long as possible. The following result takes care of it.

If $C(X)$ is a primitive polynomial, then each of the $2^L - 1$ nonzero initial states of the LFSR $\langle L, C(X) \rangle$ produces an output sequence with maximum possible period $2^L - 1$.

If $C(X) \in Z_2[X]$ is a primitive polynomial of degree L, then $\langle L, C(X) \rangle$ is called a maximum-length LFSR.

13.2.5 Linear Complexity

Linear complexity is a very important concept for the study of randomness of sequences. The linear complexity of an infinite binary sequence s, denoted $L(s)$, is defined as follows:

(i) If s is the zero sequence $s = 0, 0, 0, ...$, then $L(s) = 0$.
(ii) If no LFSR generates s, then $L(s) = \infty$.
(iii) Otherwise, $L(s)$ is the length of the shortest LFSR that generates s.

The linear complexity of a finite binary sequence $s^{(n)}$, denoted $L(s^{(n)})$, is the length of the shortest LFSR that generates a sequence having $s^{(n)}$ as its first n terms.

13.2.6 Properties of Linear Complexity

Let s and t be binary sequences.

(i) For any $n \geq 1$, the linear complexity of the subsequence $s^{(n)}$ satisfies $0 \leq L(s^{(n)}) \leq n$.
(ii) $L(s^{(n)}) = 0$ if and only if $s^{(n)}$ is the zero sequence of length n.
(iii) $L(s^{(n)}) = n$ if and only if $s^n = 0, 0, 0, \ldots, 0, 1$.
(iv) If s is periodic with period N , then $L(s) \leq N$.
(v) $L(s \oplus t) \leq L(s) + L(t)$, where $s \oplus t$ denotes the bitwise XOR of s and t.
(vi) For a finite sequence of length n, linear complexity $\leq n/2$. If the linear complexity is strictly less than $\frac{n}{2}$, the sequence is not random. For a random sequence, linear complexity should be $\frac{n}{2}$. This is one of the strongest measure of randomness.

Berlekamp [1] and Massey [26] devised an algorithm for computing the linear complexity of a binary sequence.

Let us define, $s^{(N+1)} = s_0, s_1, \ldots, s_{N-1}, s_N$. The basic idea is as follows. Let $\langle L, C(X) \rangle$ be an LFSR that generates the sequence $s^{(N)} = s_0, s_1, \ldots, s_{N-1}$. Let us define the next discrepancy as

$$d_N = \left(s_N + \sum_{i=1}^{L} c_i s_{N-i} \right) \bmod 2.$$

If d_N is 0, the same LFSR also produces S^{N+1}, else the LFSR is to be modified. The detailed algorithm is stated below.

13.2.6.1 Berlekamp–Massey Algorithm

Input: a binary sequence $s^{(n)} = s_0, s_1, s_2, \ldots, s_{n-1}$.

Output: the linear complexity $L(s^{(n)})$ of $s^{(n)}$.

1 Initialize $C(X) \leftarrow 1, L \leftarrow 0, m \leftarrow -1,$
$B(X) \leftarrow 1, N \leftarrow 0.$
2 While $(N < n)$ **do**
2.1Compute $d \leftarrow (s_N + \sum_{i=1}^{L} c_i s_{N-i})$
2.2If $d = 1$ **then**
$T(X) \leftarrow C(X), C(X) \leftarrow C(X) + B(X)X^{N-m}.$
If $L \leq N/2$ **then** $L \leftarrow N + 1 - L,$
$m \leftarrow N, B(X) \leftarrow T(X).$
2.3$N \leftarrow N + 1.$
3 Return(L).

Let us illustrate the algorithm for two sequences: $s^{(n)} = 0, 0, 1, 1, 0, 1, 1, 1, 0$ and $t^{(n)} = 0, 0, 1, 1, 0, 0, 0, 1, 1, 0$. The first sequence has linear complexity 5 and an LFSR that generates it is $\langle 5, 1 + X^3 + X^5 \rangle$. The second sequence has the linear complexity 3 and and an LFSR that generates it is $\langle 3, 1 + x + x^2 \rangle$. Since linear complexity is *less than* $n/2 = 5$, the sequence is *not random*, which is also evident from the sequence.

The steps of the Berlekamp–Massey algorithms are explained in the two following tables.

s_N	d	$T(X)$	$C(X)$	L	m	$B(X)$
–	–	–	1	0	−1	1
0	0	–	1	0	−1	1
0	0	–	1	0	−1	1
1	1	1	$1 + X^3$	3	2	1
1	1	$1 + X^3$	$1 + X + X^3$	3	2	1
0	1	$1 + X + X^3$	$1 + X + X^2 + X^3$	3	2	1
1	1	$1 + X + X^2 + X^3$	$1 + X + X^2$	3	2	1
1	0	$1 + X + X^2 + X^3$	$1 + X + X^2$	3	2	1
1	1	$1 + X + X^2$	$1 + X + X^2 + X^5$	5	7	$1 + X + X^2$
0	1	$1 + X + X^2 + X^5$	$1 + X^3 + X^5$	5	7	$1 + X + X^2$

t_N	d	$T(X)$	$C(X)$	L	m	$B(X)$
–	–	–	1	0	−1	1
0	0	–	1	0	−1	1
0	0	–	1	0	−1	1
1	1	1	$1 + X^3$	3	2	1
1	1	$1 + X^3$	$1 + X + X^3$	3	2	1
0	1	$1 + X + X^3$	$1 + X + X^2 + X^3$	3	2	1
0	0	$1 + X + X^3$	$1 + X + X^2 + X^3$	3	2	1
0	0	$1 + X + X^3$	$1 + X + X^2 + X^3$	3	2	1
1	0	$1 + X + X^3$	$1 + X + X^2 + X^3$	3	2	1
1	0	$1 + X + X^3$	$1 + X + X^2 + X^3$	3	2	1
0	0	$1 + X + X^3$	$1 + X + X^2 + X^3$	3	2	1

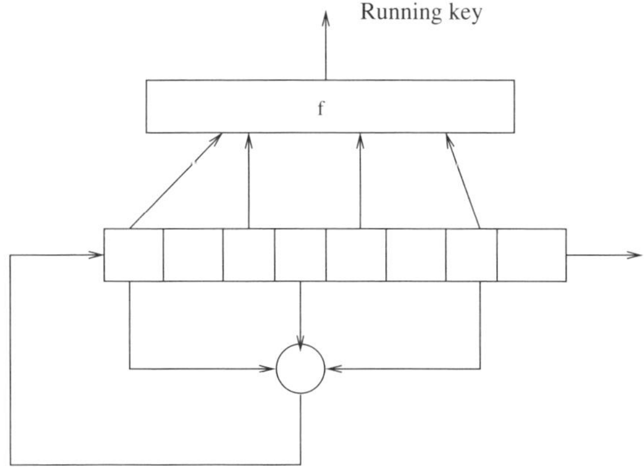

FIGURE 13.2 Nonlinear filter generator.

The running time of the algorithm for determining the linear complexity of a binary sequence of bit length n is $O(n^2)$ bit operations. For a finite binary sequence of length n, let the linear complexity be L. Then there is a unique LFSR of length L which generates the sequence iff $L \leq n/2$. For an infinite binary sequence (s) of linear complexity L, let t be a (finite) subsequence of length at least $2L$. Then the Berlekamp–Massey algorithm on input t determines an LFSR of length L which generates s.

13.2.7 Nonlinear Filter Generator

A filter generator is a running key generator for stream cipher applications. It consists of a single LFSR that is filtered by a nonlinear Boolean function f. This model has been in practical use for generating the keystream of a stream cipher. However, the strength of this model depends on the choice of the nonlinear Boolean function (Fig.13.2).

13.2.8 Synchronous and Asynchronous Stream Ciphers

There are two types of stream ciphers.

1. *Synchronous*: keys are generated before encryption process independently of the plaintext and ciphertext. Example: DES in OFB mode.
2. *Asynchronous*: encryption keys are generated using keys and a set of former ciphertext bits. Example: A5 used in GSM, DES in CFB mode (Fig.13.3).

13.2.8.1 Synchronous vs Asynchronous Stream Ciphers Attributes of synchronous stream ciphers:

- Easy to generate.
- No error propagation.
- Insertion, deletion can be detected.

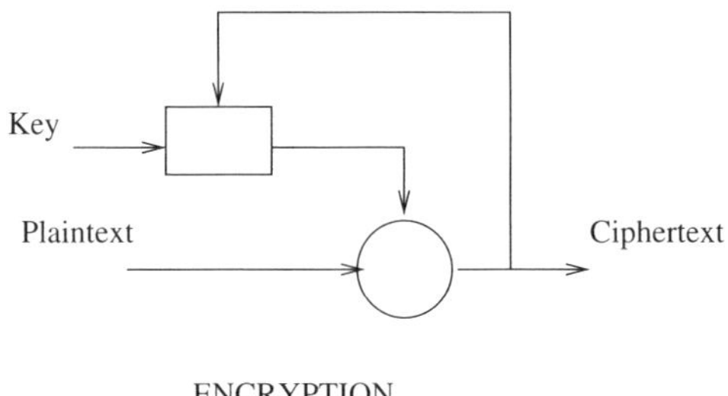

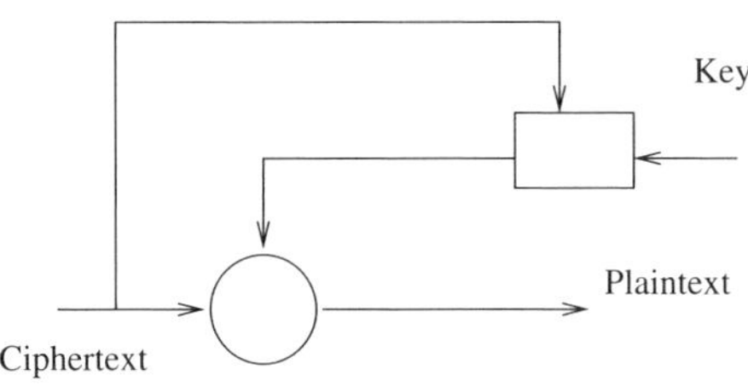

FIGURE 13.3 Asynchronous stream cipher.

- Data authentication and integrity check required.
- Synchronization required. Both the sender and receiver must be synchronized. If synchronization is lost, then decryption fails and can only be restored by resynchronization. Technique for resynchronization includes reinitialization, placing special markers at regular intervals in the ciphertext, or, if the plaintext contains enough redundancy, trying all possible keystream offsets.

Attributes of asynchronous stream ciphers.

- Self-synchronized and limited error propagation.
- More difficult to detect insertion and deletion.
- Plaintext statistics are dispersed through ciphertext.
- More resistant to eavesdropping.
- Harder to generate.

13.2.9 RC4 Stream Ciphers

RC4 was created by Rivest for RSA Securities. Inc. in 1994. Its key size varies from 40 to 256 bits. It has two parts, namely, a key scheduling algorithm (KSA) and a pseudo-random generator algorithm (PRGA). KSA turns a random key into a initial permutation S of $\{0, \cdots, N-1\}$. PRGA uses this permutation to generate a pseudo-random output sequence.

13.2.9.1 Key scheduling algorithm KSA(K)

Initialization :

For $i = 0, \ldots, N-1$ DO
  $S[i] = i$
  $j = 0$
endDo
Scrambling
  For $i = 0, \ldots, N-1$ Do
    $j = j + S[i] + K[i \bmod l]$, where l is the byte length of key
  Swap ($S[i]$, $S[j]$)
endDo

Example Let $N = 8, l = 8$, and the key

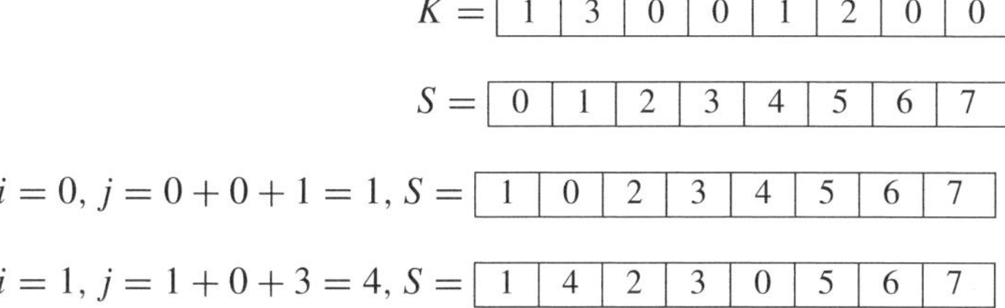

$K =$ | 1 | 3 | 0 | 0 | 1 | 2 | 0 | 0 |

$S =$ | 0 | 1 | 2 | 3 | 4 | 5 | 6 | 7 |

$i = 0, j = 0 + 0 + 1 = 1, S =$ | 1 | 0 | 2 | 3 | 4 | 5 | 6 | 7 |

$i = 1, j = 1 + 0 + 3 = 4, S =$ | 1 | 4 | 2 | 3 | 0 | 5 | 6 | 7 |

13.2.9.2 Pseudo-random Sequence Generator PRGA(K)

Initialization
  $i = 0$
  $j = 0$
Generating loop
  $i = i + 1$
  $j = j + S[i]$
Swap ($S[i]$, $S[j]$)
Output $z = S(S[i] + S[j])$

Example Let

$S =$ | 7 | 2 | 6 | 0 | 4 | 5 | 1 | 3 |

$i = 1, j = 2, S =$ | 7 | 6 | 2 | 0 | 4 | 5 | 1 | 3 |

$z = S(6 + 2) = S(0) = 7$

13.2.9.3 Weaknesses in RC4

1. The most serious weakness in RC4 was observed by Mantin and Shamir [25] who noted that the probability of a zero output byte at the second round is twice as large as expected. In broadcast applications, a practical ciphertext only attack

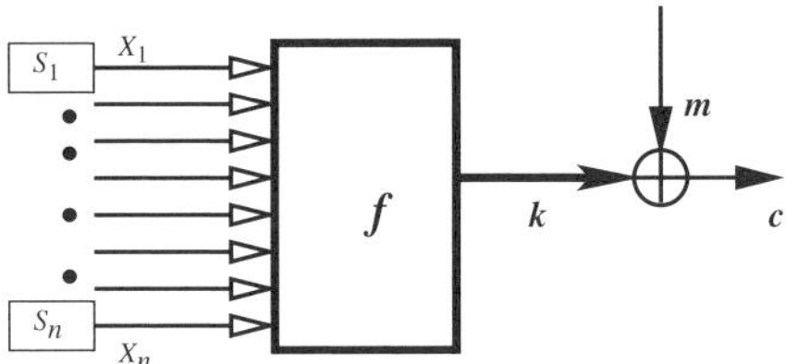

FIGURE 13.4 Nonlinear combiner model.

can exploit this weakness.

2. Fluhrer et al. [18] have shown that if some portion of the secret key is known then RC4 can be broken completely. This is of practical importance.
3. Pudovkina [31] has attempted to detect a bias, only analytically, in the distribution of the first and second output values of RC4 and digraphs under certain uniformity assumptions.
4. Paul and Preneel [30] have shown a statistical bias in the distribution of the first two output bytes of the RC4 keystream generator. They have shown that the probability of the first two output bytes being equal is $(1/N)(1 - 1/N)$. (Note that RC4 produced output bytes uniformly then the probability of that event would have been $1/N$.)

   The number of outputs required to reliably distinguish RC4 outputs from random strings using this bias is only 2^{25} bytes.

   Most importantly, the bias exists even after dropping the first N bytes and the probability of that event is $(1/N)\left(1 - 1/N^2\right)$.

13.2.10 Combiner Model

In this model, several LFSRs are considered. The output of these are combined by a Boolean function to produce the "keystream." This is one of the most commonly used stream cipher models. The strength of this model lies in the choice of the combining function.

In the next subsection, we will discuss some of cryptographic properties of Boolean function. Boolean functions play a basic role in questions of complexity theory as well as the design of circuits and chips for digital computers. In both nonlinear filter generator and nonlinear combiner model, the security depends largely on the choice of the Boolean functions. Therefore, the study of cryptographic properties of Boolean functions is extremely relevant and important (Fig. 13.4).

13.2.11 Cryptographic Properties of Boolean Function

We interpret a Boolean function $f(X_1, \ldots, X_n)$ as the output column of its *truth table* f, that is, a binary string of length 2^n, $f = [f(0, \ldots, 0), f(1, 0, \ldots, 0), f(0, 1, \ldots, 0), \ \ldots, f(1, 1, \ldots, 1)]$. $f(X_1, \ldots, X_n)$ can be written in *algebraic normal form* as $a_0 + \sum_{i=1}^{i=n} a_i X_i + \sum_{1 \le i < j \le n} a_{ij} X_i X_j +$

$\ldots + a_{12,\ldots,n}X_1X_2\cdots X_n$, where $a_0, a_{ij}, \ldots, a_{12\ldots n} \in \{0, 1\}$. The number of variables in the highest order product term with nonzero coefficient is called the *algebraic degree* of f. For example, $f(X_1, X_2, X_3) = X_3 + X_1 \cdot X_2$ is a three variable Boolean function with algebraic degree 2. Here $(+)$ and $(\cdot)$ denote addition (XOR) and multiplication (AND) over GF (2), respectively. By Ω_n, we mean the set of all Boolean functions of n variables. Functions of degree at most one are called *affine* functions. An affine function with constant term equal to zero is called a *linear* function. For example, $f(X_1, X_2, X_3) = 1 + X_1 + X_2 + X_3$ is an *affine* function. The set of all n-variable affine (respectively linear) functions is denoted by $A(n)$ respectively $L(n)$. *Nonlinearity* of an n-variable function f is

$$\mathrm{nl}(f) = \min_{g \in A(n)} (d(f, g)),$$

that is, the minimum distance from the set of all n-variable affine functions.

Example of Boolean Function

x_1	x_2	x_3	f_1	f_2	f_3	f_4	f_5	f_6	f_7	f_8	f
0	0	0	1	0	0	0	0	0	0	0	0
0	0	1	1	0	0	1	0	1	1	1	0
0	1	0	1	0	1	0	1	1	0	1	0
0	1	1	1	0	1	1	1	0	1	0	0
1	0	0	1	1	0	0	1	0	1	1	0
1	0	1	1	1	0	1	1	1	0	0	0
1	1	0	1	1	1	0	0	1	1	0	1
1	1	1	1	1	1	1	0	0	0	1	1

Function	Distance from f
f_1	6
f_2	2
f_3	2
f_4	4
f_5	6
f_6	4
f_7	4
f_8	4

Here $\mathrm{nl}(f) = 2$.

Let $X = (X_1, \ldots, X_n)$ and $\omega = (\omega_1, \ldots, \omega_n)$ both belong to $\{0, 1\}^n$ and

$$X \cdot \omega = X_1\omega_1 + \cdots + X_n\omega_n.$$

Let $f(X)$ be a Boolean function on n variables. Then the *Walsh transform* of $f(X)$ is a real-valued function over $\{0, 1\}^n$ that can be defined as

$$W_f(\omega) = \sum_{X \in \{0,1\}^n} (-1)^{f(X) + X \cdot \omega}.$$

It can be shown that

$$\mathrm{nl}(f) = 2^{n-1} - \tfrac{1}{2}\max_{\omega \in \{0,1\}^n} |W_f(\omega)|.$$

A function $f(X_1, \ldots, X_n)$ is mth order *correlation immune* (CI) iff its Walsh transform satisfies $W_f(\omega) = 0$, for $1 \le wt(\omega) \le m$. Note that f is balanced iff $W_f(0) = 0$. Balanced mth order CI functions are called *m-resilient* functions. A function $f(X_1, \ldots, X_n)$ is m-resilient iff its Walsh transform satisfies

$$W_f(\omega) = 0, \text{ for } 0 \le wt(\omega) \le m.$$

Example of Boolean Function

x_4	x_3	x_2	x_1	f
0	0	0	0	0
0	0	0	1	0
0	0	1	0	0
0	0	1	1	1
0	1	0	0	1
0	1	0	1	1
0	1	1	0	1
0	1	1	1	0
1	0	0	0	1
1	0	0	1	1
1	0	1	0	1
1	0	1	1	0
1	1	0	0	0
1	1	0	1	0
1	1	1	0	0
1	1	1	1	1

In the above example, the ANF is $f = x_4 \oplus x_3 \oplus x_2 x_1$. Also $n = 4$, algebraic degree $d = 2$, m, the order of CI is 1 and nonlinearity is 4. These are the best possible parameters for such a function.

13.2.12 Design of Boolean Function

By an (n, m, d, x) function we denote an n-variable, m-resilient function with algebraic degree d and nonlinearity x.

Tradeoffs for Design

1. Siegenthaler's inequality : $m + d \le n - 1$.
2. nlmax(n) : max nonlinearity of n-variable function.

$$\mathrm{nlmax}(n) \le 2^{n-1} - 2^{(n/2)-1}.$$

   If n is even, nlmax$(n) = 2^{n-1} - 2^{(n/2)-1}$.
3. nlr(n, m) : maximum possible nonlinearity of n-variable, m-resilient functions.

Specific construction techniques, like recursive construction, concatenation of small affine functions, are used for designing certain (n, m, d, x) functions. This is an area of active research.

Recursive Construction Basic idea proposed in the work by Siegenthalor [35]. To start with, one can consider an unbalanced function g on $n - m - 1$ variables. Next, note that the $(n - m)$-variable function $h = X_{n-m} + g(X_1, \ldots, X_{n-m-1})$ is balanced. Now consider the function f on n variables as $X_n + \ldots + X_{n-m+1} + h(X_1, \ldots, X_{n-m})$. This is an (n, m, d, x) function. We will talk about the values of d, x little later. That is, after getting the balanced function h, addition of each new variable increases the order of correlation immunity by 1. Now interpret this construction in the following way. Let $h_{k,i}$ be a k-variable resilient function of order i. Just as notation, we consider the unbalanced functions as resilient functions of order -1 and balanced non-CI functions as resilient functions of order 0. It is now clear that $X_{k+1} + h_{k,i}$ is always a $(k + 1)$-variable, $(i + 1)$-resilient function. Let us call this c (complement) operation, since the truth table of $h_{k,i}$ and its complement are concatenated to get the $(k + 1)$-variable function. Extension of this kind of construction has been discussed in the work by Camion et al. [7]. If i is even, then $(1 + X_{k+1})h_{k,i}(X_1, \ldots, X_k) + X_{k+1}h_{k,i}(1 + X_1, \ldots, 1 + X_k)$ is $(k + 1)$-variable, $(i + 1)$-resilient function. We call this as r (reverse) operation, since the truth table of $h_{k,i}$ and its reverse string are concatenated to get the $(k + 1)$-variable function. If i is odd, then $(1 + X_{k+1})h_{k,i}(X_1, \ldots, X_k) + X_{k+1}(1 + h_{k,i}(1 + X_1, \ldots, 1 + X_k))$ is $(k + 1)$-variable, $(i + 1)$-resilient function. We call this as rc (reverse and complement) operation, since the truth table of $h_{k,i}$ and its reverse and then complemented string are concatenated to get the $(k + 1)$-variable function.

Example of Recursive Construction

x_3	x_2	x_1	f
0	0	0	0
0	0	1	1
0	1	0	0
0	1	1	1
1	0	0	0
1	0	1	1
1	1	0	1
1	1	1	0

x_4	x_3	x_2	x_1	f
0	0	0	0	0
0	0	0	1	1
0	0	1	0	0
0	0	1	1	1
0	1	0	0	0
0	1	0	1	1
0	1	1	0	1
0	1	1	1	0
1	0	0	0	0
1	0	0	1	1
1	0	1	0	1
1	0	1	1	0
1	1	0	0	1
1	1	0	1	0
1	1	1	0	1
1	1	1	1	0

Note that the nonlinearity measure is bound to increase as a consequence of this kind of construction.

Advanced Recursive Construction Recursive construction by Tarannikov [38] that has been modified in the work by Pasalic et al. [29]. Given an (n, m, d, x) function, an $(n+3, m+2, d+1, 2^{n+1}+4x)$ function can be constructed. An $(n, m, d, -)$ function f is in desired form if it is of the form $f = (1+X_n)f_1 + X_n f_2$, where f_1, f_2 are $(n-1, m, d-1, -)$ functions. Let $F = f||\overline{f}||\overline{f}||f$, or written in ANF, $F = X_{n+2} + X_{n+1} + f$. Let $G = g||h||\overline{h}||\overline{g}$ where $g = f_1||\overline{f}_1$ and $h = f_2||\overline{f}_2$. In ANF, the function G is given by $G = (1 + X_{n+2} + X_{n+1})f_1 + (X_{n+2} + X_{n+1})f_2 + X_{n+2} + X_n$. We construct a function H in $n+3$ variables in the following way,

$$H = (1 + X_{n+3})F + X_{n+3}G.$$

Then, the function H constructed from f is an $(n+3, m+2, d+1, 2^{n+1}+4x)$ function in the *desired* form.

Efficient Implementation Majority of the stream ciphers are based on LFSRs. LFSR over $GF(2)$ is fast in hardware but software realization is slow. Some recent software stream ciphers such as SNOW [16] (versions 1 and 2), t-classes of SOBER [21], TURING [33] are based on word-oriented LFSRs over $GF(2^b)$. These are considerably fast in software but not time-tested. Already certain weaknesses have been found.

For resisting the known correlation attacks, following are recommended.

(i) Attack is resisted if time complexity $\geq 2^q$, q is key length.

(ii) Equivalent LFSRs

- For a given ϵ,*CJS attack* can be resisted if $L = 4q$. Required cipher bits $N \uparrow$ as $\epsilon \downarrow$, where ϵ is $(\frac{1}{2}) - p$, p being probability of some LFSR output bit being equal to the corresponding cipher bit.
- The condition $L = 4q$ resists *CT attack.*
- Consider smallest length equivalent LFSR.
- Wt. of $\psi(x)$ must be > 10 for resisting fast correlation attack [27] using sparse multiples of connection polynomial.
- Considering $L = 4q$, expected degree of the least sparse multiple (wt.5) $\approx 2^q$.

(iii) Boolean function

- Maximize nonlinearity to resist *best affine approximation attack* [12].
- The (n, m, u, x) functions with best possible nonlinearity must have three valued Walsh spectra for $m > \frac{n}{2} - 2$.
- Thereby, 2^{m-n+1} must be $\leq \epsilon$, as $\max_{\omega \in \{0,1\}^n} |W_f(\bar{\omega})| = 2^{m+2}$.

Until now, we have talked about stream ciphers. In the next section, we will discuss another very important and useful concept of private key cryptography known as block ciphers.

13.3 BLOCK CIPHER

In block ciphers, the plaintext is divided into blocks of a fixed length and encrypted into blocks of ciphertext using the same key. The mathematical definition of a block cipher is as follows:

Definition. An n-bit block cipher is a function $E : V_n \times K \rightarrow V_n$ such that for each key $k \in K$, $E(p, k)$ is an invertible mapping (encryption function for k) from V_n to V_n, written as $E_k(\ P\ )$. The inverse mapping is the decryption function, denoted $D_k(C)$. $C = E_k(P)$ denotes the ciphertext C that results from plaintext P under k.

The variable V_n is the space containing all the possible bit strings of length n.

An n-bit block cipher with a fixed key is a permutation $p : GF(2)^n \rightarrow GF(2)^n$. It would require $\log_2(2^n!)$ bits to represent the key such that all permutations p were possible, or roughly 2^n times the number of bits in a cipher block. With an ordinary block size, for example, 64 bits, this is much too big a number for practical use, therefore the key size in a practical block cipher is much smaller, typically 128 bits or 256 bits. A good encryption function must contain some nonlinear component, and this is often a substitution box S-box. An s-box is defined as a mapping $GF(2)^n \rightarrow GF(2)^m$, usually defined by a $n \times m$ lookup table. Almost all block ciphers used today are iterated block ciphers. These ciphers are based on iterating a function several times, each iteration is called a round.

In Figure 13.5, we show the process of encrypting the plaintext X_0 under a typical r-round block cipher to obtain the ciphertext X_r. Here X_i denotes the intermediate value of the block after i rounds of the encryption, so that $X_i = F_i(X_{i-1}\ , k_i)$, where $(k_1, k_2, \ldots, k_r)$ is the list of round keys which is derived from the secret key K using a policy known as KSA.

The round key is derived from the cipher key by a key schedule, which is an algorithm that expands the master key or the cipher key. Key scheduling function should be a good pseudo-random generator, however the complexity of its design is less restricted than that of the main body of the block cipher itself. This is so because in most cases a single key is used to encrypt many blocks before it is changed and thus KSA can spend more time on randomizing things than on the encryption function. Due to this reasoning, in many cases, the analysis of key scheduling function is hard. It is also hardly worth the effort since in most cases the flawed key schedule can be replaced without altering the main encryption function. An attacker may assume that subkeys are independent random variables. If the cipher is broken under this assumption, no patch of key schedule will save it. Interestingly, it is possible to avoid the need for a complex key schedule by using a fixed mixing permutation on a large set of inputs and two keys XORed at the input and at the output of the encryption function [17,34]. These keys are now called whitening keys. Many modern ciphers combine both the whitening and the key scheduling approaches.

FIGURE 13.5 A typical r-round block cipher.

The cipher key is usually between 40 and 256 bits for a block cipher, and for an r-round iterated cipher this is expanded into r-round keys. The round function is usually a combination of substitution and transposition. Substitution is when a block in the plaintext is substituted with another block by some substitution rule. Transposition is to permute the blocks or characters in the plaintext. In earlier ciphers, substitution and transposition were used on their own as a cipher, where each plaintext symbol was a block, but this proved to be insecure because of the small block size. Most modern ciphers are a combination of substitution and transposition, and are often called product ciphers [37].

Among the main building blocks of modern block ciphers are substitutions and permutations, which are primitive ciphers on their own. Substitution ciphers are known from ancient times and can be viewed simply as a change of names of the letters. For example, in a cipher attributed to Julius Caesar each letter of the alphabet is exchanged by a letter standing three positions from it (A is encrypted as D, B as E, C as F, etc.). Of course, in general, the substitution need not have a simple "shift" structure as in Caesar's cipher. However, in spite of an astronomical number of possible substitution ciphers over the English alphabet (26!), they are easily solvable, using the letter frequency analysis. As a bright illustration of this one can read Edgar Poe's fascinating story "The Golden Bug," or Conan Doyle's "The Dancing Men." A popular element of modern ciphers—a substitution box (S-box)—takes a block of m bits as its input and outputs a block of n bits (m not necessarily equals n). The S-box can perform any function on a set of its inputs: if $m = n$, it can be a permutation on a set of 2^m inputs, if $m > n$, it can be a collection of several permutations on a set of 2^n inputs. It can be a randomly chosen function or a carefully designed function with special properties. It is desirable for an S-box to perform nonlinear and nonaffine function in order for the whole cipher to be a nonlinear function. Linearity in cipher's behavior is the end of a cipher, since it essentially means that information is leaked from the plaintext to the ciphertext. Both expanding ($m < n$) and contracting ($m > n$) S-boxes can be met in modern block ciphers. Unless being calculated by a compact formula, the memory required to store an s-box grows exponentially with the linear increase in the size of its input m . Thus, the most typical sizes for S-box input are m = 4, 6, 8, and 12 bits. The second basic element—permutation (or transposition) cipher—keeps plaintext characters as they are but arranges them in a different order. One of the oldest transposition methods was used by ancient Greeks: A leather belt is tightly wound around a cylinder and a message is written on the belt across the length of the cylinder. The belt is then worn by a messenger. The message can be decrypted by a party who has a cylinder of the same diameter as was used during the "encryption." Breaking a basic permutation cipher is an easy task, especially if one knows a part of the encrypted plaintext. In modern ciphers, permutations of bits are frequently used. Although weak on their own, a line of substitutions followed by a permutation has good "mixing" properties: substitutions add to local confusion and permutation "glues" them together and spreads the local confusion to the more distant subblocks. Shannon [34] in a pioneering work "Communication Theory of Secrecy Systems," suggested the use several mixing layers interleaving substitutions and permutations. Such a design is called substitution permutation network (SPN). Figure 13.6 is an example of SPN.

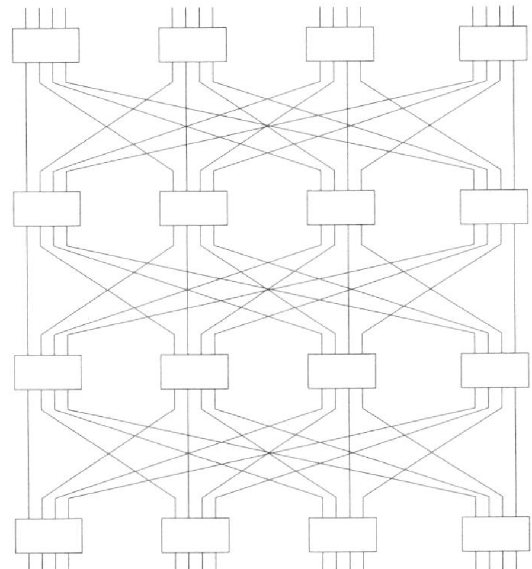

FIGURE 13.6 An example of substitution permutation network (SPN).

13.3.1 Data Encryption Standard

The DES [28] has been the most widely used iterated block cipher since it was published in 1977 by National Bureau of Standards [28] (now the National Institute of Standards and Technology, or NIST), but it is now replaced by the Advanced Encryption Standard (AES) because of too small key and block size. The DES can be seen as a special implementation of a Feistel cipher, named after Horst Feistel, where the input to each round is divided into two halves, as in the following description.

Description of DES. DES cipher is so important to the development of modern cryptanalysis that it might be worth while to describe this construction in some greater detail. It usually looks "monstrous" to the first time reader. Surprisingly, almost every

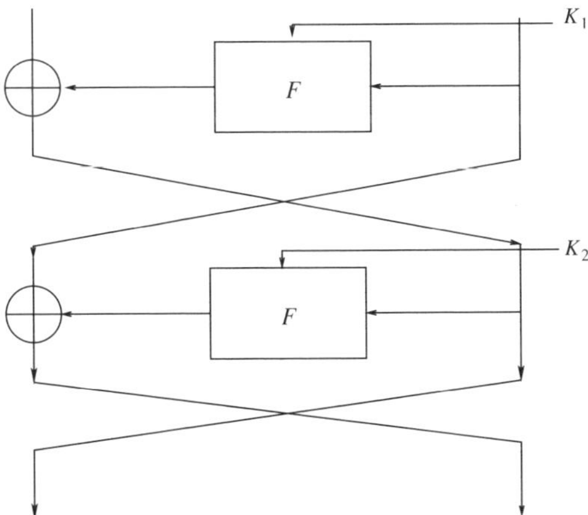

FIGURE 13.7 Two round DES.

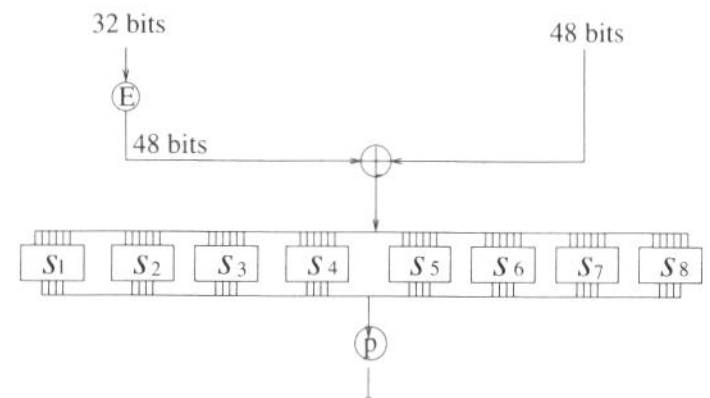

FIGURE 13.8 The F-function of DES.

bit of design in DES seems to have a security reason, and most of the changes seem to weaken the cipher considerably. Biham and Shamir [4] gave a thorough study of DES and its modifications. DES was designed by IBM crypto group from its predecessor Lucifer in early 1970 and was published in the Federal Register of March 17, 1975. DES was adopted as a standard for "unclassified" information on January 1977. Since then it has become the most widely used and the most analyzed cipher. DES is an iterative block cipher. It encrypts blocks of 64 bits into ciphertext blocks of 64 bits under control of the 56-bit secret key. DES performs 16 iterations of the round function, which is called the F-function. Figure 13.7 shows the basic structure of DES reduced to two rounds, one can see that it is a Feistel cipher. The F-function has a relatively simple structure and is based on the substitution–permutation sandwich idea of Shannon (described above).

Each round takes the 64-bit output of the previous round and divides it into two 32-bit halves–the left half L and the right half R. The F-function (described in Fig. 13.8) takes R as its input, expands it (by $E(R)$) from 32 bits into 48 bits and XORs the result with the 48-bit subkey derived from the 56-bit secret key K by the KSA. Then the result enters eight S-boxes. Each S-box takes as input six bits and outputs four bits. The 32-bit result from the row of S-boxes is permuted by the permutation P. The permuted value is the output of the F-function. In the round function, the output of the F-function, $F(R, K_i)$, is XORed with L, and the right and the left halves are swapped. Thus, the output of the ith round is ($R, L \oplus F(R; K_i)$). Note that the tables $P, S_i, i = 1, \ldots, 8, E$ are defined and fixed in the standard, so the only variable part of DES is the secret key K. The KSA of DES is as follows: The 64-bit key is permuted by the permutation $PC-1$ [37]. This permutation performs two functions: strips eight parity bits and then distributes the remaining 56 bits into two 28-bit registers C and D. On each round, 28-bit registers C and D are left shifted by one or two places (according to a fixed schedule). After the shift, the permutation $PC-2$ [37] is performed over C and D, selecting 24 bits out of each 28-bit register. These 48 bits form the subkey of the corresponding round.

AES is the successor of DES. NIST replaced DES by the new standard that is called AES in 1997. At the "First AES candidate conference" in 1998, 15 AES candidates were selected by NIST. In 1999, five of them (MARS, RC6, Rijndael, Serpent, and Twofish) were selected at the "Second AES candidate conference." Finally, Rijndael [9,10] was ultimately selected as the AES by NIST on December 4, 2001 (to be effective March 26, 2002).

13.3.2 AES

We now give a short description of AES (for details see the work by Daemen and Rijmen [10]). Rijndeal is a 128-bit block cipher with one of the three different key sizes, 128 or 192 or 256 bits. The 128-bit block is viewed as $(b_0, b_1, \ldots, b_i, \ldots, b_{15})$, where b_i is the ith byte of the block. The bytes are organized in a matrix form.

$$\begin{pmatrix} b_0 & b_4 & b_8 & b_{12} \\ b_1 & b_5 & b_9 & b_{13} \\ b_2 & b_6 & b_{10} & b_{14} \\ b_3 & b_7 & b_{11} & b_{15} \end{pmatrix}$$

This form of the input block is named "*State*." The *State* is modified by applying some transformations on it and thus 16 bytes of ciphertext is produced for every 16 bytes of plaintext. The number of rounds in AES depend on the key size. There are 10 rounds for 128 bit, 12 for 192 bit, or 14 for 256 bit key.

Each round of the cipher is composed of four different transformations. In pseudo C notation, a particular round is described as

```
Round(State, RoundKey)
{
   ByteSub(State);
   ShiftRow(State);
   MixColumn(State);
   AddRoundKey(State, RoundKey);
}
```

The final round is slightly different than other rounds. The MixColumn transformation is absent in the last round.

The inverse cipher has similar structure. The inverse cipher is as follows

```
InvRound(State, RoundKey)
{
   AddRoundKey(State, RoundKey);
   InvMixColumn(State);
   InvShiftRow(State);
   InvByteSub(State);
}
```

The initial round of the inverse cipher (corresponding to the last round of the cipher) does not have the MixColumn transformation.

In addition, an extra step of "whitening" (XORing the State with the RoundKey) is taken before the rounds begin during the cipher operation. While inverting, the last step is therefore XORing the State with the same RoundKey.

The round transformations and their inverses are described next.

1. *ByteSub and InvByteSub*: These operations act on a byte and substitute a new value in its place. A byte b consisting of bits $b_7 b_6 b_5 \cdots b_1 b_0$, is considered as a

polynomial with coefficients in {0,1}: $b_7x^7 + b_6x^6 + b_5x^5 + \cdots + b_1x + b_0$. The ByteSub operation consists of two steps. Given an irreducible polynomial $m(x) = x^8 + x^4 + x^3 + x + 1$, first the inverse of byte b in the field generated by $m(x)$ is found. The zero element is considered to be its own inverse. Next, this inverse element is operated by a bit level affine transformation $Ax + b$, where A is an 8×8 matrix binary matrix, x and b are one byte each which are considered as eight element column vectors (bit representation of the byte). This operation can be implemented with the use of a substitution table for each byte. Such a table is called the S-box.

The inverse operation can be implemented by the inverse S-box. If S-box(x) = y then Inverse-S-box(y) = x.

2. *ShiftRow and InvShiftRow*: ShiftRow transformation acts on the rows of the State array. The four bytes in row i are cyclically shifted by C_i bytes, where $C_1 = 0$, $C_2 = 1$, $C_3 = 2$, and $C_4 = 3$. That is, the first row remains unchanged, the second row, which was initially (x_1, x_2, x_3x_4), becomes (x_4, x_1, x_2x_1), and so on. The inverse of this operation shifts the bytes of the rows cyclically by $C_i^{'}$ bytes, where $C_1^{'} = 0$, $C_2^{'} = 3$, $C_3^{'} = 2$, and $C_4^{'} = 1$. It is clear that the inverse operations "undoes" the effect of the ShiftRow operation.
3. *MixColumn and InvMixColumn*: Each column (four bytes) of State array is transformed to another column in this operation. Each column is considered to be a polynomial of degree less than 4 with coefficients in $GF(2^8)$. It is multiplied by $M(x) =$ '03' x^3+ '01'x^2+ '01'$x+$ '02' and the result is taken modulo $(x^4 + 1)$. In inverse operation, first the columns are multiplied by $M^{'}(x) =$'0B' x^3+ '0D' x^2+ '09' $x+$ '0E' and then the modulo operation is performed. It is easy to see that $M(x) \cdot M^{'}(x) = 1$.
4. *AddRoundKey*: RoundKeys are generated from the given cipher key. The number of RoundKeys generated is one more than the number of rounds, and each key is of size 16 bytes. Round i of cipher operation uses the ith RoundKey. An extra RoundKey is used for "whitening" purposes. The AddRoundKey operation XORs the RoundKey with the State array. This operation is its own inverse. Thus, while deciphering, the only change needed is to change the order of the RoundKeys. That is, the last RoundKey is used in the beginning of decipher operation and so on.

For the details of the above operations and the RoundKey generation see works by Stinson [37] or Rijndael Proposal by Daemen and Rijnmen [32].

In the next section, we will discuss about one of the most widely used public key cryptosystem known as RSA cryptosystem.

13.4 PUBLIC KEY CRYPTOGRAPHY

Asymmetric cryptography, also called public key cryptography, is a relatively new field. It was invented by Diffie and Hellman in 1976. Let us briefly discuss about the motivation of Deffie–Hellman's (DH) work. For that consider the following assumption.

Assumption. One can get two functions f and g which are easily computable and $f \circ g =$ identity mapping, $g \circ f =$ identiy mapping and from f (or g), computing g (or f) is a computationally infeasibleproblem.

Protocol. Let there be n participants $P_1, P_2, \ldots, P_n$. For each participant P_i, let g_i be the public key and f_i be the private key such that $f_i \circ g_i =$ identity mapping, $g_i \circ f_i =$ identity mapping, $i = 1, 2, \ldots, n$. Now suppose, P_k wants to communicate some message M to another participant P_t, $k \neq t$, over an insecure channel. For that, first P_k will collect the public key g_t of the participant P_t and will compute $g_t(M)$. Then he/she will use his/her own private key f_k and will compute $C = f_k(g_t(M))$. Then the participant P_k will send the ciphertext C over an insecure channel to the participant P_t. On receiving the ciphertext C, the participant P_t will collect the public key g_k of P_k and first compute $g_k(C)$. After that, the participant P_t will use his/her own private key f_t to get back the original message $M = f_t(g_k(C))$, since $M = f_t(g_k(f_k(g_t(M))))$. Here, g_t is the notion of "authentication," f_k is the notion of "signature," g_k is the "signature verification" and f_t is the notion of "authorization validation."

In 1977, a year after the publication of the DH paper, three researchers at MIT developed a practical method to implement the suggested ideas. This became known as RSA, after the initials of the three developers—Ron Rivest, Adi Shamir, and Leonard Adelman—and is probably the most widely used public key cryptosystem. It was patented in the United States in 1983.

13.4.1 RSA Cryptosystem

Let n be a product of two distinct primes p and q. Let $\mathcal{P} = \mathcal{C} = \mathbf{Z}_n$. Let us define $\mathcal{K} = \{(n, p, q, e, d) : ed \equiv 1(\text{mod}\,\phi(n))\}$, where $\phi(n)$ is the number of positive integers less than n which are relatively prime to n. For each $K = (n, p, q, e, d)$, we define $e_K(x) = x^e(\text{mod}\,n)$ and $d_K(y) = y^d(\text{mod}\,n)$, where $x, y \in \mathbf{Z}_n$. The values n and e are public and the values p, q and d are used as public key.

Now we will verify that this really forms a public key cryptosystem. Suppose A wants to send a secret message to B using the public key of B. For that, first we will give algorithm for the generation of keys for B.

- B's algorithm to construct keys
  - Generate two distinct large primes p and q, each roughly of same size.
  - Compute $n = pq$ and $\phi(n) = (p-1)(q-1)$.
  - Select a random integer e with $1 < e < \phi(n)$, such that $\gcd(e, \phi(n)) = 1$.
  - Use the extended Euclidean algorithm to find the integer d, $1 < d < \phi(n)$, such that $ed \equiv 1(\text{mod}\,\phi(n))$.
  - B's public keys are n and e (i.e., known to A or C) and his private keys are p, q, and d.

- A's algorithm for encryption
  - Obtain B's public key (n, e).
  - Represent the message as an integer m in the interval $[0, n-1]$.
  - Compute $c \equiv m^e (\text{mod}\, n)$.
  - Send the ciphertext c to B.
- B's algorithm to decrypt the message
  - To obtain the plaintext message m, B uses his private key d to get $m \equiv c^d (\text{mod}\, n)$.

Proof of the decryption. It is given that $ed \equiv 1(\text{mod}\, \phi(n))$. So there must exist some integer t such that $ed = 1 + t\phi(n)$. Now we consider the following situations. If $\gcd(m, p) = 1$, then by Fermat's Theorem, $m^{p-1} \equiv 1(\text{mod}\, p) \Rightarrow m^{t(p-1)(q-1)} \equiv 1(\text{mod}\, p) \Rightarrow m^{1+t(p-1)(q-1)} \equiv m(\text{mod}\, p)$. Now if $\gcd(m, p) = p$, then also the above equality holds as both sides are equal to 0 modulo p. Hence in both the cases, $m^{ed} \equiv m(\text{mod}\, p)$. By same argument $m^{ed} \equiv m(\text{mod}\, q)$. Finally, since p and q are distinct primes, it follows that $m^{ed} \equiv m(\text{mod}\, n)$ and hence $c^d \equiv (m^e)^d \equiv m(\text{mod}\, n)$. Hence the result.

Illustration. Let us illustrate briefly the RSA algorithm with a simple example. Suppose A wants to send a secret message to B using RSA. Then A and B will follow the following algorithms.

- B's algorithm to construct keys
  - Consider two distinct primes $p = 11$ and $q = 13$.
  - Compute $n = pq = 143$ and $\phi(143) = 10 \cdot 12 = 120$.
  - Select an integer $e = 103$ with $1 < 103 < \phi(143)$, such that $\gcd(103, \phi(143)) = 1$.
  - Use the extended Euclidean algorithm to find the integer $d = 7, 1 < 7 < \phi(143)$, such that $103 \cdot 7 \equiv 1(\text{mod}\, \phi(143))$.
  - B's public key is $n = 143$ and $e = 103$ and his private key is $p = 11, q = 13$, and $d = 7$.
- A's algorithm for encryption
  - Obtain Bs public key $(n = 143, e = 103)$.
  - Represent the message as an integer m in the interval $[0, 143 - 1]$. Let $m = 7$.
  - Compute $c \equiv 7^{103}(\text{mod}\, 143) = 123$.
  - Send the ciphertext $c = 123$ to B.

- B's algorithm to decrypt the message
  - To obtain the plaintext message $m = 7$, B uses his private key $d = 7$ to get $m \equiv 123^7 (\text{mod } 143) = 7$.

Note: It is currently difficult to obtain the private key d from the public key (n, e). However, if one could factor n into p and q, then one could obtain the private key d. Thus, the security of the RSA system is based on the assumption that factoring is difficult. The discovery of an easy method of factoring would "break" RSA.

13.5 KEY AGREEMENT PROTOCOL

13.5.1 DH Key Agreement

DH proposed the first two-party single-round key agreement protocol in their seminal paper [11] that enables the users to compute a common key from a secret key and publicly exchanged information. No user is required to hold secret information before entering the protocol and each member makes an independent contribution to the common agreed key. This work invents the revolutionary concept of public key cryptography and is the most striking development in the history of cryptography.

- ***Protocol description***

  Setup: Let G be a finite multiplicative group of some large prime order q and g be a generator of G.

  Key Agreement: Assume that two entities A and B want to decide upon a common key. They perform the following steps.

  1. User A chooses a random $a \in Z_q^*$, computes $T_A = g^a$ and sends T_A to B.
  2. User B chooses a random $b \in Z_q^*$, computes $T_B = g^b$ and sends T_B to A.
  3. User A computes $K_A = T_B^a$ and similarly user B computes $K_B = T_A^b$.

  If A and B execute the above steps honestly, they will agree upon a common key $K_{\mathsf{AB}} = K_A = K_B = g^{ab}$.
- ***Assumption:*** DLP is hard.
- ***Security:*** The protocol is unauthenticated in the sense that it is secure against passive adversaries. An active adversary can mount man-in-the-middle attack.
- ***Efficiency***

  Communication: Round required is 1 and group element (of G) sent per user is 1.

  Computation: Each user computes two exponentiations.

13.5.2 Elementary Concepts on Elliptic Curves

Even though a pairing-based cryptographic primitive can be fully understood without any knowledge of elliptic curves, any implementation of such primitives will almost certainly involve the (modified) Weil or Tate pairing. We, therefore, included in the following section a brief introduction to elliptic curves that quickly leads to the definition of Weil pairing. For an elementary introduction to elliptic curves, we recommend Koblitz's book [24] and the notes by Charlap and Robbins [8]. The proofs of the results stated in this section can be found in the book by Silverman [36].

Let K be a field and $\overline{K}$ its algebraic closure. An elliptic curve over K is defined by a Weierstrass equation

$$E/K : y^2 + a_1xy + a_3y = x^3 + a_2x^2 + a_4x + a_6,$$

where $a_1, a_2, a_3, a_4, a_6 \in K$ and there are no "singular points" (singular points for a curve $f(x, y) = 0$ are those points where both the partial derivatives of f vanish). If $L \supset K$, then the set of L-rational points on E is

$$E(L) = \{(x, y) \in L \times L : y^2 + a_1xy + a_3y = x^3 + a_2x^2 + a_4x + a_6\} \cup \{\mathcal{O}\}.$$

Here $\mathcal{O}$ is an identified element, called point at infinity. If $L \supset K$, then $E(L) \supset E(K)$. We denote $E(\overline{K})$ by E. Simplified Weierstrass equation is as follows.

Case 1. If char$(K) \neq 2, 3$, then the equation simplifies to $y^2 = x^3 + ax + b, a, b \in K$ and $4a^3 + 27b^2 \neq 0$.

Case 2. If char$(K) = 2$, then the equation simplifies to

$$y^2 + xy = x^3 + ax^2 + b,\ a, b \in K, b \neq 0, \quad \text{nonsupersingular},$$

or

$$y^2 + cy = x^3 + ax + b,\ a, b, c \in K, c \neq 0, \quad \text{supersingular}.$$

For any $L \supset K$, the set $E(L)$ is an abelian group under the "chord-and-tangent law" [24] explained below: If $P \neq \mathcal{O}$, $Q \neq \mathcal{O}$, $Q \neq -P$, then $P + Q = -R$, where R is the third point of intersection of the line PQ (or tangent PQ in case $P = Q$) with the curve E.

Consider $E/K : y^2 = x^3 + ax + b$. Addition formulae are as follows:

1. $P + \mathcal{O} = \mathcal{O} + P = P$, for all $P \in E(L)$.
2. $-\mathcal{O} = \mathcal{O}$.
3. If $P = (x, y) \in E(L)$, then $-P = (x, -y)$.
4. If $Q = -P$, then $P + Q = \mathcal{O}$.

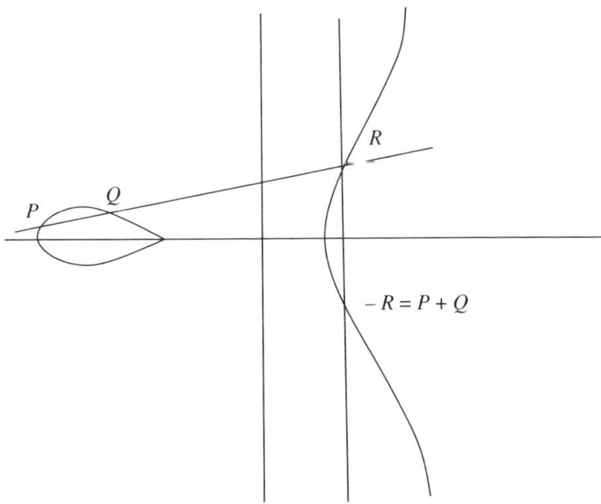

FIGURE 13.9 Elliptic curve addition ("chord-and-tangent law").

5. If $P = (x_1, y_1) \in E(L)$, $Q = (x_2, y_2) \in E(L)$, $P \neq -Q$, then $P + Q = (x_3, y_3)$, where $x_3 = \lambda^2 - x_1 - x_2$, $y_3 = \lambda(x_1 - x_3) - y_1$, and

$$\lambda = \begin{cases} \dfrac{y_2 - y_1}{x_2 - x_1} & \text{if } P \neq Q; \\ \dfrac{3x_1^2 + a}{2y_1} & \text{if } P = Q. \end{cases}$$

Note that if $P \neq \mathcal{O}$, $Q \neq \mathcal{O}$, $Q = -P$, then $P + Q = \mathcal{O}$, that is, $\mathcal{O}$ is the third point of intersection of any vertical line through P (or Q) with the curve E. Any vertical line through P (or Q) meets the curve E at infinity. This is why $\mathcal{O}$ is called point at infinity. $\mathcal{O}$ serves as the identity of the abelian group $E(L)$.

For the purpose of cryptography, assume henceforth that $K = \mathbb{F}_q$, that is, the finite field of characteristic p and of order q and $\overline{K} = \cup_{m \geq 1} \mathbb{F}_{q^m}$. The following are three important results on the group order of elliptic curve groups (Fig. 13.9).

Theorem 1 (Hasse's Theorem) $\#E(\mathbb{F}_q) = q + 1 - t$, $|t| \leq 2\sqrt{q}$. *Consequently,* $\#E(\mathbb{F}_q) \approx q$.

Theorem 2 (Schoof's Algorithm) $\#E(\mathbb{F}_q)$ *can be computed in polynomial time.*

Theorem 3 (Weil Theorem) *Let* $t = q + 1 - \#E(\mathbb{F}_q)$. *Let* α, β *be complex roots of* $T^2 - tT + q \in Z[T]$ *(where* $Z[T]$ *is the ring of polynomials in* T *with integer coefficients). Then* $\#E(\mathbb{F}_{q^k}) = q^k + 1 - \alpha^k - \beta^k$ *for all* $k \geq 1$.

The structure of elliptic curve groups is summarized by the following results.

- Let E be an elliptic curve defined over $\mathbb{F}_q$. Then $E(\mathbb{F}_q) \cong Z_{n_1} \oplus Z_{n_2}$, where $n_2 | n_1$ and $n_2 | (q - 1)$.

- $E(\mathbb{F}_q)$ is cyclic if and only if $n_2 = 1$.
- $P \in E$ is an n-torsion point if $nP = \mathcal{O}$ and $E[n]$ is the set of all n-torsion points.
- If $\gcd(n, q) = 1$, then $E[n] \cong Z_n \oplus Z_n$.

13.5.2.1 Supersingular Elliptic Curves An elliptic curve $E/\mathbb{F}_q$ is supersingular if $p|t$ where $t = q + 1 - \#E(\mathbb{F}_q)$.

Theorem 4 (Waterhouse) *$E/\mathbb{F}_q$ is supersingular if and only if $t^2 = 0, q, 2q, 3q$ or $4q$. The group structure is given by the following result.*

Theorem 5 (Schoof) *Let $E/\mathbb{F}_q$ be supersingular with $t = q + 1 - \#E(\mathbb{F}_q)$. Then*

1. *If $t^2 = q, 2q$ or $3q$, then $E(\mathbb{F}_q)$ is cyclic.*
2. *If $t^2 = 4q$ and $t = 2\sqrt{q}$, then $E(\mathbb{F}_q) \cong Z_{\sqrt{q}-1} \oplus Z_{\sqrt{q}-1}$.*
3. *If $t^2 = 4q$ and $t = -2\sqrt{q}$, then $E(\mathbb{F}_q) \cong Z_{\sqrt{q}+1} \oplus Z_{\sqrt{q}+1}$.*
4. *If $t = 0$ and $q \not\equiv 3 \bmod 4$, then $E(\mathbb{F}_q)$ is cyclic.*
5. *If $t = 0$ and $q \equiv 3 \bmod 4$, then $E(\mathbb{F}_q)$ is cyclic or $E(\mathbb{F}_q) \cong Z_{(q+1)/2} \oplus Z_2$.*

13.5.3 Cryptographic Bilinear Maps

Let G_1, G_2 be two groups of the same prime order q. We view G_1 as an additive group and G_2 as a multiplicative group. A mapping $e : G_1 \times G_1 \rightarrow G_2$ satisfying the following properties is called a cryptographic bilinear map:

Bilinearity	$e(aP, bQ) = e(P, Q)^{ab}$ for all $P, Q \in G_1$ and $a, b \in Z_q^*$.
Nondegeneracy	If P is a generator of G_1, then $e(P, P)$ is a generator of G_2.
Computability	There exists an efficient algorithm to compute $e(P, Q)$.

Modified Weil Pairing [5] and Tate Pairing [2,19] are examples of cryptographic bilinear maps.

13.5.3.1 Decision Hash Bilinear Diffie–Hellman (DHBDH) Problem Let (G_1, G_2, e) be as in Section 13.5.3. We define the following problem. Given an instance (P, aP, bP, cP, r) for some $a, b, c, r \in_R Z_q^*$ and a one-way hash function $H : G_2 \rightarrow Z_q^*$, to decide whether $r = H(e(P, P)^{abc}) \bmod q$. This problem is termed DHBDH problem as defined in the work by Barua et al. [3] and is a combination of the bilinear Diffie–Hellman (BDH) problem and a variation of the hash Diffie–Hellman (HDH) problem.

The DHBDH assumption is that there exists no probabilistic, polynomial time, 0/1-valued algorithm that can solve the DHBDH problem with nonnegligible probability of success.

13.5.4 Tree-based Group Key Agreement Using Pairing

Barua et al. [3], present a ternary tree-based unauthenticated key agreement protocol by extending the basic Joux's protocol [22] to multiparty setting and provide a proof of security against passive adversaries. In the work by Dutta et al. [14], a provably secure authenticated tree-based group key agreement from the unauthenticated protocol of Barua et al. [3] is proposed with the security analysis in the model formalized by Bresson et al. [6]. The dynamic case of the scheme in the work by the Dutta et al. [14] is further considered in the work by Dutta and Barua [15] that enables a user to join or leave the group at his desire retaining the tree structure with minimum key updates. We will present here the basic unauthenticated scheme in the work by Barua et al. [3].

- ***Protocol description***

  Setup: Suppose a set of n users $\mathcal{P} = \{U_1, U_2, \ldots, U_n\}$ wish to agree upon a secret key. Let US be a subset of users. Quite often, we identify a user with its instance during the execution of a protocol. In case US is a singleton set, we will identify US with the instance it contains. Each user set US has a representative Rep(US) and for the sake of concreteness we take Rep(US) $= U_j$, where $j = \min\{k : \Pi_{U_k}^{d_k} \in \mathsf{US}\}$. We use the notation $A[1, \ldots, n]$ for an array of n elements $A_1, \ldots, A_n$ and write $A[i]$ or A_i to denote the ith element of array $A[\,]$. Let $G_1 = \langle P \rangle, G_2$ (groups of prime order q) and $e(,)$ be as described in Section 13.5.3. We choose a hash function $H : G_2 \to Z_q^*$. The public parameters are params $= (G_1, G_2, e, q, P, H)$. Each user $U_i \in \mathcal{P}$ chooses $s_i \in Z_q^*$ at random which it uses as its ephemeral key. These keys are session specific and determine the final common key for the users in a session.

  Key agreement: Let $p = n/3$ and $r = n \bmod 3$. The set of users participating in a session is partitioned into three user sets $\mathsf{US}_1, \mathsf{US}_2, \mathsf{US}_3$ with respective cardinalities being p, p, p if $r = 0$; $p, p, p+1$ if $r = 1$; and $p, p+1, p+1$ if $r = 2$. This top-down recursive procedure is invoked for further partitioning to obtain a ternary tree structure (*cf.* Section 13.11). The lowest level 0 consists of singleton users having a secret key. CombineTwo, a key agreement protocol for two user sets, and CombineThree, a key agreement protocol for three user sets are invoked in the key tree thus obtained. These two procedures are demonstrated in Figure 13.10.

  All communications are done by representatives and users in each user set have a common agreed key. In CombineThree, a, b, c, respectively, are the common agreed key of user sets A, B, C. Representative of user set A sends aP to both the user sets B, C. Similarly, representative of B sends bP to both A, C and representative of C sends cP to both A, B. After these communications, each user can compute the common agreed key $H(e(P, P)^{abc})$. In CombineTwo, users in user set A have common agreed key a, users in user set B have common agreed key b. Representative of A sends aP to user set B and representative of B sends bP to user set A.

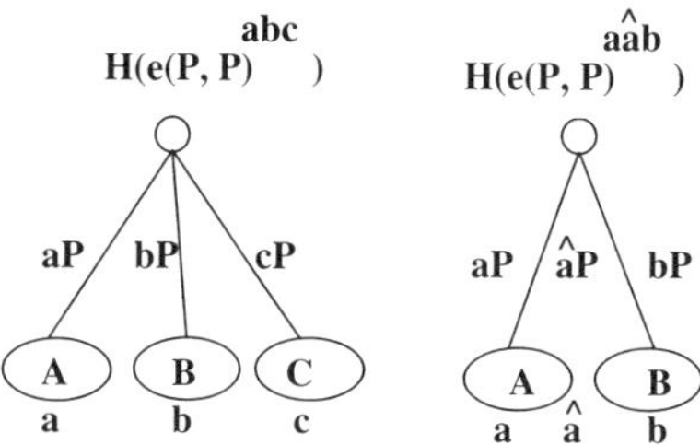

FIGURE 13.10 **Procedure** CombineThree and **procedure** CombineTwo.

Moreover, representative of user set A generates a random key $\hat{a} \in Z_q^*$ and sends $\hat{a}P$ to all the users in both A, B. After these communications, each user can compute the common agreed key $H(e(P, P)^{a\hat{a}b})$ (Fig. 13.11). The formal description of the protocol is given below.

procedure KeyAgreement(l, US$[i+1, \ldots, i+l]$, S$[i+1, \ldots, i+l]$)

if $(l = 2)$ **then**

call CombineTwo(US$[i+1, i+2]$, S$[i+1, i+2]$);

return;

end if

if $(l = 3)$ **then**

call CombineThree(US$[i+1, i+2, i+3]$, S$[i+1, i+2, i+3]$);

return;

end if

$p_0 = 0$; $p_1 = \lfloor l/3 \rfloor$; $p_3 = \lceil l/3 \rceil$; $p_2 = l - p_1 - p_3$;

$n_0 = 0$; $n_1 = p_1$; $n_2 = p_1 + p_2$;

for $j = 1$ to 3 **do in parallel**

$\widehat{\text{US}}_j = \text{US}[i + n_{j-1} + 1, \ldots, i + n_{j-1} + p_j]$;

if $p_j = 1$, **then** $\widehat{\text{S}}_j = S[i + n_{j-1} + 1]$;

else

call KeyAgreement(p_j, $\widehat{\text{US}}_j$, S$[i + n_{j-1} + 1, \ldots, i + n_{j-1} + p_j]$);

Let $\widehat{\text{S}}_j$ be the common agreed key among all members of $\widehat{\text{US}}_j$;

end if;

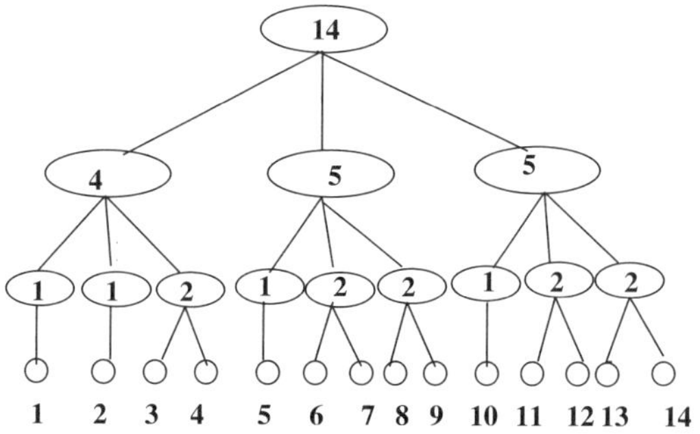

FIGURE 13.11 **Procedure** KeyAgreement for $n = 14$.

18. **end for**;
19. **call** CombineThree($\widehat{\mathsf{US}}[1, 2, 3]$, $\widehat{\mathsf{S}}[1, 2, 3]$);
end KeyAgreement

procedure CombineTwo(US[1, 2], S[1, 2])
1. **do** Steps 2 and 3 **in parallel**
2. US_1 generates $\mathsf{S} \in_R Z_q^*$ and sends SP and S_1P to US_2;
3. US_2 sends S_2P to US_1;
4. **end do**;
5. **do** steps 6 and 7 **in parallel**
6. US_1 computes $H(e(\mathsf{S}_2P, \mathsf{S}P)^{\mathsf{S}_1})$;
7. US_2 computes $H(e(\mathsf{S}_1P, \mathsf{S}P)^{\mathsf{S}_2})$;
8. **end do**;
end CombineTwo

procedure CombineThree(US[1, 2, 3], S[1, 2, 3])
1. **for** $i = 1$ to 3 **do in parallel**
2. Let $\{j, k\} = \{1, 2, 3\} \setminus \{i\}$;
3. Rep(US_i) sends S_iP to all members $\mathsf{US}_j \cup \mathsf{US}_k$;
4. **end for**;
5. **for** $i = 1$ to 3 **do in parallel**
6. let $\{j, k\} = \{1, 2, 3\} \setminus \{i\}$;
7. each member of US_i computes $H(e(\mathsf{S}_j\mathsf{P}, \mathsf{S}_k\mathsf{P})^{\mathsf{S}_i})$;
8. **end for**;
end CombineThree

The start of the recursive protocol KeyAgreement is made by the following statements:

start main
1. $\mathsf{US}_j = \{U_j\}$ for $1 \leq j \leq n$;
2. User j chooses a secret $s_j \in_R Z_q^*$;
3. User j sets $\mathsf{S}[j] = \mathsf{s}_j$;
4. **call** KeyAgreement(n, US[1, ..., n], S[1, ..., n]).
end main

The values $s_1, \ldots, s_n$ are session specific and determine the final common key for the users. Note that CombineTwo is invoked only for two individual users (i.e., $|\mathsf{US}_1| = |\mathsf{US}_2| = 1$), whereas CombineThree is invoked for three individual users as well as for three groups of users. In CombineThree the common agreed key of user sets US_1, US_2, US_3 is $H(e(P, P)^{\mathsf{S}_1\mathsf{S}_2\mathsf{S}_3})$ and in CombineTwo the common agreed key of the two users in the singleton sets US_1, US_2 is $H(e(P, P)^{\mathsf{S}_1\mathsf{S}_2\mathsf{S}})$.

The protocol described above allows $U_1, \ldots, U_n$ to agree upon a common key. The same protocol can be used by an arbitrary subset of $\{U_1, \ldots, U_n\}$ to agree upon a common key.

- ***Assumption:*** Decision hash bilinear Diffie–Hellman (DHBDH) problem is hard.
- ***Security:*** Secure against passive adversary under the assumption that DHBDH problem is hard.
- ***Efficiency***

  Communication: Round required is $\lceil \log_3 n \rceil$ and group element (of G_1) sent per user is $n\lceil \log_3 n \rceil$.

  Computation: Each user computes $< \frac{5}{2}(n-1)$ elliptic curve scalar multiplications, $n\lceil \log_3 n \rceil$ pairings, $n\lceil \log_3 n \rceil$ group exponent in G_2, and $n\lceil \log_3 n \rceil$ hash function (H) evaluation.

REFERENCES

1. Berlekamp ER. Algebraic Coding Theory. New York: McGraw-Hill; 1968.
2. Barreto PSLM, Kim HY, Scott M. Efficient algorithms for pairing based cryptosystems. Proceedings of Crypto 2002. LNCS. Volume 2442. Springer-Verlag; 2002. p 354–368. Also available at http://www.iacr.org/2002/008
3. Barua R, Dutta R, Sarkar P. Extending Joux Protocol to Multi Party Key Agreement. Proceedings of Indocrypt 2003. LNCS. Volume 2904. Springer-Verlag; 2003. p 205–217. Also available at http://eprint.iacr.org/2003/062
4. Biham E, Shamir A. Differential Cryptanalysis of the Data Encryption Standard. Springer-Verlag; 1993.
5. Boneh D, Franklin M. Identity-based encryption from Weil pairing. Proceedings of Crypto 2001. LNCS. Volume 2139. Springer-Verlag; 2001. p 213–229.
6. Bresson E, Chevassut O, Pointcheval D. Dynamic group Diffie–Hellman key exchange under standard assumptions. Proceedings of Eurocrypt 2002. LNCS. Volume 2332. Springer-Verlag; 2002. p 321–336.
7. Camion P, Carlet C, Charpin P, Sendrier N. On correlation-immune functions. Advances in Cryptology—Crypto 1991. Lecture Notes in Computer Science. Volume 576. Springer-Verlag; 1992. p 86–100.
8. Charlap L, Robbins D. An elementary Introduction to Elliptic Curves. CRD Expository Report No. 31. Institute for Defence Analysis, Princetona, December 1988.
9. Daemen J, Rijmen V. AES Proposal: Rijndael. Submitted to the Advanced Encryption Standard (AES) contest; 1998.
10. Daemen J, Rijmen V. The Design of Rijndael. 1st ed. Berlin: Springer; 2001.
11. Diffie W, Hellman M. New directions in cryptography. IEEE Trans Inform Theor 1976;IT-22(6):644–654.
12. Ding C, Xiao G, Shan W. The stability theory of stream ciphers. Lecture Notes in Computer Science. Volume 561. Berlin: Springer-Verlag; 1991.
13. Dutta R, Barua R, Sarkar P. Authenticated multi-party key agreement: a provably secure tree based scheme using pairing. Proceedings of National Workshop on Cryptology; October. 2004; Kerala, India; 2004.
14. Dutta R, Barua R, Sarkar P. Provably secure authenticated tree based group key agreement. Proceedings of ICICS 2004. LNCS. Volume 3269. Berlin: Springer-Verlag; 2004. p 92–104. Also available at http://eprint.iacr.org/2004/090

15. Dutta R, Barua R. Dynamic group key agreement in tree-based setting. Proceedings of ACISP 2005. LNCS. Volume. 3574. Berlin: Springer-Verlag; 2005. p 101–112. Also available at http://eprint.iacr.org/2005/131

16. Ekdahl P, Johansson T. A new version of the stream cipher SNOW. Selected Areas in Cryptography, SAC 2002. Lecture Notes in Computer Science. Volume 2595. Berlin: Springer-Verlag; 2003. p 47–61.

17. Even S, Mansour Y. A construction of a cipher from a single pseudorandom permutation. J Cryptol 1997;10:151–161.

18. Fluhrer SR, Mantin I, Shamir A. Weaknesses in the key scheduling algorithm of RC4. Selected Areas in Cryptography, SAC 2001. Lecture Notes in Computer Science. Volume 2259. Berlin: Springer-Verlag; 2001. p 1–24.

19. Galbraith S, Harrison K, Soldera D. Implementing the Tate Pairing. Proceedings of Algorithm Number Theory Symposium—ANTS V. LNCS. Volume 2369. Berlin: Springer-Verlag; 2002. p 324–337.

20. Gong G. Sequence Analysis. Lecture Notes for CO739x; Winter 1999. Available at website http://www.comsec.uwaterloo.ca/˜ ggong (last accessed on July 4th 2006).

21. Hawkes P, Rose G. Primitive specification and supporting documentation for sober-t32 submission to nessie. Proceedings of the First NESSIE Workshop; 2000, Belgium.

22. Joux A. A One Round Protocol for Tripartite Diffie–Hellman. Proceedings of ANTS 4. LNCS. Volume 1838. Berlin: Springer-Verlag; 2000. p 385–394.

23. Kahn D, The Codebreakers. New York: Macmillan; 1967.

24. Koblitz N. A Course in Number Theory and Cryptography. Berlin: Springer-Verlag; 1987.

25. Mantin I, Shamir A. A practical attack on broadcast RC4. Fast Software Encryption, FSE 2001. Lecture Notes in Computer Science. Volume. 2355. Springer-Verlag; 2001. p 152–164.

26. Massey JL. Shift-register synthesis and BCH decoding. IEEE Trans Inform Theor 1969; IT-15:122–127.

27. Meier W, Staffelbach O. Fast correlation attacks on certain stream ciphers. J Cryptol 1989;1(3):159–176.

28. National Bureau of Standards. Data Encryption Standard. U.S. Department of Commerce, FIPS publications 46, 1977.

29. Pasalic E, Maitra S, Johansson T, Sarkar P. New constructions of resilient and correlation immune Boolean functions achieving upper bounds on nonlinearity. Workshop on Coding and Cryptography—WCC 2001; 2001 Jan 8–12; Paris. Amsterdam: Elsevier; 2001.

30. Paul S, Preneel B. A new weakness in the RC4 keystream generator and an approach to improve the security of the cipher. Fast Software Encryption, FSE 2004. Lecture Notes in Computer Science. Volume 3017. Berlin: Springer-Verlag; 2004. p 245–259.

31. Pudovkina M,. Statistical weaknesses in the alleged RC4 keystream generator. Cryptology ePrint Archive: Report No. 2002/171, Available at:http://eprint.iacr.org/2002/171

32. Rijndael, Daemen J, Rijmen V. AES Proposal; 1999.

33. Rose GG, Hawkes P. Turing: A fast stream cipher. Fast Software Encryption, FSE 2003. Lecture Notes in Computer Science. Volume 2887. Berlin: Springer-Verlag; 2003. p 290–306.

34. Shannon CE. Communication theory of secrecy systems. Bell Sys Tech J 1949; 28(4):656–715.

35. Siegenthaler T. Correlation-immunity of nonlinear combining functions for cryptographic applications. IEEE Trans Inform Theor 1984;IT-30(5):776–780.
36. Silverman J. The Arithmetic of Elliptic Curves. Berlin: Springer-Verlag; 1986.
37. Stinson D, Cryptography: Theory and Practice. Boca Raton: CRC Press; 1995.
38. Tarannikov YV. On resilient Boolean functions with maximum possible nonlinearity. Progress in Cryptology—Indocrypt 2000. Lecture Notes in Computer Science. Volume 1977. Berlin: Springer Verlag; 2000. p 19–30.

CHAPTER 14

Secure Communication in Distributed Sensor Networks (DSN)

SUBHAMOY MAITRA and BIMAL ROY

14.1 GENERAL OVERVIEW OF DISTRIBUTED SENSOR NETWORK (DSN) AND ITS LIMITATIONS

In this chapter we will study the issues of implementing cryptographic primitives on a sensor node. The basic premise in this regard is the hardware capability of a sensor node is limited (i.e., CPU of lower speed, less amount of memory, and limited availability of power sources). Given some high complexity primitives generally used in cryptosystems, there is a need to look at the implementability of such primitives on a sensor node.

Before proceeding further, let us present a brief introduction to wireless sensor networks. A wireless sensor network consists of a number of inexpensive sensor devices spread across a geographical area. Each sensor is capable of wireless communication using the radio frequency (RF). The sensor nodes also have some limited computing capability. Let us first list a few applications of sensor networks.

1. Military sensor networks to detect and gain as much information as possible about enemy movements, explosions, and other phenomena of interest.
2. Sensor networks to detect and characterize chemical, biological, radiological, nuclear, and explosive (CBRNE) attacks and material.
3. Sensor networks to detect and monitor environmental changes in plains, forests, oceans, and so on.
4. Wireless traffic sensor networks to monitor vehicle traffic on highways or in congested parts of a city.
5. Wireless surveillance sensor networks for providing security in shopping malls, parking garages, and other facilities.
6. Wireless parking lot sensor networks to determine which spots are occupied and which are free.

Handbook of Applied Algorithms: Solving Scientific, Engineering and Practical Problems
Edited by Amiya Nayak and Ivan Stojmenović

The above list suggests that wireless ad hoc sensor networks offer certain capabilities and enhancements in operational efficiency in civilian applications as well as in assisting the national effort to increase alertness to potential terrorist threats.

Two ways to classify wireless ad hoc sensor networks are whether or not the nodes are individually addressable, and whether the data in the network are aggregated. The sensor nodes in a parking lot network should be individually addressable, so that one can determine the locations of all the free spaces. This application shows that it may be necessary to broadcast a message to all the nodes in the network. If one wants to determine the temperature in a corner of a room, then addressability may not be so important. Any node in the given region can respond. The ability of the sensor network to aggregate the data collected can greatly reduce the number of messages that need to be transmitted across the network.

The basic goals of a wireless ad hoc sensor network generally depend upon the application, but the following tasks are common to many other networks.

1. *Determine the value of some parameters at a given location:* In an environmental network, one might want to know the temperature, atmospheric pressure, amount of sunlight, and the relative humidity at a number of locations. This example shows that a given sensor node may be connected to different types of sensors, each with a different sampling rate and range of allowed values.
2. *Detect the occurrence of events of interest and estimate parameters of the detected event or events:* In the traffic sensor network, one would like to detect a vehicle moving through an intersection and estimate the speed and direction of the vehicle.
3. *Classify a detected object:* Is a vehicle in a traffic sensor network a car, a minivan, a light truck, a bus, and so on.
4. *Track an object:* In a military sensor network, track an enemy tank as it moves through the geographic area covered by the network.

In these four tasks, an important requirement of the sensor network is that the required data are to be disseminated to the proper end users. In some cases, there are fairly strict time requirements for the communication. For example, the detection of an intruder in a surveillance network should be immediately communicated to the police so that action can be taken.

Wireless ad hoc sensor network requirements include the following:

1. *Large number of (mostly stationary) sensors:* Aside from the deployment of sensors on the ocean surface or the use of mobile, unmanned, robotic sensors in military operations, most nodes in a smart sensor network are stationary. Networks of 10,000 or even 100,000 nodes are envisioned, so scalability is a major issue.
2. *Low energy usage:* Since in many applications the sensor nodes will be placed in a remote area, reworking on a node may not be possible. In this case, the

lifetime of a node may be determined by the battery life, thereby requiring the minimization of energy expenditure.

3. *Network self-organization:* Given the large number of nodes and their potential placement in hostile locations, it is essential that the network be able to self-organize; manual configuration is not feasible. Moreover, nodes may fail (either from lack of energy or from physical destruction), and new nodes may join the network. Therefore, the network must be able to periodically reconfigure itself so that it can continue to function. Individual nodes may become disconnected from the rest of the network, but a high degree of connectivity must be maintained.
4. *Collaborative signal processing:* Yet another factor that distinguishes these networks from MANETs is that the end goal is detection/estimation of some events of interest, and not just communications. To improve the detection/estimation performance, it is often quite useful to fuse data from multiple sensors. This data fusion requires the transmission of data and control messages, and so it may put constraints on the network architecture.
5. *Querying ability:* A user may want to query an individual node or a group of nodes for information collected in the region. Depending on the amount of data fusion performed, it may not be feasible to transmit a large amount of the data across the network. Instead, various local sink nodes will collect the data from a given area and create summary messages. A query may be directed to the sink node nearest to the desired location.

14.2 MODELS FOR SECURE COMMUNICATION

Given the availability of low cost, short-range radios along with advances in wireless networking, it is expected that wireless sensor networks will become commonly deployed. In these networks, each node may be equipped with a variety of sensors such as acoustic, seismic, infrared, still/motion video camera, and so on. These nodes may be organized in clusters such that a locally occurring event can be detected by most of, if not all, the nodes in a cluster. Each node may have sufficient processing power to make a decision, and it will be able to broadcast this decision to the other nodes in the cluster. One node may act as the cluster master, and it may also contain a longer range radio using a protocol such as IEEE 802.11 or Bluetooth.

14.2.1 Security Issues

Let us point out the fundamental difficulties in providing security to a sensor network.

1. The issue of taking the advantage of asymmetric cryptography is a real challenge in this area since the sensor devices have constraints in terms of computation, communication, memory, and energy resources. RSA algorithm or Diffie–Hellman key agreement protocol are difficult to implement, whereas

the symmetric solutions like Advanced Encryption Standard (AES) block cipher and HMAC-SHA-1 message authentication code are faster and easier to compute for the sensor nodes.

2. The nodes may be physically captured. Usually one should not assume that the hardware in each node is tamper resistant. Compromised nodes may behave arbitrarily, possibly in collusion with other compromised nodes.
3. Since the communication channel is wireless in sensor networks environment, eavesdropping and injection of malicious messages could be easier.
4. The sensor network security protocols should be amenable to scalability. Usually the network is often required to be scaled up to cater to several sensor nodes.
5. Lack of fixed infrastructure.
6. Unknown network topology prior to deployment.

There are different attack models. If the attacker is not an authorized participant of the network, it is called an *outsider attack*. For example, a passive eavesdropper, packet spoofer, or signal jammer may launch an outsider attack. Also physical destruction of nodes (may be intentional, climatic, or resulting from depletion of energy sources) is a form of outsider attack. Benign node failure is to be considered as a security problem since it is indistinguishable from an attack resulting into disabling a node.

On the other hand, an *insider attack* means the compromise of one or more sensor node(s). A compromised node may run some malicious code to steal some secret from the network and in turn that may disrupt the normal functioning of the complete network. If standard encryption and authentication protocols are implemented in the network, the compromised node should have some valid secret keys that enable it to join the secret and authenticated communications.

If the base station is assumed to be a trusted server that is never compromised, the problem of key distribution finds a ready solution. The base station serves as the trusted intermediary and distributes a key to each pair of nodes that need to communicate. However, for a network of very large size, the nodes in the immediate vicinity of the base station will have to continuously relay the key setup messages and very soon deplete the energy source. Also the base station will have to set up $n(n-1)/2$ keys in the worst case and becomes inefficient in case of large n.

The basic idea is to make the network resistant to outsider attacks and resilient to insider attacks (while maintaining a realistic notion of security). The former may be achieved by standard cryptographic primitives and maintaining some redundancy in the network. The network protocols should be capable of identifying the failed nodes in real time and update themselves according to the updated topology.

For the latter, the ideal situation is to detect the compromised node and revoke the keys contained therein. It is not always possible and perhaps the way out is to design protocols resilient to node capture so that the performance of the network gracefully degrades with the compromise of a small fraction of nodes. Depending on the application and sensitivity of the collected data, the security level may be relaxed or beefed up. Let us now list a few specific requirements.

1. *Authentication:* It is usually in two forms, namely source authentication and data authentication. The verification of the origin of a message/packet is known as source authentication and the condition that the data are unchanged during the transmission is known as data authentication. Though authentication prevents outsider attacks like injecting/spoofing of packets, a compromised node can authenticate itself to the network since it is in possession of valid secret keys.
2. *Secrecy:* Using standard cryptographic techniques and shared secret keys between the communicating nodes may not be sufficient to maintain secrecy because an eavesdropper can analyze the network traffic and obtain some sensitive meta data. Access control has to be exercised in order to protect the privacy of the collected data. An insider attack may defeat this purpose since the data can be revealed or the communication between two nodes may be eavesdropped by a compromised node.
3. *Availability:* Availability means the functioning of the devices for the entire lifetime. Denial of service (DoS) attacks result in a loss of availability. Both outsider and insider attacks may cause nonavailability.
4. *Integrity of service:* In the application layer, the protocols may be required to provide service integrity in the face of malfunctioning (compromised) nodes. As an example, the data aggregation service should be able to filter out the erroneous readings provided by the compromised nodes.

Secrecy and authentication may be protected from outsider attacks (like packet spoofing/modification and eavesdropping) using standard cryptographic techniques. Two sensor nodes can set up a secret and authenticated link through a shared secret key. The problem of setting up the secret key between a pair of nodes is known as the key establishment problem. There are various solutions available to this problem. Among them, the most naive one is to use a single master key for the entire network. The moment a single node is compromised, the entire network becomes insecured. At the other extreme, if one uses different keys for each pair of nodes, it will be extremely secure. This scheme is not viable because each node has to store several keys, which is not achievable due to memory constraint in sensor nodes. This solution does not scale well with the increase in the size of the network. The other solution may be obtained using public key cryptography. This is computation intensive, and one of the most important recent challenges is to implement such primitives in low end hardware. It should be noted that the public key solution is also susceptible to DoS attacks. Availability may be disrupted through DoS attacks [44] and may take place in different parts of the protocol stack.

Many sensor network protocols use broadcast and multicast, one cannot use digital signatures for the verification of the messages since public key cryptography is difficult in sensor networks. As a possible solution, in the work by Perrig et al. [35], the μTesla protocol has been proposed. A notion of asymmetry is introduced into symmetric key cryptography by the use of one-way function key chains and delayed key disclosures.

At the physical layer, jamming may be tried by propagating interfering RF signals. The other form of jamming may be by injection of irrelevant data or wastage of battery power at the reception node. The solution to this problem is discussed in the work by Pickholtz et al. [37], where frequency hopping and spread spectrum communication have been suggested. The jamming may also take place in the link layer by inducing malicious collisions or obtaining an unfair share of the radio resource. This can be resisted by carefully designing secure MAC protocols as described in the work by Wood and Stankovic [44]. If the jamming is attempted at the networking layer through the injection of malicious data packets, one can use authentication to detect such packets and nonces to detect replayed packets.

There is another kind of attack called the Sybil attack [11,34]. In this case, a malicious node claims multiple identities. The affected node can claim a major part of the radio resource. The attacker will succeed to achieve a selective forwarding and to create a sinkhole so that the affected node can capture a large amount of data [25]. The defense mechanisms have been detailed in the work by Newsome et al. [34] leveraging the key distribution strategy.

There may be different kinds of attacks like denying a message to the intended recipient, dropping of packets, and selective forwarding [25]. Multipath routing solves this problem [9,17]. Some other attacks like spreading bogus routing information, creating sinkholes or wormholes, and Hello flooding have been described [25].

Service integrity may be at stake if the attacker launches a stealthy attack in order to make the network accept a false data value. It may be achieved in different ways like compromising an aggregator node, a Sybil attack by a compromised node to affect the data value, a DoS attack to legitimate nodes to stop them reporting to the base station, and so on. The stealthy attack in data aggregation context and Secure Information Aggregation (SIA) Protocol have been proposed in the work by Przydatek et al. [38]. For an excellent and brief reading in this area we refer to the work by Perrig et al. [36].

14.3 LOW COST KEY AGREEMENT ALGORITHMS

Before starting a secure communication, the parties need to settle on one or more secret keys. In 1976, Diffie and Hellman proposed [10] a one round bipartite key agreement protocol based on the hardness of the discrete log problem in any cyclic group. Let G be a cyclic group of some large prime order p and let g be a generator of G. Suppose two entities A and B want to establish a common key between themselves. A chooses some random $a \in Z_p^*$, computes g^a, and sends it to B; while B chooses some random $b \in Z_p$, computes g^b, and sends it to A. On receiving g^a, B computes the common key as $K_{AB} = (g^a)^b$, while on receiving g^b, A computes $K_{AB} = (g^b)^a$. This protocol is secure assuming that the discrete logarithm problem (DLP) is hard over G. The DLP over G is: given g^a it should be computationally hard to obtain a. Later, in 2001, Joux [24] proposed a one round tripartite key agreement protocol based on bilinear pairing.

Both these protocols were later extended to multiparty setting, which is referred to as group key agreement [5]. Researchers also considered the dynamic scenario [40]; that is, participants are allowed to join or leave the group at any time. A general approach of group key agreement is to arrange the participants in a tree structure — for n participants this requires around n rounds. Constant round group key agreement protocol is also available in the literature [4].

The group key agreement protocols are usually implemented over elliptic curve groups [26]. Those requiring bilinear pairing further use modified Weil or Tate pairing [15] over elliptic curve groups. Operations over elliptic curve groups and implementation of bilinear pairing are computationally quite intensive. This severely restricts their application in smaller devices, especially sensor networks, though some encouraging result is available for elliptic curve cryptography in 8-bit processors [20].

For application is sensor networks, the target is to achieve some kind of optimum trade-off between computational and memory costs and communication bandwidth. In the absence of any trusted central authority, contributory group key agreement (CGKA) protocols that provide some kind of verifiable trust relationship has been suggested [31] for this kind of situation. Some recent works [29] are available in this direction based on the tree-based approach of group key agreement in the elliptic curve settings.

14.4 KEY PREDISTRIBUTION

Consider a scenario where N number of sensor nodes are dropped from an airplane in the battlefield. Thus, the geographical positioning of the nodes cannot be decided *a priori*. However, any two nodes in RF range are expected to be able to communicate securely. One option is to maintain different secret keys for each of the pairs. Then each of the nodes needs to store $N - 1$ keys. Given (i) the huge number of sensor nodes generally deployed, (ii) the memory constraint of the sensor nodes, this solution is not practical. On the other hand, online key exchange needs further research as implementation of public key framework demands processing power at the higher end. Hence, key predistribution to each of the sensor nodes before deployment is a thrust area of research and the most used mathematical tool for key predistribution is combinatorial design. Each of the sensor nodes contains M keys and each key is shared by Q nodes, (thus fixing M and Q) such that the encrypted communication between two nodes may be decrypted by at most $Q - 2$ other nodes if they fall within the RF range of the two communicating nodes. Similarly, one node can decrypt the communication between any two of at most $M(Q - 1)$ nodes if it lies within the RF range of all the nodes who share a key with it.

Let us present an exact example from the work by Lee and Stinson [28]. Take $N = 2401$, $M = 30$, $Q = 49$. The parameters are obtained using a transversal design (TD; for a basic introduction to TD, refer the work by Street and Street [43, p 133] or Section 14.4.1). It has been shown that two nodes share either 0 or 1 key. In this case, $M(Q - 1)$ gives the number of nodes with which one node can communicate. The expected number of keys that is common between any two nodes is

$M(Q-1)/N-1=0.6$. This is called the probability that two nodes share a common key [28]. Further, it can be checked that if two nodes do not share a common key, then they may communicate via another intermediate node. Let nodes ν_i, ν_j do not share a common key, but ν_i, ν_k share a common key and ν_k, ν_j share a common key, i, j, k are all distinct. Hence, the secret communication between ν_i and ν_k needs a key (encrypted by ν_i, decrypted by ν_k) and that between ν_k and ν_j needs another secret key (encrypted by ν_k, decrypted by ν_j). It has been shown that the communication between two nodes is possible in almost 0.99995 proportion of cases [28]. However, the following problems are immediate:

1. Communication between any two nodes in 60 Percent of the cases will be in one step (no involvement of any other node), but the communication between any two of them needs two steps for the rest 40 Percent of the cases, making the average of 1.4 steps in each communication. This is an overhead. Thus, we need a design where we can guarantee that there is a common key between any two nodes.
2. The direct communication between any two nodes can be decrypted by at most $Q-2$ other nodes. However, if one takes the help of a third intermediate node, then the communication can be decrypted by at most $2(Q-2)$ nodes. Thus, any communication can be decrypted by at most $1.4(Q-2)$ nodes on an average.
3. In an adversarial situation, if s nodes are compromised, it has been shown that $1-(1-(Q-2/N-2))^s$ proportion of links becomes unusable. In this specific design, for $s=10$, out of 2401 nodes, the proportion of unusable links becomes as high as 17.95 Percent.

The solution to all these problems is based on the fact that we need to increase the number of common keys between any two nodes. The issues at this point are as follows:

1. The number of keys to be stored in each node will clearly increase. So one needs to decide the availability of storage space. It has been commented that storing 150 keys in a sensor node may not be practical [28, p. 4]. On the other hand, scenarios have been described with 200 keys in the works by Du et al. [12, p. 17] and Lee and Simon [27, section 5.2]. If one considers 4 Kbytes of memory space for storing keys in a sensor node, then choosing 128-bit key (16 byte), it is possible to accommodate 256 keys.
2. It is not easy to find out combinatorial designs with prespecified number of common keys (say, e.g., 5) among any two nodes for key predistribution [8,42]. Consider the following technique. Generally, a sensor node corresponds to a block in combinatorial design [6,28]. Here one can merge a few blocks to get a sensor node. Thus, the key space at each node gets increased and the number of common keys between any two nodes can also be increased to the desired level. This technique provides a much better control over the design parameters in key predistribution algorithms.

3. Further using such a random merging strategy, one gets more flexible parameters than those given in the work by Lee and Stinson [28].

In the paper by Chakrabarti et al. [7] a randomized block merging based design strategy is used that originates from TD. The computation to find out a common key is also shown to be of very low time complexity under this paradigm as explained in Section 14.4.3.6. Note that Blom's scheme [3] has been extended in recent works for key predistribution in wireless sensor networks [12,27]. The problem with these kinds of schemes is the use of several multiplication operations (as example see the work by Du et al. [12, Section 5.2]) for key exchange.

The randomized key predistribution is another strategy in this area [14]. However, the main motivation is to maintain the connectivity (possibly with several hops) in the network. As example [14, Section 3.2], a sensor network with 10,000 nodes has been considered and to maintain the connectivity it has been calculated that it is enough if one node can communicate with only 20 other nodes. Note that the communication between any two nodes may require a large number of hops. However, as we discussed earlier, only the connectivity criterion (with too many hops) may not suffice in an adversarial condition. Further in such a scenario, the key agreement between two nodes requires exchange of the key indices.

The use of combinatorial and probabilistic design (also a combination of both—termed as hybrid design) in the context of key distribution has been proposed in the work by Camtepe and Yener [6]. In this case also, the main motivation was to have low number of common keys as in the work by Lee and Stinson [28]. On the other hand, the work by Chakrabarti et al. [7] proposes the idea of good number of common keys between any two nodes. The novelty of this approach is to start from a combinatorial design and then to apply a probabilistic extension in the form of random merging of blocks to form the sensor nodes and in this case there is good flexibility in adjusting the number of common keys between any two nodes.

First the block merging strategy is applied in a completely randomized fashion. In such a case there is a possibility that the constituent blocks (which are merged to form a sensor node) may share common keys among themselves. This is a loss in terms of the connectivity in the designed network as no shared key is needed since there is no necessity for "intranode communication." Thus, a cleverer merging strategy is used toward minimizing the number of common keys among the blocks that are being merged. A heuristic is presented for this and it works better than the random merging strategy. The scheme is a hybrid one as combinatorial design is followed by a heuristic.

14.4.1 Basics of Combinatorial Design

Let A be a finite set of subsets (also known as blocks) of a set X. A *set system* or *design* is a pair (X, A). The degree of a point $x \in X$ is the number of subsets containing the point x. If all subsets/blocks have the same size k, then (X, A) is said to be uniform of rank k. If all points have the same degree r, (X, A) is said to be regular of degree r.

A regular and uniform set system is called a $(v, b, r, k) - 1$ design, where $|X| = v$, $|A| = b$, r is the degree, and k is the rank. The condition $bk = vr$ is necessary and sufficient for existence of such a set system. A $(v, b, r, k) - 1$ design is called a (v, b, r, k) configuration if any two distinct blocks intersect in zero or one point.

A (v, b, r, k, λ) BIBD is a $(v, b, r, k) - 1$ design in which every pair of points occurs in exactly λ blocks. A (v, b, r, k) configuration having deficiency $d = v - 1 - r(k - 1) = 0$ exists if and only if a $(v, b, r, k, 1)$ BIBD exists.

Let g, u, k be positive integers such that $2 \le k \le u$. A group-divisible design of type g^u and block size k is a triple $(X, \mathcal{H}, \mathcal{A})$, where X is a finite set of cardinality gu, $\mathcal{H}$ is a partition of X into u parts/groups of size g, and $\mathcal{A}$ is a set of subsets/blocks of X. The following conditions are satisfied in this case:

1. $|H \bigcap A| \le 1 \; \forall H \in \mathcal{H}, \; \forall A \in \mathcal{A}$,
2. every pair of elements of X from different groups occurs in exactly one block in $\mathcal{A}$.

A TD (k, n) is a group-divisible design of type n^k and block size k. Hence, $H \bigcap A = 1 \; \forall H \in \mathcal{H}, \; \forall A \in \mathcal{A}$.

Let us now describe the construction of a TD. Let p be a prime power and $2 \le k \le p$. Then there exists a TD(k, p) of the form $(X, \mathcal{H}, \mathcal{A})$ where $X = \mathbb{Z}_k \times \mathbb{Z}_p$. For $0 \le x \le k - 1$, define $H_x = \{x\} \times \mathbb{Z}_p$ and $\mathcal{H} = \{H_x : 0 \le x \le k - 1\}$.

For every ordered pair $(i, j) \in \mathbb{Z}_p \times \mathbb{Z}_p$, define a block $A_{i,j} = \{x, (ix + j) \bmod p : 0 \le x \le k - 1\}$. In this case, $\mathcal{A} = \{A_{i,j} : (i, j) \in \mathbb{Z}_p \times \mathbb{Z}_p\}$. It can be shown that $(X, \mathcal{H}, \mathcal{A})$ is a TD(k, p).

Now let us relate a $(v = kr, b = r^2, r, k)$ configuration with sensor nodes and keys. X is the set of $v = kr$ number of keys distributed among $b = r^2$ number of sensor nodes. The nodes are indexed by $(i, j) \in \mathbb{Z}_r \times \mathbb{Z}_r$ and the keys are indexed by $(i, j) \in \mathbb{Z}_k \times \mathbb{Z}_r$. Consider a particular block $A_{\alpha,\beta}$. It will contain k number of keys $\{(x, (x\alpha + \beta) \bmod r) : 0 \le x \le k - 1\}$. Here $|X| = kr = v$, $|\mathcal{H}_x| = r$, the number of blocks in which the key (x, y) appears for $y \in \mathbb{Z}_r$, $|A_{i,j}| = k$, the number of keys in a block. For more details on combinatorial design refer the works by Lee and Stinson [28] and Street and Street [28,43].

Note that if r is a prime power, one cannot get an inverse of $x \in \mathbb{Z}_r$ when $\gcd(x, r) > 1$. This is required for key exchange protocol (see Section 14.4.3.6). So basically one should consider the field $GF(r)$ instead of the ring $\mathbb{Z}_r$. However, there is no problem when r is a prime by itself. One may generally use $\mathbb{Z}_r$ if r is considered to be a prime.

14.4.2 Lee–Stinson Approach [28]

Consider a (v, b, r, k) configuration (which is in fact a (rk, r^2, r, k) configuration). There are $b = r^2$ sensor nodes, each containing k distinct keys. Each key is repeated in r nodes. Also v gives the total number of distinct keys in the design. One should note that $bk = vr$ and $v - 1 > r(k - 1)$. The design provides 0 or 1 common key between two nodes. The design $(v = 1470, b = 2401, r = 49, k = 30)$ has been used as an example in the work by Lee and Stinson [28]. The important parameters of the design are as follows:

1. *Expected number of common keys between two nodes:* It is $p_1 = k(r-1)/b-1 = k/r+1$ and in this example $p_1 = 30/49+1 = 0.6$.
2. *Consider an intermediate node:* There is a good proportion of pairs (40 Percent) with no common key, and two such nodes will communicate through an intermediate node. Assuming a random geometric deployment, the example shows that the expected proportion such that two nodes are able to communicate either directly or through an intermediate node is as high as 0.99995.
3. *Resiliency:* Under adversarial situation, one or more sensor nodes may get compromised. In that case, all the keys present in those nodes cannot be used for secret communication any longer, that is, given the number of compromised nodes, one needs to calculate the proportion of links that cannot be used further. The expression for this proportion is

$$\text{fail}(s) = 1 - \left(1 - \frac{r-2}{b-2}\right)^s,$$

where s is the number of nodes compromised. In this particular example, $\text{fail}(10) \approx 0.17951$. That is, given a large network comprising as many as 2401 nodes, even if only 10 nodes are compromised, almost 18 Percent of the links become unusable.

14.4.3 Chakrabarti–Maitra–Roy Approach [7]

14.4.3.1 Merging Blocks in Combinatorial Design. Let us present the concept of merging blocks to form a sensor node. Note that all the following materials of this section are taken from the work by Chakrabarti et al. [7]. Initially no specific merging strategy is considered and that blocks are merged randomly.

Theorem 1 *Consider a (v, b, r, k) configuration with $b = r^2$. Merge z randomly selected blocks to form a sensor node. Then*

1. *There will be $N = \lfloor b/z \rfloor$ sensor nodes.*
2. *The probability that any two nodes share no common key is $(1-p_1)^{z^2}$, where $p_1 = k/(r+1)$.*
3. *The expected number of keys shared between two nodes is $z^2 p_1$.*
4. *Each node will contain M distinct keys, where $zk - \binom{z}{2} \leq M \leq zk$. The average value of M is $\hat{M} = zk - \binom{z}{2}k/(r+1)$.*
5. *The expected number of links in the merged system is*

$$\hat{L} = \left(\binom{r^2}{2} - \left\lfloor \frac{r^2}{z} \right\rfloor \binom{z}{2}\right) \frac{k}{r+1} - (r^2 \bmod z)k.$$

6. *Each key will be present in Q nodes, where $\lceil r/z \rceil \leq Q \leq r$. The average value of Q is*

$$\hat{Q} = \frac{1}{kr}\left(\left\lfloor \frac{b}{z} \right\rfloor\right)\left(zk - \binom{z}{2}\frac{k}{r+1}\right).$$

Proof. The first item is easy to see.

Since the blocks are merged randomly, any two sensor nodes will share no common key if and only if none of the keys in z blocks constituting one sensor node are available in the z blocks constituting the other sensor node. Thus, there are z^2 cases where there are no common keys. As we have considered random distribution in merging z blocks to form a node, under reasonable assumption (corroborated by extensive simulation studies), all these z^2 events are independent. Note that p_1 is the probability that two blocks share a common key. Hence, the proof of the second item.

The number of common keys between two blocks approximately follows binomial distribution. The probability that two blocks share i common keys is given by $\binom{z^2}{i} p_1^i (1-p_1)^{z^2-i}$, $0 \le i \le z^2$. Thus, the mean of the distribution is $z^2 p_1$, which proves the third item.

For the fourth item, note that each block contains k distinct keys. When z blocks are merged, then there may be at most $\binom{z}{2}$ common keys among them. Thus, the number of distinct keys M per sensor node will be in the range $zk - \binom{z}{2} \le M \le zk$. The average number of common keys between two nodes is $k/(r+1)$. So the average value of M is $zk - \binom{z}{2}k/(r+1)$.

Consider that z blocks are merged to form a node, that is, given a $(v = rk, b = r^2, r, k)$ configuration we get $\lfloor r^2/z \rfloor$ sensor nodes. The total number of links was $\binom{r^2}{2}k/(r+1)$ before the merging of blocks. For each of the nodes (a node is z blocks merged together), $\binom{z}{2}k/(r+1)$ links become intranode links and totally, there will be a deduction of $\lfloor r^2/z \rfloor \binom{z}{2}k/(r+1)$ links (to account for the intranode links) on an average. Further as we use $\lfloor r^2/z \rfloor$ sensor nodes, we discard $(r^2 \bmod z)$ number of blocks, which contribute to $(r^2 \bmod z)k$ links. There will be a deduction for this as well. Thus the expected number of links in the merged system is

$$\left(\binom{r^2}{2} - \left\lfloor \frac{r^2}{z} \right\rfloor \binom{z}{2}\right)\frac{k}{r+1} - (r^2 \bmod z)k.$$

This proves the fifth item.

Note that a key will be present in r blocks. Thus, a key may be exhausted as early as after being used in $\lceil r/z \rceil$ sensor nodes. On the other hand, a key may also be distributed to a maximum of r different nodes. Hence, the number of distinct nodes Q corresponding to each key is in the range $\lceil r/z \rceil \le Q \le r$. Now we try to find out the average value of Q, denoted by $\hat{Q}$. Total number of distinct keys in the merged design does not change and is also kr. Thus, $\hat{Q} = N\hat{M}/kr = (1/kr)(\lfloor b/z \rfloor)\left(zk - \binom{z}{2}(k/(r+1))\right)$. This proves the sixth item. ■

The expression fail(s), the probability that a link become unusable if s nodes are compromised, has been calculated in the following way in the work by Lee and Stinson [28]. Consider that there is a common secret key between the two nodes N_i, N_j. Let

N_h be a compromised node. Now the key that N_i, N_j share is also shared by $r-2$ other nodes. The probability that N_h is one of those $r-2$ nodes is $r-2/b-2$. Thus, the probability that compromise of s nodes affect a link is approximately $1-(1-(r-2)/(b-2))^s$. Given the design ($v=1470, b=2401, r=49, k=30$) and $s=10$, fail (10) ≈ 0.17951.

We calculate this expression in a little different manner. Given $b=r^2$ nodes, the total number of links is $\binom{r^2}{2}k/(r+1)$. The compromise of one node reveals k keys. Each key is repeated in r nodes, that is, it is being used in $\binom{r}{2}$ links. Thus, if one key is revealed, it disturbs the following proportion of links:

$$\frac{\binom{r}{2}}{\binom{r^2}{2}\frac{k}{r+1}} = \frac{1}{kr}.$$

Now s nodes contain $ks - \binom{s}{2}k/(r+1)$ distinct keys on an average. This is because there are $\binom{s}{2}$ pairs of nodes and a proportion of $k/r+1$ of them will share a common key. Thus, in our calculation, on an average

$$\text{Fail}(s) = \frac{ks - \binom{s}{2}\frac{k}{r+1}}{kr} = \frac{s}{r}\left(1 - \frac{s-1}{2(r+1)}\right).$$

Note that to distinguish the notation we use Fail(s) instead of fail(s) in the work by Lee and Stinson [28]. Note that considering the design ($v=1470, b=2401, r=49, k=30$), we tabulate the values of fail(s), Fail(s) and experimental data (average of 100 runs for each s) regarding the proportion of links that cannot be used after compromise of s nodes. The results look quite similar. However, it may be pointed out that our approximation is in better conformity with the experimental values than that of Lee and Stinson [28], which looks a bit underestimated.

Now we present the calculation of Fail(s) when more than one blocks are merged. Let N_a and N_b be two given nodes. Define two events E and F as follows:

1. E: N_a and N_b are disconnected after the failure of s number of nodes
2. F: N_a and N_b were connected before the failure of those s nodes

The sought for quantity is

$$\text{Fail}(s) = P(E|F) = \frac{P\left(E \bigcap F\right)}{P(F)}.$$

Let X be the random variable denoting the number of keys between N_a and N_b and following the proof of Theorem 1(2), we assume that X follows $B(z^2, k/(r+1))$. Thus,

$$P(F) = P(X>0) = 1 - P(X=0) = 1 - \left(1 - \frac{k}{r+1}\right)^2.$$

Next define two sets of events:

1. E_{1i}: i number of keys (shared between N_a and N_b) are revealed consequent upon the failure of s nodes,
2. E_{2i} : i number of keys are shared between N_a and N_b.

Let $E_i = E_{1i} \bigcap E_{2i}$ for $i = 1, 2, \ldots, z^2$. So, $E_i \bigcap E_j = \emptyset$ for $0 \leq i \neq j \leq z^2$. As

$$E \bigcap F = \bigcup_{i=1}^{z^2} E_i, \text{ we have } P\left(E \bigcap F\right) = P\left(\bigcup_{i=1}^{z^2} E_i\right)$$

$$= \sum_{i=1}^{z^2} P(E_i) = \sum_{i=1}^{z^2} P(E_{1i}|E_{2i})P(E_{2i}) \text{ and also}$$

$$P(E_{2i}) = \tbinom{z^2}{i}\left(\frac{k}{r+1}\right)^i\left(1 - \frac{k}{r+1}\right)^{z^2-i}.$$

Now we estimate $P(E_{1i}|E_{2i})$ by hypergeometric distribution. Consider the population (of keys) of size kr and γ number of defective items (the number of distinct keys revealed). We shall draw a sample of size i (without replacement) and we are interested in the event that all the items drawn are defective.

Note that γ is estimated by the average number of distinct keys revealed, that is,

$$\gamma = szk\left(1 - \frac{sz - 1}{2(r+1)}\right).$$

So $P(E_{1i}|E_{2i}) = \tbinom{\gamma}{i}/\tbinom{kr}{i}, i = 1, 2, \ldots, z^2$.

Finally

$$P(E|F) = \frac{P(E \bigcap F)}{P(F)}$$

$$= \frac{\sum_{i=1}^{z^2} \frac{\binom{\gamma}{i}}{\binom{kr}{i}}\binom{z^2}{i}\left(\frac{k}{r+1}\right)^i\left(1 - \frac{k}{r+1}\right)^{z^2-i}}{1 - \left(1 - \frac{k}{r+1}\right)^2}.$$

The estimate γ is a quadratic function of s and hence is not an increasing function (though in reality, it should be an increasing function of s $\forall s$). That is why Fail(s) increases with s as long as γ increases with s. Given $\gamma = szk\,(1 - (sz - 1)/2(r + 1)))$, it can be checked that γ is increasing for $s \leq (2r + 3)/2z$. As we are generally interested in the scenarios where a small proportion of nodes are compromised, this constraint on the number of compromised nodes s is practical.

Based on the above discussion, we have the following theorem.

Theorem 2 *Consider a (v, b, r, k) configuration. A node is created by random merging of z nodes. For $s \leq (2r + 3)/2z$,*

$$\text{Fail}(s) \approx \frac{\displaystyle\sum_{i=1}^{z^2} \frac{\binom{\gamma}{i}}{\binom{kr}{i}} \binom{z^2}{i} \left(\frac{k}{r+1}\right)^i \left(1 - \frac{k}{r+1}\right)^{z^2-i}}{1 - \left(1 - \dfrac{k}{r+1}\right)^2},$$

where $\gamma = szk\,(1 - ((sz - 1)/2(r + 1))$.

It may be mentioned that while estimating $P(E_{1i}|E_{2i})$ by $\binom{\gamma}{i}/\binom{kr}{i}$, we are allowing a higher quantity in the denominator. The number of distinct keys revealed is under the restriction that the keys are distributed in s distinct blocks. However, the denominator is the expression for choosing i number of distinct keys from a collection of kr keys without any restriction. As a consequence, the resulting probability values will be under estimated.

Note that in Theorem 2, there is a restriction on s. Next we present another approximation of Fail(s) as follows where such a restriction is not there. However, the approximation of Theorem 3 is little further than that of Theorem 2 from the experimental results.

Theorem 3 *Consider a $(v = kr, b = r^2, r, k)$ configuration. A node is prepared by merging $z > 1$ nodes. Then in terms of design parameters,*
Fail(s) $\approx$

$$\frac{1}{1 - (1 - \frac{k}{r+1})^{z^2}} \sum_{i=1}^{z^2} \binom{z^2}{i} \left(\frac{k}{r+1}\right)^i \left(1 - \frac{k}{r+1}\right)^{z^2-i} \pi^i,$$

where

$$\pi = szk\left(1 - \frac{sz-1}{2(r+1)}\right)\frac{\hat{Q}(\hat{Q}-1)}{2\hat{L}}$$

.

Proof. Compromise of one node reveals $\hat{M}$ keys on an average. Thus, there will be $s\hat{M}$ keys. Further, between any two nodes, $z^2(k/(r + 1))$ keys are common on an average. Thus, we need to subtract $\binom{s}{2}z^2k/r + 1$ keys from $s\hat{M}$ to get the number of distinct keys. Thus, the number of distinct keys in s merged nodes is

$$= s\hat{M} - \binom{s}{2}z^2\frac{k}{r+1} = s\left(zk - \binom{z}{2}\frac{k}{r+1}\right) - \binom{s}{2}z^2\frac{k}{r+1} = szk\left(1 - \frac{sz-1}{2(r+1)}\right).$$

We have $N = \lfloor b/z \rfloor$ sensor nodes, and $\hat{L} = \left(\binom{r^2}{2} - \lfloor \frac{r^2}{z} \rfloor \binom{z}{2}\right) \frac{k}{r+1} - (r^2 \bmod z)k$ average number of total links. Each key is repeated in $\hat{Q}$ nodes on an average, that is, it is being used in $(\hat{Q})(\hat{Q} - 1)/2$ links. Thus, if one key is revealed that disturbs $(\hat{Q}(\hat{Q} - 1))/2\hat{L}$ links on an average. Hence, compromise of 1 key disturbs $(Q(Q - 1))/2/\hat{L}$ proportion of links. Hence, compromise of s nodes disturbs

$$\pi = szk\left(1 - \frac{sz - 1}{2(r + 1)}\right) \frac{\hat{Q}(\hat{Q} - 1)}{2\hat{L}}$$

proportion of links on an average. Thus, we can interpret π as the probability that one link is affected after compromise of s merged nodes.

Now the probability that there are i links between two nodes given at least one link exists between them is

$$\frac{1}{1 - \left(1 - \frac{k}{r+1}\right)^{z^2}} \binom{z^2}{i} \left(\frac{k}{r + 1}\right)^i \left(1 - \frac{k}{r + 1}\right)^{z^2 - i}.$$

Further the probability that all those i links will be disturbed due to compromise of s nodes is π^i. Hence

$$\text{Fail}(s) = \frac{1}{1 - \left(1 - \frac{k}{r+1}\right)^{z^2}} \sum_{i=1}^{z^2} \binom{z^2}{i} \left(\frac{k}{r + 1}\right)^i \left(1 - \frac{k}{r + 1}\right)^{z^2 - i} \pi^i. \blacksquare$$

The following example illustrates our approximations vis-a-vis the experimental results. Consider a $(v = 101 \cdot 7, b = 101^2, r = 101, k = 7)$ configuration and merging of $z = 4$ blocks to get a node. Thus, there will be 2550 nodes. In such a situation we present the proportion of links disturbed if s $(1 \leq s \leq 10)$ nodes are compromised; that is, this can also be seen as the probability that two nodes get disconnected, which were connected earlier (by one or more links).

14.4.3.2 Comparison with the work by Lee and Stinson In the example presented in the work by Lee and Stinson [28], the design $(v = 1470, b = 2401, r = 49, k = 30)$ has been used to get $N = 2401$, $M = 30$, $Q = 49$, $p_1 = 0.6$, $1 - p_1 = 0.4$.

Now we consider the design $(v = 101 \cdot 7 = 707, b = 101^2 = 10201, r = 101, k = 7)$. Note that in this case $p_1 = (k)/(r + 1) = (7)(102)$. We take $z = 4$. Thus, $N = \lfloor 10201/4 \rfloor = 2550$. Further, the probability that two nodes will not have a common key is $(1 - (7/102))^{16} = 0.32061$. Note that this is considerably lesser (better) than the value 0.4 presented in the work by Lee and Stinson [28] under a situation where the number of nodes is greater $(2550 > 2401)$ and number of keys per node is

TABLE 14.1 Comparison with an Example Presented in the Work by Lee and Stinson [28]

Comparison	Random merging (Section 14.4.3.1)	Heuristic (Section 14.4.3.4)	Lee and Stinson [28]
Number of nodes	2550	2550	2401
Number of keys per node	≤ 28	≤ 28	30
Probability that two nodes do not share a common key	0.320555	0.30941	0.4
Fail(s), for $s = 10$	0.222167	0.218968	0.185714

lesser ($28 < 30$) in our case. Thus, our strategy is clearly more efficient than that of Lee and Stinson [28] in this respect. On the other hand, the Fail(s) value is worse in our case than what has been achieved in the work by Lee and Stinson [28]. In Table 14.4.3.2, for our approaches, we present the experimental values that are average over 100 runs. For the time being let us concentrate on the comparison between our contribution in this section (Section 14.4.3.1) and the idea presented in the work by Lee and Stinson [28]. In the next section (Section 14.4.3.4), we will present a better idea and the result of that is also included in Table 14.1 for brevity.

The comparison in Table 14.1 is only to highlight the performance of our design strategy with respect to what is described in the work by Lee and Stinson [28] and that is why we present a design with average number of common keys between any two nodes ≤ 1. However, we will present a practical scenario in the next subsection where there are more number (≥ 5) of common keys (on an average) between any two nodes and consequently the design achieves much less Fail(s) values.

One more important thing to mention is that we consider the average case analysis for our strategy. The worst-case situation will clearly be worse than the average case, but that is not of interest in this context as we will first try to get a merging configuration that is close to the average case. As this is done in preprocessing stage, we may go for more than one attempts for the configuration and it is clear that in a few experiments, we will surely get a configuration matching the average case result. On the other hand, it is very important to identify the best case as this will provide a solution better than the average case. However, this is open at this point of time.

The strength of our scheme is in the presence of several common keys between two nodes, which in fact makes it more resilient. Of course, this is at the cost of an obvious increase in number of keys in each node by a factor of z. The examples presented in Sections 14.4.3.2 and 14.4.3.3 illustrate this fact. In Section 14.4.3.2, we deliberately allowed a very low number of common keys (so that the node size is comparable to that of Lee and Stinson [28]) and hence the negative resiliency measure Fail(s) increased slightly. In what follows, we demonstrate that with an increase in the node capacity, the negative resiliency measure Fail(s) assumes a negligible value.

14.4.3.3 A Practical Design with More Than One Keys (On Average) Shared Between Two Nodes We start with the idea that a node can contain 128 keys and as we like to compare the scenario with the work by Lee and Stinson [28], we will consider the number of sensor nodes ≥ 2401, as it has been used in the examples in the work by Lee and Stinson [28].

Consider a $(v = rk, b = r^2, r = 101, k = 32)$ configuration. If one merges $z = 4$ blocks (chosen at random) to construct a node, the following scheme is obtained (refer to Theorems 1 and 2).

1. There will be $\lfloor 10201/4 \rfloor = 2550$ sensor nodes.
2. The probability that two nodes do not share a common key is approximately $(1 - 32/102)^{16} = 0.0024$.
3. Expected number of keys shared between two nodes $= (16 \cdot 32/102) \geq 5$.
4. Each node will contain on an average $\hat{M} = 4 \times 32 - \binom{4}{2}(32/102) \approx 126$ distinct keys and at most 128 keys.
5. $\text{Fail}(10) = 0.019153 \approx 2$ percent and $\text{Fail}(25) = 0.066704 \approx 7$ percent.

This example clearly uses more keys (≤ 128) per sensor node than the value 30 in the example of Lee and Stinson [28]. Note that directly from a (v, b, r, k) configuration, it is not possible to have $k > r$. However, in a merged system that is always possible. Moreover, the average number of keys shared between any two nodes is ≈ 5. It is not easy to get a combinatorial design [43] to achieve such a goal directly. This shows the versatility of the design proposed by us.

14.4.3.4 A Heuristic: Merging Blocks Attempting to Minimize the Number of Intra Node Common Keys So far we have used the concept of merging blocks to form a sensor node without any constraints on how the blocks will be chosen to form a node. Now we add the constraint that the blocks that will be merged to form a node such that the number of common keys between two blocks of the same node is minimized (the best case is if the number is zero). For this we present the following heuristic.

Heuristic 1

1. *flag = true; count = 0; all the blocks are marked as unused;*
2. *an array node[...] is available, where each element of the array can store z blocks;*
3. *while(flag){*
   (a) *choose a random block, mark it as used and put it in node[count];*
   (b) *for $(i = 1; i < z; i++)${*
      (i) *search all the unused blocks in random fashion and put the first available one in node[count] that has no common key with the existing blocks already in node[count];*
      (ii) *mark this block as used;*
      (iii) *if such a block is not available then break the for loop and assign flag = false;*

(c) *}(end for)*
(d) *if flag = true then count = count + 1;*

4. *} (end while)*
5. *report that count nodes are formed such that there is no intranode connectivity.*
6. *for rest of the ($r^2 - count \cdot z$) blocks, merge z blocks randomly to form a node (they may have intranode connectivity) to get ($\lfloor r^2/z \rfloor - count$) many extra nodes; this constitutes the initial configuration.*
7. *assign the initial configuration to current configuration and run step 8 for i iterations.*
8. *make m* ***moves*** *(explained below) on the current configuration and choose the one that gives rise to the maximum increase in connectivity; update the current configuration with this chosen one.*

We define a **move** as follows:

1. start **move**;
2. copy the current configuration in a temporary configuration and work on the temporary configuration;
3. from the list of pairs of nodes sharing more than one common keys, select one pair of nodes randomly; call them a and b;
4. from the list of pairs of nodes sharing no common key, select one pair of nodes randomly; call them c and d.
5. select one block each from a and b (say block α from node a and block β from node b) and remove them such that α and β intersect each other and nodes a and b are still connected after the removal of α, β, respectively; if this condition is not satisfied then go to step 9;
6. select one block each from nodes c and d and remove them; let the removed blocks be γ and δ respectively;
7. put γ in a, δ in b, α in c, and β in d;
8. store this temporary configuration in some container;
9. end **move**.

In Heuristic 1 we use a simple hill climbing technique and for experimental purposes we took $m = 100, i = 100$. It will be encouraging to apply more involved metaheuristic techniques in step 8 of Heuristic 1. This we recommend for future research.

It is very clear that given (v, b, r, k) configuration with $b = r^2$, if one merges z blocks to get each node then the maximum possible nodes that are available could be $N \leq \lfloor b/z \rfloor$. However, it is not guaranteed that given any configuration one can really achieve the upper bound $\lfloor b/z \rfloor$ with the constraint that the blocks constituting a node cannot have any common key among themselves. Using Heuristic 1 up to step 5, one can use all the blocks in some cases, but sometimes it may not be possible also. That is the reason we go for step 6 for merging the rest of the blocks where we remove the constraints that no two blocks of a node can have a common key.

The following example illustrates the experimental results. Consider a $(v = 101 \cdot 7, b = 101^2, r = 101, k = 7)$ configuration and merging of $z = 4$ blocks to get a node. Thus, there will be 2550 nodes. In such a situation we present the proportion of links disturbed if s $(1 \leq s \leq 10)$ nodes are compromised; that is, this can also be seen as the probability that two nodes get disconnected, which were connected earlier (by one or more links).

Let us refer to Table 14.1 for the comparison. As usual, we consider the $(v = 101 \cdot 7 = 707, b = 101^2 = 10201, r = 101, k = 7)$ configuration to attain a comparable design after merging. Note that in this case $p_1 = k/(r+1) = 7/(102)$. We take $z = 4$. Thus, $N = \lfloor 10201/4 \rfloor = 2550$. Considering the binomial distribution presented in Theorem 1(3), the theoretical probability that two nodes will not have a common key is $(1 - (7/102))^{16} = 0.32061$. Experimentally with 100 runs we find the average value as 0.30941, which is less (better) than the theoretically estimated value and also the experimental value 0.320555 as explained in Section 14.4.3.1 under the same experimental setup. Note that this is considerably lesser than the value 0.4 presented in the work by Lee and Stinson [28]. The average number of common keys between any two nodes is $z^2 p_1 = z^2 k/(r+1) = 16 \cdots 7/102 = 1.098039$. Experimentally with 100 runs we get it as 1.098362 on an average, which is a higher (improved) value than the theoretical estimate and also the experimental value 1.098039 as given in Section 14.4.3.1 under the same experimental setup.

14.4.3.5 More Keys Shared Between Two Nodes As in Section 14.4.3.3, consider a $(v = rk, b = r^2, r = 101, k = 32)$ configuration. If one merges $z = 4$ blocks to construct a node according to Heuristic 1, the following scheme is obtained.

1. There are $\lfloor 10201/4 \rfloor = 2550$ sensor nodes.
2. The probability that two nodes do not share a common key is approximately $(1 - 32/102)^{16} = 0.002421$. The experimental value on an average is 0.002094 with 100 runs, which is lesser (better) than the theoretically estimated value.
3. Expected number of keys shared between two nodes $= \frac{16 \cdot 32}{102} \geq 5.019608$. The experimental value with 100 runs is 5.021088 on an average, little better than the theoretically estimated value.

14.4.3.6 Key Exchange In this section, we present the key exchange protocol between any two nodes. First we present the key exchange protocol (as given in the work by Lee and Stinson [28]) between two blocks N_a, N_b having identifiers (a_1, a_2) and (b_1, b_2), respectively. We take a $(v = kr, b = r^2, r, k)$ configuration. Thus, the identifier of a block is a tuple (a_1, a_2) where $a_1, a_2 \in \{0, \ldots, r-1\}$ and the identifier of a key is a tuple (k_1, k_2) where $k_1 \in \{0, \ldots, k-1\}, k_2 \in \{0, \ldots, r-1\}$.

1. Consider two blocks N_a, N_b having identifiers (a_1, a_2) and (b_1, b_2), respectively.
2. If $a_1 = b_1$ (and hence $a_2 \neq b_2$), then N_a and N_b do not share a common key.

3. Else $x = (b_2 - a_2)(a_1 - b_1)^{-1} \bmod r$. If $0 \leq x \leq k - 1$, then N_a and N_b share the common key having identifier $(x, a_1x + a_2)$. If $x \geq k$, then N_a and N_b do not share a common key.

They can independently decide whether they share a common key in $O(\log_2^2 r)$ time as inverse calculation is used [41, Chapter 5].

In the proposed system, a node comprises of z number of blocks. Since each block has an identifier (which is an ordered pair $(x, y) \in Z_r \times Z_r$), a node in the merged system has z number of such identifiers, which is maintained in a list.

1. For the tth block in the node N_a, $t = 1, \ldots, z$
   (a) send the identifier corresponding to the tth block to the other node N_b;
   (b) receive an identifier corresponding to a block in N_b;
   (c) compare the received identifier from N_b with each of the z identifiers in it (i.e., N_a) using Algorithm 14.4.3.6;
   (d) if a shared key is discovered acknowledge N_b and terminate;
   (e) if an acknowledgment is received from N_b that a shared key is discovered then terminate;
2. Report that there is no shared key;

Since N_a and N_b participate in the protocol at the same time, the above algorithm is executed by N_a and N_b in parallel. There will be $O(z)$ amount of communications between N_a and N_b for identifier exchange and the decision whether they share a common key. At each node at most z^2 inverse calculations are done (each identifier of the other node with each identifier of the node), which gives $O(z^2 \log_2^2 r)$ time complexity.

14.5 LOW COST SYMMETRIC CIPHERS FOR ACTUAL COMMUNICATION

Once the secret key(s) between the communicating parties are settled, actual symmetric ciphers are required for the secured communication. There are two major areas in symmetric cipher design, one is block cipher and another is stream cipher. The most well-known block cipher of recent time is the (AES), also known as Rijndael [48]. Here we leave the detailed study regarding the implementation of AES on low end hardware (see http://www.iaik.tugraz.at/research/krypto/AES/ and http://www2.mat.dtu.dk/people/Lars.R.Knudsen/aes.html for more details), but mainly concentrate on a few well-known stream ciphers for low end applications.

Stream ciphers have important applications in cryptography. A private or secret key between two communicating nodes is fixed earlier and it is supposed that the key is not known to any other person. This key is used as a seed in a pseudorandom bit generator. The generator outputs a stream of pseudorandom bits based on the initial key (the seed) called the keystream. The message bits are bitwise XORed with the keystream bits to generate the ciphertext bits. These ciphertext bits are communicated over a public channel. It is expected that if the stream cipher design is proper, then it is hard (in practical sense impossible) to extract the message from the cipher without knowing the secret key. From cryptanalysis point of view, it is assumed that the attacker will know everything about the encryption and decryption algorithm, and the only unknown parameter will be the secret key itself. We present one of the most famous stream cipher RC4 and point out briefly its implementation on a low end device.

14.5.1 RC4

The RC4 stream cipher has been designed by Ron Rivest for RSA Data Security in 1987, and was a propriety algorithm until 1994. It uses an S-Box $S = S_0, \ldots, S_{255}$ of length 256, with each location of 8 bits. It is initialized as $S_i = i$ for $0 \leq i \leq 255$. Another array $KEY = KEY_0, \ldots, KEY_{255}$ is used, where each location is of 8 bits. The minimum key size is 40 bits, that is, in this case $KEY_0, \ldots, KEY_{40}$ will be filled by the key and then that is repeated number of times to fill up the entire array KEY. Initially, an index j is set to 0 and the following code is executed for key scheduling.

$$\text{for } (i = 0;\ i < 256;\ i++)\{j = (j + S_i + KEY_i) \bmod 256;\ \text{Swap } S_i \text{ and } S_j;\}$$

The following code is used to generate a random byte.

$$i = j = 0; i = (i + 1) \bmod 256;\ \ j = (j + S_i) \bmod 256;\ \text{Swap } S_i \text{ and } S_j;$$

$$t = (S_i + S_j) \bmod 256;\ \text{keyByte} = S_t;$$

The keyByte is XORed with the message byte to generate the cipher byte at the sender end and again the keyByte is XORed with the cipher byte to generate the message byte at the receiver end.

An exact implementation of RC4 for ATMega 163L has been presented in the work by Sheshadri et al. [39]. ATMega 163L is a high performance, low power 8-bit microcontroller working at a clock speed of 4 MHz. It has 130 instructions and 32 general purpose 8-bit registers. For detailed description of the hardware see http://www.chipdocs.com/datasheets/datasheet-pdf/Atmel-Corporation/ATMEGA163.html [45]. The implementation of RC4 takes only eight machine instructions on ATMega 163L microcontroller to generate one byte of keystream after every 13 cycles.

One may refer to www.cosic.esat.kuleuven.be/ecrypt/stream/ [46] to get the complete details on state of the art research in the area of stream ciphers. There are a few well-known stream ciphers that are proposed keeping in mind that one may implement them in low end devices and keeping that in mind we will discuss E0, A5/1, and Grain.

14.5.2 E_0

E_0 is a stream cipher used in the Bluetooth protocol for link encryption (see Bluetooth SIG(2001) [47]). The encryption function in Bluetooth has a variable key size that is decided upon during the manufacturing stage and never changed. Each device uses a PIN code, which can be supplied to the device by the user. This PIN code has a variable length, from 1 to 16 bits. In addition, each unit has a unique address, BD_ADDR (Bluetooth device address), which is a publicly known 48-bit value.

Firstly, for a point-to-point communication setup, a 128-bit initialization key is derived in both units based on the PIN and the BD_ADDR of the claimant unit. This key is used for a few transaction to establish a new 128-bit key called the link key K_{link}. From the link key, the cipher key K_c is derived. The link key is only used for the authentication and is not as strictly regulated as the encryption keys, thus K_{link} is always 128 bits.

The cipher key K_c, together with a 48-bit BD_ADDR, a 128-bit publicly known random value, and the 26 least significant bits from the master clock are used as initialization values for the link encryption algorithm E_0. This is a stream cipher with linear feedback shift registers (LFSRs) feeding a finite state machine (FSM). The binary output from the state machine is the keystream, which is Xored to the plaintext to form the ciphertext. There are four LFSRs having lengths 25, 31, 33, and 39 bits. The cipher is shown in Figure 14.1.

The boxes labeled z^{-1} are delay elements holding two bits each. T_1 and T_2 are two different linear bijections over F_2^2, $T_1(x_1, x_0) \rightarrow (x_1, x_0)$ and $T_2(x_1, x_0) \rightarrow (x_0, x_1 \oplus x_0)$. Let x_t^i denotes the output from the $LFSR_i$ at time t. The output from the keystream generator z_t is given by

$$z_t = x_t^1 \oplus x_t^2 \oplus x_t^3 \oplus x_t^4 \oplus c_t^0.$$

The following relations also hold:

$$s_{t+1} = (s_{t+1}^1, s_{t+1}^0) = \left\lfloor \frac{y_t + c_t}{2} \right\rfloor,$$

$$y_t = x_t^1 + x_t^2 + x_t^3 + x_t^4,$$

$$c_{t+1} = (c_{t+1}^1, c_{t+1}^0) = (s_{t+1}^1, s_{t+1}^0) \oplus T_1(c_t) \oplus T_2(c_{t-1}).$$

Since the addition operations are over integers, we have the possible values $y_t \in \{0, 1, 2, 3, 4\}$ and $s_t \in \{0, 1, 2, 3\}$. Furthermore, (s_t^1, s_t^2) is the binary vector representation of s_t with the natural mapping $0 \rightarrow (0, 0)$, $1 \rightarrow (0, 1)$, and so on.

The four feedback polynomials used in the LFSRs are given in Table 14.2. The LFSR output x_t^i is not taken from the end of the shift registers but from the taps as shown in Table 14.2.

The key initialization in E_0 is somewhat more complicated and involves a premixing of the initially loaded key material, the details of which are available in the

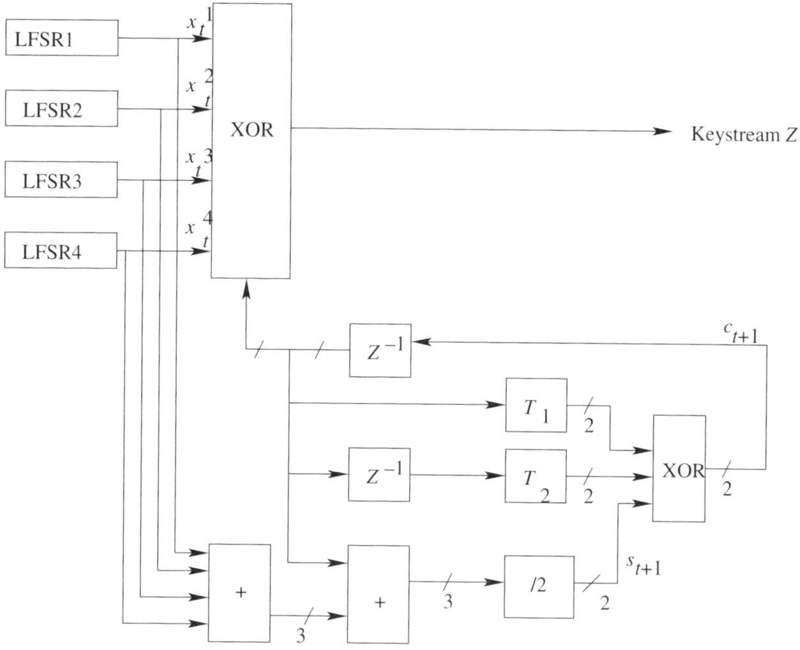

FIGURE 14.1 Bluetooth stream cipher E_0.

Bluetooth documentation [47]. However, it is important that the initial values of the LFSRs are dependent on the master clock, and that the registers are reinitialized and premixed for each frame. Two consecutive frames with the little difference in the master clock will not generate initial states with little difference due to premixing.

The first attack was presented in 1999 by Hermelin and Nyberg [23]. Their attack can recover the initial state of the shift registers with a given keystream length of 2^{64} and a computational complexity of 2^{64}.

In 2001, Fluhrer and Lucks [16] found a theoretical attack with 2^{80} operations precalculation and key search space of complexity of about 2^{65} operations. Fluhrer's attack is an improvement upon the earlier work by Golic et al. [19] who devised a 2^{70} operations attack on E_0.

In 2005, Lu et al. [30] published a cryptanalysis of E_0 based on a conditional correlation attack. Their result required the first 24 bits of the $2^{23.8}$ frames and 2^{38} computations to recover the key.

TABLE 14.2 Feedback Polynomials used in the LFSRs

LFSR	Feedback polynomial	Output tap
1	$t^{25} + t^{20} + t^{12} + t^8 + 1$	24
2	$t^{31} + t^{24} + t^{16} + t^{12} + 1$	24
3	$t^{33} + t^{28} + t^{24} + t^4 + 1$	32
4	$t^{39} + t^{36} + t^{24} + t^4 + 1$	32

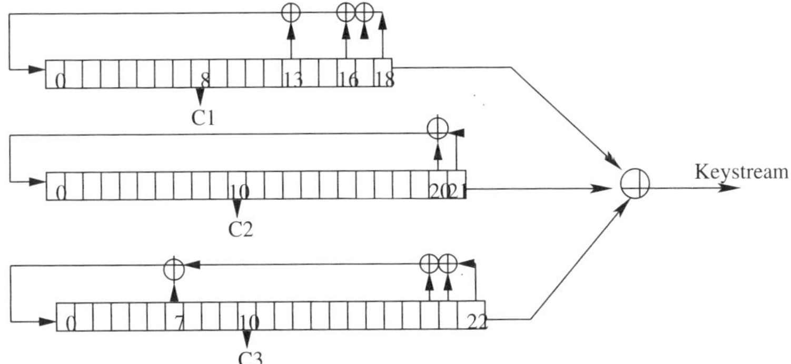

FIGURE 14.2 The A5/1 stream cipher.

14.5.3 A5/1

A5/1 is a stream cipher used to provide over-the-air voice privacy in the GSM cellular telephone standard. A GSM conversation is sent as a sequence of frames, where one frame is sent every 4.6 ms. Each frame contains 114 bits representing the communication from the mobile station (MS) to the base transceiver station (BTS), and another 114 bits in the other direction. A5/1 is used to produce 228 bits of keystream, which is XORed with the frame. A5/1 is initialized using a 64-bit key together with a publicly known 22-bit frame number.

A5/1 consists of three short binary LFSRs of lengths 19, 22, and 23 denoted by R1, R2, and R3, respectively. All these three LFSRs have primitive feedback polynomials (see Table 14.3). The keystream of A5/1 is the XOR of the outputs of these three LFSRs, as shown in Figure 14.2.

The LFSRs are clocked in an irregular fashion. It is a type of stop/go clocking with majority rule as follows: each register has a certain clocking tap, denoted by C1, C2, and C3, respectively. Each time the LFSRs are clocked, the three clocking taps C1, C2, and C3 determine which of the LFSRs is to be clocked, according to Table 14.4. At each step at least two LFSRs are clocked.

TABLE 14.3 Primitive Feedback Polynomials for LFSRs

LFSR number	Length in bits	Characteristic polynomial	Clocking bit
1	19	$x^{19}+x^5+x^2+x+1$	8
2	22	$x^{22}+x+1$	10
1	23	$x^{23}+x^{15}+x^2+x+1$	10

TABLE 14.4 Register Clocking Taps

Conditions			Registers clocked		
C1	$= C2$	$= C3 \oplus 1$	R1	R2	
C1	$= C2 \oplus 1$	$= C3$	R1		R3
$C1 \oplus 1$	$= C2$	$= C3$		R2	R3
C1	$= C2$	$= C3$	R1	R2	R3

First, the LFSRs are initialized to zero. Then for 64 cycles, the 64-bit secret key is mixed in accordance to the following scheme: in cycle $0 \leq i \leq 64$, the ith key bit is added to the least significant bit of each register using XOR as follows:

$$R[0] = R[0] \oplus K[i],$$

where R is a register and K is the key. Each register is then clocked (ignoring the irregular clocking).

In the second step, the three registers are clocked for 100 additional clock cycles with irregular clocking, but ignoring the output. Then finally, the three registers are clocked for 228 additional clock cycles with the irregular clocking, producing the 228 bits that form the keystream. As A5/1 is an additive stream cipher, the keystream is XORed to the plaintext to form the ciphertext. The keystream output is denoted as $z = z_1, z_2, \ldots, z_{228}$.

In 1997, Golic [18] described two attacks on A5/1. The first is an attack by solving the system of linear equations that requires about 2^{40} operations. The second attack is a time–memory trade-off one that can find the initial state of the ciphers using a precomputed table of 2^{42} 128-bit entries, and probing the table with about 2^{22} queries during the active phase of the attack.

In 2000, Biryukov, et al. [2] refined the attack of Golic. They presented two attacks both based on highly optimized and cipher-specific search algorithms. One needs encrypted voice data for 2 s for this attack and the attack itself requires around 2 min. However, the preprocessing time requires 2^{48} steps and 150 GB of data storage.

The same year Biham and Dunkelman [1] published an attack in A5/1 with a total work complexity of $2^{39.91}$ clockings of the cipher, given $2^{20.8}$ bits of known plaintext. The attack requires 32 GB of data storage after a precomputation time complexity of 2^{38}.

Ekdahl and Johannson [13] published an attack in 2003, based on initialization procedure that breaks A5/1 by observing 2–5 min of encrypted conversation. No preprocessing stage is required. Maximov et al. [33] improved this requiring less than 1 min.

14.5.4 Grain

Grain [21] is a stream cipher primitive that is designed to be accommodated in low end hardware. It is based on two shift registers and a nonlinear filter function. The key size is 80 bits. Grain is a bit-oriented stream cipher.

The cipher is presented as in Figure 14.3. It contains three main building blocks, namely an LFSR, a nonlinear feedback shift register (NFSR), and a filter function. The LFSR guarantees a minimum period for the key stream and it provides balancedness in the output. The NFSR, together with the nonlinear filter, introduces nonlinearity.

Both the shift registers are 80 bits in size. The contents of the LFSR are denoted by $s_i, s_{i+1}, \ldots, s_{i+79}$ and the contents of the NFSR are denoted by $b_i, b_{i+1}, \ldots, b_{i+79}$. The

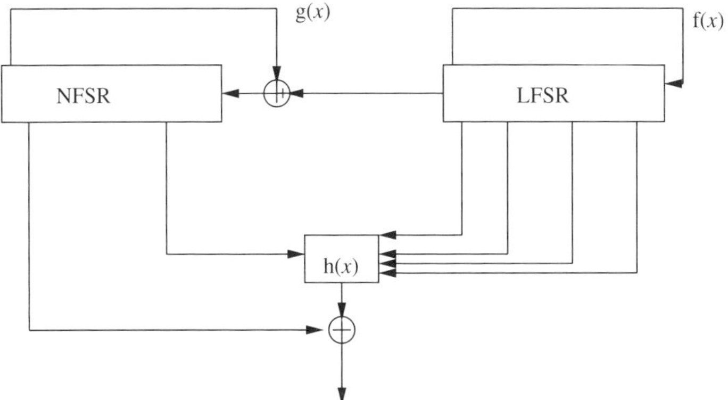

FIGURE 14.3 The grain cipher.

feedback polynomial of the LFSR, $f(x)$, is a polynomial of degree 80 and is defined as

$$f(x) = 1 + x^{18} + x^{29} + x^{42} + x^{57} + x^{67} + x^{80}.$$

Thus, the update function of the LFSR is

$$s_{i+80} = s_{i+62} + s_{i+51} + s_{i+38} + s_{i+23} + s_{i+13} + s_i\ .$$

The feedback polynomial of the NFSR, $g(x)$, is defined as

$$\begin{aligned} g(x) = {} & 1 + x^{17} + x^{20} + x^{28} + x^{35} + x^{43} + x^{47} + x^{52} + x^{59} + x^{65} + x^{71} + x^{80} \\ & + x^{17}x^{20} + x^{43}x^{47} + x^{65}x^{71} + x^{20}x^{28}x^{35} + x^{47}x^{52}x^{59} + x^{17}x^{35}x^{52}x^{71} \\ & + x^{20}x^{28}x^{43}x^{47} + x^{17}x^{20}x^{59}x^{65} + x^{17}x^{20}x^{28}x^{35}x^{43} + x^{47}x^{52}x^{59}x^{65}x^{71} \\ & + x^{28}x^{35}x^{43}x^{47}x^{52}x^{59}. \end{aligned}$$

The update function has the bit s_i masked with the input and can be defined as

$$\begin{aligned} b_{i+80} = {} & s_i + b_{i+63} + b_{i+60} + b_{i+52} + b_{i+45} + b_{i+37} + b_{i+33} + b_{i+28} + b_{i+21} \\ & + b_{i+15} + b_{i+9} + b_i + b_{i+63}b_{i+60} + b_{i+37}b_{i+33} + b_{i+15}b_{i+9} \\ & + b_{i+60}b_{i+52}b_{i+45} + b_{i+33}b_{i+28}b_{i+21} + b_{i+63}b_{i+45}b_{i+28}b_{i+9} + \\ & b_{i+60}b_{i+52}b_{i+37}b_{i+33} + b_{i+63}b_{i+60}b_{i+21}b_{i+15} + b_{i+63}b_{i+60}b_{i+52}b_{i+45}b_{i+37} \\ & + b_{i+33}b_{i+28}b_{i+21}b_{i+15}b_{i+9} + b_{i+52}b_{i+45}b_{i+37}b_{i+33}b_{i+28}b_{i+21}. \end{aligned}$$

The contents of the two shift registers represent the state of the cipher. From this state, five variables are taken as input to a Boolean function, $h(x)$. This filter function is chosen to be balanced, first-order correlation immune and with algebraic degree 3. The nonlinearity is the highest possible for five-variable functions, namely 12. The function $h(x)$ is defined as

$$h(x) = x_1 + x_4 + x_0x_3 + x_2x_3 + x_3x_4 + x_0x_1x_2 + x_0x_2x_3 \\ + x_0x_2x_4 + x_1x_2x_4 + x_2x_3x_4.$$

where the variables x_0, x_1, x_2, x_3, and x_4 correspond to the tap positions s_{i+3}, s_{i+25}, s_{i+46}, s_{i+64}, and b_{i+63}, respectively. The output of the filter function is masked with the bit b_i from the NFSR to produce the keystream. The ciphertext can then be obtained simply by XORing the plaintext bits with the keystream bits.

Before generating the keystream, the cipher must be initialized with the key and an initialization vector (IV). Let the bits of the key k be denoted by k_i, $0 \leq i \leq 79$, and the bits of the IV be denoted by IV_i, $0 \leq i \leq 63$. First the NFSR bits are loaded with the key bits $b_i = k_i$, for $0 \leq i \leq 79$, and the first 64 bits of the LFSR are loaded with the IV, $s_i = \mathrm{IV}_i$, $0 \leq i \leq 63$, and the remaining bits of the LFSR are filled with ones. The cipher is clocked 160 times without producing any running key. Instead, the output of the filter function, $h(x)$, is fed back and XORED with the input, both to the LFSR and to the NFSR as shown in Figure 14.4.

The exhaustive key search attack requires the complexity 2^{80}. It is known that an LFSR with degree d and having a primitive connection polynomial produces an output with period $2^d - 1$. Because of the NFSR and the fact that the input is masked with the output of the LFSR, the exact period will depend on the key and the IV used.

Both the shift registers are regularly clocked so that the cipher will output 1 bit per clock. However, the speed can be increased at the expense of more hardware. This is done by just implementing the feedback functions $f(x)$, $g(x)$ and filtering function $h(x)$ several times. The last 15 bits of the shift registers, s_i, $65 \leq i \leq 79$, and b_i, $65 \leq i \leq 79$, are not used in the feedback functions or inputs to the filter function.

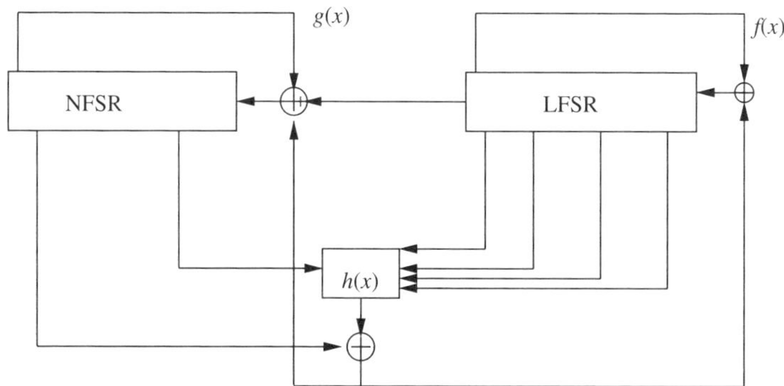

FIGURE 14.4 The key initialization.

This increases the speed up to 16 times if sufficient hardware is available. An example of implementation is shown in Figure 3 in the work by Hell et al. [21]. Moreover, the shift registers also need be implemented such that each bit is shifted t steps instead of one thus increasing the speed by a factor of t. By increasing the speed 16 times, the cipher outputs 16 bits/clock. Since in key initialization, the cipher is clocked 160 times, the possibilities to increase the speed is limited to factors less than or equal to 16 that are divisible by 160. The number of clockings used in the key initialization is then $160/t$. Since the filter and feedback functions are quite small, the throughput can be increased in this way.

Grain was implemented in hardware based on standard FPGA architectures [21]. The whole design was described in VHDL. The ALTERA MAX 3000A family was chosen because MAX 3000A uses flash memory as storage for programming data. This is persistent and no loading procedure is necessary as with RAM-based FPGA. Using the ALTERA Quartus design tool, a place/rout and postlayout timing analysis was done. It has been found that $t \leq 4$ fits into the EPM3256, leading to a usage of about 90 Percent of the 256 available macrocells.

The maximum clock frequency is in the range of 35–50 MHz, depending on the operating mode and the output interface. Also $t = 8$ fits into the chips, but the maximum clock frequency is then limited to 30 Hz. The number of output bits per second is t times the clock frequency. Design on other FPGA families, namely the ALTERA MAX II and ALTERA Cyclone, allowed the cipher to be clocked at higher speed and it also allowed an implementation when the speed is increased by a factor of 16, that is, $t = 16$.

Given certain attacks on Grain [32], a new cipher in a similar direction, called Grain-128 is proposed in the work by Hell et al. [22]. In the hardware implementation of the Grain family of stream cipher, the gate count is very low and it is expected that these ciphers can be accommodated in very low end hardwares, even in RFID tags.

14.6 CONCLUSION

In this chapter we have discussed the issues of secure communication in sensor network environment. One should note that in general the sensor nodes have limited processing power and restricted amount of memory. We have first described a basic introduction of security issues in distributed wireless sensor networks. Toward low cost key agreement algorithms, we note that using a public key kind of situation may not be recommendable in low end hardware platforms. Thus, key predistribution issues are discussed in detail. Further we study some specific stream ciphers those are possible to implement in low end hardware.

ACKNOWLEDGMENTS

The authors like to acknowledge Dibyendu Chakrabarty as some introductory material and the key predistribution issues were identified during his doctoral work. Further, we like to thank Sanjit Chatterjee, Deepak Kumar Dalai, and Sushmita Ruj for detailed discussion on this chapter.

REFERENCES

1. Biham E, Duneklman O. Cryptanalysis of the A5/1 GSM stream cipher. Progress in Cryptology—INDOCRYPT 2000. Lecture Notes in Computer Science. Volume 1977. Springer-Verlag; 2000. p 43–51.
2. Biryukov A, Shamir A, Wagner D. Real time cryptanalysis of A5/1 on a PC. Fast Software Encryption—FSE 2000. Lecture Notes in Computer Science. Volume 1978. Springer-Verlag; 2000. p 1–13.
3. Blom R. An optimal class of symmetric key generation systems. Proceedings of Eurocrypt 1984; LNCS. Volume 209. 1985. p 335–338.
4. Boyd C, González Nieto JM. Round-optimal contributory conference key agreement. Public Key Cryptography, PKC 2003. Lecture Notes in Computer Science. Volume 2567. Springer-Verlag; 2003. p 161–174.
5. Burmester M, Desmedt Y. A secure and efficient conference key distribution system (extended abstract). Proceedings of Eurocrypt 1994; 1994. p 275–286.
6. Camtepe S, Yener B. Combinatorial design of key distribution mechanisms for wireless sensor networks. Proceedings of ESORICS 2004. LNCS. Volume 3193. 2004. p 293–308.
7. Chakrabarti D, Maitra S, Roy B. Clique size in sensor networks with key pre-distribution based on transversal design. Int J Distrib Sensor Netw 2005;1(4:345–354.
8. Colbourn CJ, Dinitz JH. The CRC Handbook of Combinatorial Designs. Boca Raton: CRC Press; 1996.
9. Deb B, Bhatnagar S, Nath B. ReInForM: reliable information forwarding using multiple paths in sensor networks. Proceedings of 28th Annual IEEE International Conference on Local Computer Networks (LCN '03); October, 2003; p 406–415.
10. Diffie W, Hellman ME. New directions in cryptography. IEEE Trans Inform Theor 1976;22:644–654.
11. Douceur JR. The Sybil attack. Proceedings of the IPTPS02 Workshop. LNCS. Volume 2429. Cambridge, MA; March 2002. p 251–260.
12. Du W, Ding J, Han YS, Varshney PK. A pairwise key pre-distribution scheme for wireless sensor networks. Proceedings of the 10th ACM Conference on Computer and Communications Security; ACM CCS 2003. p 42–51.
13. Ekdahl P, Johansson T, Another attack on A5/1. IEEE Trans Inform Theor 2003;49(1):284–289.
14. Eschenauer L, Gligor VB. A key-management scheme for distributed sensor networks. Proceedings of the 9th ACM Conference on Computer and Communications Security; ACM CCS 2002. p 41–47.
15. Eisenträger K, Lauter K, Montgomery PL. Improved Weil and Tate pairings for elliptic and hyperelliptic curves. Algorithmic Number Theory, 6th International Symposium, ANTSVI, Burlington, VT, USA, June 13–18, 2004. Lecture Notes in Computer Science. Volume 3076. Springer-Verlag; 2004. p 169–183.

16. Fluhrer SR, Lucks S. Analysis of E_0 encryption system. Selected Areas in Cryptology—SAC 2001. Lecture Notes in Computer Science. Volume 2259. Springer-Verlag; 2001. p 38–48.
17. Ganesan D, Govindan R, Shenker S, Estrin D. Highly resilient, energy-efficient multipath routing in wireless sensor networks. Mobile Comput Commun Rev 2001; 5(5):11–25.
18. Golic JD. Cryptanalysis of alleged A5 stream cipher. Advances of Cryptology—EUROCRYPT'97. Lecture Notes in Computer Science. Volume 1233. Springer-Verlag; 1997. p 239–255.
19. Golic JD, Bagini V, Morgari G. Linear cryptanalysis of Bluetooth stream cipher. Advances in Cryptology—EUROCRYPT 2002. Lecture Notes in Computer Science. Volume 2332. Springer-Verlag 2002. p 238–255.
20. Gura N, Patel A, Wander A, Eberle H, Shantz SC. Comparing elliptic curve cryptography and RSA on 8-bit CPUs, CHES, 2004. Lecture Notes in Computer Science. Volume 3156. Springer-Verlag; 2004. p 119–132.
21. Hell M, Johansson T, Meier W. Grain—A stream cipher for constrained environments. Int J Wirel Mobile Comput. (Special Issue on Security of Computer Network and Mobile Systems) 2006.
22. Hell M, Johansson T, Maximov A, Meier W. A stream cipher proposal: Grain-128. Proceedings of the ISIT; Seattle, USA; 2006.
23. Hermelin M, Nyberg K. Correlation properties of Bluetooth combiner. Information Security and Cryptology—ICISC'99. Lecture Notes in Computer Science. Volume 1787. Springer-Verlag; 2000. p 17–29.
24. Joux A. A one round protocol for tripartite Diffie-Hellman. J Cryptol 2004;17(4): 263–276.
25. Karlof C, Wagner D. Secure routing in wireless sensor networks: attacks and countermeasures. Elsevier's Ad Hoc Netw J (Special Issue on Sensor Network Applications and Protocols) 2003;1(2–3):293–315.
26. Koblitz N. Elliptic curve cryptosystem. Math Comput 1987;48:203–209.
27. Lee J, Stinson D. Deterministic key pre-distribution schemes for distributed sensor networks. Proceedings of SAC 2004. LNCS. Volume 3357. 2004. p 294–307.
28. Lee J, Stinson D. A combinatorial approach to key pre-distribution for distributed sensor networks. IEEE Wireless Computing and Networking Conference (WCNC 2005); New Orleans, LA, USA; 2005.
29. Liao L, Manulis M. Tree-based group key agreement framework for mobile ad-hoc networks. Proceedings of the 20th International Conference on Advanced Information Networking and Applications (AINA 2006), April 18–20, 2006; Vienna, Austria. IEEE Computer Society; 2006. p 5–9. http://doi.ieeecomputersociety.org/10.1109/AINA.2006.336
30. Lu Y, Meier W, Vaudenay S. The conditional correlation attack: a practical attack on Bluetooth encryption. Advances in Cryptology—CRYPTO 2005. Lecture Notes in Computer Science. Volume 3621. Springer-Verlag; 2001. p 97–117.
31. Manulis M. Contributory group key agreement protocols, revisited for mobile ad hoc groups. Proceedings of the MASS; 2005.
32. Maximov A. Cryptanalysis of the "Grain" family of stream ciphers. ACM Symposium on Information, Computation and Communications Security (ASI–ACCS '06); 2006. p 283–288.

33. Maximov A, Johansson T, Babbage S. An improved correlation attack on A5/1. Selected Areas in Cryptography (SAC 2004). Lecture Notes in Computer Science. Volume 3357. Springer-Verlag; 2004. p 1–18.
34. Newsome J, Shi E, Song D, Perrig A. The Sybil attack in sensor networks: analysis and defenses. Proceedings of the IEEE International Conference on Information Processing in Sensor Networks; April 2004.
35. Perrig A, Szewczyk R, Wen V, Culler D, Tygar J. SPINS: security protocols for sensor networks. Wirel Netw J 2002;8(5):521–534.
36. Perrig A, Stankovic J, Wagner D. Security in wireless sensor networks. Commun ACM 2004;47(6):53–57.
37. Pickholtz RL, Schilling DL, Milstein LB. Theory of spread spectrum communications: a tutorial. IEEE Trans Commun 1982;COM-30(5):855–884.
38. Przydatek B, Song D, Perrig A. SIA: security information aggregation in sensor networks. Proceedings of the 1st ACM International Conference on Embedded Networked Sensor Systems; 2003. p 255–265.
39. Seshadri A, Perrig A, v Doorn L, Khosla P. SWATT: software-based AT-testation for embedded devices. IEEE Symp Secur Privacy; 2004.
40. Steiner M, Tsudik G, Waidner M. Diffie–Hellman key distribution extended to group communication. ACM Conference on Computer and Commun Secur; 1996. p 31–37.
41. Stinson D. Cryptography: Theory and Practice. 2nd ed. Chapman & Hall, CRC Press; 2002.
42. Stinson D. Combinatorial Designs: Constructions and Analysis. New York: Springer; 2003.
43. Street AP, Street DJ. Combinatorics of Experimental Design. Oxford: Clarendon Press; 1987.
44. Wood A, Stankovic J. Denial of service in sensor networks. IEEE Comput 2002; 35(10):54–62.
45. http://www.chipdocs.com/datasheets/datasheet-pdf/Atmel-Corporation/ATMEGA163.html (accessed on Aug 25, 2006)
46. https://www.cosic.esat.kuleuven.be/ecrypt/stream/ (accessed on Aug 25, 2005)
47. SIG Bluetooth. Bluetooth specification. Available at http://www.bluetooth.com (accessed on Aug 25, 2006)
48. http://csrc.nist.gov/CryptoToolkit/aes/rijndael/
49. http://www.iaik.tugraz.at/research/krypto/AES/(accessed on Aug 25, 2006)
50. http://www2.mat.dtu.dk/people/Lars.R.Knudsen/aes.html(accessed on Aug 25, 2006)

CHAPTER 15

Localized Topology Control Algorithms for Ad Hoc and Sensor Networks

HANNES FREY and DAVID SIMPLOT-RYL

15.1 INTRODUCTION

Ad hoc networks are formed by portable devices that are communicating wirelessly without using a stationary network infrastructure. Such networks may be desired when users are collaborating via mobile devices or may be of great importance in case of disaster control whenever infrastructure-based communication is no longer available. In addition, multihop ad hoc networking techniques can be used in order to extend the limited range of wireless access points, thus extending the area from where wireless nodes can access the Internet. Finally, ad hoc networking can be used as an alternative communication platform to existing wired network infrastructures. For instance, in urban regions, specific wireless routing nodes installed on top of some selected buildings might span a high speed wireless communication network. Installation and maintenance of such wireless *rooftop networks* is less expensive and time consuming compared to their fiber- or copper-based counterparts.

A specific networking scenario that received significant attention within the past years is *sensor networks*. The idea is to combine sensor, processing, and communication capabilities in small wireless network nodes that perform a measurement in a collaborative way. In general, a single sensor node plays no significant role in the whole measurement. Individual measurements are aggregated along a path to selected data sinks. In this way, individual measurements emerge to a global picture of the observed physical phenomenon. Compared to installing a set of hard-wired sensors, sensor networks enable a rapid deployment of sensors at the measured phenomenon. Moreover, since sensor nodes are small in general, and, in fact, not tied together by cables, sensor networks have only a minimal influence on the whole measurement. It is expected that in the near future, sensor networks will lead us significantly beyond scale, precision, and detail compared to what we can measure today.

To enable the receiver to decode the received signal correctly, any wireless communication requires the signal to be received above a certain minimum signal strength. More precisely, the relationship between the received signal strength and the noise

Handbook of Applied Algorithms: Solving Scientific, Engineering and Practical Problems
Edited by Amiya Nayak and Ivan Stojmenović

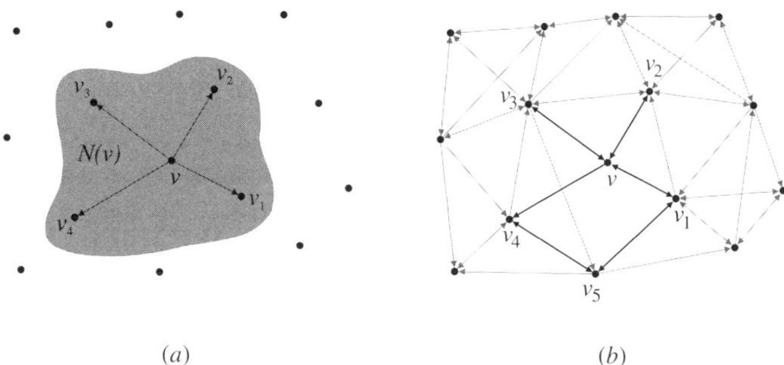

FIGURE 15.1 (*a*) The neighbor set of node v. (*b*) The topology of the entire node set.

at the receiver has to be above a certain threshold referred as *signal-to-noise ratio*. In general, the received signal strength significantly decreases with the distance between sending and receiving devices. In addition, due to its limited energy resources and physical constraints, any sending device may only issue signals that are below or equal to a maximal possible signal strength. Consequently, the potential destinations of a message transmission are limited to spatial close network nodes (see Fig. 15.1a). The nodes that are immediately reachable from a node v are denoted as its *neighbors* $N(v)$. The size of this set is the *degree* of node v. The *topology* of an ad hoc network can be defined as an extension of the neighbor relation by referring to the network graph (V, E) consisting of all nodes V and all edges (v, w) that satisfy that w is a neighbor of v (see Fig. 15.1b).

Generally speaking, *topology control* refers to all methods that, starting from a given network topology (V, E), construct a graph (V', E') that satisfies a desired property. More specifically, existing topology control mechanisms can be distinguished according to the following three subclasses. *Neighbor elimination schemes* (NES) maintain the entire set of network nodes while reducing each node's neighbor set; that is, the resulting topology (V', E') satisfies $V' = V$ and $E' \subseteq E$ with additional properties like connectivity. In contrast, *backbone construction schemes* remove nodes from the original node's neighbor set while keeping all edges from the original graph that connect the remaining nodes. This corresponds to constructing a graph (V', E') that satisfies $V' \subseteq V$ and $E' = \{(v, w) : v, w \in V' \text{ and } (v, w) \in E\}$. Finally, the class of *overlay topology control schemes* utilizes the original network topology in order to construct a virtual graph consisting of nodes and edges that are not contained in V and E of the original network topology, respectively.

Owing to limited communication ranges, ad hoc and sensor networks are inherently *decentralized*; that is, in general there exists no node that has at once an entire view on the global network topology. Thus, algorithms that rely on a global network view require the local views of individual nodes being collected and transferred to one, all, or a set of dedicated nodes that are responsible for performing the topology control. Moreover, in most cases an ad hoc network topology will vary over the time. This might be due to varying external noises, changes in the environment, unrelated message transmissions, and device mobility. In such a dynamic environment, nodes that are in charge of the topology control have to receive periodic update messages reflecting the most current view of the entire network topology. Consequently, any centralized method may end up with a scalability problem when the number of devices is increasing.

A promising solution that does not suffer from this scalability problem is *localized methods* where network nodes obtain only a local view on the constructed topology. In addition, the topology construction is performed by using local knowledge only. More precisely, local knowledge of a node v refers to a constant amount of information about the network nodes that are able to send a message to v by using a well-defined maximum number of intermediate forwarding nodes. In the simplest setting this constant is equal to 1; that is, topology control executed at each node v is based on a constant amount of information about the nodes for which v is an immediate neighbor.

In this chapter we focus on the basic ideas and results about the most prominent neighbor elimination-based localized topology control mechanisms. First of all, the different objectives that can be followed by topology control mechanisms are discussed in Section 15.2. Possible model assumptions and elementary mechanisms are subsequently presented in Section 15.3. With the elementary terms and definitions introduced, basic NESs are presented in Section 15.4. For more details concerning the classes backbone construction and overlay topologies the reader can refer to the works by Simplot-Ryl et al.[43] and by other authors [10–12], respectively. Finally, Section 15.5 concludes the chapter by discussing future research directions.

15.2 TOPOLOGY CONTROL OBJECTIVES

In the most general definition, topology control refers to transferring the underlying network topology to a graph having a desired property. This section will substantiate this definition by listing some of the most frequently pursued objectives. Depending on the application of scenario topology control, only some of the listed objectives might be desired ones. Moreover, some of the objectives may be conflicting ones and cannot be followed at the same time. Whenever two conflicting goals are of interest at the same time, pro and contra of both methods have to be assessed in order to find the right trade-off between them.

15.2.1 Connectivity

In general, a constructed topology is intended to serve as the basis for supporting communication within a network. Thus, *connectivity* is one of the most intrinsic properties that should be satisfied by any topology control method. More precisely, the existence of a path connecting two nodes v and w in the original topology (V, E) should always imply that sending a message from v to w is as well possible by utilizing a path in the constructed topology (V', E').

However, some schemes sacrifice connectivity of the resulting topology putting main emphasis on other objectives like reducing node degree. Moreover, some approaches may not result in a connected topology in all possible cases but may show good performance regarding connectivity in the typical network scenarios they are designed for. Finally, for some schemes that do not guarantee connectivity it might be possible to prove that they produce a connected topology with some high probability.

15.2.2 Energy Consumption

The lifetime of a wireless network depends on the battery capacity of its individual nodes. As long as nodes are not recharged, energy will permanently dissipate, initially resulting in some nodes to fail, and eventually resulting in malfunction of the entire network. Thus, energy efficient protocol design is vital for the practical applicability of wireless ad hoc and sensor networks.

The energy requirement $f(v, w)$ for message transmission along a single link (v, w) can be extended to the energy requirement for an entire path $p = v_1 v_2 \ldots v_n$ by setting $f(p) = f(v_1, v_2) + \ldots + f(v_{n-1} v_n)$. According to this definition, a topology control mechanism might aim to construct a subtopology that *supports all energy optimal paths* from the original network topology. More precisely, a path p connecting two nodes v and w is energy optimal if any other path q connecting these two nodes satisfies $f(p) \leq f(q)$. For each energy minimal path p, there has to be a path p' in the subtopology that connects the same end points of p and that satisfies $f(p') = f(p)$.

A less restrictive form of this definition is topology control mechanisms that *support energy efficient paths up to a constant factor*; that is, there exists a specific constant $c \geq 1$ such that for each energy minimal path p in the original topology, there exists a path p' in the subtopology that connects the same end points and that satisfies $f(p') \leq cf(p)$. A topology with this property is also denoted as a *spanner* with respect to the considered edge weight function.

Energy optimality can as well be expressed in terms of optimizing the *transmission power assignment*, which refers to the minimum power needed at a node in order to reach all its neighbors in the constructed topology. Two objectives have been considered so far, *min-max* and *min-total assignments*. The min-max assignment problem is to find a transmission power assignment such that the topology is connected and the maximum over the transmission power assigned to each node is minimal. The min-total assignment problem tries to find a transmission power assignment such that the resulting topology is connected and the sum over the transmission powers assigned to each node is minimal.

The transmission power assignment problem has some similarity to the minimum energy broadcasting problem. Any broadcasting can be seen as a directed tree T rooted at the broadcasting initiator. The cost of a broadcasting tree can be calculated as the sum over the cost of each node in the tree. Leaf nodes require no further transmission and are assigned energy cost 0. Under the assumption that the communication hardware supports transmission power adjustments, intermediate nodes are assigned the minimum power required in order to reach all neighbor nodes in the broadcasting tree. Finding an optimal broadcasting tree is not possible in polynomial time under $P \neq NP$ [7]. However, a possible topology control objective might be to support energy minimal broadcasting up to a constant factor, that is, when using the neighbor relation of the topology, independent from the broadcasting initiator should consume only a constant factor more energy than the optimal broadcasting tree.

15.2.3 Node Degree

A further objective that is considered by many topology control mechanisms is to keep each node's degree small. A desirable property is that a topology control mechanism guarantees the degree of any network node limited by a certain constant from above. Using such a subtopology may support scalability of protocols that rely on neighborhood information since there is less amount of information that has to be kept up to date. For an example, consider a protocol that requires two-hop neighbor information; that is, each node needs information about the neighbors a neighbor node is able to reach. Suppose this information is periodically provided by each node to its neighbors by sending control messages over the wireless communication media. The size of these control messages depends on the degree of the nodes. In densely deployed networks without topology control, such control messages might get arbitrarily large sizes. Thus, control message exchange in this case will consume a significant amount of energy and might congest the wireless network.

Topology control that reduces a network node's set of neighbor nodes may as well support network throughput due to spatial reuse of the communication media. The reduced set of neighbor nodes might extend over a smaller area than it does for the entire neighbor set from the original network topology. If the communication hardware supports signal strength adjustments, the maximum transmission power needed is less than or equal to the one required in order to reach all neighbor nodes in the original topology. Communications that involve only nodes from the reduced neighbor set might disturb less other nodes that are communicating at the same time.

15.2.4 Planarity

Constructing a *planar* topology is an important ingredient of planar graph routing schemes [4,20,22]. In this context, the notion planar refers to a two-dimensional geometric graph with no intersecting edges. The general idea of this routing scheme can be described as follows. A planar graph partitions the plane into faces that are made up of the polygons described by the graph edges. Beginning with the face containing the starting node, planar graph routing accomplishes message forwarding by following a sequence of faces that provide general progress toward the final destination node. Exploration of a single face and deciding the right sequence of faces can be done in a pure localized manner. Each forwarding node needs information about its immediate neighbors only.

A wireless network consisting of nodes that are deployed on a plane defines a geometric graph in a natural way. In general, the resulting graph is not planar (see Fig. 15.1b, for instance). From a global point of view, planarity can simply be obtained by repeatedly removing one of the two intersecting edges until no intersection remains. However, without any further structural assumptions on the underlying topology, it is easy to construct an example where connectivity and planarization by edge removal are conflicting goals. Moreover, under this general network setting, the described global planarization scheme cannot be applied in a localized manner. The edge end point of two intersecting edges might be connected by only one path of length n. Thus, detecting the intersection requires message exchange along n communication hops while n can be arbitrarily large.

15.2.5 Symmetry

Under a given topology (V, E), two nodes v and w are connected by a *symmetric link* if both edges (v, w) and (w, v) are present in E. Otherwise, if only one of both edges is present in E, the connection will be referred as *asymmetric* or *unidirectional*. In general, a topology may contain unidirectional links. For instance, in the network topology depicted in Figure 15.1, it is possible that node v is able to reach node v_4 but v_4 is not able to reach node v. In this situation, node v is able to send a message to node v_4 but node v_4 is not able to send a direct reception acknowledgment to v. Thus, the message gets lost in case of a transmission failure. The objective of symmetric topologies is to maintain only symmetric connections that provide reliable communications due to direct link acknowledgments.

Given an arbitrary topology T, a symmetric topology can be constructed by removing all unidirectional edges or by introducing a backward edge for each unidirectional edge. The resulting topology will be denoted as *symmetric subtopology* T^- and *symmetric supertopology* T^+, respectively. Symmetric sub- and supertopologies can be constructed in a localized way due to the fact that an asymmetric link can be detected at the sending node due to missing acknowledgments from the receiver. A symmetric subtopology is obtained when each sending node removes all potential receivers from which it has not received an acknowledgment. The symmetric supertopology requires that the sender informs all unidirectional connected receivers to introduce the backward edge. This can be obtained by increasing the receiver node's transmission power, for instance.

15.3 MODEL ASSUMPTIONS AND BASIC MECHANISMS

Different topology control mechanisms may have similar requirements on the hardware capabilities and the structure of the underlying network topology. Moreover, even schemes that are highly different in the applied topology construction rule may have some elementary building blocks in common. The following lists some of the most fundamental model assumptions and elementary building blocks that are often used in the literature.

15.3.1 Energy Models

In the most general form, the energy that is consumed by two communicating nodes can be described as a function $f : V \times V \to \mathbb{R}^+$, that is, each communication requires a positive amount of energy. In the simplest setting, the energy required for signal transmission between two nodes v and w might be assumed as $f(v, w) = c$ with $c > 0$. This model is a reasonable choice whenever the communication hardware cannot adapt the signal strength and thus always sends with full transmission power.

When a sender can adjust its power to the minimum needed in order to reach the message receiver, the energy model requires a closer look. In general, the energy required for a communication between two close nodes will be less than the energy required for a communication between two distant ones. Under the assumption of omnidirectional and unobstructed signal dispersion, it is reasonable to define f as a nondecreasing function depending only on the distance between two nodes. This will be denoted as a *distance-based* energy model in the following.

A well-established distance-based energy model is the *exponential path loss model* [40] that defines the power required for two communicating nodes v and w as $f(v, w) = t|vw|^\alpha + c$ for appropriate $\alpha > 1$, $t > 0$, and $c \geq 0$. The parameter α is typically set to 2 or 4. $|vw|^\alpha$ reflects the signal attenuation along the transmission path. The value of t that is sometimes assumed as 1 can be used as a normalizing constant depending on the hardware parameters and the utilized energy unit. Finally, the constant c that is sometimes neglected is used in order to take energy requirements for message processing in the sender and receiver devices into account. This value is assumed to be independent of the distance between the communicating devices.

The energy considerations described so far refer to energy efficient unicast communications between two devices. Multicasting and broadcasting are important communication paradigms as well, which differ in the way that a single message transmission might have more than one recipient. In order to take multiple recipients into account, the general energy consumption model can be extended to a mapping $f : V \times P(V) \to \mathbb{R}^+$ with $P(V)$ being the power set of V. Constant energy consumption in each message transmission can be extended to $f(v, \{v_1, \ldots, v_n\}) = a + n \cdot b$. In this connection, a reflects the constant amount of energy spent at the sending device and b the constant amount of energy spent at the receiving devices. In the same way, the exponential path loss model can be extended to $f(v, \{v_1, \ldots, v_n\}) = a + td^\alpha + n \cdot b$ with $d = \max\{|vv_1|, |vv_2|, \ldots, |vv_n|\}$; that is, the power is set to the value required to reach the recipient that is most distant from v.

15.3.2 Geometric Data

The class of *geographic topology control* schemes needs additional information about the nodes' physical location. This might be available by GPS [19], a local positioning infrastructure [15], or relative positioning based on signal strength estimations [5]. In contrast, schemes that only need information about the current reachability between neighbor nodes can be denoted as *link-based topology control*.

Some geographic topology schemes require weaker geometric information in the form of the direction or the distance to the signal sender. Such methods can be denoted as *direction-based* and *distance-based* topology control, respectively. Whenever nodes can determine their physical location, direction or distance can simply be computed by using the sending and receiving node positions. The position of a message sender can be made available by piggybacking this information on each transmitted message.

Direction- and distance-based topology control does not necessarily require a localization mechanism. Directional information can be made available if the communication hardware uses more than one directional antenna [21] in order to determine the *angle of arrival*; that is, relatively to its own orientation a receiving node is able to determine the direction of a sending node. Distance information might be inferred by measuring the signal strength of a received message. The reception hardware might provide this always existent information to upper protocol layers. Assuming an adequate distance-based energy model, the receiver can compute the distance to the sender by applying the inverse function of the power model on the sent and received signal strength. The signal strength used in the sending device might be either known in advance or transmitted as additional information within each message.

15.3.3 Neighbor Discovery

From a global point of view, determination of the neighbors of a given node is obvious. For instance, under the topology depicted in Figure 15.1(b), one can see that the neighbors of node v are $\{v_1, v_2, v_3, v_4\}$. However, by exploiting local information only, how is node v able to determine that it is able to reach node v_4? It might send a hello message with full signal strength and thus might be able to reach v_4 but any direct reply from node v_4 is not possible under this topology. On the contrary, node v_4 might use path $v_4 v_5 v_1 v$ (wherever it gets this information from) in order to send a reply message to v.

In general, such a backward path might be arbitrarily long. Consequently, without any further network assumptions determining the entire set of neighbors a node might be able to reach is not always possible in a localized way. There has been a general discussion about the usefulness of unidirectional links. Some works like those by Pearlman et al. [35] and Ramasubramanian et al. [39] suggest that protocols might benefit from treating such links as bidirectional; for example, node v_4 in Figure 15.1b might introduce a virtual link (v_4, v) that is mapped on the backward path $v_4 v_5 v_1 v$. Whenever node v_4 receives a message from v it might send an acknowledgment along the virtual link, implicitly using the assigned backward path. However, maintaining such virtual links requires some control overhead when the network is changing over the time. The required control overhead might outweigh the usefulness of such virtual links [32].

A localized neighbor discovery protocol that simply ignores such unidirectional links can basically be implemented by the following request reply protocol. A node v sends a broadcast message by using a desired transmission power p (this might be the full signal strength, for instance). All nodes receiving the request message will reply with appropriate signal strength. After the node v has received all reply messages, it knows about the set of bidirectional connected neighbors it is able to reach with power p.

When the network is changing dynamically, the request reply protocol has to be repeated periodically. As an alternative solution, each node might periodically broadcast a hello message including the list of all nodes it recently received a hello message from. Whenever a node is an element in a hello message it received from a node w, it knows that there exists a bidirectional link between itself and w.

Neighbor discovery is sometimes considered under the simplified assumption that the underlying topology is undirected; that is, two nodes v and w are either mutually in their communication ranges or cannot communicate directly at all. In this case, sending a plain hello message containing the sending node ID is sufficient since reception of a hello message from w implies that w is able to receive messages from node v as well.

As an alternative to the described *active* neighbor discovery scheme, neighbor detection might as well be performed in a *passive* manner. Such schemes do not require any hello message exchange but rely on the fact that communication among two nodes can be overheard by nearby nodes running their transceivers in *promiscuous mode*, passing as well messages to upper protocol layers that were not addressed to this node. In such a scheme it is assumed that other networking protocols are producing control and data messages. A neighbor is any node from which a message was overheard recently.

Finally, an alternative passive scheme might just utilize neighbor information that is already maintained within another protocol. For instance, proactive routing schemes periodically exchange hello messages in order to keep their routing tables up to date. In order to avoid unnecessary additional control overhead, a topology control scheme can utilize the information that is already available at the routing layer.

15.3.4 Unit Disk Graphs (UDGs)

Assuming a distance-based energy model has an important implication on the resulting network topology. When issuing a specific signal strength p the recipients that a node v is able to reach are exactly the nodes located within a circle centered at v, while the circle radius $r(p)$ depends on the signal strength p. Assuming that each node utilizes a uniform signal strength p implies a topology consisting of exactly those edges (v, w) that satisfy $|vw| \leq r(p)$. A topology with this structural property is referred to as a UDG. A generalization of this concept is quasi-unit disk graphs that allow the communication range to vary within a minimum and a maximum transmission radius. More precisely, the communication range may be any shape boundary lying outside a minimum circle with center v and radius $R_{\min}$, and inside a maximum circle with center v and radius $R_{\max}$.

UDGs are an important class of network topologies that are often used in order to simplify the theoretical analysis of a given topology control scheme. Moreover, some topology control schemes require that the underlying topology has the UDG property. This assumption might be justified if the network is deployed in a well-tempered medium like sensor nodes flowing in a large water basin. Moreover, it might be possible to construct a unit disk graph by throwing away all long edges. More precisely, in many scenarios network nodes are able to communicate with each other whenever their distance is not larger than a critical distance parameter d. When the distance between two devices is larger than d, communication might no longer be possible. Whenever nodes are able to precisely determine the distance of a message sender, a unit disk graph can be constructed locally by simply ignoring all messages that were transmitted over a distance larger than d. However, connectivity of the original topology might get lost under this construction.

15.3.5 Power Control

Topology control mechanisms based on *power control* assume that each network node is able to adjust its transmission power. Power control can be used as a general mechanism to control the network topology. A node that transmits a signal with reduced signal strength will be visible to less or at most the same set of nodes that will see this node in the original network topology. Power control is sometimes used as a synonym for topology control. However, when defining topology control in the general sense, power control refers to a specific subclass of the possible topology control mechanisms.

The general idea of localized power control is that each node attempts to find its own optimal power level such that the set of neighbors it discovers with this power level satisfies a desired property. This general idea can further be classified according to the way the optimal power setting is found. This may either be a *direct* or a *feedback scheme*.

Under a direct power control scheme each node uses its maximum possible transmission range in order to discover the set of neighbor nodes first and determine the right power setting afterward. Setting the power level in one step requires an appropriate signal propagation model that enables a node to estimate the subset of neighbor nodes it is able to reach after reducing the power level. The calculation might rely on either neighbor node positions or signal strength measurements.

A feedback-based adjustment refers to all methods that starting from an initial power level successively adapt a node current power level in order to reach an optimal one. Within each step, the current power level p_i is used for the neighbor discovery procedure. The neighbor information detected with the current power level p_i is used in order to calculate the next power level p_{i+1}. A feedback scheme might either be *transient* in the sense that it finds an optimal power level after a finite number of steps, or it might be applied *permanently* in order to keep the power level optimal with regard to possible changing network parameters.

15.4 NEIGHBOR ELIMINATION SCHEMES

The following lists the basic ideas of the most prominent localized NESs. In general, neighbor elimination can be obtained either in a direct way by eliminating some elements from the list of all currently known neighbor nodes or in an indirect way by reducing a node maximum transmission range such that neighbor discovery will find less neighbor nodes. A reduced neighbor set may be a by-product when using topology control to reduce a node's maximum transmission range. On the contrary, explicitly reducing a node's neighbor set may be used in order to reduce a node's maximum transmission range as well. In other words, power control and explicit neighbor elimination accompany each other.

15.4.1 Relay Regions and Enclosures

The *minimum energy communication network (MECN)* algorithm that was introduced by Rodoplu and Meng [41] is a sophisticated geographic topology control mechanism that keeps all energy efficient paths from the original network topology. The algorithm was further improved by Li and Halpern [24]. The improvement—referred as *small minimum energy communication network (SMECN)*—constructs a subgraph of the topology obtained by MECN while maintaining its minimum energy property.

Both algorithms are based on the concept of *relay regions*, which for two given nodes u and v describes the region of node positions where message transmission from node u via node v consumes less power than transmitting the message directly. In other words, the relay region $R(u, v)$ is defined as the point set $\{w : p(u, v) + p(v, w) < p(u, w)\}$, while $p(u, v)$ denotes the power required to send a message from u to v. Refer to Figure 15.2a for a typical shape of a relay region resulting from the exponential path loss model.

Basically, in both algorithms each node v explores its surrounding with a successively increasing broadcast range. In each step, v determines the *enclosure* of its already discovered neighbors. For a given set of nodes N, the enclosure of a node u defines the region where a direct message transmission is less expensive than sending it over a relay node. This can be defined as the intersection of the nonrelay regions of the neighbor nodes in N, that is, $\cap_{v \in N} R(u, v)^c$. For instance, Figure 15.2b depicts the enclosure of u defined by the discovered nodes v_1, v_2, and v_3. Neighbor exploration stops when the broadcast area becomes a superset of the node's discovered enclosure.

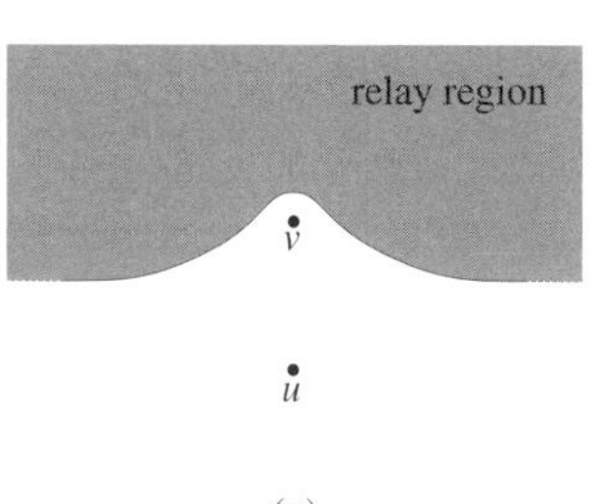

(*a*)

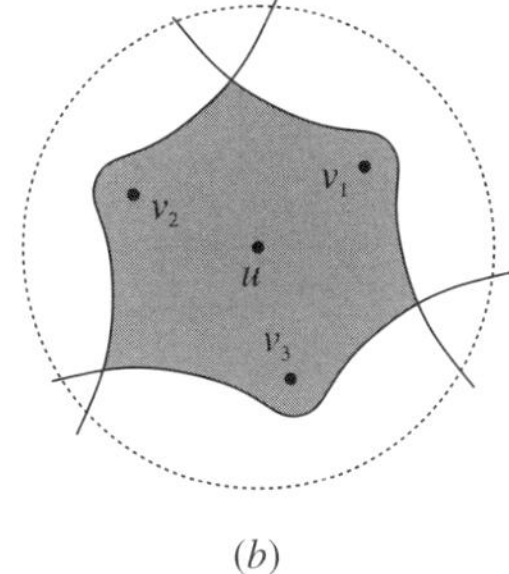

(*b*)

FIGURE 15.2 (*a*) The relay region for node u and its neighbor node v. (*b*) The enclosure for node u obtained due to its neighbor nodes v_1, v_2, and v_3.

For instance, in Figure 15.2b neighbor exploration will stop since the current broadcast range (the dashed circle) contains the entire enclosure formed by the already discovered neighbor nodes v_1, v_2, and v_3.

Refer to Figure 15.2b to follow the intuition behind the described power control scheme. Network nodes might be located either within or without the broadcast range of node u. The nodes within the broadcast range are the discovered nodes v_1, v_2, and v_3 that are located within the enclosure $E(u)$ of node u. Minimum power consumption is achieved when addressing them directly (in general, minimum energy communication within the enclosure $E(u)$ might as well require an appropriate relay node within $E(u)$). Sending a message to any other node located outside the broadcast range of u and thus outside the enclosure $E(u)$, always consumes less power when it is sent via the right relay node in $\{v_1, v_2, v_3\}$. Thus, regarding energy minimal paths, further neighbor discovery beyond the enclosure is not required.

Enclosure-based topology control requires knowledge about the geometry of relay and broadcast regions. Although this is not a necessary condition for the key idea described by MECN and SMECN, the algorithms are introduced under the exponential path loss model assuming omnidirectional free space radio propagation. In this case, by using the power model parameters it is possible to compute the broadcast region, which is a circle centered at u, and the relay region, which is a bell-shaped curve as depicted in Figure 15.2a.

15.4.2 The Cone-Based Approach

The *cone-based topology control (CBTC)* mechanism by Wattenhofer et al. [47] is a directional topology control mechanism that provides a parameter α that can be used in order to control the energy efficiency and node degree of the resulting topology. For a given angle α, each node v running CBTC determines the minimum broadcast range that satisfies that any cone with angle α centered at v (see Fig. 15.3a) contains at least one neighbor node. Obviously, the condition is equivalent to finding the power setting such that the angular distance between two successive neighbors, according to their direction, is less than α.

The original work [47] and a subsequent publication [25] give a precise analysis of the parameter α regarding network connectivity and energy efficient communication

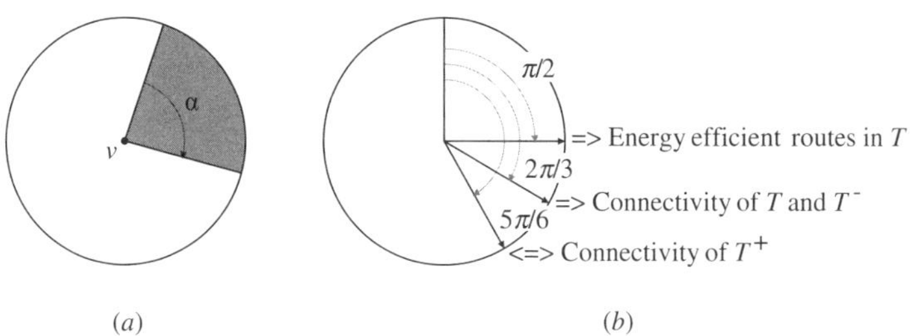

FIGURE 15.3 (*a*) Every cone with angle α must contain at least a neighbor node. (*b*) The structural properties of CBTC depending on the parameter α.

paths. Refer to Figure 15.3b for a summary on the results. It is shown that $\alpha \leq 2\pi/3$ is a sufficient condition for connectivity of the resulting topology. It can be observed that the topology T resulting from CBTC might contain asymmetric links. Thus, the parameter α is as well analyzed with respect to the symmetric sub- and supertopology of T. It holds that $\alpha \leq 2\pi/3$ is even a sufficient condition for connectivity of the symmetric subtopology T . For the symmetric supergraph T^+ the parameter $\alpha \leq 5\pi/6$ is a necessary and sufficient condition for connectivity. Energy efficiency is analyzed under the assumption that the power $p(u, v)$ required to send a message from u to v is lower and upper bounded by $cd^x \leq p(u, v) \leq czd^x$, with $d = |uv|$, $z \geq 1$, $c > 0$, and $x \geq 2$. It is shown that the topology obtained for $\alpha \leq \pi/2$ preserves power optimal routes up to a constant factor.

15.4.3 Counting-Based Methods

A direct approach to reduce the average node degree of a given topology is to set each node's communication range to the minimum power that is required to obtain a given number of neighbor nodes. Assuming a distance power model, this approach is similar to setting each node's communication range to the minimum that is required in order to reach the k-nearest-neighbor node. In the following, the topology resulting from connecting each node with its k closest neighbors will be denoted as a *k-neighbor graph G_k*.

Assuming that network nodes are uniformly distributed on a given square, it has been proved by Xue and Kumar [48] that for an increasing number of nodes n there exists a constant c_1 such that the symmetric supertopology of $G_{c_1 \log n}$ tends to be connected with probability 1. Furthermore, there exists as well a constant $c_2 < c_1$ such that the symmetric supertopology $G_{c_2 \log n}$ tends to be disconnected with probability 1. Based on this result, Blough et al. [2] proved the same property even for the symmetric subtopology, that is, even when all asymmetric links are removed from $G_{c_1 \log n}$ the resulting graph tends to be connected with probability 1.

This theoretical result shows that topology control that considers the k closest neighbors is a reasonable approach to control the node degree while preserving network connectivity with a high probability. The critical part with this approach is that the value of k depends on $\log n$, that is, on the total number of network nodes. Thus, the number of nodes has to be known in advance in order to adjust the right k value at each node.

A counting-based approach can be implemented in many variants. A straightforward method is to actively search for neighbor nodes with increasing power adjustment until the desired number of neighbors has been discovered. If nodes are able to estimate their mutual distances, the k closest neighbors can be determined directly without the need for successively increasing node current power setting. This approach is followed by the *k-neighborhood protocol (k-Neigh)* by Blough et al. [2]. Basically, each node sends its node ID with the maximum possible signal strength. After a certain time-out delta each node sends again with full signal strength its own ID and the IDs of the k-nearest-neighbor nodes it has heard about. With this information each node is able to decide the neighbors it is bidirectionally connected with.

Each node sets its maximum power to the one that is needed in order to reach the most distant bidirectional connected neighbor.

Passive realizations of the counting-based approach have been proposed as well. In the *MobileGrid* approach by Liu and Li [31] the current number of neighbor nodes is estimated by listening for control and data messages issued by neighboring devices. After a certain time-out interval the algorithm checks whether the number of discovered neighbors is within a certain interval, which is an external parameter tuned according to the network characteristics. If the number of neighbors is this interval, the device's own power level is proportionally increased or decreased. Passive neighbor discovery and power level adaption are performed periodically such that the power level gradually approaches the desired one. A similar approach is taken by the *Local Information No Topology (LINT)* protocol described by Ramanathan and RosalesHain [38]. LINT is intended to run in combination with a routing protocol where each node keeps a neighbor table used for routing. This already available information is exploited by LINT in order to determine whether the current number of neighbors is within a certain minimum and maximum threshold. If not satisfied, the node transmission power is gradually modified according to a function of the current and the desired node degrees.

15.4.4 Gabriel and Relative Neighborhood Graph

Gabriel graphs (GGs) [13] and *relative neighborhood graphs (RNGs)* [45] have extensively been studied in conjunction with planar graph routing. Both methods assume an underlying UDG topology and that each node is aware of the positions of its immediate neighbor nodes. A Gabriel graph is obtained by removing all UDG edges (u, v) that satisfy that at least one neighbor node w lies in the circle $U(u, v)$ with diameter $|uv|$ and passing through nodes u and v (see Fig. 15.4a). A relative neighborhood graph is obtained in the same way but using a broader area to detect such a node w. The edge (u, v) is removed whenever a further neighbor node w lies within the intersection of the circles $U(u)$ and $U(v)$ with radius $|uv|$ and centers u and v, respectively (see Fig. 15.4b).

Both graph constructions result in a planar graph and maintain connectivity of the underlying UDG. Moreover, it is obvious that relative neighborhood graph is a subtopology of Gabriel graph. Bose et al. investigated that RNG and GG are not network spanners with respect to the Euclidean distance metric [3]. In the worst case

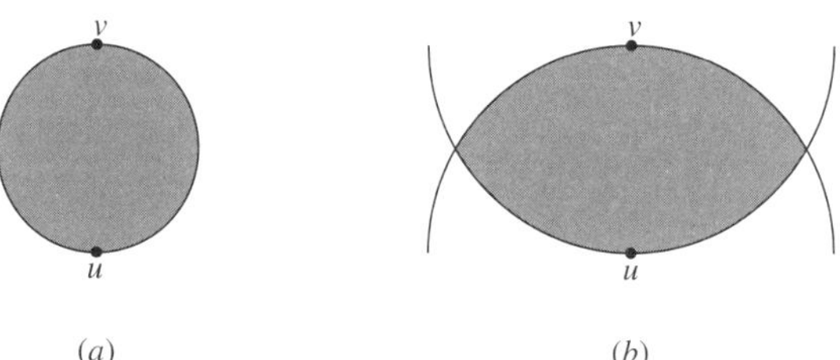

FIGURE 15.4 (*a*) The Gabriel graph criterion. (*b*) The relative neighborhood graph criterion.

the shortest path in a Gabriel graph might be $O(\sqrt{n})$ times longer than the shortest path in the UDG. Worse still, in a relative neighborhood graph the shortest path might be $O(n)$ times longer.

It was pointed out by Li [27] that a node in a relative neighborhood graph might have an arbitrarily large node degree. This observation is based on the fact that in the original RNG construction an edge (u, v) is preserved even if a node w is lying on the boundary of $U(u) \cap U(v)$. The work proposes a modified version of RNG, denoted as RNG$'$, which takes such boundary nodes w as well into account. Removing all edges (u, v) that have a node on the boundary of $U(u) \cap U(v)$ might result in a disconnected topology. For this reason, the RNG$'$ variant utilizes a total ordering to decide if such an edge has to be removed or not. More precisely, besides the original RNG criterion an edge (u, v) is removed as well if there exists a node w on the boundary of $U(u) \cap U(v)$ and one of the following three conditions is satisfied: w is closer to v than to u and $\text{id}(w) < \text{id}(v)$, w is closer to u than to v and $\text{id}(w) < \text{id}(u)$, or w has the same distance to u and v and $\text{id}(w) < \min\{\text{id}(u), \text{id}(v)\}$.

The UDG assumption is inevitable for the described Gabriel graph and relative neighborhood graph constructions. For arbitrary graphs the resulting topology can neither guarantee planarity nor connectivity. Barriere et al. [1] described an extension of the Gabriel graph construction that can be applied in quasi-UDG that satisfies that the ratio between the maximum and the minimum transmission radius is lower than or equal to $\sqrt{2}$. As it was pointed out as well by Kuhn et al., this ratio forms a sharp transition with respect to locally detectable edge intersections [23]. In the Gabriel graph-based robust topology control mechanism described in the work by Barriere et al.[1], each node u determines for all its nonprocessed outgoing edges (u, v) if there exists a neighbor node w that is located within $U(u, v)$. Any such detected neighbor node w is announced to node v. Node v adds w as an unprocessed node to its neighbor set, if node w was not a known to v so far. In addition, node v stores a virtual edge (v, w) that is mapped to the relay node u; that is, a message from v to w is first sent to node u and from there sent to node w. Finally, when all nodes have completed the virtual edge construction step, local Gabriel graph construction is applied on this extended neighbor set. It is proved that the resulting topology is connected and planar if the quasi-unit disk graph satisfies $r_{\max}/r_{\min} \leq \sqrt{2}$ [1].

Note that the virtual edge concept is a recursive structure; that is, the edges (v, u) and (u, w) used for the virtual edge (v, w) might be virtual edges as well. It is shown that the path corresponding to a virtual edge might be arbitrarily long [1]. Thus, without any further provision the algorithm is not localized. However, it is shown that for graph families providing a minimum Euclidean distance between any two network nodes (sometimes denoted as *civilized graphs* or $\omega(1)$-*model*) the total length of a virtual edge's corresponding path is bounded by a constant. Moreover, it is proved that the length of the corresponding path is as well bounded by a constant if the quasi-unit disk graph has a bounded node degree [1]. The work proposes a localized two-phase scheme that first applies a topology control mechanism that limits the node degrees but maintains the quasi-unit disk graph property. Afterward, the virtual edge topology control scheme and localized Gabriel graph construction is applied on this backbone structure.

15.4.5 Localized Delaunay Triangulation

Given a point set V, the *Delaunay triangulation* is obtained by all triangles $T = (u, v, w) \in V^3$ that satisfy that there exists no further node $x \in V$ that is contained in the disk $U(u, v, w)$ passing through the nodes u, v, and w. The geometric graph formed by a Delaunay triangulation is planar and is known to be a spanner with respect to the Euclidean distance metric [9,18]. It is thus an ideal candidate planar graph routing schemes might be applied on. However, Delaunay triangulation requires knowledge of the entire node set, and, moreover, might contain arbitrarily long edges, that is, edges that are longer than the communication range of its end points.

It has been observed that the intersection of the Delaunay triangulation and UDG over a node set V, which is referred as *unit Delaunay triangulation $UDel(V)$*, preserves the spanning property; that is, with respect to the Euclidean distance metric for two given nodes the shortest path in $UDel(V)$ is at most a constant longer than the shortest path in the UDG [14,28]. Thus, Delaunay triangulation is an interesting candidate for constructing planar, spanning topologies in a localized manner.

In the method described by Gao et al. [14], each node locally constructs a Delaunay triangulation over all its one-hop neighbor nodes and announces this triangulation to its one-hop neighbors. Based on this information a node u checks for each incident Delaunay triangulation edge (u, v) if there exists a one-hop neighbor w that is connected to v but does not contain the edge (u, v) in its local Delaunay triangulation. In this case, the edge (u, v) is removed at node u. It is shown that this topology construction method always produces a planar graph that preserves all edges of $UDel(V)$ [14]. The topology is thus a spanner as well.

Li et al. [28] introduced the concept of *k-localized Delaunay triangulation $LDel^k(V)$*, which denotes the topology obtained by preserving each node Gabriel graph edges and edges of all *k-localized Delaunay triangles*. The latter refers to all triangles (u, v, w) that satisfy that u, v, and w can reach each other in the underlying UDG and that the disk $U(u, v, w)$ does not contain any k-hop neighbor of u, v, or w. It is observed that $LDel^k(V)$ may be nonplanar for $k = 1$ while it is always planar for $k > 2$. Moreover, it is shown that $LDel^k(V)$ is a spanner. $LDel^k(V)$ can be used for localized topology control since it requires only local neighborhood information. However, the communication cost will be high for $k > 1$. For this reason the *planarized $LDel^1(V)$ method $PLDel(V)$* described in the work by Li et al. [28] first locally constructs the nonplanar topology $LDel^1(V)$ and then removes intersecting edges by the following scheme. In a previous step, a node u removes a triangle (u, v, w) from $LDel^1(V)$ if one edge of a neighbor nodes triangle (x, y, z) is lying in the circle $U(u, v, w)$. Afterward, node u keeps all incident edges that are either a Gabriel graph edge or an edge from a triangle (u, v, w) that was kept by each triangle nodes u, v, and w.

The *partial Delaunay triangulation PDT* described by Li et al. [29] employs an alternative definition of Delaunay triangulation in a localized way. It is easy to show that the edges (u, v) of a Delaunay triangulation are exactly those edges that satisfy that there exists a circle having u and v on its boundary that does not contain any other nodes. A node u running the PDT method keeps an edge (u, v) if the empty circle rule is satisfied for the specific circle $U(u, v)$ that is used for Gabriel graph construction; that is, node u keeps the edge if the circle $U(u, v)$ is empty. Two cases arise if the circle is not empty. Node u removes the edge (u, v) if all nodes found in the circle are located on both sides of the line segment uv. If these nodes are lying on one side only, the node w maximizing the angle $\angle uwv$ is considered. The edge is kept iff each neighbor node x of u and v satisfies $\angle uvw + \angle uxv < \pi$.

15.4.6 Explicit Planarization

An inherent property of a UDG is that for any intersection between two edges (a, b) and (c, d) there exists at least one edge end point (for instance, node c in Fig. 15.5a) that is connected to the remaining nodes. This will be referred as the *redundancy property* in the following. Any undirected graph satisfying the redundancy property supports local detection of intersecting edges. More precisely, when the network nodes exchange two-hop neighbor information the intersection between two edges (a, b) and (c, d) is visible to all nodes in $\{a, b, c, d\}$.

Whenever edge intersection can be detected locally it is a natural approach to explicitly remove only an edge whenever it intersects with another one. This, however, raises the question which of both edges has to be removed. As depicted in Figure 15.5a, removing both edges or even removing only the wrong one (edge (c, d) in this case) might result in a disconnected subtopology. Moreover, even if a localized construction method always selects the right edge the question remains if the resulting topology remains connected. More precisely, from a local point of view, removing edge (a, b) in Figure 15.5a will not cause disconnection since messages sent along the edge (a, b) can as well be relayed along the intermediate node c by using edge (a, c) at first and edge (c, b) afterward. However, during some construction step these edges might have been removed as well. Thus, in order to keep the resulting topology connected any direct planarization scheme has to ensure that there remains at least a path from a to c and a path from c to b that do not use the removed edge (a, b).

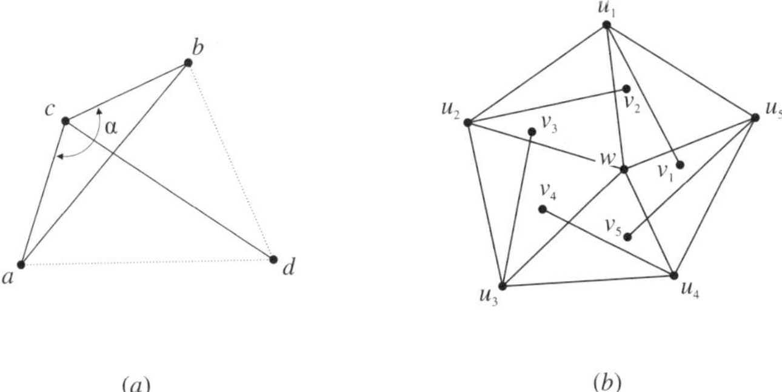

FIGURE 15.5 (*a*) The redundancy property. (*b*) A graph that loses connectivity when planarizing it by edge removal.

It turns out that the redundancy property alone does not provide edge removal based construction of a planar and connected topology. Refer to Figure 15.5b for an example. The depicted graph satisfies the redundancy property. Planarity can be obtained by removing either the edges (u_i, v_i) or the edges (u_i, w). However, in both cases either the nodes v_i or the node w get isolated from the remaining network. On the contrary, the fact that any UDG can be planarized (for instance, by applying the Gabriel graph method) reveals that these graphs have besides the redundancy an additional structural property that assures that an example like depicted in Figure 15.5b cannot be constructed. An explicit formal graph theoretic definition of this property is still missing.

A localized rule in order to decide which one of the two intersecting edges (a, b) or (c, d) has to be removed can be described as follows. For edge (a, b) compute the maximum over the angles $\angle acb$ and $\angle adb$. For edge (c, d) compute the maximum over the angles $\angle cad$ and $\angle cbd$. The edge with the larger maximum value is removed. For instance, in Figure 15.5a the angle $\alpha = \angle acb$ is the maximum one in the quadrilateral (a, b, c, d). Thus, the edge (a, b) will be removed. Ties, that is, both edges have the same maximum value, are broken by using a given edge ordering and removing the "smaller" edge. In the following, the described construction method will be referred as *angle-based direct planarization (ABDP)*.

Since for each intersection at least one edge is removed, it is obvious that the resulting topology is planar. Moreover, it is easy to show that a Gabriel graph is a subtopology of the topology obtained by ABDP. Consequently, the resulting topology preserves connectivity of the underlying UDG. The intuition behind this approach is to maintain shortest paths up to a constant factor. An edge is removed only if it intersects, and from both edges exactly the one is removed whose alternative path introduces the smallest increment in terms of the Euclidean distance. However, it is not known by now if the resulting topology is a spanner with respect to the Euclidean distance metric.

15.4.7 Minimum Spanning Trees

For an undirected and connected graph $G = (V, E)$ a *spanning tree* T is an acyclic subset of E, which connects any two nodes in V. For any edge weight function $f : E \to \mathbb{R}_0^+$ the *weight of a tree* $f(T)$ is defined as the sum over its edge weights. A *minimum spanning tree* $T_{\min}$ denotes a spanning tree that satisfies $f(T_{\min}) \leq f(T)$ for any other spanning tree T of G. Under the assumption that the nodes are deployed on the plane and that the edge weight function is the Euclidean distance, the degree of each node in the resulting *Euclidean minimum spanning tree* is limited by 6 [33]. Thus, when aiming at small node degrees a minimum spanning tree might be a desired topology.

Standard minimum spanning tree constructions like those by Prim [36] require global knowledge about the entire network and are thus limited in scale when the network is dynamically changing. The geographic topology construction method by Li et al. [26] referred as *local minimum spanning tree (LMST)* applies an independent localized minimum spanning tree construction at each node. More precisely, by using the full available signal strength the nodes exchange position information among their one-hop neighbors and apply a standard Euclidean minimum spanning tree construction on all detected one-hop neighbors. The transmission power of each node is adjusted to the minimum power that is required to reach the most distant node that is adjacent in the locally constructed spanning tree.

It is shown that the topology G_0 resulting from all locally preserved edges—the edges (v, w) that satisfy that w is an immediate neighbor of v in the spanning tree constructed at node v—inherits the same node degree bound that is observed for Euclidean minimum spanning trees [26]. Moreover, it is shown that topology G_0 is connected (assuming that the original network topology is connected). However, the topology G_0 might contain directed links. Thus, when an undirected topology is desired, either the symmetric subtopology G_0^- or the symmetric supertopologies G_0^+ can be constructed afterward. The theoretical analysis reveals that G_0^- being a subgraph of G_0 preserves connectivity of G_0, and that G_0^+ being a supergraph of G_0 remains bounded in degree; that is, each node still has at most six neighbor nodes in G_0^+.

Minimum spanning trees are as well of special interest for energy minimal broadcasting. Under the exponential path loss model $|uv|^\alpha$ (i.e., the constant c is ignored here), it was shown by Wan et al. that a minimum spanning tree supports energy minimal broadcasting up to a constant factor [46]. Thus, localized topology control based on minimum spanning trees is a desired topology to support energy efficient broadcasting. In order to assess the quality of a topology with respect to minimum spanning trees, Li introduced the concept of *low weight graphs* that denotes all graphs whose total edge length is within a constant factor of the total edge length of the minimum spanning tree [27]. In the same work it is shown as well that localized construction of a localized minimum weighted graph requires some two-hop information. This implies that, for instance, LMST although based on minimum spanning trees is not a low weight graph in general. However, by allowing some two-hop neighbor information the basic idea of LMST can be extended to a low weight graph construction. Li et al. [30] introduce two such methods that are based on the UDG assumption. The *two-hop* LMST (LMST_2) employs the idea of LMST but each node utilizes two-hop information in order to construct its local minimum spanning tree. The *incident MST and RNG (IMRG)* graph is basically a combination of RNG′ and LMST. By using one-hop neighbor information, a RNG′ topology is constructed first. The incident edges of RNG′ are then broadcast to all one-hop neighbors. The LMST construction is then applied on this partial two-hop neighbor information. For both structures LMST_2 and IMRG it is shown that they are planar, connected, limited by 6 in node degree, and that they have the low weight graph property. The latter construction method is suggested as the favored one since construction has a significantly reduced message complexity due to the fact that only two-hop neighbor information with respect to the constructed RNG′ topology has to be exchanged.

It is important to note that a low weight graph does not imply that this graph supports energy efficient broadcasting up to a constant factor. However, it is shown in the work by Li et al. [30] that IMRG improves the RNG method that has as well been applied as a broadcasting topology [42]. More precisely, it is pointed out there that RNG can consume about $O(n^{\alpha})$ (with α being the path loss exponent) times the energy used by the optimal broadcasting method. This is improved by IMRG by the factor $O(n)$; that is, it is shown that IMRG consumes up to a constant factor of $O(n^{\alpha-1})$ more energy than the optimal broadcast tree. Another example of RNG and LMST for broadcasting can be found in the work by Ingelrest et al. [17] where authors propose a variation of NES for broadcasting by limiting surveillance to RNG or LMST neighbors and by reducing transmission power to noncovered monitored neighbors.

15.4.8 Redundant Edges

Generally speaking, an edge (u, w) can be denoted as *redundant* whenever for some intermediate node v the edges (u, v) and (v, w) do exist and communication along (u, v) and (v, w) is "cheaper" than direct communication along (u, w). In terms of a given edge weight $f : E \rightarrow \mathbb{R}_0^+$ this condition can be expressed as $f(u, v) + f(v, w) < f(u, w)$. For instance, in Figure 15.6a (u, w) is a redundant edge due to (u, v) and (v, w). It is obvious that removing a redundant edge from a given topology T will end up into a topology T' that preserves all paths in T that are minimal regarding the given edge weight function.

The concept of redundant edges can be used either as a topology control mechanism on its own or as a subsequent refinement to further reduce the subset of neighbors already constructed by a given topology control mechanism. For instance, in the CBTC method [47], after the minimum required power setting for each node has been found, in the second phase node degrees are reduced by removing distant neighbors that can be reached by intermediate nodes. The described NES assumes that the required transmission power $p(v, w)$ is a nondecreasing function depending on the distance between two communicating nodes v and w. In addition, neighbor nodes have to perform a local exchange of the minimum transmission power they need in order to reach their neighbor nodes. By inspecting the neighbor nodes with increasing power distance, node u will check for the current neighbor node w if there exists a neighbor node v that satisfies $p(u, v) \leq p(u, w)$ and $p(u, v) + p(v, w) \leq q \cdot p(u, w)$. In this case, the node w will be removed from the neighbor list. In other words, node w is removed whenever there exists an alternative closer relay node v and the power required to send a message along the relay node v is lower than a constant factor more than the power required in order to transmit the message directly to w. The

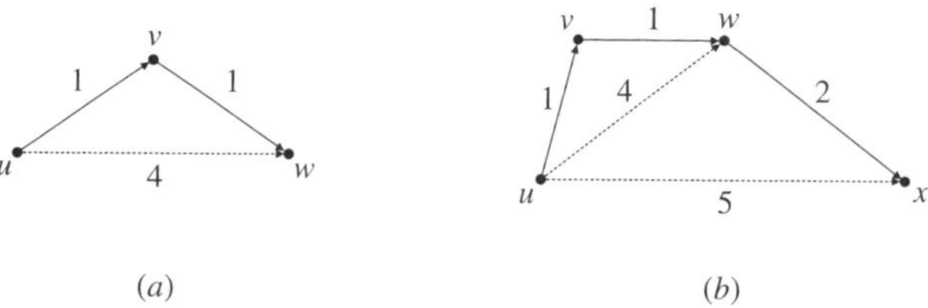

FIGURE 15.6 (*a*) A redunant edge in terms of single relay node. (*b*) A redundant edge in terms of a sequence of relay nodes.

constant q is introduced to adjust CBTC between energy optimal paths and reduced node degree that is forming a trade-off. Obviously, for $q_1 \leq q_2$ the neighbor set constructed for q_2 will be a subset of the neighbor set constructed for q_1. A large q value may result in relay nodes that require significantly more energy consumption.

When assuming an arbitrary power model, reducing all redundant edges might still leave some more room for improvement. For an example refer to Figure 15.6b. The edge (u, x) with edge weight 5 is not redundant since using the only possible relay node w will produce energy consumption of $4 + 2 = 6$. However, the sequence of the relay nodes v and w will produce energy consumption of only $1 + 1 + 2 = 4$. The pruning stage of the k-Neigh protocol [2] can be seen as an improved version of the basic redundant edge elimination scheme that implicitly takes such sequences of relay nodes into account. Considering all neighbor nodes in ascending edge order, the neighbor w is removed if it is redundant. In this case, the algorithm looks for the relay node v that produces the minimum energy expenditure. The edge of the removed node w is considered in all further steps; however, its weight is reduced from $f(u, w)$ to $f(u, v) + f(v, w)$. For instance, in Figure 15.6b the edge weight of (u, w) will be reduced from 4 to $1 + 1 = 2$. Thus, when inspecting the next edge (u, x) the relay node w will consume power $2 + 2 = 4$; that is, edge (u, x) is a redundant one as well.

The removal of redundant nodes can be used in order to reduce the information that is disseminated in the network. For instance, in OLSR protocol [6], each node broadcasts information about topology (edges) to all nodes in the network. In the original proposition, the reduced graph $G' = (V', E')$ has to satisfy shortest paths, for all couple of nodes (u, v) that are not neighbors in G, the hop distance between u and v has to be the same in G and in G'. In order to compute locally G', each node selects a subset of its neighbors called multipoint relays (MPRs) [37]. Ideally, this subset is the smallest subset of one-hop neighbors, which covers two-hop neighbors. Since this problem is shown to be NP-complete, Viennot et al. propose a greedy heuristic. For QoS purpose, Moraru et al. [34] propose to replace advertisement of MPR links by links that preserve widest paths (bandwidth) or quickest path (delay).

15.5 PERSPECTIVES

In this chapter we have presented last recent results of topology control in the domain of neighbor elimination. The emphasis was on presenting the basic mechanisms in this domain. The result of a topology control mechanism A applied on a network graph is again a graph (with some desired properties). If the structural properties of this result is compatible with the input domain of a further topology control mechanism B, concatenation of mechanism B after mechanism A can be applied as a combined topology control mechanism.

This intrinsic property opens a wide spectrum of possible combined schemes, which can be subject of future research. The combination of two basic topology control building blocks could result in a scheme that supports the objectives of both schemes. For instance, producing a topology that supports cost efficient routing paths and applying a planar graph construction mechanism that supports face routing is a prominent example of such a combined scheme [8].

Foundation of many works in topology control is the so-called UDG, which is a simplified version of wireless communication in real world. While in theory this model plays an important role, in future research a majority of well established mechanisms could be investigated with focus on more realistic models [16,44].

15.6 EXERCISES

1. Assume neighbor discovery based on periodic hello messages including the sending node's ID only. What is the main problem when applying this scheme under an arbitrary topology? Argue why this scheme might, however, be applied in this case and discuss possible solutions.
2. Design a localized protocol that maintains two-hop neighbor information under arbitrary topologies. More precisely, each node should know about the node to which itself and its neighbors are connected. Estimate the message complexity of your scheme.
3. Suppose a simple neighbor discovery protocol where a requesting node sends out a broadcast message and each node receiving this broadcast immediately sends a neighbor discovery reply. What is the problem with this approach? Describe possible solutions to this problem.
4. Assume that communication between two nodes v and w consumes energy according to the exponential path loss model $|vw|^{\alpha} + c$. Show that there are an optimal number of equidistant intermediate forwarding nodes that require the minimum amount of energy.
5. Neighbor nodes can be discovered passively by listening for control and data messages issued by the surrounding nodes. Discuss the advantage and disadvantage of this approach.
6. Show by an example that CBTC may produce a directed topology.
7. Suppose that network nodes are deployed on a rectangular area D. Assume the enclosure of a node u and its neighbors N defined as $\bigcap_{v \in N} R(u, v)^c$.
   (a) Show by an example that with this definition nodes at the boundary of D will often set their broadcast range to the maximum possible one.
   (b) Describe an improved definition of enclosure, which avoids this problem.
8. Show by an example that CBTC does not necessarily preserve connectivity for $\alpha = 5\pi/6 + \epsilon$ for any $\epsilon > 0$.
9. Show by an example that the topology derived under LMST might be asymmetric.
10. Construct an example network that loses connectivity when planarizing it by edge removal.
11. From a global point of view, graph planarization can be obtained by simply removing one of two intersecting edges until no intersection remains. Discuss if such a scheme can be implemented in a localized way.

12. Show that there exists no localized algorithm that constructs a topology so that the maximum node power based on this structure is within a constant factor of that based on a minimum spanning tree.
13. Show by an example that Gabriel and relative neighborhood graphs are not bounded in degree.
14. Show by an example that Gabriel and relative neighborhood graph construction might produce disconnection and nonplanarity in arbitrary graphs.
15. Let G be a quasi-UDG with minimum transmission range $r_{\min}$ and maximum transmission range $r_{\max}$.
    (a) Show that for $r_{\max}/r_{\min} \leq \sqrt{2}$ an edge intersection can always be detected locally, that is, at least one end point of the first edge is connected to an end point of the second edge.
    (b) Show that for $r_{\max}/r_{\min} > \sqrt{2}$ local detection of an intersection is not always possible.
    (c) Construct an example that shows that the path corresponding to a virtual edge of the robust Gabriel graph construction might be arbitrarily long.
16. Prove that for any intersecting edges in a UDG at least one of the edge end points is connected to all other nodes.
17. Show that under the assumption of a connected UDG the ABDP method results in a planar and connected subtopology.
18. Show by an example that $LDel^1(V)$ is not planar in general.

REFERENCES

1. Barriere L, Fraigniaud P, Narajanan L, Opatrny J. Robust position-based routing in wireless ad hoc networks with unstable transmission ranges. Proceedings of the 5th ACM International Workshop on Discrete Algorithms and Methods for Mobile Computing and Communications (DIAL M 01); 2001. p 19–27.
2. Blough D, Leoncini M, Resta G, Santi P. The K-neigh protocol for symmetric topology control in ad hoc networks. Proceedings of the 4th ACM International Symposium on Mobile Ad Hoc Networking and Computing (MOBIHOC-03); June 1–3; 2003; New York: ACM Press; 2003. p 141–152.
3. Bose P, Devroye L, Evans W, Kirkpatrick D. On the spanning ratio of Gabriel graphs and beta-skeletons. Proceedings of the Latin American Theoretical Informatics (LATIN02); April 3–6, 2002; Cancun, Mexico; 2002.
4. Bose P, Morin P, Stojmenovic I, Urrutia J. Routing with guaranteed delivery in ad hoc wireless networks. Proceedings of the 3rd ACM International Workshop on Discrete Algorithms and Methods for Mobile Computing and Communications (DIAL M 99); August 20, 1999; Seattle, WA; 1999. p 48–55.
5. Capkun S, Hamdi M, Hubaux J-P. GPS-free positioning in mobile ad-hoc networks. Cluster Comput J 2002;5(2):118–124.
6. Clausen T, Jacquet P. Optimized link state routing protocol (olsr). RFC 3626 (Experimental); 2003. Available at: http://www.ietf.org/rfc/rfc3626.txt

7. Clementi A, Crescenzi P, Penna P, Rossi G, Vocca P. On the complexity of computing minimum energy consumption broadcast subgraphs. Proceedings of the 18th Annual Symposium on Theoretical Aspects of Computer Science (STACS 2001); 2001. p 121–131.
8. Datta S, Stojmenovic I, Wu J. Internal node and shortcut based routing with guaranteed delivery in wireless networks. Cluster Comput 2002;5(2):169–178.
9. Dobkin DP, Friedman SJ, Supowit KJ. Delaunay graphs are almost as good as complete graphs. Discrete Comput Geom 1990.
10. Frey H, Geographical cluster based routing with guaranteed delivery. 2nd IEEE International Conference on Mobile Ad-hoc and Sensor Systems (MASS 2005); November 7–10, 2005; Washington, DC, USA.
11. Frey H, Görgen D. Planar graph routing on geographical clusters. Ad Hoc Networks (Special Issue on Data Communication and Topology Control in Ad Hoc Networks) 2005;3(5):560–574.
12. Frey H, Görgen D. Geographical cluster based routing in sensing-covered networks. IEEE Trans Parallel Distrib Syst (Special Issue on Localized Communication and Topology Protocols for Ad Hoc Networks) 2006;17(4).
13. Gabriel KR, Sokal RR. A new statistical approach to geographic variation analysis. Syst Zool 1969;18:259–278.
14. Gao J, Guibas LJ, Hershberger J, Zhang L, Zhu A. Geometric spanner for routing in mobile networks. Proceedings of the Second ACM International Symposium on Mobile Ad Hoc Networking and Computing MobiHoc'01; October 2001; Long Beach, CA, USA; 2001. p 45–55.
15. Hightower J, Borriella G. Location systems for ubiquitous computing. IEEE Comput 2001;34(8):57–66.
16. Ingelrest F, Simplot-Ryl D. Maximizing the probability of delivery of multipoint relay broadcast protocol in wireless ad hoc networks with a realistic physical layer. Proceedings of the 2nd International Conference on Mobile Ad-hoc and Sensor Networks (MSN 2006); 2006; Hong Kong, China.
17. Ingelrest F, Simplot-Ryl D, Stojmenovic I. Optimal transmission radius for energy efficient broadcasting protocols in ad hoc networks. IEEE Trans Parallel Distrib Syst 2006;17(6):536–547.
18. Keil JM, Gutwin CA. Classes of graphs which approximate the complete Euclidean graph. Discrete Comput Geom 1992; 7.
19. Kaplan ED. Understanding GPS: Principles and Applications. Boston, MA: Artech House; 1996.
20. Karp B, Kung HT. GPSR: greedy perimeter stateless routing for wireless networks. Proceedings of the 6th ACM/IEEE Annual International Conference on Mobile Computing and Networking (MOBICOM-00); Aug 6–11, 2000; NY: ACM Press; 2000. p 243–254.
21. Krizman K, Biedka TE, Rappaport TS. Wireless position location: fundamentals, implementation strategies, and source of error. Proceedings of the IEEE 47th Vehicular Technology Conference; Volume 2; 1997. p 919–923.
22. Kuhn F, Wattenhofer R, Zhang Y, Zollinger A. Geometric ad hoc routing: of theory and practice. Proceedings of the 22nd ACM International Symposium on the Principles of Distributed Computing (PODC); July 13–16, 2003; Boston, MA, USA; 2003. p 63–72.

23. Kuhn F, Wattenhofer R, Zollinger A. Ad hoc networks beyond unit disk graphs. ACM DIALM-POMC Joint Workshop on Foundations of Mobile Computing; September 2003; San Diego; 2003. p 69–78.

24. Li L, Halpern JY. Minimum-energy mobile wireless networks revisited. Proceedings of IEEE International Conference on Communications ICC 2001; Volume 1; June 2001. p 278 283.

25. Li L, Halpern JY, Bahl P, Wang Y-M, Wattenhofer R. Analysis of a cone-based distributed topology control algorithm for wireless multi-hop networks. Proceedings of the 20th ACM SIGACT-SIGOPS Symposium on Principles of Distributed Computing (PODC 2001); 2001.

26. Li N, Hou JC, Sha L. Design and analysis of an MST-based topology control algorithm. Proceedings of the 22nd Annual Joint Conference of the IEEE Computer and Communications Societies (INFOCOM 2003); 2003.

27. Li X-Y. Localized construction of low weighted structure and its applications in wireless ad hoc networks. Wireless Netw 2005;11(6):697–708.

28. Li X-Y, Calinescu G, Wan P-J. Distributed construction of a planar spanner and routing for ad hoc wireless networks. Proceedings of the 21st Annual Joint Conference of the IEEE Computer and Communications Society (INFOCOM'02); Volume 3; June 23–27, 2002; Piscataway, NJ, USA: IEEE Computer Society; 2002. p 1268–1277.

29. Li X-Y, Stojmenovic I, Wang Y. Partial Delaunay triangulation and degree limited localized Bluetooth scatternet formation. IEEE Trans Parallel Distrib Syst 2004;15(4): 350–361.

30. Li X-Y, Wang Y, Wan P-J, Frieder O. Localized low weight graph and its applications in wireless ad hoc networks. Proceedings of the 23rd Conference of the IEEE Communications Society (INFOCOM 2004); 2004.

31. Liu J, Li B. Mobilegrid: Capacity-aware topology control in mobile ad hoc networks. Proceedings of the 11th IEEE International Conference on Computer Communications and Networks (ICCCN 2002); October 14–16, 2002; Miami, FL, USA; 2002. p 570–574.

32. Marina MK, Das SR. Routing performance in presence of unidirectional links in multihop wireless networks. Proceedings of the Third ACM International Symposium on Mobile Ad Hoc Networking and Computing (MobiHoc 2000); 2002 p 12–23.

33. Monma C, Suri S. Transitions in geometric minimum spanning trees. Proceedings of the Annual ACM Symposium on Computational Geometry; 1991 North Convay, NH, 1991. 239–249.

34. Moraru L, Simplot-Ryl D. QoS preserving topology advertising reduction for olsr routing protocol for mobile ad hoc networks. Proceedings of the 3rd Wireless On Demand Network Systems and Services (WONS 2006); 2006 Les Ménuires, France.

35. Pearlman MR, Haas ZJ, Manvell BP. Using multi-hop acknowledgements to discover and reliably communicate over unidirectional links in ad hoc networks. Proceedings of the IEEE Wireless Communications and Networking Conference (WCNC 2000); Volume 2; 2000. p 532–537.

36. Prim R. Shortest connection networks and some generalizations. Bell Syst Technol J 1957;36:1389–1401.

37. Qayyum LVA, Laouiti A. Multipoint relaying for flooding broadcast messages in mobile wireless networks. Proceedings of the 35th Hawaii International Conference on System Sciences; 2002; Hawaii, USA.

38. Ramanathan R, Hain R. Topology control of multihop wireless networks using transmit power adjustment. Proceedings of the 2000 IEEE Computer and Communications Societies Conference on Computer Communications (INFOCOM-00); March 26–30, 2000; Los Alamitos: IEEE; 2000. p 404–413.

39. Ramasubramanian V, Chandra R, Mosse D. Providing a bidirectional abstraction for unidirectional ad hoc networks. Proceedings of the 21st IEEE Conference on Computer Communications (INFOCOM 2002); June 23–27, 2002; New York, USA; 2002. p 1258–1267.

40. Rappaport TS. Wireless Communications: Principles and Practice. Prentice Hall; 2002.

41. Rodoplu V, Meng TH. Minimum energy mobile wireless networks. IEEE J Selected Areas Commun 1999;17(8):1333–1344.

42. Seddigh M, Solano-González J, Stojmenovic I. RNG and internal node based broadcasting algorithms for wireless one-to-one networks. Mobile Comput Commun Rev 2001;5(2):37–44.

43. Simplot-Ryl D, Stojmenovic I, Wu J. Energy efficient broadcasting, and area coverage in sensor networks. In: Stojmenovic I editor. Handbook of Sensor Network Algorithms and Architecture. New York; Wiley; 2005.

44. Stojmenovic I, Nayak A, Kuruvila J. Design guidelines for routing protocols in ad hoc and sensor networks with a realistic physical layer. IEEE Commun Magazine (Ad Hoc and Sensor Networks Series) 2005;43(3):101–106.

45. Toussaint G. The relative neighborhood graph of a finite planar set. Pattern Recog 1980;12(4):261–268.

46. Wan P-J, Calinescu G, Li X, Frieder O. Minimum-Energy broadcast routing in static ad hoc wireless networks. Proceedings of the Twentieth Annual Joint Conference of the IEEE Computer and Communications Societies (INFOCOM-01); April 22–26, 2001; Los Alamitos, CA; IEEE Computer Society; 2001. p 1162–1171.

47. Wattenhofer R, Li L, Bahl P, Wang Y-M. Distributed topology control for power efficient operation in multihop wireless ad hoc networks. Proceedings of the 20th Annual Joint Conference of the IEEE Computer and Communications Societies; 2001. p 1388–1397.

48. Xue F, Kumar PR. The number of neighbors needed for connectivity of wireless networks. Wireless Netw 2004;10(2):169–181.

CHAPTER 16

A Novel Admission Control for Multimedia LEO Satellite Networks

SYED R. RIZVI, STEPHAN OLARIU, and MONA E. RIZVI

16.1 INTRODUCTION

Terrestrial wireless networks provide mobile communication services with limited geographic coverage since they are economically infeasible in areas of rough topography or inadequate user population [1]. In order to provide global information access, a number of satellite systems have been proposed. These satellite networks are well suited for worldwide communication services and to complement the terrestrial wireless networks because they can support not only the areas with terrestrial wireless networks but also the areas that lack a wireless infrastructure. Among the satellite systems, Low Earth Orbit (LEO) satellite systems play an important role in the near future of communication services. The satellite system could interact with the terrestrial wireless network to absorb the instantaneous traffic overload of these networks. In other words, it is possible to route a connection using intersatellite links (ISL) without relying on terrestrial resources. However, a number of mobility problems that did not exist in terrestrial systems should be solved in order to have feasible implementations of the LEO systems.

In response to the demand for truly global coverage by personal communication services (PCS), a new generation of mobile satellite networks intended to provide *anytime–anywhere* communication services has emerged [4,5]. LEO satellite networks, deployed at altitudes ranging from 500 to 2000 km, are well suited to handle bursty Internet and multimedia traffic and to offer anytime–anywhere connectivity to mobile hosts (MH). LEO satellite networks offer numerous advantages over terrestrial networks including global coverage and low cost-per-minute access to MHs equipped with handheld devices. Because LEO satellite networks are expected to support real-time interactive multimedia traffic, they must be able to provide their users with quality-of-service (QoS) guarantees for metrics that includes bandwidth, delay, jitter, call dropping, and call blocking probability [8].

Handbook of Applied Algorithms: Solving Scientific, Engineering and Practical Problems
Edited by Amiya Nayak and Ivan Stojmenović

16.2 LEO SATELLITE NETWORKS AND MAIN QoS PARAMETERS

Although providing significant advantages over their terrestrial counterparts, LEO satellite networks present protocol designers with an array of daunting challenges, including handoff, mobility, and location management [8]. Because LEO satellites are deployed at low altitude, Kepler's third law implies that these satellites must traverse their orbits at a very high speed. We assume an orbital speed of about 26,000 km/h. As can be seen in Figure 16.1, the coverage area of a satellite—a circular area of the surface of the Earth—is referred to as its *footprint*. For spectral efficiency reasons, the satellite footprint is partitioned into slightly overlapping cells, called *spotbeams*. As their coverage area changes continuously, in order to maintain connectivity, MHs must switch from spotbeam to spotbeam and from satellite to satellite, resulting in frequent intra and intersatellite handoffs. Identical frequencies can be reused in different spotbeams if the spotbeams are geographically separated to limit the interference. In this chapter, we focus on intrasatellite handoffs, referred to, simply, as handoffs.

Due to the large number of handoffs experienced by a typical connection during its lifetime, resource management and connection admission control are very important tasks if the system is to provide fair bandwidth sharing and QoS guarantees. In particular, a reliable handoff mechanism is needed to maintain connectivity and to minimize service interruption to on-going connections, as MHs. In fact, one of the most important QoS parameters for LEO satellite networks is the call dropping probability (CDP), quantifying the likelihood that an ongoing connection will be force-terminated due to an unsuccessful handoff attempt. The call blocking probability (CBP) expresses the likelihood that a new call request will not be honored at the time it is placed. The extent to which the existing bandwidth in a spotbeam is efficiently used is known as bandwidth utilization (BU). The main goal of a network designer becomes to provide acceptably low CDP and CBP while, at the same time,

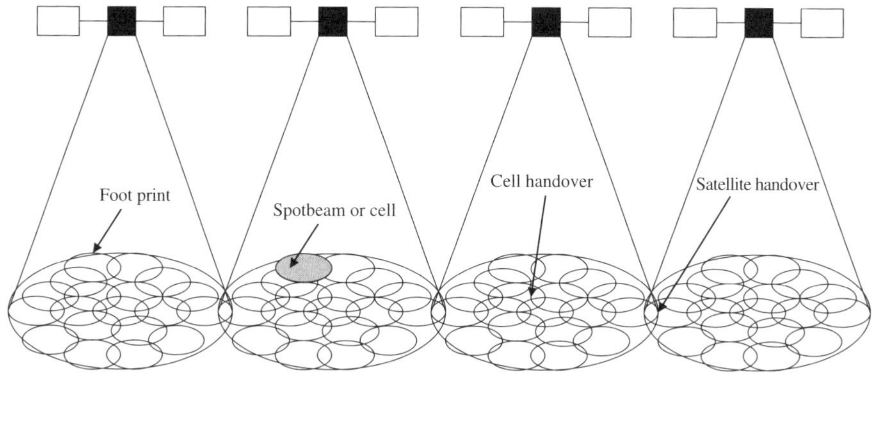

FIGURE 16.1 Illustration of satellite footprint.

maximizing BU [8,11]. This chapter describes in detail four recent resource allocation strategies in multimedia LEO satellite networks that use a novel call admission control concept. The performance of these schemes is compared and simulation results show that they offer low CDP, providing for reliable handoff of on-going calls, good CBP for new call requests, while maintaining high BU.

16.3 BACKGROUND AND RELATED WORK

In this section, we briefly outline a number of call admission algorithms proposed in the literature. One noticeable prioritization scheme is the handoff with queueing (HQ) technique [3]. This scheme outlines the overlapped area between two cells where the handoff takes place. When an MH is in an overlapped area, the handoff process is initiated. If a channel is available in the new cell, it is allocated to the MH; otherwise, the handoff request is queued. When a channel becomes available, one of the calls in the queue is served. A handoff call is blocked if no channel is allocated for the call in the new cell when the power level received from the current cell falls below the minimum power level that is required for a successful data transfer. The HQ scheme reduces the handoff call dropping; however, its performance depends on the new call arrival rate and the size of the overlapped area. In the worst case, high call arrival rates or small overlapped areas would result in a high value of handoff CDP.

Later, Markhasin et al. [6] introduced two different mobility models for satellite networks. In the first model, only the motion of the satellite is taken into account, whereas in the second model, other motion components such as the rotation of the Earth and user mobility are considered. To design a call admission control algorithm for mobile satellite systems, the authors introduced a new metric called mobility reservation status, which provides the information about the current bandwidth requirements of all active connections in a specific spotbeam in addition to the possible bandwidth requirements of mobile terminals currently connected to the neighboring spotbeams. A new call request is accepted in the spotbeam where it originated, say m, if there is sufficient available bandwidth in the spotbeam, and the mobility reservation status of particular neighboring spotbeams have not exceeded a predetermined threshold $TNewCall$. If a new call is accepted, the mobility reservation status of a particular number S of spotbeams will be updated. A handoff request is accepted if bandwidth is available in the new spotbeam and the handoff threshold is not exceeded. The key idea of the algorithm is to prevent handoff dropping during a call by reserving bandwidth in a particular number S of spotbeams into which the call is likely to move. The balance between new call blocking and handoff call dropping depends on the selection of predetermined threshold parameters for new and handoff calls. However, during simulation implementation, we found that this scheme has a problem determining threshold points in the case of LEO satellite networks.

Uzunalioglu [14] proposed a call admission strategy based on the MH location. In his scheme, a new call is accepted only if the handoff CDP of the system is below the target dropping rate at all times. Thus, this strategy ensures that the handoff dropping probability averaged over the contention area is lower than a target handoff dropping probability PQoS (QoS of the contention area). The system always traces the location of all the MHs in each spotbeam and updates the MH's handoff dropping parameters. The algorithm involves high processing overhead to be handled by the satellite, and seems therefore to be unsuitable for high-capacity systems where a satellite footprint consists of many small-sized spotbeams, each having many active MHs. Cho [1] employs MH location information as the basis for adaptive bandwidth allocation for handoff resource reservation. In a spotbeam, bandwidth reservation for handoff is allocated adaptively by calculating the possible handoffs from neighboring spotbeams. A new call request is accepted if the spotbeam where it originated has enough available bandwidth for new calls. This reservation mechanism provides a low handoff dropping probability compared to the fixed reservation strategy. However, the use of location information in handoff management suffers from the disadvantage of updating locations, which then results in a high processing load for the onboard handoff controller, thereby increasing the complexity of terminals. The method seems suitable for only fixed users. El-Kadi et al. [2] proposed a probabilistic resource reservation strategy for real-time services. They introduced a call admission algorithm where real-time and non-real-time service classes are treated differently. The novel concept of a sliding window is proposed in order to predict the necessary amount of reserved bandwidth for a new call in its future handoff spotbeams. For real-time services, a new call request is accepted if the spotbeam where it originated has available bandwidth, and resource reservation is successful in future handoff spotbeams. For non-real-time services, a new call request is accepted if the spotbeam where it originated satisfies its maximum required bandwidth. Handoff requests for real-time traffic are accepted if the minimum bandwidth requirement is satisfied. Non-real-time traffic handoff requests are honored if there is some residual bandwidth available in the cell.

This chapter describes four recent QoS provisioning strategies for multimedia LEO satellite networks that perform admission control by using the concept of a sliding window, which was first proposed by El-Kadi et al. [2].

16.4 MOBILITY MODEL AND TRAFFIC PARAMETERS

Although several mobility models exist for LEO satellites [7,8], it is customary to assume a one-dimensional mobility model where the MHs move in straight lines and at a constant speed, essentially the same as the orbital speed of the satellite [7]. Since the speed of users (even in fast moving vehicles) is negligible compared to the satellite's speed and the Earth's rotation, MH speed can be ignored. For example, users in fast vehicles move with a maximum speed of 80 m/s, while a LEO satellite's ground track speed is more than 5700 m/s and the speed of the rotation of the Earth at the equatorial level is nearly 460 m/s. For simplicity, all the spotbeams (also referred to as cells) are identical in shape and size. Although each spotbeam is, in reality, circular, the use of squares to approximate spotbeams is justifiable. Some authors use regular hexagons instead of squares. We assume an orbital speed of 26,000 km/h. The width of a cell is taken to be 425 km. Thus, the time t_S it takes an end-user to cross a cell is, roughly, 65 s. Referring to Figure 16.2, the MH remains in the cell where the

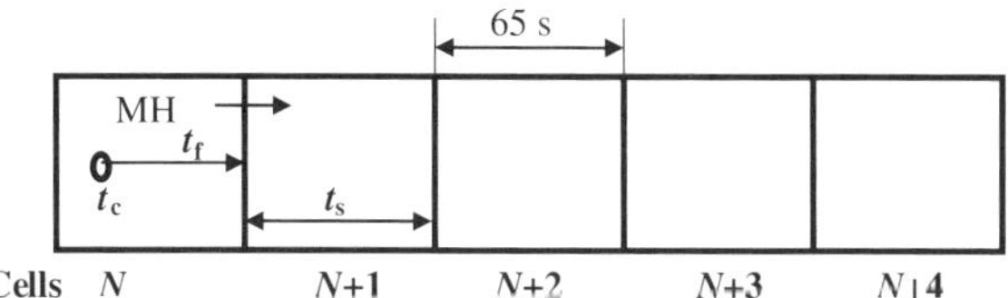

FIGURE 16.2 Illustration of some of the mobility and cell parameters.

connection was initiated for t_f time, where t_f is uniformly distributed between 0 and t_s. Thus, t_f is the time until the first handoff request, assuming that the call does not end in the original cell. After the first handoff, a constant time t_s is assumed between subsequent handoff requests until call termination.

As illustrated in Figure 16.2, when a new connection C is requested in cell N, it is associated with a trajectory, consisting of a list $N, N+1, N+2, \ldots, N+k, \ldots$ of cells that the connection may visit during its lifetime.

The traffic offered to the satellite may be real-time multimedia traffic, such as interactive voice and video applications, and non-real-time data traffic, such as email or ftp. Thus, traffic offered to the satellite system is classified as

- Class I traffic—real-time multimedia traffic, such as interactive voice and video applications.
- Class II traffic—non-real-time data traffic, such as email or ftp.

When a mobile user requests a new connection C in a given cell, it provides the following parameters:

- The desired class of traffic for C (either I or II).
- M_C the desired amount of bandwidth for the connection.

If the request is for a Class I connection, the following parameters are also specified:

1. m_C, the minimum acceptable amount of bandwidth, that is, the smallest amount of bandwidth that the source requires in order to maintain acceptable quality, for example, the smallest encoding rate of its codec.
2. θ_C, the largest acceptable CDP that the connection can tolerate.
3. $1/\mu_C$, the mean holding time of C.

16.5 A NOVEL CALL ADMISSION CONTROL USING THE SLIDING WINDOW CONCEPT

Connection admission control is one of the fundamental tasks performed by the satellite network at call setup time in order to determine if the connection request can be accepted into the system without violating prior QoS commitments. The task is nontrivial because the traffic offered to the system is heterogeneous due to new call attempts and handoff requests. El-Kadi et al. [2] proposed the following two novel call admission criteria.

- The first call admission criterion, which is local in scope, applies to both Class I and Class II connections, and attempts to ensure that the originating cell has sufficient resources to provide the connection with its desired amount of bandwidth.
- The second admission control criterion, which is global in scope, applies to Class I connections only, and attempts to minimize the chances that, once accepted, the connection will be dropped later due to a lack of bandwidth in some cell into which it may handoff. The second criterion is inspired by the sliding window criterion first proposed by El-Kadi et al. [2].

Consider a request for a new Class I connection C in cell N at time t_C and let t_f be the estimated residence time of C in N. Referring to Figure 16.3, the key observation that inspired the second criterion is that when C is about to handoff into cell $N + 1$, the connections resident in $N + 1$ are likely to be those in region A of call N and those in region B of cell $N + 1$. More precisely, these regions are defined as follows:

- A connection is in region A if at time t_C its residual residence time in cell N is less than or equal to t_f.
- A connection is in region B if at time t_C its residual residence time in cell $N + 1$ is larger than or equal to t_f.

In general, the satellite does not know the exact position of a new call request in generic cell N. This makes the computation of the bandwidth committed to connections in areas A and B difficult to assess. Some schemes rely on a MH location database by utilizing global positioning system (GPS). While GPS-enabled devices will become ubiquitous in the future, at present the use of GPS in call admission and handoff management schemes for LEO satellite networks has many disadvantages. For one thing, in order for GPS localization to be effective, three or more satellites

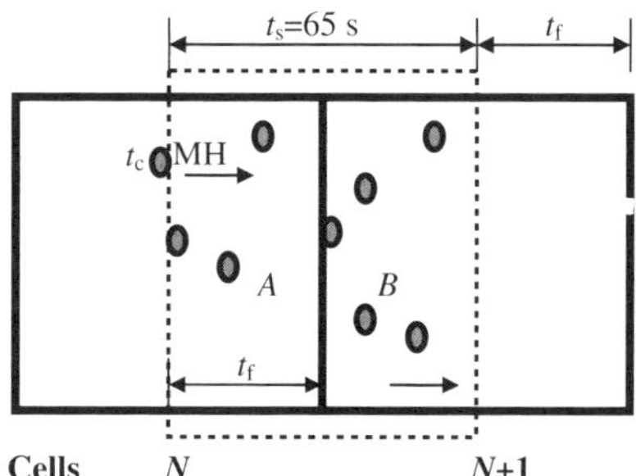

FIGURE 16.3 Illustration of the sliding window concept for call admission.

must be visible to the GPS receiver. This limits the applicability of GPS in urban areas where the buildings may obstruct line of sight to the satellites. By the same token, GPS localization does not work indoors. Likewise, GPS does not work well in poor atmospheric conditions. The protocols discussed here do not use GPS to determine the MH location.

16.5.1 SILK—A Selective Look-Ahead Allocation Scheme

The main goal of this section is to spell out the details of SILK, a selective look-ahead bandwidth admission control and handoff management scheme.

16.5.1.1 SILK—The Basic Idea SILK [13] admission policies distinguish between real-time (known as Class I) and non-real-time (known as Class II) connections. As in [11], Class I handoffs are admitted only if their minimum bandwidth requirements can be met. However, Class II handoff requests will be accepted as long as there is some residual bandwidth left in the cell. Thus, bandwidth reservation pertains only to Class I handoffs. The key idea of SILK is to allocate bandwidth to each accepted Class I connection in a look-ahead horizon of k cells along its trajectory. Here, k is referred to as the depth of the look-ahead horizon. The intuition for this concept is provided by the fact that the deeper the horizon, the smaller the likelihood of a handoff failure, and the smaller the CDP. Because at setup time the connection C specifies the CDP it can tolerate, it implicitly specifies the depth k of the corresponding look-ahead horizon. Thus, for each connection C, SILK looks ahead just enough to ensure that the CDP of θ_C can be enforced. Thus, in SILK, the look-ahead allocation is determined by the negotiated QoS.

Let p_{h} denote the handoff failure probability of a Class I connection, that is, the probability that a handoff request is denied for lack of resources. Let S_k denote the event that a Class I connection C admitted in cell N goes successfully through k handoffs and will, therefore, show up in cell $N+k$. It is easy to confirm that the probability of S_k is

$$\Pr[S_k] = p_{\mathrm{f}}(1 - p_{\mathrm{h}})[p_{\mathrm{s}}(1 - p_{\mathrm{h}})]^{k-1}$$

where $p_{\mathrm{f}}(1 - p_{\mathrm{h}})$ is the probability that the first handoff request is successful and $[p_{\mathrm{s}}(1 - p_{\mathrm{h}})]^{k-1}$ is the probability that all subsequent $k-1$ handoff requests are also successful.

Likewise, let D_{k+1} be the event that C will be dropped at the next handoff attempt. Thus, we have

$$\begin{aligned}\Pr[D_{k+1}] &= \Pr[S_k]p_{\mathrm{s}}p_{\mathrm{h}} = p_{\mathrm{f}}p_{\mathrm{s}}p_{\mathrm{h}}(1 - p_{\mathrm{h}})[p_{\mathrm{s}}(1 - p_{\mathrm{h}})]^{k-1} \\ &= p_{\mathrm{f}}p_{\mathrm{h}}.[p_{\mathrm{s}}(1 - p_{\mathrm{h}})]^{k-1}\end{aligned}$$

as $p_{\mathrm{s}}p_{\mathrm{h}}$ is the probability that the connection will attempt but fail to secure the $(k+1)$th handoff.

Now, assuming that the connection C has negotiated a CDP of θ_C, it follows that

$$\Pr[D_{k+1}] = p_\mathrm{f}\, p_\mathrm{h}[p_\mathrm{s}(1 - p_\mathrm{h})]^{k-1} = \theta_C,$$

which implies:

$$k = \frac{\log \frac{\theta_C}{p_\mathrm{f}\, p_h}}{\log[p_\mathrm{s}(1 - p_h)]}. \tag{16.1}$$

There are a number of interesting features of Equation (16.1), which computes the value of k. First, the only variable parameter in the equation is p_h. All the others are known beforehand. Todorova et al. [13] argued that the satellite maintains p_h as the ratio between the number of unsuccessful handoff attempts and the total number of handoff attempts. Second, since p_h may change with the network conditions, the depth k of the look-ahead horizon will also change accordingly. This interesting feature shows that SILK is indeed adaptive to traffic conditions. Finally, k is dynamically maintained by the satellite either on a per-connection or, better yet, on a per-service class basis, depending on the amount of onboard resources and network traffic.

As it turns out, the above computed value of k is at the heart of SILK. The details are spelled out as follows:

- In anticipation of its future handoff needs, bandwidth is allocated for connection C in a number k of cells corresponding to the depth of its look-ahead horizon; no allocation is made outside this group of cells.
- For $1 \leq i \leq k$, allocate in cell $N + i$ an amount of bandwidth equal to $\mathrm{B}_{N+i} = m_C \Pr[S_i]$.
- This amount of bandwidth will be allocated for connection C during the time interval

$$I_{N+i} = [t_C + t_f + (i - 1)t_s, t_C + t_f + it_s]$$

where t_C is the time connection C was admitted into the system.

As pointed out by Todorova et al. [13], SILK is lightweight. Indeed, the mobility parameters t_f and t_s are readily available and the look-ahead horizon k is maintained by the satellite for each service class. Similarly, since the trajectory of connection C is a straight line, the task of computing for every $1 \leq i \leq k$ the amount of bandwidth B_{N+i} to allocate, as well as the time interval I_{N+i} during which B_{N+i} must be available is straightforward and can be easily computed by the satellite using its onboard capabilities.

16.5.1.2 SILK—The Call Admission Strategy

SILK's call admission strategy involves two criteria mentioned earlier. However, unlike [2], SILK only looks at the first k cells on C's trajectory. The connection satisfies the second criterion if all these k cells have sufficient bandwidth to accommodate C, that is, for every i, $(1 \leq i \leq k)$, the amount of residual bandwidth in the cell during the time interval I_{N+i} must not be less than B_{N+i}. The motivation for this second criterion is very simple: if the residual bandwidth available in cell $N + i$ is less than the projected bandwidth needs of connection C, it is very likely that C will he dropped. To avoid such a situation, connection C is not admitted into the system. Thus, the second admission criterion acts as an additional safeguard against a Class I connection to be accepted, only to be dropped at some later point.

16.5.2 Q-WIN—A Predictive Allocation and Management Scheme

The main goal of this section is to discuss in full detail the Q-WIN protocol proposed in the study of Olariu et al. [9]. A key ingredient of Q-WIN is a novel predictive resource allocation protocol. Q-WIN involves some processing overhead. However, as it turns out, this overhead is transparent to the MHs, being absorbed by the onboard processing capabilities of the satellite. Consequently, Q-WIN is expected to scale and to accommodate a large population of MHs.

16.5.2.1 Q-WIN—The Data Structures A Class I connection C in a generic cell N is said to be

- *Regular* if C has confirmed bandwidth reservations in cells $N + 1$ and $N + 2$. The regular connections in cell N are maintained in the queue $R(N)$.
- *One-short* if C has confirmed bandwidth reservation in cell $N + 1$ but not in cell $N + 2$. The one-short connections in cell N are maintained in the queue $S1(N)$.
- *Two-short* if C has no confirmed reservation in cells $N + 1$ and $N + 2$. The 2-short connections in cell N are maintained in the queue $S2(N)$.
- Finally, we note that Class II connections in cell N are maintained in a separate queue $Q(N)$.

From the above classification, observe that two-short connections are liable to be dropped at the next handoff attempt, while one-short connections are in no imminent danger of being dropped. The stated goal of our bandwidth allocation scheme is to minimize the likelihood of a connection being dropped. It is widely acknowledged that priority should be given to calls-in-progress versus primary call requests. The intuition in prioritizing handoff calls are that voice users are bothered more by a dropped call than had the call never been accepted. (Note, this is not necessarily true for data traffic where users may be satisfied to transfer some of their files during a short connection time.) The principle vehicle for achieving this goal is a judicious priority-based bandwidth allocation strategy.

16.5.2.2 Q-WIN—The Call Admission Strategy Consider a request for a new connection C in cell N. Very much like SILK [13], Q-WIN [9] bases its connection admission control on a novel scheme that combines the following two criteria:

- *Local availability*: The first call admission criterion, which is *local* in scope, ensures that the originating cell N has sufficient resources to provide the connection with its desired amount of bandwidth M_C. Both Class I and Class II connections are subject to this first admission criterion. A Class II connection request that satisfies the first admission criterion is accepted into the system and placed into the queue $Q(N)$ of Class II connections currently in cell N. On the contrary, if the first admission criterion is not satisfied, the connection request is immediately rejected.
- *Short-term guarantees*: The second admission control criterion, which is *non-local* in scope, applies to Class I connections only, attempting to minimize the chances that, once accepted, the connection will be dropped later due to a lack of bandwidth in some cell into which it may handoff.

In general, the satellite does not know the exact position of a new call request in generic cell N. This makes the computation of the bandwidth committed to connections in areas A and B difficult to assess (see Fig. 16.3). In what follows, we describe a heuristic that attempts to approximate the bandwidth held by the connections in A and B. For this purpose, we partition the union of cells N and $N+1$ into $m+1$ virtual windows $W_0, W_1, \ldots, W_m$ each of width t_s. In this sequence, W_0 is the *base* window, and its left boundary is normalized to 0. For every i, $0 \leq i \leq m$, window W_i stretches from

$$\frac{i \times t_s}{m} \text{ to } t_s + \frac{i \times t_s}{m} \tag{16.2}$$

In particular, by Equation (16.2), window W_0 coincides with cell N, and window W_m with cell $N+1$. We refer the reader to Figure 16.4 for an illustration, with $m = 5$. All the virtual windows have the exact shape and size of a cell (shown with different sizes in Fig. 16.4).

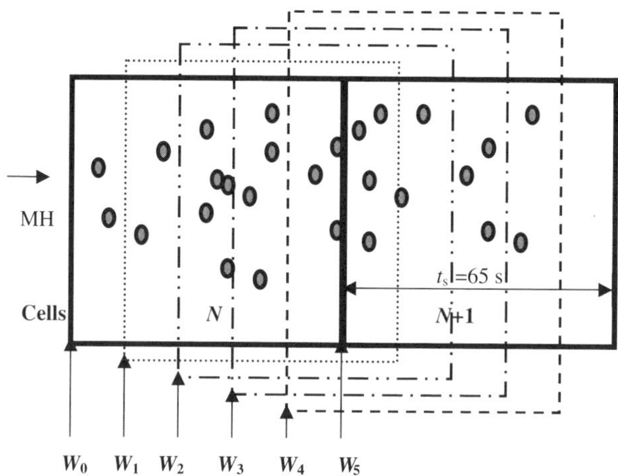

FIGURE 16.4 Illustration of the virtual windows.

For later reference, we partition a generic window W_i into a left subwindow W_i^N and a right subwindow W_i^{N+1} denoting, respectively, the intersection of W_i with cells N and $N+1$.

We distinguish between mobile hosts that have experienced a handoff (referred to as *old*) from those that have not (referred to as *new*). As we are about to describe, mobile hosts may or may not be assigned *timers*. Specifically, each old mobile host is assigned a timer θ; no timer is assigned to new mobile hosts. Upon entering a new cell, θ is set to t_s (the time it takes to traverse a cell). Every time unit, θ is decremented by 1, making it close to zero by the time the MH is about to reach the next handoff. For illustration purposes, we note that in Figure 16.4, since $m = 5$, W_1^N contains the old users in cell N with $\theta \leq 65 - 65/5 = 52$; likewise, W_1^{N+1} contains the old users in cell $N+1$ with $\theta > 52. W_2^N$ contains the old users in cell N with $\theta \leq 65 - 2 \times 65/5 = 39$, and so on.

Let B_i and D_i denote, respectively, the total amount of bandwidth in use by the old and new mobile hosts in window W_i. Notice that the amount of bandwidth B_i is easy to compute by the satellite since, by virtue of timers, the position of old mobile hosts, up to the granularity of a virtual window, is known.

The location of new mobile hosts defined earlier is unknown. It is, therefore, difficult to determine D_i exactly. However, it is reasonable to assume that, within each of the cells N and $N+1$, these mobile hosts are uniformly distributed. Notice that this does not imply a uniform distribution of new mobile hosts across the union of cells N and $N+1$. Let n_N and n_{N+1} stand, respectively, for the number of new mobile hosts in cells N and $N+1$. As illustrated in Figure 16.5, the assumption of uniform distribution of new mobile hosts in cell N implies that the expected number of mobile hosts W_i^N is $n_N(1 - i/m)$. Likewise, since the new mobile hosts are uniformly distributed in cell $N+1$, the expected number of new mobile hosts W_i^{N+1} is $n_{N+1} \times i/m$. Thus, by a simple computation we obtain the following approximation for D_i:

$$D_i = n_N + \frac{i}{m}[n_{N+1} - n_N]. \tag{16.3}$$

Let M stand for the total bandwidth capacity of a cell. Using B_i and the value of D_i from Equation (16.3), the virtual window W_i determines the residual bandwidth $R_i = M - B_i - D_i$. If $R_i \geq M_C$, W_i *votes* in favor of accepting the new request C with desired bandwidth M_c; otherwise it *votes* against its admittance. After counting the votes, if the majority of the virtual windows had *voted* in favor of admittance, the new connection request is admitted into the system. Otherwise, it is rejected. Once admitted, the desired bandwidth of connection C is reserved in the current cell, and the connection is placed in queue $S2(N)$.

16.5.3 OSCAR: An Opportunistic Resource Management Scheme

The main idea behind OSCAR [10] is a multiple virtual window call admission protocol and average line mechanism based on dynamic channel reservation for handoff calls for multimedia LEO satellite networks. The essence of this predictive resource

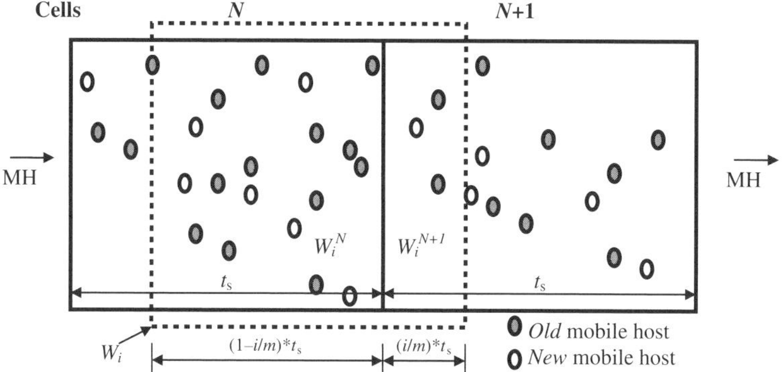

FIGURE 16.5 Illustration of the computation of D_i.

allocation protocol is that it achieves results comparable to those of Q-WIN but eliminates the queues. Even though it uses up more processing time, the overhead of maintaining queues during heavy traffic is avoided and hence makes this algorithm simpler and less dependent on buffers. Moreover, the processing time is transparent to the MH, being absorbed by the onboard processing capabilities of the satellite. Consequently, OSCAR scales to a large number of users.

Consider a request for a new connection C in cell N. Very much like SILK and Q-WIN, OSCAR bases its connection admission control on a novel scheme that combines the two call admission criteria. However, unlike both SILK and Q-WIN that either look at a distant horizon or maintain rather complicated data structures, OSCAR looks ahead only one cell. Surprisingly, simulation results indicate that this short horizon works well when supplemented by an opportunistic bandwidth allocation scheme. OSCAR's second admission criterion relies on a novel idea that is discussed in full detail below.

16.5.3.1 OSCAR—The Average Load Line Concept OSCAR implements the predictive strategy combined with an opportunistic handoff management scheme. In OSCAR, handoff calls fall into one of the two types discussed below:

- *Type 1*: those that are still not assigned a timer, that is, newly admitted calls that are about to make their first handoff.
- *Type 2*: those that are assigned a timer, that is, the calls that have already made one or more handoffs.

It is important to observe that by virtue of OSCAR's call admission scheme that is looking at both the originating cell and the next one along the MH's path, handoffs of Type 1 succeed with high probability. We will, therefore, show only how to manage Type 2 handoffs. The details of this scheme are discussed below.

Each cell in the network dynamically reserves a small amount of bandwidth specifically for handoffs of Type 2. When a Type 2 handoff request is made, the algorithm

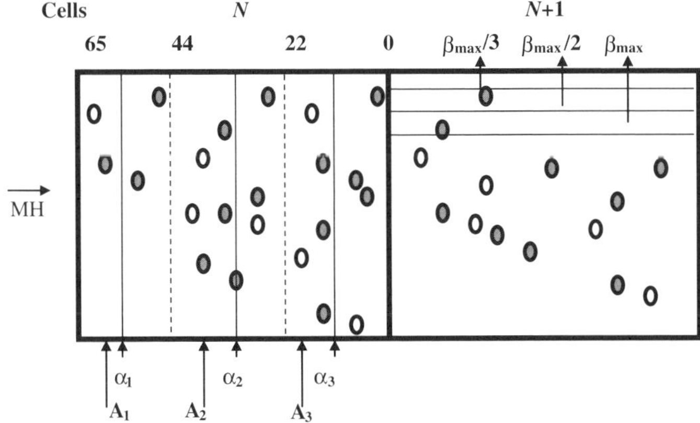

FIGURE 16.6 Illustration of the handoff scheme in OSCAR.

will first try to satisfy the request by allotting the bandwidth from the reserved amount. If the reserved bandwidth has already been used, the request will be allotted the bandwidth from the remaining available bandwidth of the cell. Otherwise, the handoff request is dropped.

Let the maximum amount of bandwidth that could be reserved be $\beta_{\max}$ (a small percentage of total available bandwidth). The amount of bandwidth reserved for Type 2 handoffs dynamically varies between 0 and $\beta_{\max}$ depending on the relative position of the *average load line* in the previous neighboring cell.

To explain the concept of average load line, consider a cell N, and refer to Figure 16.6. Assume that cell $N-1$ contains k Type 2 handoff calls with residual residence times in cell $N-1$ denoted by $t_1, t_2, \cdots, t_k$ such that $t_1 \leq t_2 \leq \cdots \leq t_k$ and let the corresponding amounts of bandwidth allocated to the calls be by $b_1, b_2, \cdots, b_k$. Let B be the sum total of b_j where j ranges from 1 to k. The average load line L is defined as the average of t_i and t_{i+1} where i is the smallest subscript for which the inequality below holds.

$$\sum_{j=1}^{i} b_j \left\lceil \frac{B}{2} \right\rceil .$$

We note that, from a computational standpoint, determining the average load line L is a simple instance of the prefix sums problem and can be handled easily by the satellite.

16.5.3.2 OSCAR—The Dynamic Reservation Scheme The dynamic bandwidth reservation scheme in cell N can be explained as follows. Since cell N knows about its neighbors, it can track all the Type 2 handoff calls in cell $N-1$ as shown in Figure 16.6. $A1$, $A2$, and $A3$ represent equal-sized areas of a cell $N-1$. The average load line L will always fall into one of these three areas depending upon the distribution of Type 2 handoff calls. The bandwidth for Type 2 calls in cell N is reserved depending upon the position of the average load line as detailed below:

- If the position of average load line L is at α_1 in area $A1$, then it can be inferred that roughly half of the bandwidth required by Type 2 handoff calls is concentrated in area $A1$. Since L is relatively far from cell N, an amount $\beta_{\max}/3$ of bandwidth is reserved for Type 2 handoff calls in cell N as shown, in such a way that more bandwidth is available for other call requests.
- If the average load line L is at α_2 in $A2$, then an amount $\beta_{\max}/2$ of bandwidth is reserved in cell N.
- If the average load line L is at α_3 in $A3$, then an amount $\beta_{\max}$ of bandwidth is reserved in cell N.

16.5.4 RADAR: A Refined Call Admission Control Strategy

A key ingredient in RADAR [12] is a novel predictive resource allocation protocol. This scheme was named RADAR because the absence region detection technique refines the similar call admission control scheme used in Q-WIN [9], and consequently the bandwidth utilization was increased. RADAR overcomes the problem faced by Q-WIN and OSCAR in their call admission scheme where they have assumed a uniform distribution of the MHs that are newly accepted and have not experienced any handoff. As in Q-WIN, all the mobile hosts can be divided into two types:

- Those that have experienced a handoff (referred to as *old*).
- Those that have not experienced a handoff (referred to as *new*).

Unlike in Q-WIN and OSCAR, in the RADAR protocol, all the mobile hosts are assigned *timers*. Each old mobile host is assigned a timer θ; whereas a timer α is assigned to the new mobile hosts. This timer α, assigned to the new mobile hosts, is an essential element of the RADAR scheme because this timer α helps to detect the *absence region* for thenew mobile host. This unique characteristic, the absence region detection, of the RADAR scheme is explained in the following paragraphs. Similar to Q-WIN, when a MH enters a new cell, θ is set to t_s (the time it takes to traverse a cell). Every time unit, θ is decremented by 1, making it close to zero by the time the MH is about to reach the next handoff. Similarly, as soon as a MH is accepted into the system and is in its new MH state, that is, the MH has not yet experienced a handoff, α is set to 0. Every time unit, α is incremented by 1.

When a new MH is accepted into the system, that is, bandwidth is allocated to it, the major problem encountered is to determine its relative location with respect to the current cell and neighboring cells. This is not in case for old MHs because the timer θ helps determine its relative position with respect to the current cell and neighboring cells. Once accepted into the system, a new MH could be located anywhere in its cell of origin. As shown in Figure 16.7, let a new MH z originate in cell $N + 1$. For z, with timer α_z, the rectangle *cdmn* such $cn = dm = \alpha_z$ forms the *absence region* where it is impossible for z to be present. This is concluded by taking into consideration the

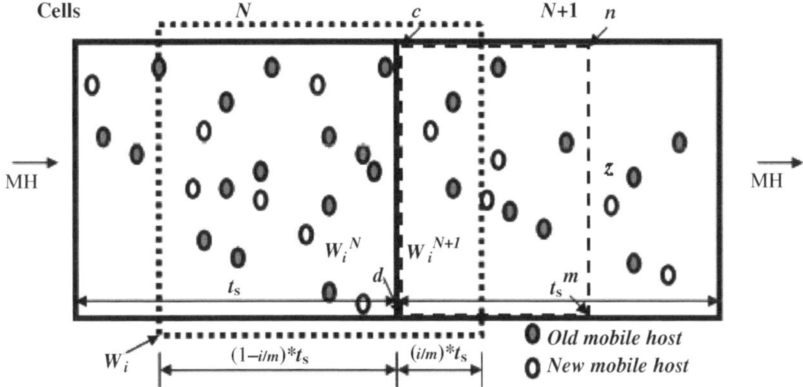

FIGURE 16.7 Illustration of the computation of D_i.

worst case scenario for z that it originated from the extreme left side cd of the cell $N + 1$. In other words, in time α_z, even if z originated from the extreme left side cd, it would have traveled the distance $cn = dm = \alpha_z$, hence be absent from the area $cdmn$. Let B_i and D_i denote, respectively, the total amount of bandwidth in use by the old and new mobile hosts in window W_i. Notice that the amount of bandwidth B_i is easy to compute by the satellite since, by virtue of timers, the position of old mobile hosts, up to the granularity of a virtual window, is known.

The location of new mobile hosts (i.e., newly accepted ones that have not yet experienced their first handoff) is unknown. It is, therefore, difficult to determine D_i exactly. Q-WIN and OSCAR assumed that, within each of the cells N and $N + 1$, these mobile hosts are uniformly distributed. Notice that this does not imply a uniform distribution of new mobile hosts across the union of cells N and $N + 1$. No doubt this assumption makes the computation of D_i simple, but it might calculate an estimated D_i different from the actual D_i, in the particular case of highly variable and heavy loads. As we are about to see, the RADAR scheme uses the absence region detection method to find the MH absence region for one part of the virtual window along with the uniform distribution assumption for the rest of the virtual window, thus making the MH's distribution nonuniform across the virtual window. Let d n_N and dn_{N+1} stand, respectively, for the sum total of bandwidth consumed by the new mobile hosts in cells N and $N + 1$. As illustrated in Figure 16.7, the assumption of uniform distribution of new mobile hosts in cell N implies that the consumed bandwidth of new mobile hosts W_i^N is d$n_N(1 - i/m)$. Likewise, since the new mobile hosts are uniformly distributed in cell $N + 1$ except for the new MHs that are absent in the right side of the virtual window; the expected bandwidth of new mobile hosts W_i^{N+1} is d$n_{N+1} \times i/m$ − (dn_{N+1} with $\alpha > i/m$). Thus, by a simple computation, we obtain the following approximation for D_i:

$$D_i = n_N + \frac{i}{m}[n_{N+1} - n_N] - (\mathrm{d}n_{N+1}\ \text{with}\ \alpha > i/m) \qquad (16.4)$$

Notice that the RADAR scheme uses the absence region detection method to determine the region where the MH will be absent, thus the computation for the consumed bandwidth is done with partial nonuniform distribution across the virtual window for the new users.

Let M stand for the total bandwidth capacity of a cell. Using B_i and the value of D_i from Equation (16.4), the virtual window W_i determines the residual bandwidth $R_i = M - B_i - D_i$. If $R_i \geq M_C$, W_i votes in favor of accepting the new request C with desired bandwidth M_c; otherwise it votes against its admittance. After counting the votes, if the majority of the virtual windows have voted in favor of admittance, the new connection request is admitted into the system. Otherwise, it is rejected. Once admitted, the desired bandwidth of connection C is reserved in the current cell, and the connection is placed in queue $S2(N)$.

16.6 PERFORMANCE EVALUATION

16.6.1 Simulation Model

Based on the description of the queue management, virtual windows, and the call admission control algorithm in Sections 16.4 and 16.5, we have developed a simulation program based on the one used in the work by El-Kadi et al. [2] to evaluate the performances of the proposed scheme.

16.6.1.1 Server Functions The server functions implemented are

- To monitor the amount of available bandwidth in the spotbeam.
- To reserve bandwidth required by future connections.
- To accept or reject new call requests.
- To accept or reject handoff requests.

The system parameters used in our simulation experiments are described in Table 16.1 and are based on the well-known Iridium satellite system [3]. New call arrival rate follows a Poisson distribution and connection duration is exponentially distributed. We define six types of services with different QoS requirements and assume equal mean arrival rate for each service type and a fixed bandwidth in each spotbeam [11].

16.6.2 Simulation Results

The simulation results are shown in Figures 16.8–16.12. We compare the CDP, CBP, and BU performances of the following schemes:

- Q-WIN [9] with 13 virtual windows.
- SILK [13].

TABLE 16.1 Simulation Parameters

Spotbeam parameters						
1.0	Radius	212.5				
1.0	Capacity	30,000 kbit				
1.0	Speed	26,000 km/h				
Service parameters	Class 1	Class2				
	Type1	Type2	Type3	Type1	Type2	Type3
Mean duration (s)	180	300	600	30	180	120
Maximum bandwidth (kbps)	30	256	6000	20	512	10,000
Minimum bandwidth (kbps)	30	256	1000	5	64	1000

- OSCAR [10] with three zones within each cell to compute the average load line.
- RADAR [12] with 13 virtual windows.

The results of our simulation, summarized in Figures 16.8 and 16.9 show that the CDP for Class I connections in RADAR is better than in SILK and OSCAR, since RADAR takes into account the well-determined horizon limited to two cells. Also the CDP for Class II gains better performance in RADAR.

Figure 16.10 shows that the CBP for all traffic in RADAR is better than in SILK, Q-WIN, and OSCAR, since RADAR involves the refined admission technique with multiple virtual window approach for new call admissions.

Figure 16.12 shows that the bandwidth utilization with RADAR is the best out of all the competing schemes. It is well known that the goals of keeping the CDP low and that of keeping the bandwidth utilization high are conflicting. It is easy to ensure a low CDP at the expense of bandwidth utilization and similarly, it is easy to ensure high bandwidth utilization at the expense of CDP. The challenge, of course, is to come up with a handoff management protocol that strikes a sensible balance between the two.

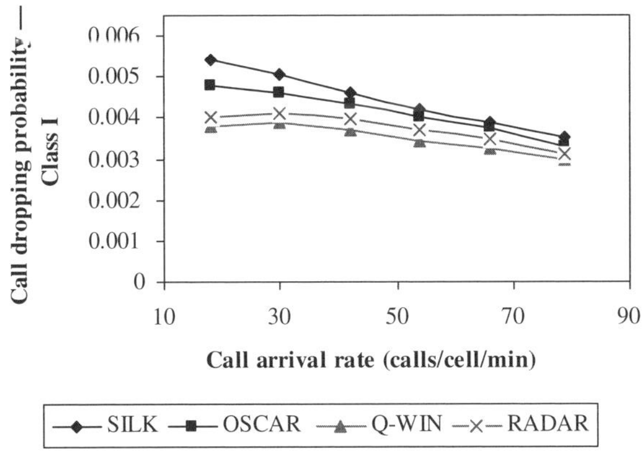

FIGURE 16.8 CDP of Class I traffic.

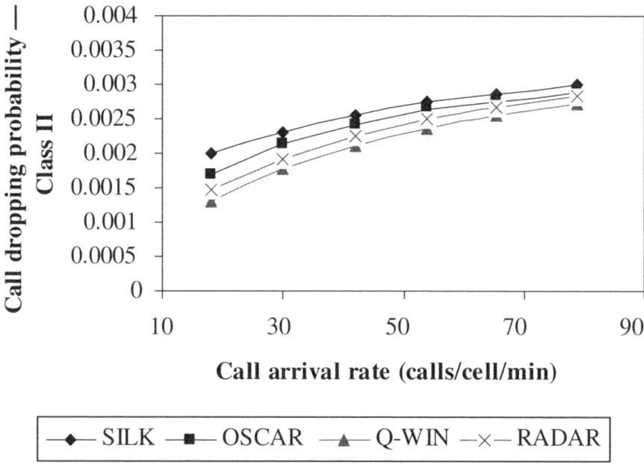

FIGURE 16.9 CDP of Class II traffic.

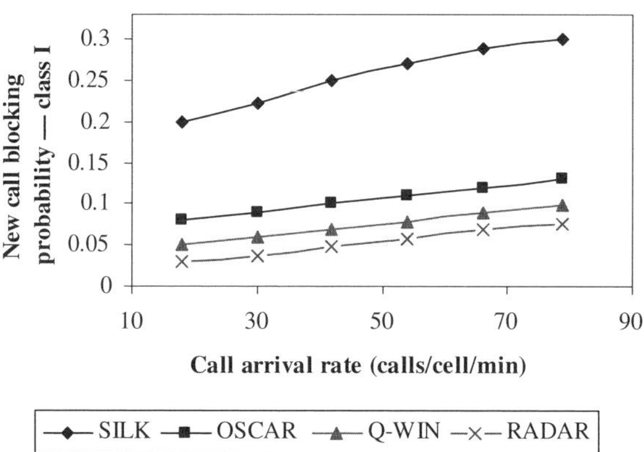

FIGURE 16.10 New CBR of Class I traffic.

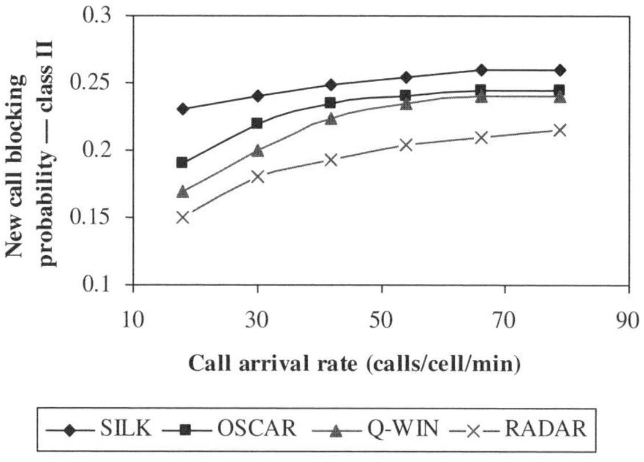

FIGURE 16.11 New CBR of Class II traffic.

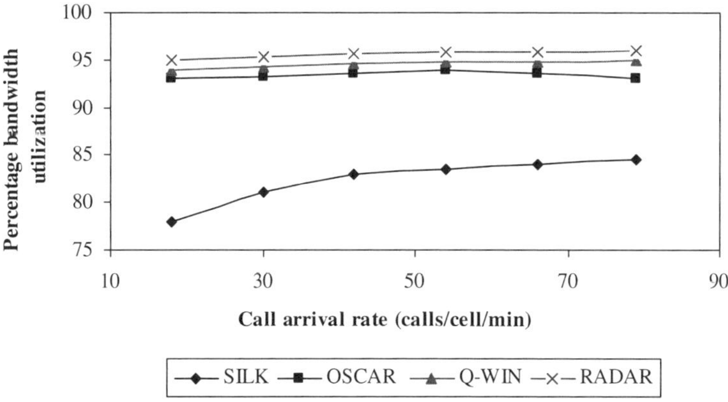

FIGURE 16.12 Bandwidth utilization.

16.7 CONCLUDING REMARKS

LEO satellites are expected to support multimedia traffic and to provide their users with the appropriate QoS. However, the limited bandwidth of the satellite channel, satellite rotation around the Earth, and mobility of MHs makes QoS provisioning and mobility management a challenging task. In this chapter we have surveyed four recent resource management protocols for LEO satellite networks that use a novel call admission control based on a sliding widow concept. These protocols are specifically tailored to meet the QoS needs of multimedia connections, as real-time and non-real-time service classes are differently treated. Also, they do not use GPS for MH locations. Each of these protocols features a different philosophy of bandwidth management. But, in a sense, they complement each other since the solutions they offer may each appeal to a different set of applications, or to different specific network configurations or network performance goals.

We have implemented these protocols and have evaluated their performance by simulation. Our simulation results expose the differences in performance due to design decisions. In summary, these protocols are well suited for QoS provisioning in multimedia LEO satellite networks.

ACKNOWLEDGMENTS

The authors are grateful to Nam Nguyen, Rajendra Shirhatti, and Petia Todorova for many insightful discussions on QoS provisioning in LEO satellite networks.

REFERENCES

1. Cho I. Adaptive dynamic channel allocation scheme for spotbeam handover in LEO satellite networks. Proc IEEE VTC 2000;1925–1929.

2. El-Kadi M, Olariu S, Todorova P. Predictive resource allocation in multimedia satellite networks. Proceedings of the IEEE GLOBECOM; November 2001; San Antonio.
3. Fantacci R, Del Re E, Giambene C. Efficient dynamic channel allocation techniques with handover queuing for mobile satellite networks, IEEE J Sel Area Commun 1995;13(2):397–405.
4. Jamalipour A, Tung T. The role of satellites in global IT: trends and implications. IEEE Per Commun 2001;8(3):5–11.
5. Luglio M. Mobile multimedia satellite communications. IEEE Multimedia 1999;6:10–14.
6. Markhasin A, Olariu S, Todorova P. An overview of QoS oriented MAC protocols for future mobile applications. In: KosrowPour M, editor. Encyclopedia of Information Science and Technology. Hershey, PA: Idea Group; 2005.
7. Nguyen HN, Olariu S, Todorova P. A novel mobility model and resource allocation strategy for multimedia LEO satellite networks. Proceedings of the IEEE WCNC; 2002; Orlando, FL.
8. Nguyen HN. Routing and Quality-of-Service in Broadband LEO Satellite Networks. Boston: Kluwer Academic; 2002.
9. Olariu S, Rizvi SR, Shirhatti R, Todorova P. QWIN—A new admission and handoff management scheme for multimedia LEO satellite networks. Telecommun Sys 2003;22(1–4):151–168.
10. Olariu S, Shirhatti R, Zomaya AY. OSCAR: An opportunistic call admission and handoff management scheme for multimedia LEO satellite networks. Proceedings of the International Conference on Parallel Processing, ICPP'2004; Montreal, Canada.
11. Oliviera C, Kim JB, Suda T. An adaptive bandwidth reservation scheme for high-speed multimedia wireless networks. IEEE J Sel Area Commun 1998;16:858–874.
12. Rizvi SR, Olariu S, Rizvi ME. RADAR—A novel call admission and handoff management scheme for multimedia LEO satellite networks. Proceedings of the IEEE MILCOM 2006; October 2006; Washington DC.
13. Todorova P, Olariu S, Nguyen HN. SILK—A selective look-ahead bandwidth allocation scheme for reliable handoff in multimedia LEO satellite networks. Proceedings of the ECUMN2002; April 2002; Colmar, France.
14. Uzunalioglu H. A connection admission control algorithm for LEO satellite networks. Proceedings of the IEEE ICC; 1999; p 1074–1078.

CHAPTER 17

Resilient Recursive Routing in Communication Networks

COSTAS C. CONSTANTINOU, ALEXANDER S. STEPANENKO, THEODOROS N. ARVANITIS, KEVIN J. BAUGHAN, and BIN LIU

17.1 INTRODUCTION

The function of routing in communication networks is to determine a consistent set of local switching decisions at all the nodes such that data can be transported from any source to any destination. In general, routing algorithms can be loosely classified in many ways, for example, unicast versus multicast, centralized versus distributed, proactive versus reactive, single-path versus multipath, and so on, but in practice, routing algorithms can fall in between such simplistic classifications whose discussion is beyond the scope of this chapter. Furthermore, routing is frequently cast as an optimization problem, which can be either static or dynamic in nature (although in some instances routing can and is formulated as a constraint satisfaction problem).

This chapter will concentrate on a dynamic, unicast, proactive, link-state routing algorithm only. The aim of the algorithm is to achieve a scalable approach to the representation and exploitation of path diversity in communication networks. By "scalable" we here mean that the number of message updates needed to support adaptation to changes in the state of the network scales well (i.e., as a polynomial) with respect to the number of nodes and links in the network. After a brief critique of well-established routing algorithms and their application to communication networks, we discuss the desirable properties of adaptive routing protocols. We then introduce a graph-theoretic framework on which a dynamic routing protocol can be constructed in a scalable fashion. This framework is a recursive abstraction of the physical network topology that can be also employed in analyzing the network path diversity, as well as the applicability of various types of dynamic routing protocols to a network belonging to a specific topology class. Finally, we present our routing protocol, called resilient recursive routing, which is built upon this framework, and demonstrate through simulations that it meets the desirable properties of adaptive routing protocols identified earlier.

Handbook of Applied Algorithms: Solving Scientific, Engineering and Practical Problems
Edited by Amiya Nayak and Ivan Stojmenović

The chapter concludes with presenting a collection of open problems that arise from both the network abstraction and the routing protocol itself.

17.2 OVERVIEW AND CRITIQUE OF CURRENT ROUTING PROTOCOLS

One of the cornerstones of routing algorithms is to ensure that data are correctly delivered to its destination by following a path that is loop-free. We exclude from our discussion exceptional cases such as deflection routing in optical networks where looping is employed to compensate for the fact that there are no optical buffers that can be used to "hold" data during localized congestion events. Frequently, the underlying physical network possesses a rich topology and many loop-free paths exist. The role of the routing protocol is to compute one such path to the destination. In essence, the routing protocol takes as an input the physical topology of the network, that is, a mathematical graph, and for every node, reduces this to a spanning tree, routed at this node. As a tree is a loop-free structure by construction, there cannot be any looping of data traffic once the protocol has converged. Furthermore, at a practical level a spanning tree can be trivially implemented as a set of unique routing table entries to all destinations. Multipath routing protocols are often computed as a collection of distinct trees so that alternative paths to the same destination consist of edges that are disjoint.

There is always an implicit assumption that protocol freedom from data loops can only be guaranteed if the spanning subgraph employed in constructing routing tables is itself loop-free; that is, it is a tree. As we shall see shortly, this is an assumption that can in fact be relaxed and still result in routing protocols that are loop-free in their operation.

The spanning trees used to construct routing tables have to generate unique paths between all pairs of nodes in the network to ensure loop-free data forwarding operation. The choice of these paths is made unique by imposing some optimality criteria, for example, having the smallest number of hops, or edge-weighted hops, thus yielding a shortest path tree (SPT).

Several algorithms exist to compute the SPT for a network graph. The two most widespread methods are based either on Dijkstra's [1] or the Bellman–Ford [2,3] algorithm. Both algorithms work by computing some minimal spanning tree at each node that contains a consistent set of shortest paths between any pair of nodes.

Dijkstra's algorithm requires that the complete graph of the network is known in advance at each node and that the costs of edges between nodes are nonnegative. Dijkstra's algorithm has (computation) time complexity $O(m + n \log n)$, where n is the number of nodes in the graph and m is the number of edges. In the case of a completely connected graph, $m = n(n - 1)/2$ (for a review cf. the work by Zwick [4]), it yields a worst-case performance of $O(n^2)$. However, Dijkstra's algorithm incurs a significant communication overhead (or communication complexity) in order to disseminate the topology information through a flooding procedure that does not scale well with increasing n, whereby all nodes advertise the weights of the links to their neighbors through networkwide broadcasts.

On the other hand, the Bellman–Ford algorithm has a time complexity of $O(mn)$, which for a completely connected graph yields a worst-case performance of $O(n^3)$ [4], albeit often at a much lower communication overhead cost, as messages are restricted to immediate neighbor exchanges of SPTs. The lower overheads make the algorithm scalable to large networks, but at the expense of convergence delay. This delay arises from the iterative nature of the algorithm and the exchange of local information only.

The above considerations focus on the relative merits of the two algorithms during network initialization. Another important issue is the performance of a routing algorithm in response to a change in the network topology (e.g., the failure of a node, or the addition of a new link): Dijkstra's algorithm in fully distributed nonhierarchical networks requires the complete dissemination of updated topology information to be flooded throughout the network (called link-state advertisements), which is expensive in terms of communication overhead and does not scale well with increasing network size, but is fast. The computation overhead of Dijkstra's algorithm can be further improved if an incremental version is employed [5]. The Bellman–Ford algorithm has a re-convergence time that is highly topology dependent and in some cases infinite, as is evident from the count-to-infinity problem [6]. Furthermore, during re-convergence, both types of routing protocols can loop and possibly drop data.

A generalization of the Bellman–Ford algorithm that eliminates many of the problems associated with re-convergence is EIGRP [7]. This exploits the concept of diffusing computations [8,9] to enable the algorithm to compute shortest paths distributively and as quickly as link-state routing protocols based on flooding while maintaining loop-free operation at all times. However, these operational properties presuppose the presence of a transport mechanism used to exchange update messages amongst routers that is not only reliable but also guarantees ordered delivery [7]. A detailed discussion of EIGRP is beyond the scope of this chapter.

A further class of routing protocols of interest here avoids global topology change information dissemination. This is achieved by implementing local restoration algorithms and thus computing suboptimal paths to destinations once a re-convergence is necessitated [10,11] through node or link failure, or link cost change. Naturally, if the shortest path to each destination needs to be computed, it is still possible to avoid global flooding, but the number of nodes involved in the re-convergence procedure increases significantly [12].

Irrespective of which of the above-described algorithms is employed in a routing protocol, changes to the network topology always necessitate protocol re-convergence. Some of the more advanced algorithms referred to above can avoid data looping. However, data may become nonroutable during the topology information update and shortest path algorithm re-computation, which then results in packets being dropped. The only way of endowing networks with resilience to failures is to compute more than one disjoint path to each destination and either make use of both paths simultaneously (thus also providing a load balancing capability in the network) or switch over to the second path immediately after the first one fails. A number of such multipath routing schemes have been proposed, the most widely adopted one being the equal cost multipath (ECMP) extension to link-state routing protocols [13,14].

Hitherto we have discussed routing protocols that assume that the average time between topology updates or changes is much longer than the routing protocol re-convergence time, which includes the time necessary for sending topology updates to relevant nodes, as well as the time required to perform the SPT re-computation (whether distributed or not). If the network state varies on a shorter timescale, not only does the re-computation become very expensive, but data losses become unacceptably high as well. A class of networks that is susceptible to frequent changes is mobile *ad hoc* networks (MANETs), which are decentralized wireless networks where each node is both router and host [15]. In such networks, dynamic routing protocols that discover paths to a destination *on demand* (i.e., reactively rather than proactively) tend to be favored. Two examples of such routing protocols are the dynamic source routing (DSR) and the *ad hoc* on demand distance vector (AODV) routing protocols [16]. As expected, MANET routing protocols not only discover routes dynamically but sometimes also adopt a local route restoration mechanism to cope with a rapidly changing network topology.

All of the above protocols determine an optimal or near-optimal SPT for every source. However, when all sources are considered simultaneously, the overall solution is not necessarily optimal for the entire network in terms of traffic load distribution, as some links or nodes could become congested. In this sense, the "optimality" of SPTs is not network oriented when traffic is taken into account. One approach of making such solutions optimal for the network as a whole is to make link weights change dynamically in response to traffic loading. However, this needs to be done on a slower timescale than the time taken for the information on the changing network link weights to propagate across the entire network. Provided this is the case, the expensive (in terms of communication and computation overheads) process of re-convergence of all the SPTs can occur repeatedly until an overall optimal solution for the network is reached. This assumes that the external offered traffic to the network does not change significantly during the re-convergence time, which is often not the case.

At the root of this problem lies a fundamental issue: Adaptation requires path choices to be available without delay and SPTs eliminate such choices by decimating the complete network graph into a tree. Since SPTs are global structures, their recalculation takes time, which, in turn, hampers the adaptation process.

In order to build dynamic routing protocols that optimize the operation of a network as a whole, we first need to understand the relevant timescales of all the underlying dynamical processes and their interrelations. The relevant timescales are (1) the timescale for network topology discovery and dissemination, (2) the timescale for topology change, (3) the timescale for external offered traffic change, (4) the timescale for route discovery (*route* is defined henceforth to mean the collection of paths to a destination that a routing protocol can admit), (5) the timescale for a path selection from a route, and (6) the timescale for making switching decisions (we take this to be the shortest characteristic timescale in the network).

Existing routing protocols often force a number of these timescales to be either identical or of the same magnitude, which impacts the scalability of the routing protocol, its convergence properties, and performance. As an example, we cannot apply conventional distance-vector or link-state routing protocols to MANETs, as the topology change timescale is comparable to the topology discovery timescale and such protocols assume that the topology change timescale is much larger than that for topology discovery and dissemination.

17.2.1 Desirable Dynamic Routing Protocol Properties

A dynamic routing protocol should (1) avoid the creation of congestion hot spots and adapt to changes in offered traffic, (2) make maximum use of underlying network capacity according to some optimality criteria that may or may not be global, (3) adapt to topology changes in the network, and (4) be scalable with respect to increasing network size.

It should be evident from the discussion of the preceding subsection that global optimality in network operation is desirable, but may not be attainable given the range of timescales characterizing all the relevant dynamic processes. In such a case, local optimality needs to be considered instead. However, this raises an important question: What is locality in this context? It is our contention that a topological locality must be associated with an elementary routing protocol function, which, in turn, is ascribed an operational timescale. As we have a range of different timescales, this implies that we should consider a hierarchy of localities, and routing protocol adaptation must occur at both the appropriate timescale and its associated appropriate locality. For example, since we require the routing protocol to react fast to link failures, the pertinent locality must relate to the immediate "neighborhood" of the failure and must contain a local restoration path in order to be able to select this very quickly.

17.3 LOGICAL NETWORK ABRIDGEMENT PROCEDURE

The first challenge to be met is to define a graph-theoretic framework for considering a hierarchy of localities in networks.

When considering the connection diversity and thus resiliency in a network, it is important to quantify the number of distinct paths between any pair of end nodes. The loss of one path is insignificant if numerous other paths exist. At the other extreme, if only a single path exists, loss of any of its component nodes or links results in the network becoming disconnected into two disjoint subnetworks. The simplest and most elementary form of diversity is when two disjoint paths connect two nodes; that is, these nodes belong to a ring, or cycle in graph-theoretic terminology (see Fig. 17.1). We shall refer to such a topological relation, as a simple neighborhood (elementary locality), and all nodes belonging to the same cycle are thus neighbors.

Every cycle can be represented by an incidence vector of its constituent edges. Given an exhaustive enumerated list of all m edges in a graph, the incidence vector is an m-dimensional vector of binary elements, where a 1 denotes an edge belonging to the cycle and a 0 if it does not. These incidence vectors form an algebra relative to the binary addition operation. The binary addition (or symmetric difference) of two vectors (and thus their associated cycles) is the set of edges, which are in either cycle, but not in both [17]. This operation is the set-theoretic equivalent of the XOR operation

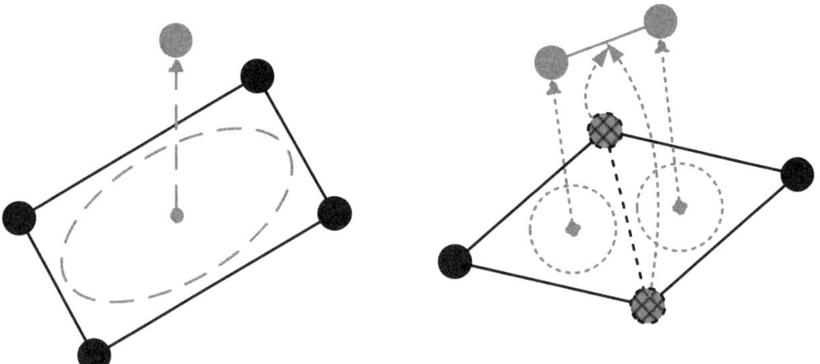

FIGURE 17.1 Simplest form of path diversity in a graph is a simple cycle, which is abstracted to a logical node (left). Definition of a connection (adjacency) between two logical nodes (right).

in Boolean logic. Cycle independence is then defined as a linear independence of associated incidence vectors in this space [17]. Any connected graph with n nodes and m edges has $\nu = m - n + 1$ independent cycles, where ν is defined as the cyclomatic number of the connected graph [17]. A maximal set of independent cycles forms a basis from which all the remaining cycles can be generated. The choice of a basis set of cycles is not unique as we shall see shortly.

Every independent cycle or neighborhood of nodes can be abstracted to a logical node (e.g., the gray node in Fig. 17.1), intended to represent a diversity unit. In the context of a communication network, this logical node represents shared routing state information among all the nodes that belong to this cycle. Two cycles are defined to be adjacent (in a diversity sense) if they share at least one common edge (e.g., the edge and its incident nodes highlighted in dotted black in Fig. 17.1). This can be justified since two adjacent cycles have at least two nodes in common and are thus connected diversely. The nodes incident to the common edge are gateway nodes between the two cycles, and in the context of a communication network, they are responsible for the exchange of the routing information between these two logical nodes. Connecting logical nodes (e.g., the gray nodes in Fig. 17.1) with their associated logical edges (e.g., the gray edge in Fig. 17.1), we can construct the next logical level graph that is an abstraction of the physical network. Any connected linear set of nodes ending in a leaf node is implicitly eliminated from the next level abstracted graph, as this is tantamount to the logical collapsing of such subtrees into their root node, which is a member of a cycle. The reason for this is that there is no path choice (i.e., no path diversity) on a subtree. If the abstracted logical level description of the network contains cycles, we can repeat the above procedure as many times as required, or until it terminates in a highest-level loop-free logical network structure at logical level $\ell = L$. In Figure 17.2, we have the original physical level (level $\ell = 0$) and logical levels $\ell = 1$ and $\ell = L = 2$ (the latter being trivially a single logical node rather than a tree). We call this recursive procedure logical network abridgment, or LNA for short. We label nodes as $\ell.n$, where ℓ denotes the level of abstraction and n is the node number at that level. Thus, 1.2 is node 2 at level 1 (identified with the

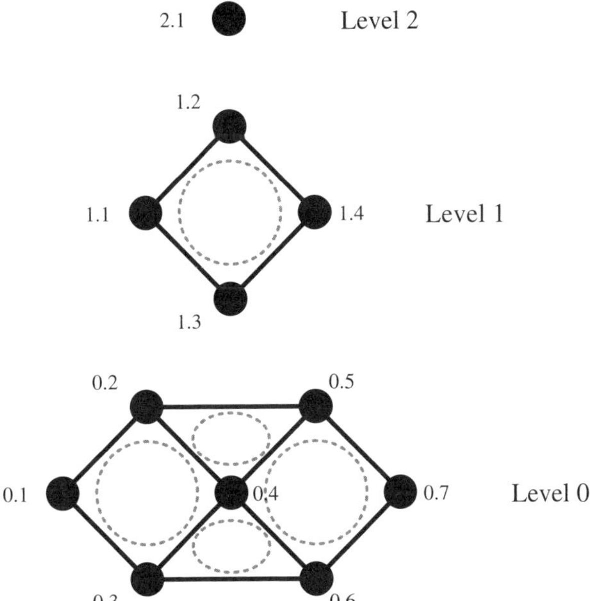

FIGURE 17.2 Logical network abridgement (LNA) procedure applied to a simple network. The LNA abstraction is the ensemble of levels 0, 1, and 2. Physical cycles at level 0 are identified as logical nodes at level 1; common links between cycles at level 0 correspond to logical links at level 1; the abstraction is iterated until a highest level 2 (loop-free) graph is arrived at. The labeling of nodes has two parts: the first one corresponding to the level of abstraction and the second enumerating the node at this level.

cycle $0.2 - 0.4 - 0.5 - 0.2$ at level 0 in Fig. 17.2). It is worth pointing out that when we refer to the LNA abstraction, we signify the entire *ensemble* of levels.

The choice of basis cycle set is usually far from unique [18]: The presence of a K_4 subgraph is sufficient to destroy uniqueness. In weighted graphs, it is possible to ensure uniqueness through a slight perturbation of edge weights [19]. However, it is known that the number of different choices of minimal cycle bases for a graph can be potentially exponential with respect to the size of the graph [20].

Since the choice of basis cycle set is not unique, it follows trivially that the LNA procedure is also not unique, as it is dependent on this choice at each step of recursion. Additional criteria suited to the problem or application at hand need to be employed to make the choice of basis cycle set unique. The number of logical nodes at level $\ell + 1$ is determined by the cyclomatic number, ν_ℓ, of level ℓ. However, the number of logical links at level $\ell + 1$ is determined by our nonunique choice of the set of independent cycles at level ℓ. For the purposes of our discussion, we choose to minimize the number of logical links at the next level of abstraction, as this will not only speed up the convergence of the LNA, but will also minimize the amount of control information overhead incurred in routing. To the best of our knowledge, there is no polynomial algorithm currently in existence that can be used to determine the basis set of cycles that minimizes the number of logical links at the next logical level of abstraction.

The problem of minimizing the number of logical levels of abstraction, L, is even more complex because it implies a global minimization procedure across all levels of abstraction.

In the absence of a polynomial complexity algorithm discussed above, we currently use the minimal cycle basis of a reduced graph, determined as follows: We first remove all nodes of degree 1, repeatedly, until no such nodes remain. We subsequently "eliminate" transient nodes of degree 2 by contracting [17] either of the edges incident on each such node (this is equivalent to removing the transient node and inserting a new edge between the nodes adjacent to the removed one). Finally, we remove all parallel edges, as they constitute trivial cycles that can be reinserted later. The computational complexity of the basic Horton's minimum cycle basis algorithm that can be used is $O(m^3n)$ (cf. [18]). However, improved versions have been reported [18], especially for sparse graphs. The above operations still do not yield a unique cycle basis, but significantly reduce the algorithm's running time by reducing the size of the problem, as well as the number of different cycle bases that often helps minimize the number of logical links at the next level.

The convergence of the LNA procedure to a loop-free graph in a finite number of steps is guaranteed for finite planar graphs. The reason for this lies in the fact that for a particular embedding of a planar graph (with basis cycle set chosen as the set of faces of this embedding), the LNA procedure is tantamount to finding the modified dual of a graph minus the exterior node and ignoring parallel edges and loops. As two consecutive dual graph transformations yield the original graph, the LNA in this case will always give a smaller planar graph, thus guaranteeing convergence in a finite number of steps. We conjecture that for sparse nonplanar graphs, the procedure will also converge, while the question of how many steps it takes to reach convergence still remains. This is supported by numerous applications for the LNA procedure to nonplanar sparse graphs derived from actual Internet service provider core networks. For arbitrarily large, densely connected graphs, such as fully connected graphs (cliques), the LNA convergence remains an open question.

Every level of abstraction conveys summarized path diversity information for the previous level, which can aid both the visualization and analysis of this diversity. The summarization is not done on an arbitrary clustering basis, but is dictated by the underlying network topology and introduces a natural measure for the network diversity, $\mathcal{L} \equiv \min[L]$. The minimum is taken over all choices of sets of independent cycles across all levels. This is an open graph-theoretic problem that merits further study. Clearly, the bigger the $\mathcal{L}$, the more intrinsic path diversity exists in a network. If the graph at any level of abstraction becomes disconnected, it indicates the existence of a path diversity bottleneck at the previous level. An example of the application of the LNA procedure to a graph illustrating the above point is shown in Figure 17.3.

17.4 NETWORK DIVERSITY

We now consider the application of the LNA to routing in communication networks, and specifically to routing in packet-switched networks such as the Internet. The

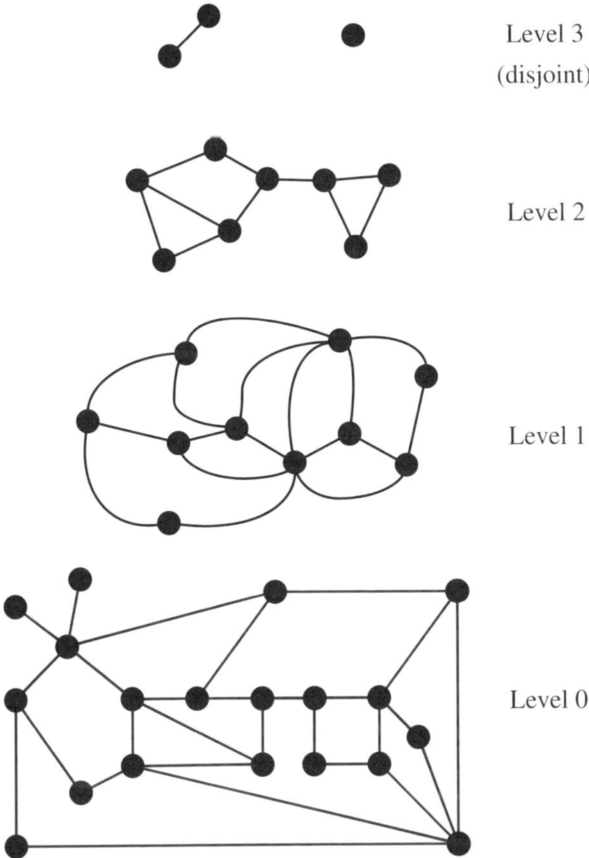

FIGURE 17.3 Logical network abridgment abstraction of a graph that results in a disjointed logical level 3. The disjoint nature of logical level 3 is a characteristic signature of reduced path diversity between more highly connected clusters in the physical level 0 network (i.e., path diversity is not homogeneous across the network) and must not be confused with the absence of connectivity.

network diversity index, $\mathcal{D}$, is the simplest global measure of diversity in a network and can be defined as $\mathcal{D} \equiv \mathcal{L}/n$. We note that any graph consisting of n nodes can be derived from the completely connected graph K_n by removing a number of edges. Therefore, we can conclude that the diversity index of any graph is bounded as follows: $0 \leq \mathcal{D} \leq \mathcal{D}_{K_n}$. The lower bound arises trivially if the original network is a forest (i.e., loop-free). The upper bound arises as the removal of edges from K_n results in the reduction of both the numbers of logical nodes and edges at the higher levels of abstraction. Unfortunately, we do not have at present any exact results or conjectures for $\mathcal{D}_{K_n}$, which could even be infinite for sufficiently large n.

Irrespective of the actual value of $\mathcal{D}_{K_n}$, the diversity index $\mathcal{D}$ can be used to determine the type of routing protocol best suited to the network. If $\mathcal{D} \gtrsim 0$, the network is dominated by trees and a shortest path type protocol is highly scalable and efficient. An adaptive routing protocol in this case will not bring any benefits, but will simply incur internal communication overheads. At the other extreme, if $\mathcal{D} \lesssim \mathcal{D}_{K_n}$, the network is very close to being fully meshed and random deflection routing is scalable (has very low internal communication overheads), robust, and sufficient, because if a destination is not reachable directly there is a high probability that it can be reached through any one of the neighboring nodes chosen at random.

Away from these two extreme cases, a shortest path type protocol fails to exploit the underlying network diversity and will take time to re-converge if congestion or failures arise, while on the other hand random deflection routing is unlikely to result in the successful delivery of data to its intended destination, as nodes are likely to be separated by many hops. To exploit the underlying network diversity, a dynamic, adaptive routing protocol is then required.

In cases where the path diversity in a network is inhomogeneous (e.g., a typical well-engineered ISP network tends to have a highly meshed core of nodes), then a global measure such as $\mathcal{D}$ fails to capture this fact. A local version of the diversity index can in principle be defined on a set of subgraphs of the original graph. We can then analyze the local diversity index of each subgraph to determine the most pertinent type of routing protocol, which may then lead us to the conclusion that a nonuniform routing procedure is required. For example, if a network contains a number of cliques, K_c where $c > 3$, then we could abstract each clique to a logical node with its own internal routing procedure and then apply the LNA to such a modified network. Such an approach would lead to a faster LNA convergence and smaller internal communication overheads.

17.5 RESILIENT RECURSIVE ROUTING

The LNA can be augmented with a number of forwarding rules to create a resilient recursive routing (R^3) protocol. Here we consider the high-level generic features of such a protocol that adheres to the properties discussed in Section 17.2.1. There can be more than one specific implementation of the generic algorithm, and we shall describe our specific choice that we have proceeded to simulate in Section 17.5.2.

17.5.1 Generic R^3 Algorithm

The routing algorithm must operate recursively at each level of abstraction of the network either to route a packet around a single cycle or along a tree. Routing information on a tree is a trivial exercise in the sense that all forwarding decisions are deterministic, and we shall not discuss this any further. The fundamental algorithm must route a packet from a source to a destination, both of which are members of the same level 1 logical node and thus are members of the same cycle at level 0 (hereafter referred to as level 0 neighbors). The algorithm must be capable of (i) loop-free data routing across the cycle, (ii) load balancing across the cycle, and (iii) fast reaction to link or node failures in the cycle of nodes. A specific implementation of the fundamental routing algorithm will be discussed in the next section.

If the source and destination are members of the same level 2 logical node (i.e., they belong to the same level 1 cycle and are thus level 1 neighbors), the fundamental routing algorithm should be applied iteratively twice, once at level 1 and once at the current (local) level 0 cycle.

For source and destination nodes that are level ℓ neighbors, the fundamental routing algorithm needs to be applied $\ell + 1$ times iteratively, from the current highest level ℓ down to the local level 0 cycle.

If at some level of abstraction ℓ' the LNA graph of the network is disjoint (in Fig. 17.3, e.g., $\ell' = 3$), the fundamental routing algorithm cannot find a level 3 cycle or tree across some source and destination pairs. In this case, the algorithm must drop down to level $\ell' - 1$, where at least one cut-node (in the case of Fig. 17.3 two cut-nodes and a cut-edge) needs to be traversed *deterministically* at the $\ell' - 1$ level of abstraction, just as routing on a tree needs to operate. This implies that cut-nodes need to exchange reachability information about their corresponding bi-connected parts of the network.

The routing methodology embodied in the generic algorithm enables us to route a packet in a loop-free manner, while performing load balancing and enabling failure recovery across the network. The iterative nature of the algorithm though does not on its own guarantee the scalability of all the properties of the fundamental routing algorithm to the entire network. The first condition necessary for the scalability of the routing protocol is the need to have the minimum number of levels of abstraction $\mathcal{L}$ to be significantly smaller than the number of nodes n in the original network, as the size of the network grows, that is, $\mathcal{L} \ll n$, or equivalently $\mathcal{D} \ll 1$. The second condition for scalability relates to the characteristic reaction times of the fundamental routing algorithm to congestion and failures at the higher levels of abstraction. The higher levels must use summarized information, for example, for congestion along their logical cycles, over longer timescales to reflect the summarized nature of this higher-level neighborhood. For example, if for a sufficiently sparse class of network graphs it were to turn out that $\mathcal{L} \sim \log n$, as $n \to \infty$, it would be natural to select adaptation/update time intervals, τ_ℓ, for higher levels that grow exponentially, $\tau_\ell \approx \tau_0 \, b^\ell, \;\; \ell = 0, \ldots, \mathcal{L}$, for some base $b > 1$ that depends on the sparsity of the graph and a desirable fastest adaptation time, τ_0, at physical level $\ell = 0$.

Naturally, the adaptation can be "terminated" prematurely at an earlier level of abstraction and the higher-level iterations of the fundamental routing algorithm can become static, if the network operation is deemed to be sufficiently adaptive by the protocol designer.

It should be noted that the proposed scheme bears some similarities to routing in networks based on abstraction hierarchies (see e.g., works by other authors [21–23]), but differs fundamentally in that both the number of hierarchy levels and their clustering structure are not determined *a priori*, or through extrinsic criteria to the network, but arise naturally from the topology itself.

17.5.2 A Specific Implementation of the R^3 Algorithm

We now proceed to discuss a practical implementation of R^3 through a simple example. Even though we can devise a topology discovery and destination host advertisement mechanism based on R^3, we choose to adopt for simplicity a standard link-state routing protocol such as IS–IS [25] or OSPF [13] to achieve both of these network functions *the initialization stage only*. This is done in order to concentrate on developing the routing function of the protocol alone.

Routing is achieved by employing labels hereafter called *circulation vectors*, which are also implemented recursively (i.e., they are nested in the header of each packet). For a level ℓ destination, each circulation vector describes a local level 0 simple path that is a subgraph of the local level 0 cycle (i.e., loop segment or arc on the local level 0 cycle) toward the destination, a level 1 arc on the local level 1 cycle toward the destination, and so on, all the way up to a "local" level ℓ arc on the "local" level ℓ cycle, containing the destination.

Note that this routing scheme is not the same as source routing [16], as it does not specify a precise path to the destination, but rather a progressively abstracted route (in the sense of an ensemble of many physical level 0 paths defined in Section 17.2) to the destination. This provides a connectionless service that gives specific physical path selection on the shortest timescale of a level 0 neighborhood, but as a result of the increasing levels of abstraction provides more flexibility in subsequent physical path selections across any remaining higher-level neighborhoods, on a longer timescale. This retained flexibility is then used at subsequent nodes to make local forwarding decisions in order to overcome any congestion and failure situations that might arise.

A selected cycle segment at level ℓ requires that the packet be forwarded from one node to an adjacent node using a link, all at level ℓ. Each node at level ℓ is in fact a representation of a neighborhood/cycle at level $\ell - 1$. Therefore, the link at level ℓ is in fact a representation of the nodes held in common between two adjacent neighborhoods at level $\ell - 1$. These common nodes are gateways and thus represent an intermediate destination at level $\ell - 1$ of a selected path at level ℓ. Nodes receiving a packet will forward the packet so as to *maintain* its given direction of circulation on the designated cycle, until it reaches the gateway. Once the packet reaches a gateway, the circulation vectors of all completed arcs are removed and new ones are added, based on more recent information regarding congestion and even failures, until the packet is routed to its final destination.

The above procedure can be best illustrated using the simple two-level network of Figure 17.4. In sending a packet from a host A connected directly to node 0.1 to a host B connected directly to node 0.11, host A generates a packet with destination address B. Node 0.1 will have knowledge of the existence of B through the advertisement protocol (borrowed unchanged from IS–IS for this particular implementation) only as a level 2 destination attached to the level 2 node 2.3. As the level 2 network description is a simple tree, the routing on it is deterministic and we omit the use of level 2 circulation vectors in our discussion for simplicity and clarity. The omitted circulation vector lists the deterministic hops to the destination node 2.3.

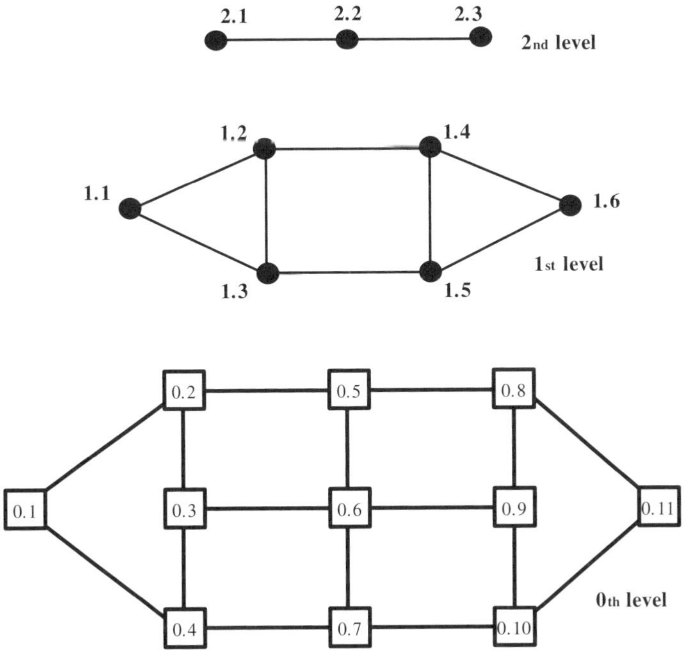

FIGURE 17.4 Routing on a simple network.

The routing required at level 2 is from neighborhood 2.1, to which node 0.1 belongs, to neighborhood 2.3, to which node 0.11 belongs, with the immediate next hop at level 2 being 2.1 → 2.2. The link from 2.1 to 2.2 is represented by the nodes that cycles 2.1 and 2.2 have in common at the next lower level, that is, gateway nodes 1.2 and 1.3. At level 1 there is path diversity, as node 0.1 may send the packet either clockwise around cycle 2.1 to gateway 1.2 or anticlockwise around cycle 2.1 to gateway 1.3. Node 0.1 must then select one of these two paths, for example, cycle 2.1 clockwise to gateway 1.2, based on summarized performance information around the level 1 cycle 2.1 on a longer timescale, and attaches an inner label containing the selected circulation vector to the packet.

The routing required at level 1 is to forward the packet from neighborhood 1.1 to neighborhood 1.2. The link from cycle 1.1 to cycle 1.2 is represented by the nodes that 1.1 and 1.2 have in common as the next lower level, that is, nodes 0.2 and 0.3. At level 1 there is, therefore, path diversity, as node 0.1 may send the packet either clockwise around cycle 1.1 to gateway 0.2 or anticlockwise around cycle 1.1 to gateway 0.3. Node 0.1 selects one of these two paths, for example, cycle 1.1 clockwise to gateway 0.2, based on measured performance information around the level 0 cycle 1.1 on the shortest timescale, and attaches an outer label containing the selected circulation vector to the packet.

The routing required at level 0 is now to forward the packet from node 0.1 to node 0.2, according to the attached circulation vectors. As the link from 0.1 to 0.2 corresponds to a physical link between these nodes, there is no further path

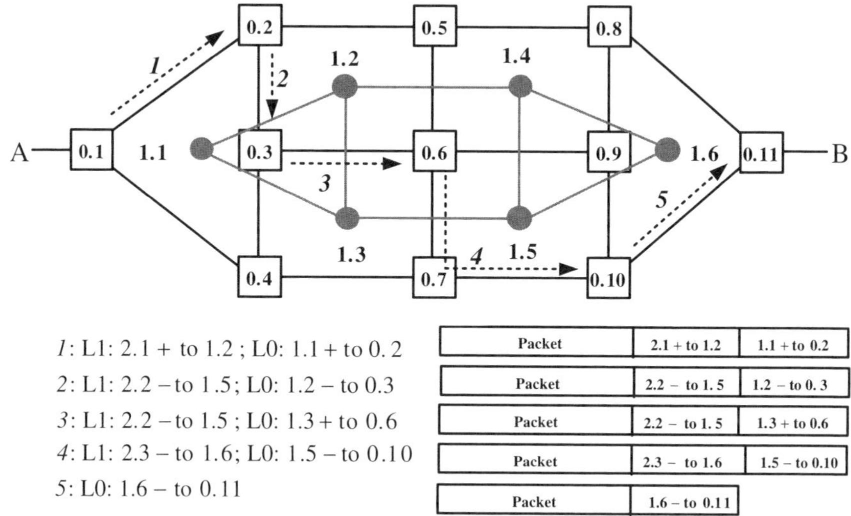

FIGURE 17.5 Set of routing decisions on the network of Figure 17.4.

diversity that can be exploited and the packet is forwarded along the physical link to node 0.2.

In each cycle, there are always two possible circulation directions. The sense of circulation direction, denoted by positive or negative, does not need to be defined globally, but needs to be unambiguously defined only for the member nodes of the cycle. In our planar graph example of Figure 17.4, we denote positive (negative) circulation around a cycle to be clockwise (anticlockwise) for ease of illustration. A possible packet structure corresponding to the first routing decision, shown in Figure 17.5, would be label 1 (inner label): 2.1+ to 1.2 and label 0 (outer label): 1.1+ to 0.2. When this packet arrives at the level 0 node 0.2, this node identifies itself as being 0.2, the destination gateway, of the outer label, and so strips the outer label. It also identifies itself as being a member of the neighborhood 1.2, the destination gateway of the inner label, and so strips the inner label as well.

This occurs because neither of the circulation vectors is required in addition to the destination host address B to ensure deterministic routing. Indeed, it is quite acceptable to adopt a policy of penultimate node label stripping, so that labels are stripped if the adjacent node that the packet is being sent to is in fact the label destination. Labels are, therefore, only needed in order to ensure that packets are correctly transited through intermediate nodes at all levels in the abstraction. Implicit in this statement is the fact that all transit nodes must respect the circulation vector at their relevant level, unless there is a failure. In our example, there was no need to add any labels to the packet leaving node 0.1. However, for clarity, all labels will continue to be shown throughout this example.

Node 0.2 follows the same process of establishing the associated path diversity and then making path selections based on performance information associated with each level in the LNA. The packet leaves node 0.2 toward 0.3, for example, with an inner label, abbreviated as L1, of 2.2− to 1.5 and an outer label, L0, of 1.2− to 0.3.

Upon reaching node 0.3, the outer L0 label reaches its destination and is removed, but the inner L1 label has not and so it is retained and the next level 0 path is selected. Node 0.3 must maintain the circulation at level 1 of 2.2− to 1.5 and as node 0.3 belongs to neighborhood 1.3, it must forward the packet along 2.2− from gateways 1.3 to 1.5. As the common nodes between cycles 1.3 and 1.5 are gateways 0.6 and 0.7, node 0.3 can forward the packet on either 1.3+ to 0.6, or 1.3− to 0.7. Node 0.3 selects one of these two paths, for example, 1.3+ to 0.6 using a single label L0, based on the most recent level 0 performance (e.g., congestion) information.

Upon reaching node 0.6, both the outer and inner labels have reached their destination and are thus removed. New labels are inserted following the same process that occurred at node 0.2.

Upon reaching node 0.7, neither the outer label L0 nor the inner label L1 gateway destinations have been reached, and 0.7 simply maintains both circulation vectors and the packet is forwarded without choice to node 0.10 without any change to the labels.

Upon reaching node 0.10, the outer label L0 and the inner label L1 have both reached their destination and are removed. Node 0.10 has knowledge of the existence of B through the advertisement protocol as a level 0 destination as nodes 0.10 and 0.11 are both members of neighborhood 1.6. Node 0.10, therefore, follows the same process but only has to consider whether to send the packet on either 1.6+ to 0.11 or 1.6− to 0.11. In this example, the packet is forwarded on 1.6− to 0.11 based on the most recent level 0 performance information.

The simplest performance information we employ in our protocol is the measured cumulative delay that a modified "hello" packet experiences per hop in traversing each loop in each of the circulation directions, approximately every 100 ms. Each router time stamps a cycle-specific "hello" packet upon processing it, together with its router address and forwards this to the next router in the cycle. Each node then computes the associated delay to each other node on the same cycle in both directions of circulation. For a given destination node on the same cycle, new packets are forwarded using the direction of circulation that is currently experiencing the lowest delay.

Higher-level summarized performance information is computed from the average cycle delay in both directions and is disseminated through restricted flooding on progressively longer timescales: A router that is not a member of the logical abstracted node for which the summarized performance information is intended simply discards the packet. In our implementation, level 1 summarized performance information was updated on the order of ~ 1 s, while higher levels were static.

We want to stress here that labels are not path specific but network specific. The labels are determined by the LNA and thus are tied to the network topology. Here, we need to introduce a further refinement in our terminology of paths. A level ℓ path is an arc on a level ℓ cycle, which in turn contains a set of level $\ell - 1$ paths. Routes (the ensemble of paths), which are instantiated as circulation vectors, are thus fixed, or at least determined on the same long timescale as the topology itself. The choice of a particular physical path to a destination is not determined in advance but is done progressively as the packet is forwarded through the network, based on up-to-date, local congestion information.

As data flows can suffer from jitter in any multipath routing scheme, a further improvement in our protocol can be to perform per flow routing (using hash tables), with lower quality of service (QoS) flows being switched to alternative paths earlier than higher QoS flows if the need arises, say due to congestion.

17.5.3 Simulations

R^3 has been implemented in the discrete-event network simulator OPNET® [24] and has been used in extensive simulation experiments, which fall into three large sets: First, we have simulated congestion that might arise at the BGP gateway of a core IP network of a medium-sized European Internet Service Provider, examining the end-to-end delay and packet-loss-rate characteristics of the network with ECMP IS–IS [14,25] as a baseline routing protocol for comparison. A second set of simulations concentrated on the study of similar metrics in scale-free networks [26] suffering asymmetric attacks (i.e., single as well as multiple highly connected node attrition). ECMP IS–IS was also used as a benchmark in this set of simulations. Finally, we have also studied the impact of unstable links and restricted endhost mobility in a tactical network example under many simulation scenarios, comparing the performance of R^3 against ECMP IS–IS as well as AODV [16]. In all three cases, R^3 outperformed the remaining protocols, as it indeed managed to spread the data traffic load evenly across all available network resources and was capable of operating in "broken" networks on a reduced set of circulation vectors without the need for any re-convergence whatsoever. For highly mobile networks where the rate of link breakage and formation is large, R^3 cannot at present outperform any of the current MANET protocols. In order to improve its performance, we need to fundamentally rethink the physical level abstraction, as wireless networks cannot be appropriately abstracted by simple graphs. This discussion is beyond the scope of this chapter.

As we can see from Table 17.1 the good performance of R^3 was achieved without a complete implementation of its adaptation functionality.

Here we discuss only one simulation example due to space limitations. We concentrate on asymmetric attacks on scale-free networks with multiple highly connected node failures.

Scale-free networks arise naturally in many contexts, including the Internet, when new nodes attach themselves preferentially to the existing highly connected nodes. This makes the network efficient in terms of routing, as it limits the number of end-to-end hops significantly (the diameter of scale-free networks is small). Also, scale-free networks are quite robust to random node failures. However, when highly connected nodes are preferentially targeted by an attacker, such networks can be easily compromised.

The scale-free network model we used has been developed on the basis of measurements of the Internet. The Albert–Barabási algorithm [26] outlined below summarizes how to generate a scale-free network: When new nodes are to be connected to an

TABLE 17.1 Implementation Versions of R^3

<table>
<tr><th></th><th>R^3v4</th><th>R^3v5</th><th>R^3v6</th><th>R^3v7</th><th>R^3v8</th><th>R^3v9</th></tr>
<tr><td>R^3 routes initialization</td><td colspan="6">Dynamic</td></tr>
<tr><td>R^3 route labeling</td><td colspan="6">All levels</td></tr>
<tr><td>R^3 path selection</td><td colspan="6">Dynamic</td></tr>
<tr><td>R^3 stub collapsing</td><td colspan="3">Level 0</td><td colspan="3">All levels</td></tr>
<tr><td>Node/link failure adaptation</td><td colspan="2">Level 0 nodes</td><td>Level 0 nodes and links</td><td colspan="3">All levels nodes and links</td></tr>
<tr><td>Traffic congestion adaptation</td><td colspan="3">Level 0</td><td colspan="3">Levels 0 and 1</td></tr>
<tr><td>LNA</td><td colspan="5">Static/scripted</td><td>Partially dynamic</td></tr>
</table>

existing network nucleus of nodes, they connect each of their q available links to an existing network node i with probability $P(q_i) = q_i / \sum_j q_j$, where q_i is the current node degree for the already existing nodes in the network.

Such a model creates networks where the node degree (connectivity) distribution has a power-law behavior, with most nodes having a low degree and a very small proportion of the nodes being highly connected. Consequently, such a network has the advantage of providing highly efficient communication through small number of key, highly connected nodes that act as hubs.

We generated a 120-node, 117-link scale-free network using the Albert–Barabási algorithm [26], starting with a core of nine highly meshed nodes. We subsequently simplified this network in order to speed up the simulation by removing stubs and purely transit nodes that do not play a role in routing, as all their switching decisions are trivial. To further simplify the LNA so as to aid graph visualization, we removed 17 nonplanar links. This made the network more vulnerable to the loss of highly connected nodes, as these links provided alternative distant connections across the network. The resulting modified scale-free network had 39 nodes and 70 links, and its R^3 levels of abstraction are shown in Figure 17.6.

For the purposes of our simulation analysis, we started with realistic values for bandwidth, packet size, and so on, but used *bit and time scaling* in order to speed up the rather lengthy simulations. The *scaled* simulation parameters were the following: All links had a bandwidth of 1 Mb/s; we employed 4000-bit-long constant-length packets; the node buffers had a capacity of 2000 packets per outgoing link (incoming links were nonblocking); the packet generation rate was 142 packets/s per node; the packets generated at each node had a stochastic destination address with probability proportional to the destination node degree; the level 0 adaptation time was 1 s; and

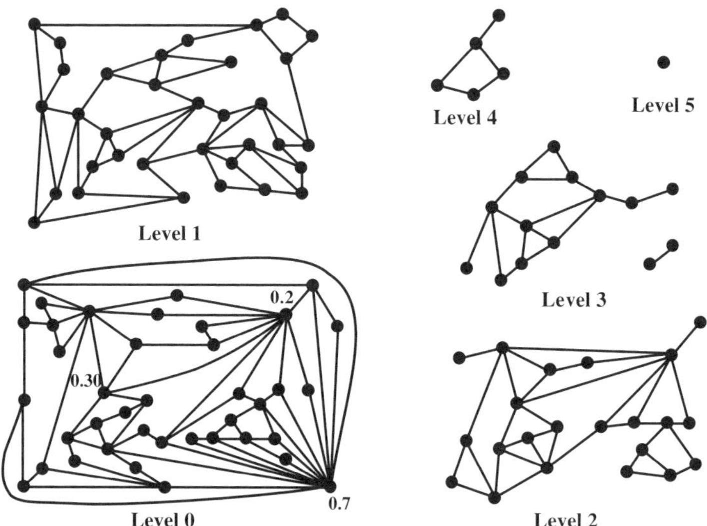

FIGURE 17.6 LNA of modified scale-free network used in the simulations.

the level 1 adaptation time was 5 s. The simulation was run for 100 s, with node 0.2 (with degree 10—the second most connected node) failing at 30 s and node 0.30 (with degree 5—the fourth most connected node) failing at 60 s. Once a node failed, the data traffic destined for that node was dropped in order to avoid simulating artificial, meaningless losses. It is also worth pointing out that the cascade failures of nodes 0.2 and 0.30 result in a level 1 logical link failure, which is a relatively severe test for R^3.

We avoided simulating the most highly connected node failing (node 0.7), as this represented 20 percent of all links failing, making this an excessively compromised network. This decision was supported by earlier simulations that demonstrated quantitatively that excessive attrition of links and nodes in scale-free networks often left little or no scope for adaptation in routing.

The results of the simulation are shown in Figures 17.7–17.9. The network traffic generation rate was chosen in such a way that the network is originally neither lightly loaded nor congested, but had average buffer occupancies at around 1 percent of their capacity and maximum buffer occupancies at around 25 percent as can be seen in Figure 17.8. Once node 0.2 fails, IS–IS dropped a few hundreds of packets during the re-convergence period (see Fig. 17.7), whereas R^3 immediately rerouted data around the failure. Prior to this first node failure, IS–IS had the lowest average buffer occupancies and shortest end-to-end packet transport delays (Fig. 17.8), as it always selected the shortest paths, and all the nodes along these paths were not congested.

The severely reduced number of available paths in the network resulted in a gentle increase in traffic at the nodes along the surviving paths as time progressed, and at around 50 s both IS–IS ECMP and R^3v4, which adapt and thus perform load balancing only at the physical level (i.e., level 0), started experiencing the onset of congestion and hence dropped packet at hot spot nodes along their chosen routes. As R^3v4 exploits more level 0 paths than IS–IS ECMP, its loss rate (i.e., the slope of the

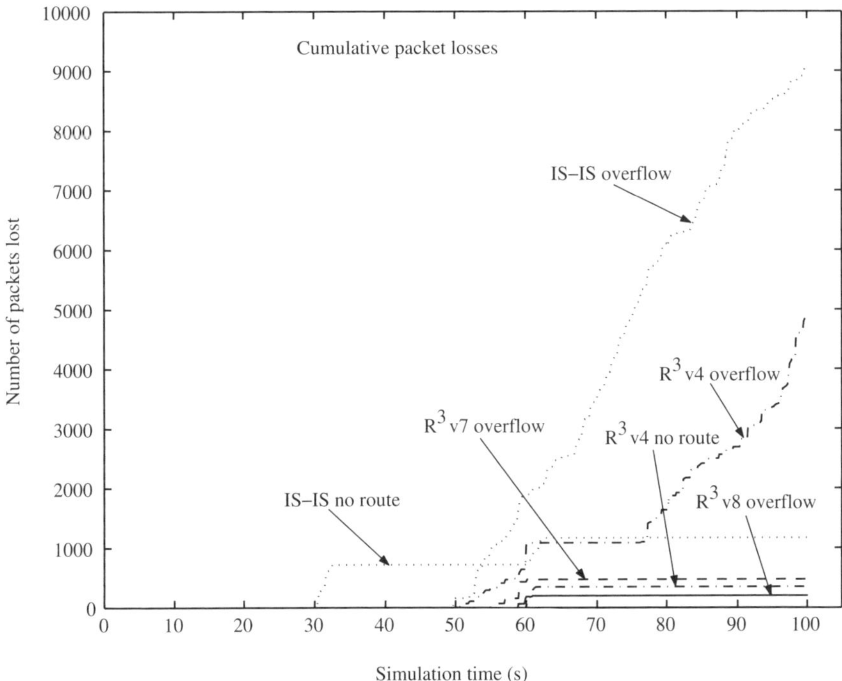

FIGURE 17.7 Comparative cumulative packet loss rate of different R^3 versions and IS–IS ECMP operating on the scale-free network of Figure 17.6 when node 0.2 fails at 30 s and node 0.30 fails at 60 s (R^3 versions 7 and 8 have zero no route losses).

graph in Fig. 17.7) is about half that of IS–IS. R^3v7 adapts to logical link failures but fails to perform higher-level load balancing and was also beginning to show signs of congestion, albeit with some delay. However, R^3v8, which adapts both to logical link failures and performs load balancing at levels 0 and 1, did not experience losses due to congestion.

When the node 0.30 also failed, the data packets destined for this node were dropped and R^3v7 came out of congestion, whereas R^3v4 temporarily came out of congestion, but through physical level adaptation only re-developed it a short time later at around 78 s. Being unable to exploit the inherent path diversity in the decimated network, IS–IS ECMP continued to experience packet losses. In contrast to all these protocols, R^3v8 only experienced a small transient packet loss during the time it needed to adapt to the failure of a logical link at level 1, as adaptation at this level was slower than adaptation at level 0 by design, in order to make the protocol scalable. The final observation worth making, was that in this failing scale-free network, the end-to-end packet transfer delays of all the protocols were comparable, even though their corresponding path hop counts were significantly different, as can be seen in Figure 17.9. This was the case because the shorter hop-count paths were always chosen first and thus experienced congestion earlier. The significant difference between the

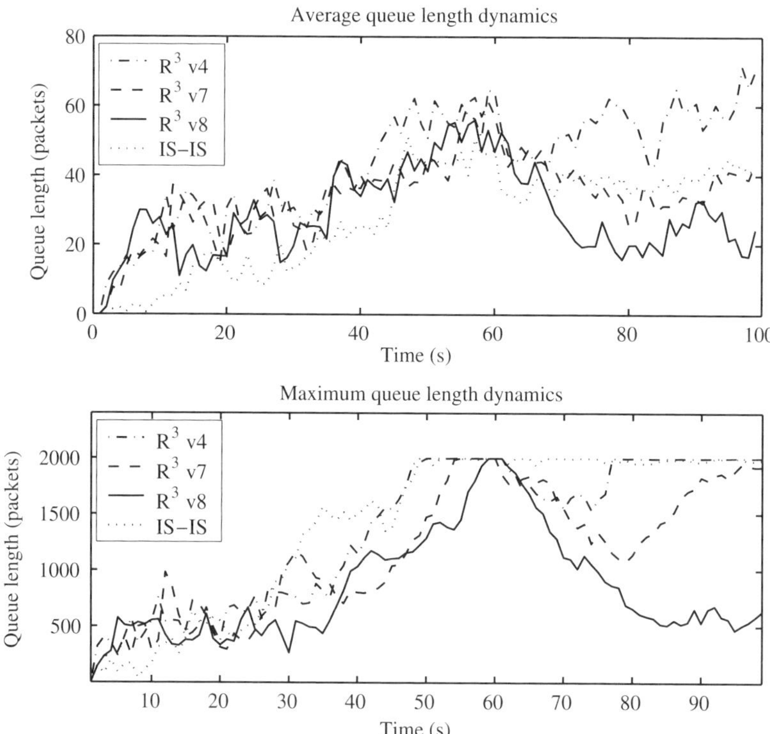

FIGURE 17.8 Comparative queue buffer occupancy analysis for the same simulation scenario as Figure 17.7.

protocols was in their ability to deliver successfully significantly different numbers of packets in this simulation scenario.

17.6 CONCLUDING REMARKS AND OPEN PROBLEMS

The simulation results of Section 17.5.3 demonstrate that routing protocols that exhibit only local adaptation to failures and congestion will often fail to yield the desired result in achieving better network operation. Global adaptation on the other hand will work if it is capable of exploiting all the paths in a network, but at a high overhead cost, which will not scale well for very large networks. The approach taken by R^3 is to adapt to changes on appropriate range of localities and timescales for the specific network topology, thus reaping the desired benefits while minimizing the adaptation overhead costs and maintaining reasonable scalability.

The routing protocol presented here is only one implementation of a very general and novel class of LNA-based protocols. Variants of our implementation of R^3 might employ different ways of summarizing information to be used by the adaptation mechanism at different levels of abstraction. Furthermore, the relative timescales

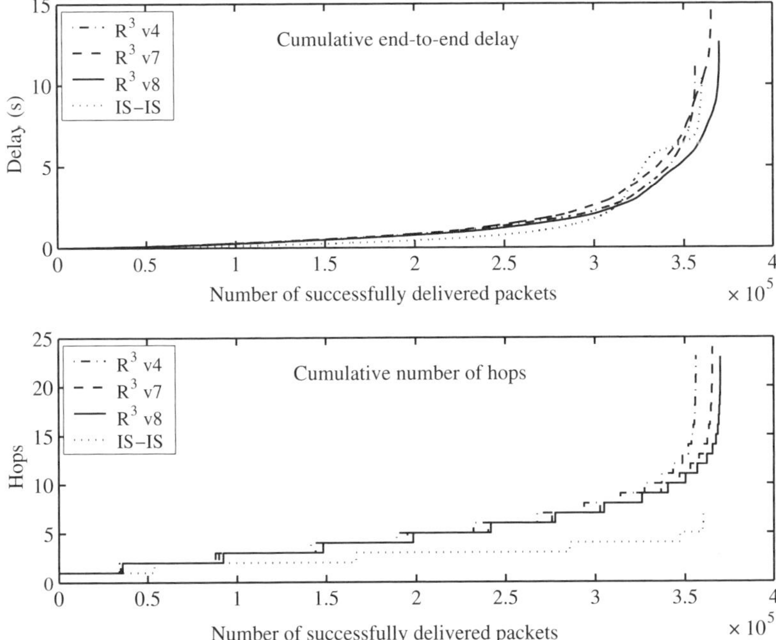

FIGURE 17.9 Comparative cumulative end-to-end packet transfer delay and cumulative hop count for the same simulation scenario as Figure 17.7.

of adaptation at different logical levels were chosen on the basis of a plausibility argument that merits a much more thorough investigation.

The practical implementation issues of R^3 are also an interesting subject for further study, as our chosen mechanism of employing labels to instantiate circulation vectors carries a significant overhead. For example, only circulation vectors that are not changed in traversing a node need to be explicitly declared. However, employing explicitly circulation vectors at all times endows the protocol with the ability to perform data backtracking if the network links are particularly unstable, which may be a desirable property for some types of networks and/or applications. The current implementation simply drops packets rather than employing backtracking, which gives pessimistic performance results for R^3.

Most importantly, there is scope for significant further work on the graph theoretic aspects of the LNA, as well as the corresponding algorithms. As mentioned in this chapter, the choice of cycle basis is not unique and requires additional, possibly application-specific, criteria. The sensitivity of the LNA and its convergence to such additional criteria is an open question.

The computational complexity of the LNA algorithm as a whole is also intimately dependent on the cycle basis choice criteria. The detailed study of the computational complexity of the LNA algorithms is also an important topic for further work. However, this needs to be preceded by a rigorous study of the convergence of the LNA procedure on a number of graph families, such as large densely connected graphs, sparse graphs, Erdös–Rényi, and scale-free graphs.

ACKNOWLEDGMENTS

We wish to thank the European Office of Aerospace Research and Development, Air Force Office of Scientific Research, United States Air Force Research Laboratory for supporting this work, in part, through grants FA8655-04-M-4037 and FA8655-05-M-4026.

REFERENCES

1. Dijkstra EW. A note on two problems in connexion with graphs. Numerische Math 1959;1:269–271.
2. Ford LR. Jr. Network flow theory. Technical Report No. P-923, Santa Monica, CA: The Rand Corporation, August 1956.
3. Bellman R. On a routing problem. Quarterly Appl Math 1958;16(1):87–90.
4. Zwick U. Exact and approximate distances in graphs—a survey. Lecture Notes in Computer Science (LNCS2161). Berlin: Springer-Verlag; 2001. p 33–48.
5. Ramalingam G, Reps TW. An incremental algorithm for a generalization of the shortest-path problem. J Algor 1996; 21(2): 267–305.
6. Perlman R. Interconnections. 2nd ed. Reading, MA: Addison-Wesley; 1999.
7. Albrightson B, Garcia-Luna-Aceves JJ, Boyle J. EIGRP—a fast routing protocol based on distance vectors. Proceedings of Networld/Interop; 1994.
8. Dijkstra EW, Scholten CS. Termination detection for diffusing computations. Inform Process Lett 1980;11(1):1–4.
9. Garcia-Luna-Aceves JJ. Loop-free routing using diffusing computations. IEEE/ACM Trans Netw 1993;1(1):130–141.
10. Narvaez P, Siu K-Y, Tzeng H-Y. Fault-tolerant routing in the Internet without flooding. In: Aversky D, editor. Dependable Network Computing. Drodrecht: Kluwer Academic; 2000. p 193–206.
11. Wu J, Dai F, Lin X, Cao J, Jia W. An extended fault-tolerant link-state routing protocol in the Internet. IEEE Trans Comput 2003;52(10):1298–1311.
12. Narvaez P, Siu K-Y, Tzeng H-Y. New dynamic algorithms for shortest path tree computation. IEEE/ACM Trans Netw 2000;8(6):734–746.
13. Moy J. OSPF Version 2 IETF, RFC 2328, 1998 .
14. Hopps C. Analysis of an equal-cost multi-path algorithm. IETF, RFC 2992, 2000.
15. Basagni S, Conti M, Giordano S, Stojmenović I, editors. Mobile Ad Hoc Networking. Wiley-IEEE Press; 2004.
16. Perkins CE, editor. Ad Hoc Networking. Addison-Wesley; 2001.
17. Diestel R. Graph Theory. Graduate Texts in Mathematics. Volume 173, Berlin: Springer; 2000.
18. Berger F. Minimum Cycle Bases in Graphs. Shaker-Verlag; 2004

19. Mardon R. Optimal cycle and cut bases in graphs. Ph.D. thesis, Northwestern University, 1990.
20. Vismara P. Union of all the minimum cycle bases of a graph. Electr J Comb 1997; 4(1):R9.
21. Kleinrock L, Kamoun F. Optimal clustering structures for hierarchical topological design of large computer networks. Networks 1980;10(3):221–248.
22. McQuillan JM, Richer I, Rosen EC. The new routing algorithm for the arpanet. IEEE Trans Commun 1980;COM-28:711–719.
23. Tsai WT, Ramamoorthy CV, Tsai WK, Nishiguchi O. An adaptive hierarchical routing protocol. IEEE Trans Comput 1989;38(8):1059–1075.
24. http://www.opnet.com/
25. Callon R. Use of OSI IS–IS routing in TCP/IP and dual environments. IETF 1990;RFC 1195.
26. Albert R, Barabási A-L. Statistical mechanics of complex networks. Rev Mod Phys 2002;74:47–97.

CHAPTER 18

Routing Algorithms on WDM Optical Networks

QIAN-PING GU

18.1 INTRODUCTION

The bandwidth of an optical fiber is about four orders of magnitude higher than a peak electronic data rate of a few Gbps. A bottleneck in realizing the huge bandwidth of optical fibers is this opto-electronic bandwidth mismatch. Wavelength division multiplexing (WDM) is the current favorite technology to eliminate this bottleneck. In WDM networks, the transmission spectrum of an optical fiber is partitioned into multiple wavelengths, each wavelength supports a channel that is usually operated at a peak electronic data rate. The bandwidth of a wavelength channel may be further shared by multiple low-rate traffic demands. A WDM network consists of network nodes connected by point-to-point optical links. A network node provides the switching between optical links connected to it and the interface between end users (at electronic domain) and the optical network. An optical link consists of an optical fiber (or multiple parallel fibers) that carries optical signals from one node to another. To transmit data from a source node s to a destination node t, electronic data are converted to optical signals at s, the optical signals are transmitted to t via a sequence of optical links (an optical path), and converted to electronic data at t. The transmission is called *all-optical* or *one-hop of optical routing*. The opto-electronic conversions at s and t are also known as *add data to network* and *drop data from network* operations, respectively. Interested readers are referred to the works of Ramaswami and Sivarajan [38], Sivalingam and Subramaniyam [40], and Stern and Bala [41] for details of WDM networks.

A communication application on a WDM network can be specified by a set $D = \{(s, t, d_{st})\}$ of traffic demands, each demand (s, t, d_{st}) requires a bandwidth d_{st} from source node s to destination node t of the network. It is preferred that all traffic demands are realized by all-optical routing. However, due to some constraints such as limited wavelength channels, all-optical routing may not be available for all traffic demands. In this case, some traffic demands may have to be realized by multihop routing: data

Handbook of Applied Algorithms: Solving Scientific, Engineering and Practical Problems
Edited by Amiya Nayak and Ivan Stojmenović

are added to one wavelength channel at source, dropped at an intermediate node, and then added to another wavelength channel at the intermediate node, and the routing is repeated until data are dropped at the destination. In general, to realize a communication application specified by a set of traffic demands, algorithms are required to solve the following problems.

- *The logical topology design problem:* Given a WDM network G, define a set $R = \{(u, v)\}$ of connection requests over a set V of nodes of G such that for every pair of nodes s and t in V there is a path from s to t in R when R is viewed as a graph with a connection request as an edge. For each request $(u, v) \in R$, data from u to v are transmitted by all-optical routing in G. Usually V is the set of all nodes of G or the set of nodes in a given set D of traffic demands. Graph R is known as the *logical topology* and a connection request in R is called a *logical link* of the network.
- *The routing and wavelength assignment (RWA) problem:* Given a set R of connection requests in a WDM network G, find a routing path in G for every $(u, v) \in R$ and assign each path a wavelength such that the paths with the same wavelength do not share any common link in G. The routing path with an assigned wavelength is called a *light path.*
- *The traffic grooming problem:* Given a set D of traffic demands and a set of light paths in a WDM network G, multiplex the traffic demands of D into the light paths subject to the bandwidth constraint of the light paths. The multiplexing is realized by a device called add-drop multiplexer (ADM) at nodes of G.

A general goal in the study of the above problems is to determine the resources required to achieve a given connectivity as a function of network size and functionality of network nodes. Much of the discussions for the goal is on minimizing the number of wavelength channels and the number of ADMs for realizing a given communication application. These optimization problems have been extensively studied in both communication and graph algorithms communities for WDM networks [1,2,9,15,38,40,41]. Since the quality of the solutions for these optimization problems is critical for the performance of WDM networks, it is extremely important to have efficient algorithms with good guaranteed performance for those problems. However, it is challenging to develop such algorithms in most cases. For example, the RWA problem is NP-hard for even very simple networks like rings and trees [9,19]. Inapproximability results are known for the RWA problem in networks with more complex topologies when the routing paths are given [32]. Similar hardness results are known for the other optimization problems as well. For those problems on simple networks like rings and trees, many efficient algorithms with good guaranteed performance have been known. But for the problems on more complex networks, no effective approach has been developed and the existing algorithms are based on the integer linear programming or ad hoc heuristics. The performance of those algorithms are not guaranteed. Readers may refer to other works on the subject [1,32,38,40,41] for more details on the minimization problems on complex networks. In this chapter, we focus on reviewing algorithms with guaranteed performance for the minimization problems on networks with simple and well-used topologies.

The rest of this chapter is organized as follows. In Section 18.2, we describe the network model and give the preliminaries. The logical topology design problem is addressed in Section 18.3. Algorithms for the RWA problem and traffic grooming problem are introduced in Sections 18.4 and 18.5, respectively. The final section summarizes the chapter.

18.2 NETWORK MODEL

A WDM network consists of a set of nodes connected by a set of point-to-point optical links and can be modeled as a graph $G(V, E)$ with $V(G)$ for the nodes and $E(G)$ for the links in the network, respectively. In practice, an optical link is unidirectional and a WDM network is expressed by a directed graph. However, undirected graphs are often used as an abstract model in theoretical and algorithmic studies for WDM networks. Both directed and undirected graphs will be used in this chapter. We use (u, v) for an edge from node u to node v in a directed graph and $\{u, v\}$ for an edge between u and v in an undirected graph. Readers may refer to works by other authors [38,40,41] for more technical details on optical networks.

We assume that each optical link consists of a single optical fiber. The bandwidth of an optical link is partitioned into a number of channels, each channel is supported by a wavelength and usually has a bandwidth of a peak electronic data rate. The bandwidth of a wavelength channel may be further shared by multiple traffic streams of low data rates. In the study of WDM networks, a wavelength is often called a *color*, and these two terms are used interchangeably in this chapter. A network node provides the optical switching between optical links and the interface (opto-electronic conversion) between optical networks and end users. Major devices in a network node include demultiplexers (DEMUX), optical switches, optical add-drop multiplexers (OADM), multiplexers (MUX), and add-drop multiplexers (ADM). Figure 18.1 gives conceptual structures of network nodes. A DEMUX demultiplexes the wavelength

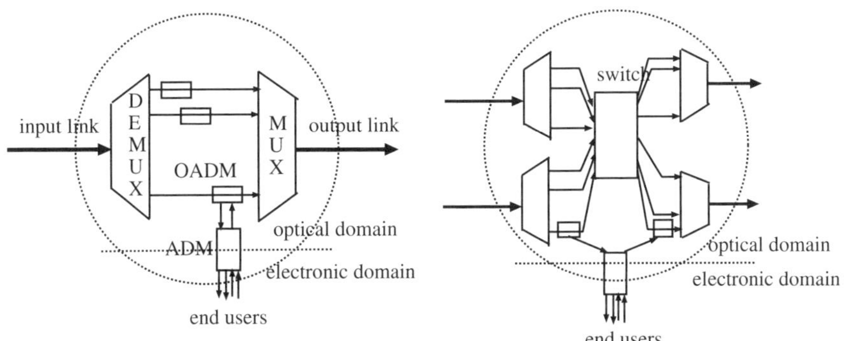

FIGURE 18.1 Conceptual structures of WDM network nodes.

channels from an input optical link. An optical switch connects a wavelength channel from a DEMUX to a channel to a MUX using the circuit switching. When an optical wavelength converter is not available, the two channels connected by the switch must be supported by the same color. An OADM on a wavelength channel may drop optical signals from the channel to an ADM, bypass the optical signals in the channel, or/and add the optical signals from an ADM to the channel. A MUX multiplexes the wavelength channels into an output optical link. An ADM may convert the optical signals from an OADM into electronic ones (drop data from network) and convert the electronic signals from end users into optical ones to an OADM (add data to network). The DEMUX, OADM, optical switch, and MUX work at optical domain. The ADM provides the interface between the optical network and end users. Since optical wavelength converters are expensive and not commonly used, we introduce algorithms for networks without such converters. In this case, a light path of all-optical routing is supported by the same color.

In practice, the bandwidth requirement of a single traffic demand is usually much smaller than the capacity provided by a wavelength channel. So multiple lowrate traffics are multiplexed to share a high-rate wavelength channel. Synchronous Optical Network (SONET) is the current transmission and multiplexing standard for high speed digital transmission on optical fibers in North America. In SONET/WDM networks, the multiplexing is known as *traffic grooming* and the maximum number of traffics that can be multiplexed into a wavelength channel is called *grooming factor*. For example, 16 OC-3 (155.52 Mbps) traffics can be multiplexed into a wavelength channel operated at OC-48 (2488.32 Mbps), giving a grooming factor of 16. In SONET/WDM networks, traffic grooming is carried out by ADMs (known as SONET ADMs or SADMs). With the current technology, SADMs dominate the cost of WDM/SONET networks.

A communication application on a WDM network is specified by a set of traffic demands. Each traffic demand is defined by three parameters: a source node, a destination node or a set of destination nodes, and a required bandwidth. In this chapter, we focus our discussion on the traffic demands with a single destination node in each demand. Such a demand is called a *one-to-one* or *unicast* demand. So a communication application on a network G can be specified by a traffic demand matrix $D = \{(s, t, d_{st})\}$, where $s \in V(G)$ is the source node, $t \in V(G)$ is the destination node, and d_{st} is the bandwidth required by the demand (usually the number of low-rate channels). A *static* or *off-line routing* problem is to connect the source–destination pairs of D after all the traffic demands of D are given. A *dynamic* or *on-line routing* problem is that the traffic demands of D arrive in sequence $(s_1, t_1, d_{s_1t_1})$, ..., $(s_i, t_i, d_{s_it_i})$, ..., and the connection for s_i and t_i is realized without information on the demands arriving after $(s_i, t_i, d_{s_it_i})$. In this chapter, we restrict our discussion on static routing problems.

The readers may refer to a graph theory book such as that by Berge [3] for basic graph definitions and terminology. For undirected graph G, the degree $\delta(u)$ for $u \in V(G)$ is the number of edges incident to u. For directed graph G, $\delta(u)$ is defined as the number of edges originated at u (out degree of u). The maximum node degree of graph G is $\Delta(G) = \max\{\delta(u) | u \in V(G)\}$. We use a *path* for a simple path in G (i.e., repetition of nodes is not allowed). Two paths in G *intersect* if they have a common link. A set of paths in G is *edge-disjoint* if any two paths in the set do not intersect. The distance $d(u, v)$ from u to v is the minimum number of edges in a path from u to v in G. The diameter of graph G is $d(G) = \max\{d(u, v)|u, v \in V(G)\}$. A clique of G is a complete subgraph of G. The number of nodes in the largest clique of G is the *clique number* of G, denoted by $\rho(G)$.

Let G be an undirected (multi)graph. The *vertex coloring* of G is to assign each node of G a color such that any pair of adjacent nodes are given distinct colors. The minimum number of colors for the vertex coloring of G is called the *chromatic number* of G, denoted by $\lambda(G)$. It is known that $\rho(G) \leq \lambda(G) \leq \Delta(G) + 1$. The *edge coloring* of G is to assign each edge of G a color such that any pair of edges incident to the same node are given distinct colors. The minimum number of colors for the edge coloring of G is called the *chromatic index* of G, denoted by $\mu(G)$. It is known that $\Delta(G) \leq \mu(G) \leq \lfloor 3\Delta(G)/2 \rfloor$. It is NP-hard to find $\lambda(G)$ and $\mu(G)$ for arbitrary graphs [23]. An edge coloring of G using at most $\lfloor 3\Delta(G)/2 \rfloor$ colors [39] and a vertex coloring using at most $\Delta(G) + 1$ colors can be found (in polynomial time).

For an NP-hard minimization problem, an algorithm is an α-approximation algorithm if for any instance of the problem, α is an upper bound on the ratio of the solution produced by the algorithm over the optimal solution. We also say the algorithm has the guaranteed performance ratio α.

Popular topologies for WDM networks include rings, trees, and trees of rings (see Fig. 18.2). We define the undirected ring network with n nodes as C_n with $V(C_n) = \{u|0 \leq u \leq n-1\}$ and $E(C_n) = \{\{u, v\}|u = v \pm 1 \mod n\}$. A directed ring $\vec{C}_n$ is defined as the graph obtained by replacing every edge in C_n by a pair of directed edges, one in each direction. Given a pair of nodes u and v in a ring network, we define the segment from node u to node v, denoted as $[u, v]$, to be the subgraph induced by the nodes from u to v in the clockwise direction in the ring. We define the undirected tree network with n nodes as T_n which is a connected undirected graph with $n - 1$ edges. A directed tree $\vec{T}_n$ is defined as the graph obtained by replacing every edge in T_n by a pair of directed edges, one in each direction. An undirected tree of rings, denoted as *TR*, is defined as follows: A single ring is a tree of rings, and the graph obtained by adding a node-disjoint ring to an existing tree of rings and then merging one node of the ring and one node of the tree of rings into one node is also a tree of rings. Similarly, we can define the directed tree of rings $\vec{TR}$.

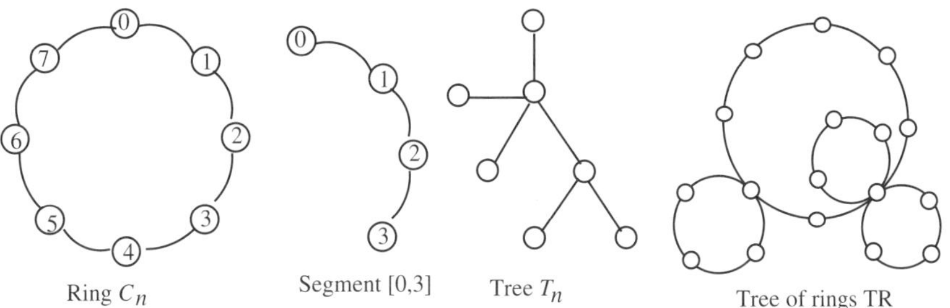

FIGURE 18.2 Ring, segment, tree, and tree of rings.

18.3 LOGICAL TOPOLOGY DESIGN PROBLEM

Given a set D of traffic demands in a WDM network G, it is preferred to realize all demands of D by all-optical routing. However, this may not be possible due to some constraints such as limited wavelength channels. In this case, some traffic demands may have to be realized by multihop routing. If data are routed via k intermediate nodes, the routing is called $(k + 1)$-hop routing. The logical topology design problem is to define a logical topology R over $V(G)$ (or the set of nodes in D) such that every pair of nodes of $V(G)$ (or the set of nodes in D) is connected by R. To realize R on G, each connection request $(u, v) \in R$ is realized by a light path in G (all-optical routing). This is also known as the embedding R to G. For a traffic demand $(s, t, d_{st}) \in D$, if s and t are connected by a logical path of length k in R then the routing from s to t is realized by k-hops of routing. A primary goal in the logical topology design is to provide the connectivity required by the nodes of G (or D) using a minimum number of colors. Since add/drop operations at intermediate nodes are the major bottleneck for data transmission, another key issue in the design is to provide the connectivity using a minimum number of hops. This is equivalent to define a topology R with small diameter. The logical topology design problem is difficult for arbitrary network G and arbitrary set D of traffic demands. The design problem may depend on the RWA problem and the traffic grooming problem. In most cases, the problem is modeled as an integer linear programming problem with specified optimization goals. Readers may refer to other works [16,41] for more details.

18.3.1 Full Connectivity on Rings

An important problem in WDM networks is to support the *full connectivity* (or *all-to-all connection*) of a network G. To do so for every pair of nodes $s, t \in V(G)$, we need to find a light path or a sequence of light paths to connect s and t such that the light paths of the same color are edge-disjoint. The full connectivity on rings have been well studied and the following results are known.

Theorem 1 (Bermond et al. [4], Ellinas and Bala [17]) *The necessary and sufficient number of colors for realizing the full connectivity on $\vec{C}_n$ by all optical routing is $(n^2 - 1)/8$ for n odd, $(n^2 + 4)/8$ for $n/2$ odd, and $n^2/8$ for $n/2$ even.*

Outline of Proof. The algorithm in Figure 18.3 is given in the work by Ellinas and Bala [17] (also see work by Stern and Bala [41]) for realizing the full connectivity on $\vec{C}_n$ for n odd. For $k = 3$, the algorithm uses $(k^2 - 1)/8 = 1$ color to realize the full connectivity on $\vec{C}_k$. Assume that the algorithm uses $(k^2 - 1)/8$ colors for $k \geq 3$. For $k + 2$, there are four sets of new paths: $P_1 = \{u \to i | 0 \leq i \leq (k-1)/2\}$, $P_2 = \{u \to v, u \to i | (k+1)/2 \leq i \leq k-1\}$, $P_3 = \{v \to i | 0 \leq i \leq (k-1)/2\}$, and $P_4 = \{v \to u, v \to i | (k+1)/2 \leq i \leq k-1\}$ (see Fig. 18.4a). Obviously every set has $(k+1)/2$ paths and for any path $p \in P_i$ and $q \in P_j$ with $i \neq j$, p and q are edge-disjoint. So the

```
Procedure Full-Connectivity_on_Ring(C⃗n)
Input: A directed ring network C⃗n.
Output: A set of light paths realizing the full-
  connectivity on C⃗n by all-optical routing.
begin
  k := 3.
  For every pair of nodes u,v ∈ V(C⃗k), connect (u,v) and
    (v,u) by the shortest paths and assign each path the
      same color.
  while (k ≤ n) {
    Insert node u between k − 1 and 0, and
      insert node v between (k − 1)/2 and (k + 1)/2 in C⃗k.
    Connect u to every i ∈ V(C⃗k) ∪ {v} by the shortest path.
    Connect v to every i ∈ V(C⃗k) ∪ {u} by the shortest path.
    Assign the paths above new colors not used for the
      full-connectivity of C⃗k s.t. the paths with the
        same color are edge-disjoint.
  k := k + 2 and relabeling the nodes of C⃗k from 0 to k − 1.
  }
end.
```

FIGURE 18.3 Algorithm for the full connectivity on $\vec{C}_n$ for n odd.

paths of P_1, P_2, P_3, and P_4 can be colored by $(k+1)/2$ colors. The total number of colors required for $\vec{C}_{k+2}$ is $(k^2-1)/8+(k+1)/2=((k+2)^2-1)/8$. This shows that the full connectivity on $\vec{C}_n$ can be realized by $(n^2-1)/8$ colors for n odd. The proof for other values of n are similar and readers may refer to Bermond et al. [4] and Ellinas and Bala [17] for details. ■

The full connectivity on $\vec{C}_n$ by all-optical routing requires about $n^2/8$ colors which could be beyond the number of available colors for even moderate value of n. There is a simple logical topology to realize the full connectivity of $\vec{C}_n$ by $\lceil (n-1)/2 \rceil$ colors in two-hops routing [36]: Select a hub node u in $\vec{C}_n$ and define $R=\{(v,u),(u,v) | v \in V(\vec{C}_n), v \neq u\}$. It is known that the full connectivity on $\vec{C}_n$ requires about $n/3$ colors by two-hops of routing [13].

Theorem 2 (Choplin et al. [13]) *The number of colors for realizing the full connectivity on $\vec{C}_n$ by two-hops of routing is at least $\lfloor (n-1)/3 \rfloor$ and at most $\lfloor (n+1)/3 \rfloor$.*

Outline of Proof. The upper bound of the theorem can be shown by designing a logical topology of diameter 2 as follows (see Fig. 18.4b): Three hub nodes u, v, and w are selected such that the three nodes cut the ring into three segments $[u, v-1]$, $[v, w-1]$, and $[w, u-1]$, with each segment having at most $\lfloor (n+1)/3 \rfloor$

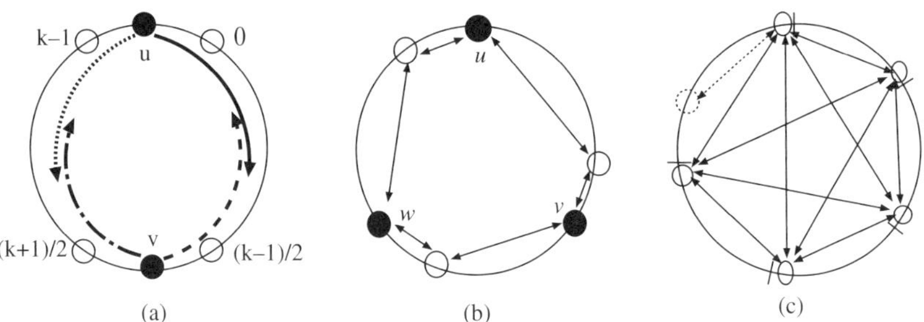

FIGURE 18.4 Full connectivity on ring networks.

nodes, where the arithmetic operations on the nodes are modulo n. For every node x in segment $[u, v-1]$, there are logical links $(x, u), (u, x), (x, v)$, and (v, x). For every node x in segment $[v, w-1]$, there are logical links $(x, v), (v, x), (x, w)$, and (w, x). For every node x in the segment $[w, u-1]$, there are logical links $(x, w), (w, x), (x, u)$, and (u, x). It is easy to see that every pair of nodes of $\vec{C}_n$ is connected by a logical path of length at most two and at most $\lfloor (n+1)/3 \rfloor$ colors are needed to realize this logical topology. Due to the limited space, we omit the proof for the lower bound of the theorem and readers may refer to the work by Choplin et al. [13] for details. ■

It is conjectured in the work by Choplin et al. [13] that $\lfloor (n+1)/3 \rfloor$ is the necessary number of colors as well.

To realize the full connectivity of $\vec{C}_n$ by the logical topology given in the proof of Theorem 2, the working load of the hub nodes u, v, and w may be much higher than that of other nodes. It is desirable that every node in the network has the same working load. To do so, we need to design a logical topology R with the same node degree for every node in R. The following result has been known on the regular logical topology for the full connectivity on $\vec{C}_n$ [27].

Theorem 3 (Gu and Peng [27]) *The full connectivity on $\vec{C}_n$ can be realized by $c_k n^{1+1/k}$ colors, where $c_k \approx 1/2^{2+1/k}$, using a regular logical topology of diameter k.*

Outline of Proof. For $k = 2$, $\vec{C}_n$ is partitioned into n/l segments, each of which has $l = (n/2)^{1/2}$ nodes. For each segment j $(1 \leq j \leq n/l)$, the nodes in the segment are labeled from j_1 to j_l. For every pair of node j_i in segment j and node j'_i in segment j' $(1 \leq j \neq j' \leq n/l, 1 \leq i \leq l)$ there are logical links (j_i, j'_i) and (j'_i, j_i), and for every pair of nodes j_i and $j_{i'}$ $(i \neq i')$ in the same segment j there are logical links $(j_i, j_{i'})$ and $(j_{i'}, j_i)$ (see Fig. 18.4c). Obviously, the logical topology, given above realizes the full connectivity on $\vec{C}_n$ and has diameter 2. It is easy to check that the logical topology can be realized by $c_2 n^{1+1/2}$ colors, where $c_2 \approx 1/2^{2+1/2}$. Readers may refer to the work by Gu and Peng [27] for the full proof. ■

18.4 RWA PROBLEM

The RWA problem is that given a set R of connection requests on a network G, find a routing path in G for every $(u, v) \in R$ and assign each path a color such that the paths with the same color are edge-disjoint. A general goal in this study is to determine the number of colors needed to achieve a given connectivity as a function of network size.

18.4.1 Lower Bounds on the Number of Colors

We first give some lower bounds on the number of colors for realizing a given set R of connection requests on a network G. A lower bound on the number of colors needed for realizing R on G can be derived from the distance $d(u, v)$ for every $(u, v) \in R$ and the number of links in G. This lower bound is called the *aggregate network capacity bound* [41], denoted by

$$W_{\text{Netcap}} \geq \sum_{(u,v)\in R} \frac{d(u, v)}{|E(G)|}.$$

Another lower bound is known as *limiting cut bound* [41]. This lower bound on the number of colors is based the maximum flow and minimum cut theorem. Partition the nodes of $V(G)$ into two subsets X and Y. Let $E_{X,Y}$ be the set of all links $(u, v) \in E(G)$ with $u \in X$ and $v \in Y$ ($E_{X,Y}$ is called a cut set). Let $R_{X,Y} = \{(u, v)|(u, v) \in R, u \in X, v \in Y\}$. Then the number of colors for realizing $R_{X,Y}$ on G is at least $|R_{X,Y}|/|E_{X,Y}|$. Taking the maximum over all cuts in G,

$$W_{\text{limcut}} = \left\lceil \max_{E_{X,Y}} . \frac{|R_{X,Y}|}{|E_{X,Y}|} \right\rceil.$$

Those lower bounds can be used to evaluate the performance of routing algorithms on WDM networks. For the bidirectional ring $\vec{C}_n$ and $R = \{(u, v)|u, v \in V(\vec{C}_n), u \neq v\}$ (full connectivity): $W_{\text{limcut}} \geq (n^2 - 1)/8$ for n odd and $W_{\text{limcut}} \geq n^2/8$ for n even. The lower bounds show that the upper bounds in Theorem 1 are optimal.

18.4.2 Wavelength Assignment and Vertex Coloring

In the RWA problem, when the set of paths is given, we only need to assign the set of paths colors to meet the distinct color assignment constraint. The color assignment is known as the *wavelength assignment* (WA) problem. Given a set $W = \{\lambda_1, \lambda_2, ...\}$ of colors and a set P of paths, a color assignment from W to P is called a *valid coloring* if each path in P is assigned a single color of W and the paths with the same color are edge-disjoint. Finding a valid coloring for P is also called *coloring* P. Given a set P of paths in G, let L be the maximum number of paths of P on any link of G, w_{opt} be the minimum number of colors for coloring P, and w_{up} denote an upper bound on the number of colors for coloring P. Then $L \leq w_{\text{opt}} \leq w_{\text{up}}$. L is also known as *link load*. The WA problem for a set P of paths in a network has a close relation with the vertex coloring problem of the *path conflict graph* $G_P(V, E)$ defined as follows: $V(G_P) = \{p_i | p_i \in P\}$ and $\{p_i, p_j\} \in E(G_P)$ if and only if paths p_i and p_j share a common link of the network. Obviously, a vertex coloring of G_P gives a valid coloring for the set P of paths and $w_{\text{opt}} = \lambda(G_P)$. From this, we have $\rho(G_P) \leq \lambda(G_P) = w_{\text{opt}} \leq \Delta(G_P) + 1$.

A well used strategy for the WA problem is the *first-fit coloring*: Given a set $W = \{\lambda_1, \lambda_2, ...\}$ of colors and a set P of paths, the paths in P are colored one by one in arbitrary order, and a path $p \in P$ is assigned a color λ_i with the smallest index i such that no path of $P \backslash \{p\}$ already colored by λ_i intersects with p. We say a set of elements is assigned distinct colors if any two different elements in the set are assigned different colors. We say a path is on a *link* (resp. a *node*) if the path contains the link (resp. the node). We say a path is on a graph (e.g., a ring) if the path contains a link of the graph. We denote W_P as the set of colors assigned to a set P of paths, and denote W_{uv} as the set of colors assigned to the paths on a link (u, v) (or $\{u, v\}$) of G.

18.4.3 RWA Problem on Rings

The ring topology is popular for optical networks due to its simple structure and symmetric property. The RWA and WA problems on ring networks have been extensively studied. In Section 18.3, the number of colors for supporting the full connectivity on $\vec{C}_n$ is given. We now introduce well-known algorithms for the WA and RWA problems for arbitrary connection requests on ring networks. We first discuss the WA problem.

18.4.3.1 WA Problem Given a set P of directed paths on $\vec{C}_n$, P can be partitioned into two subsets, one is the subset of clockwise paths and the other is the subset of counter-clockwise paths. A clockwise path only uses links of $\vec{C}_n$ in the clockwise direction and does not share a link with any counter-clockwise path. So the WA problem on $\vec{C}_n$ can be solved for each subset independently. The WA problem on $\vec{C}_n$ for each subset can be studied as the WA problem on the undirected ring C_n.

Given a set P of paths on C_n, the conflict graph G_P is a *circular arc graph* [42]. It is known that for circular arc graph G_P the clique number $\rho(G_P)$ can be computed in polynomial time [24] but finding the chromatic number $\lambda(G_P)$ is NP-hard [22]. Algorithms that use at most $2L - 1$ and $\lfloor 3\rho(G_P)/2 \rfloor$ colors are known [31,42]. To describe those algorithms, we first introduce some new notation. A path p on C_n is identified by the segment $[p^{\vdash}, p^{\dashv}]$, where $p^{\vdash}$ and $p^{\dashv}$ are end nodes of p. Given a node u, $a \leq b$ (resp. $a \geq b$) if $a \in [u, b]$ (resp. $a \in [b, u]$).

Theorem 4 (Tucker [42]) *The WA problem for a set P of paths on C_n with link load L can be solved by at most $2L - 1$ colors.*

Outline of Proof. Let $p \in P$ be an arbitrary path. Then there are at most $L - 1$ paths of P other than p that are on link $\{p^{\vdash}, p^{\vdash} + 1\}$ of C_n. We color these $L - 1$ paths by $L - 1$ colors by the first-fit coloring. For the other paths of P, none of them contains node $p^{\vdash}$ as an internal node. So the other paths of P can be viewed as a set of paths on a segment obtained by cutting C_n at node $p^{\vdash}$. The link load of the paths on the segment is at most L. The WA problem on a segment with link load L can be solved by L colors. ■

The algorithm of Tucker [42] is a 2-approximation algorithm. The following example [42] shows that the upper bound of $2L - 1$ given in Theorem 4 is tight. Let n be an odd integer and $P = \{p_i | 0 \leq i \leq n - 1\}$, where p_i is the path with $p_i^{\vdash} = i$ and $p_i^{\dashv} = (n + 2i + 1)/2$ (arithmetic operations are modulo n). It is easy to check that the load of P on C_n is $L = (n + 1)/2$ and $|P| = 2L - 1$. The conflict graph G_P is complete and the WA problem for P requires at least $2L - 1$ colors. It is conjectured in the work by Tucker [42] that the WA problem on C_n can be solved by at most $\lfloor 3\rho(G_P)/2 \rfloor$ colors and this conjecture is proved in the work by Karapetian [31].

Theorem 5 (Karapetian [31]) *The WA problem for a set P of paths on C_n can be solved by at most $\lfloor 3\rho(G_P)/2 \rfloor$ colors.*

We introduce the algorithm of Karapetian but omit the proof details due to the limited space. Readers may refer to the work by Karapetian [31] for details. The key components of the algorithm are the clockwise sweep and counter-clockwise sweep. In each run of the clockwise sweep, a set of paths that can share a same color is found. To do so, a path p is first included into a set A. Then another path that can share a color with the paths in A is searched in the clockwise direction. If there are multiple candidates then the path q with the smallest end node $q^{\vdash}$ is included into A. The process is repeated until no path can be included in A. Similarly, in each run of the counter-clockwise sweep, a set of paths that can share a same color is found. To do so, a path p is first included into a set B. Then another path that can share a color with the paths in B is searched in the counter-clockwise direction. If there are multiple candidates then the path q with the largest end node $q^{\dashv}$ is included into B. The process is repeated until no path can be included in B. The algorithm calls the clockwise sweep at most $\rho(G_P)/2$ times and calls the counter-clockwise sweep at most $\rho(G_P)$ times. The algorithm of Karapetian [31] is given in Figure 18.5.

The algorithm of Karapetian [31] is a 1.5-approximation algorithm. The upper bound $\lfloor 3\rho(G_P)/2 \rfloor$ is tight in the sense that there are instances of the WA problem on C_n that require at least $\lfloor 3\rho(G_P)/2 \rfloor$ colors. An example of such instances can be constructed as follows: Let $n = 5k$ $(k \geq 1)$ and $P = P_0 \cup P_1 \cup P_2 \cup P_3 \cup P_4$, where P_i $(0 \leq i \leq 4)$ is a set of L paths between node $i \times k$ and node $(i + 1) \times k + 1$ (arithmetic operations are modulo n). Then the clique number $\rho(G_P)$ is $2L$ and any valid coloring for P requires at least $3L$ colors.

18.4.3.2 RWA Problem For the RWA problem on the ring network, a well-used approach for paths selection is the *edge avoidance routing* in which every routing path

```
Procedure WA_on_Ring(Cn,P)
Input: A set P of paths in Cn.
Output: A valid coloring from W = {λ1, λ2, ...} to P.
begin
  Compute ρ(GP) and add dummy paths to P to make L = ρ(GP).
  Label paths of P s.t. P = {p1, ..., pL, ...},
    p1, ..., pL contain node u, and pi⊣ ≤ pj⊣ for i < j.
  R = P \ {p1, ..., pL}.
  for i = 1, ..., ⌊L/2⌋ {
    Find pi ∈ R s.t. pi⊢ is minimum.
    Ai =Clockwise-sweep (R,pi) is colored by color λL+i.
    R = R \ Ai.
  }
  R = R ∪ {p1, ..., pL}.
  for i = L, ..., 1 {
    Bi =Counter-clockwise-sweep(R, pi) is colored by color
    λi R = R \ Bi.
  }
end.
Subroutine Clockwise-sweep (R,p)
Input: A set R of paths and a path p in Cn.
Output: A set of paths which can be colored by one color.
begin
  A = {p} and Q = {q ∈ P, q ∩ p = Ø}.
  while (Q ≠ Ø) {
    Find a q̂ ∈ Q s.t. q̂⊢ is minimum;
    A = A ∪ {q̂};
    Q = Q \ {q ∈ Q, q ∩ q̂ ≠ Ø}
  }
end.
Subroutine Counter-clockwise-sweep (R,p)
Input: A set R of paths and a path p in Cn.
Output: A set of paths which can be colored by one color.
begin
  B = {p} and Q = {q ∈ P, q ∩ p = Ø};
  while (Q ≠ Ø) {
    Find a q̂ ∈ Q s.t. q̂⊣ is maximum;
    B = B ∪ {q̂};
    Q = Q \ {q ∈ Q, q ∩ q̂ ≠ Ø};
  }
end.
```

FIGURE 18.5 $\lfloor 3\rho(G_P)/2 \rfloor$ algorithm for the WA problem on C_n.

is selected in such a way that a prespecified ring edge is avoided [10]. The approach can be further generalized as the weight-based routing in which each ring edge is assigned a weight and the routing paths are selected subject to the constraint on the weights of the ring edges in the path [10]. For example, if we assign the prespecified edge weight 1 and all the other edges weight 0, and the routing paths are selected such that the weight of the ring edges in the path is 0, then it is the edge avoidance routing. After the routing paths are selected, the WA problem for the selected paths is solved by the algorithms described above. It is NP-hard to find w_{opt} for the RWA problem on ring networks [19]. The RWA problem can be solved by at most $2w_{\text{opt}}$ colors for both C_n [37] and $\vec{C}_n$ [29]. For the RWA problem on C_n, both randomized and deterministic algorithms with $w_{\text{up}} < 2w_{\text{opt}}$ have been developed [12,34].

18.4.4 RWA Problem on Trees

For any pair (u, v) of nodes in a tree network, there is a unique path from u to v in the tree. So the RWA problem in a tree network becomes the WA problem. It is NP-hard to find the w_{opt} for the WA problem on both undirected tree T_n [37] and directed tree $\vec{T}_n$ [20]. Given a set P of paths on T_n with load L, an algorithm solves the WA problem using at most $\lfloor 3L/2 \rfloor$ colors is known [37]. The idea of the algorithm is to reduce the WA problem into the edge-coloring problem of a multigraph. For an internal node u of T_n and the set of paths on u, a multigraph G_u can be constructed as follows: For every edge e_i incident to node u there is a corresponding vertex e_i in G_u. Since each path on u can be on at most two edges incident to u, for every path p on u a unique edge in G_u can be defined. To eliminate the self loops, an additional vertex f_i is introduced for every e_i. More precisely, $V(G_u) = \{e_i, f_i | e_i \text{ is an edge incident to } u\}$ and

$$\begin{aligned} E(G_u) = \quad & \{(e_i, e_j, p) | \text{path } p \text{ is on edges } e_i \text{ and } e_j\} \\ & \cup \{(e_i, f_i, p) | \text{path } p \text{ is on edge } e_i \text{ only}\}, \end{aligned}$$

where (x, y, p) is an edge between x and y with label p. Obviously, an edge coloring of G_u gives a valid coloring for the paths containing node u. Notice that $\Delta(G_u) \leq L$. To solve the WA problem on T_n, T_n is viewed as a rooted tree and the internal nodes of T_n can be processed in a breadth first search (BFS) order, starting from the root. In processing a node u, the paths on u are colored by the edge coloring of G_u. Since the edge coloring of G_u can be solved by at most $\lfloor 3\Delta(G_u)/2 \rfloor$ colors [39], $\Delta(G_u) \leq L$ for P with load L, the WA problem on T_n can be solved by $\lfloor 3L/2 \rfloor$ colors.

Theorem 6 (Raghavan and Upfal [37]) *The WA problem on T_n can be solved using at most $\lfloor 3L/2 \rfloor$ colors.*

The algorithm in the work of Ragavan and Upfal [37] is a 1.5-approximation algorithm for the WA problem on T_n. The upper bound $\lfloor 3L/2 \rfloor$ is tight in the sense that there are instances requiring at least $\lfloor 3L/2 \rfloor$ colors [37]. Here is an example

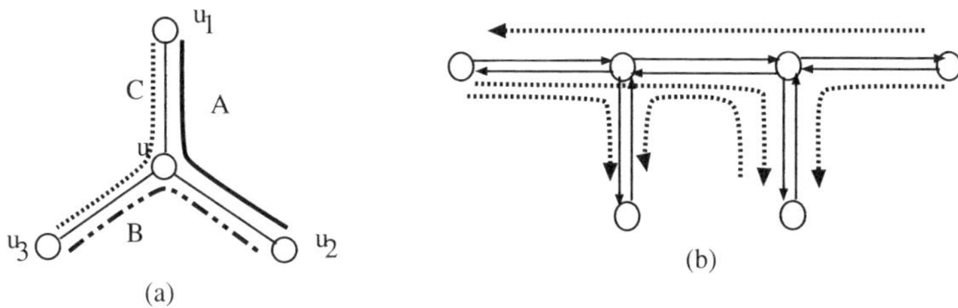

FIGURE 18.6 Instances for lower bounds for WA problem on trees.

of such instances. Let T be the tree with three edges $\{u, u_1\}$, $\{u, u_2\}$, $\{u, u_3\}$ and $P = A \cup B \cup C$, where A is a set of $L/2$ (L is even) paths on edges $\{u, u_1\}$ and $\{u, u_2\}$, B is a set of $L/2$ paths on edges $\{u, u_2\}$ and $\{u, u_3\}$, and C is a set of $L/2$ paths on edges $\{u, u_3\}$ and $\{u, u_1\}$ (see Fig. 18.6a). Then the load of P on T is L and the conflict graph of P is a complete graph of $3L/2$ nodes. At least $3L/2$ colors are needed to color P.

For the WA problem on directed trees $\vec{T}_n$, there are a number of algorithms that follow a general approach described in Figure 18.7 [8,9,29,30,33]. In the coloring procedure, a path is called *colored* if it has been assigned a color, otherwise *uncolored*. Processing a node u means coloring the uncolored paths on u. The nodes of $\vec{T}_n$ is processed in the BFS order. A number of techniques have been developed for processing a node in the above approach, giving a number of algorithms. One technique is to reduce the coloring of paths containing node u to an edge-coloring problem on a bipartite graph $G_u(U, V, E)$ [8,29,30,33]. $\vec{T}_n$ is viewed as a rooted tree and for a node u in $\vec{T}_n$, assume that v_0 is the parent and $v_1, ..., v_k$ are children of u. The graph G_u is constructed as follows: For each node v_i, there are four vertices a_i, b_i, c_i, d_i and $U = \{a_i, d_i | 0 \le i \le k\}$ and $V = \{b_i, c_i | 0 \le i \le k\}$. For a path on links (v_i, u) and (u, v_j), there is an edge $\{a_i, b_j\} \in E(G_u)$. For a path on link (v_i, u) and u is the end node of the path, there is an edge $\{a_i, c_i\} \in E(G_u)$. For each path on link (u, v_i) and u is the start node of the path, there is an edge $\{d_i, b_i\} \in E(G_u)$. It is shown that an edge coloring of $G_u(U, V, E)$ gives a valid coloring of paths on u [8,29,30,33].

Procedure `WA_Tree(`$\vec{T}_n$`, P)`
Input: `A set P of paths in` $\vec{T}_n$.
Output: `A valid coloring from` $W = \{\lambda_1, \lambda_2, ...\}$ `to P.`
begin
1. Fix a BFS (Breadth-first search) order, starting from a node (say u_0), on the nodes of $\vec{T}_n$.
2. Process the starting node u_0.
3. Process the other nodes u in the BFS order.

end.

FIGURE 18.7 A framework of algorithms for the WA problem on directed trees.

Theorem 7 (Kaklamanis et al. [30]) *The WA problem on $\vec{T}_n$ can be solved using at most $5L/3$ colors.*

The algorithm in work of Kaklamanis et al. [30] is a (5/3)-approximation algorithm for the WA problem on $\vec{T}_n$. For the WA problem on $\vec{T}_n$, there are instances requiring at least $5L/4$ colors [33]. An example of such instances is shown in Figure 18.6 b. In the figure, each set has $L/2$ paths. The load on T is L and there are $5L/2$ paths. It is easy to check that at most two paths can be given the same color. From this, at least $(5L/2)/2 = 5L/4$ colors are needed.

18.4.5 RWA on Tree of Rings

A tree of rings is another important topology for WDM networks. We first discuss the WA problem on trees of rings. Similar to the WA problem on rings, the WA problem on directed trees of rings can be studied as the WA problem on undirected trees of rings. In a tree of rings *TR*, any two rings have at most one node in common, and for any pair of nodes u and v in *TR* there are exactly two edge-disjoint paths between u and v. *TR* remains connected even if an arbitrary link fails in each ring, and thus provides a better fault tolerance than a tree network. Many research efforts have been devoted to the study of the WA problem on *TR* [6,14,18]. An important property for the paths on *TR* is that for any node $u \in V(TR)$, a path on u can be on at most two rings that contain u. For a node u in a ring of *TR*, we denote u^- as the neighbor of u in the counter-clockwise direction and u^+ as the neighbor of u in the clockwise direction in the ring (see Fig. 18.8a). Given a set P of paths on *TR* of arbitrary node degree with link load L, it is known that the WA problem can be solved by at most $3L$ colors [6]. The upper bound is tight in the sense that there are instances of the problem that require at least $3L$ colors. For the WA problem on *TR* of degree at most 6 (each node can appear in at most three rings), an algorithm that uses at most $2w_{\text{opt}}$ colors is known [6]. Both algorithms follow a same framework as shown in Figure 18.9.

At any stage of the coloring procedure, a path is called *colored* if it has been assigned a color, otherwise *uncolored*. Processing a node u means coloring the

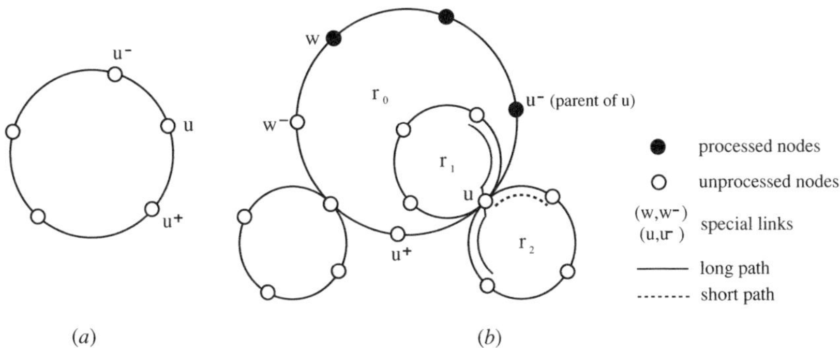

FIGURE 18.8 Illustration of some terms defined on a tree of rings *TR*.

Procedure `Framework(TR,P)`
Input: `A set P of paths in TR.`
Output: `A valid coloring from` $W = \{\lambda_1, \lambda_2, ...\}$ `to P.`
begin

```
1. Fix a DFS (depth-first search) order, starting from
   a node (say u0) of degree two, on the nodes of TR.
2. Process the starting node u0.
3. Process the other nodes u in the DFS order.
   Let r0 be the ring which contains u and the parent of u.
   3.1 Color the set P0 of uncolored paths on u and r0.
   3.2 Color the set P1 of other uncolored paths on u.
```

end.

FIGURE 18.9 A framework of algorithms for the WA problem on trees of rings.

uncolored paths on u. We call a node u *processed* if the coloring process for u has been completed, otherwise *unprocessed*. The nodes of *TR* is processed in the depth-first search (DFS) order introduced in the work by Erlebach [18]. For a node u, its *parent* is the node from which u is reached in the DFS order (see Fig. 18.8b). A link is called *special* if it connects a processed node and an unprocessed node (see Fig. 18.8b). There are either 0 or 2 special links in a ring in *TR*. A path on a special link is colored and only such a path has a possibility to intersect with an uncolored path. We assume that in Step 1, the nodes in the same ring are searched in the clockwise direction in the DFS order.

18.4.5.1 WA Problem on TR of Arbitrary Degree Algorithm A1 for the WA problem on *TR* of arbitrary degree follows the framework of Figure 18.9. In Step 2, the paths on links $\{u_0, u_0^-\}$ and $\{u_0, u_0^+\}$ are assigned distinct colors of W. In Step 3, the parent of node u in the DFS order is node u^- in some ring that is called r_0. If u appears in $k+1$ rings, the other k rings are denoted by r_i, $1 \leq i \leq k$ (see Fig. 18.8b). Let Q_0 be the set of paths on special links $\{u, u^-\}$ or $\{w, w^-\}$. In Step 3.1, P_0 is colored using the colors of $W \setminus W_{Q_0}$ by the first-fit coloring. It is easy to see that the paths of $Q_0 \cup P_0$ are given distinct colors in Step 3.1. This is critical for Step 3.2.

In Step 3.2, the path-coloring problem is converted to the edge-coloring problem of a multigraph G_u with rings r_i $(0 \leq i \leq k)$ as vertices and all paths on u as edges. Notice that a path on u is on either one ring or two rings of r_i. A path on u is called a long path if it is on two rings, otherwise a short path (see Fig. 18.8b). To eliminate self-loops, we introduce a vertex s_i for every r_i in G_u. More specifically, G_u is defined as: $V(G_u) = \{r_i, s_i | 0 \leq i \leq k\}$, and

$$E(G_u) = \{(r_i, r_j, p) | p \text{ is a long path on } r_i \text{ and } r_j, 0 \leq i < j \leq k\}$$
$$\cup \{(r_i, s_i, p) | p \text{ is a short path on } u \text{ and } r_i, 0 \leq i \leq k\},$$

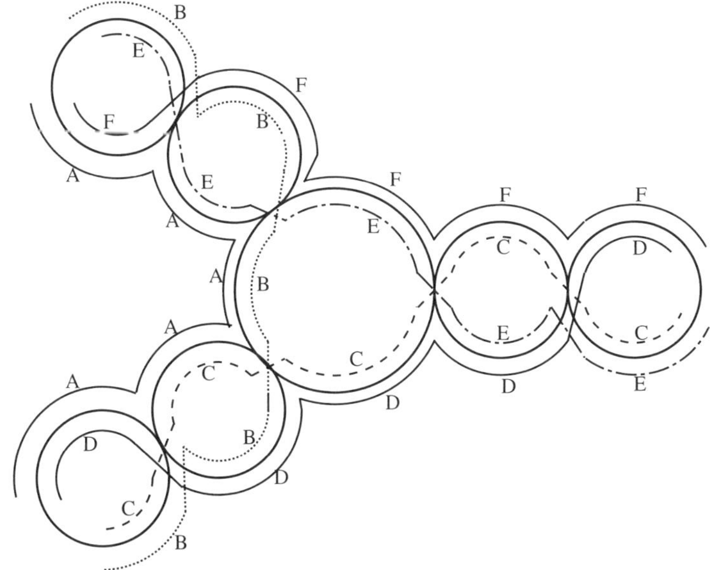

FIGURE 18.10 An instance for the $3L$ lower bound.

where (x, y, p) denotes an undirected edge between vertices x and y with label p. There is a one-to-one correspondence between the paths on u and the edges in G_u. We color the paths of P_1 by solving the edge coloring of G_u. To apply the edge coloring of G_u in Step 3.2 as shown above, it is required that $Q_0 \cup P_0$ is assigned distinct colors.

Theorem 8 (Bian et al. [6]) *Algorithm A1 solves the WA problem on TR by at most* $3L$ *colors.*

Algorithm A1 is a 3-approximation algorithm for the WA problem on *TR* of arbitrary degree. The $3L$ upper bound is tight. Below is an example that requires at least $3L$ colors [6]. Let $P = A \cup B \cup C \cup D \cup E \cup F$ be the set of paths, with each subset having $L/2$ (L is even) paths, as shown in Figure 18.10. The maximum number of paths on any link in the tree of rings is L. The conflict graph G_P is a complete graph of $3L$ nodes and thus any coloring of P requires at least $3L$ colors.

18.4.5.2 WA Problem on TR of Degree 6 Algorithm A2 for the WA problem on *TR* of degree at most 6 follows the framework of Figure 18.9 too. In Algorithm A1, to apply the edge coloring of G_u in Step 3.2, it is required that $Q_0 \cup P_0$ has been assigned distinct colors. This requirement may be too strict for solving the WA problem on *TR* since two paths in $Q_0 \cup P_0$ can have the same color if they are edge disjoint. In Algorithm A2, instead of using edge-coloring approach for Step 3.2, a different path-coloring scheme that is designed specifically for *TR* of degree 6 is used. Recall that P_0 and P_1 are the sets of paths to be colored in Step 3.1 and Step

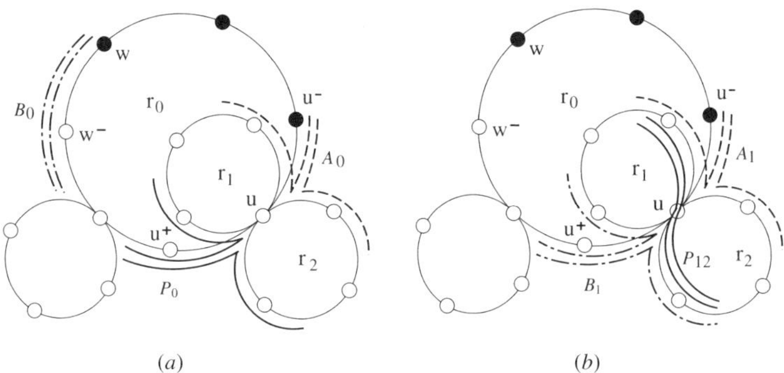

FIGURE 18.11 The sets of paths related to Schemes S31 and S32.

3.2 of the framework in Figure 18.9, respectively. We first introduce a scheme for coloring P_0 and a scheme for coloring a subset of P_1. The scheme for P_0, called S31, works as follows. Let A and B be the sets of paths on special links $\{u, u^-\}$ and $\{w, w^-\}$, respectively. Then $Q_0 = A \cup B$. Define $A_0 \subseteq A$ (resp. $B_0 \subseteq B$) to be the set of paths on link $\{u, u^-\}$ (resp. on $\{w, w^-\}$), each of which has a color in $W_A \setminus W_B$ (see Fig. 18.11a). We construct a graph G_0 with

$$V(G_0) = P_0 \cup A_0 \text{ and } E(G_0) = \{\{p, q\} \mid p \text{ and } q \text{ are edge disjoint}\}.$$

We find a maximum matching M_0 of G_0. Notice that G_0 is bipartite and for each pair $\{p, q\} \in M_0$, $p \in P_0$, and $q \in A_0$. For each pair $\{p, q\} \in M_0$, assign the color of $q \in A_0$ to p.

The second scheme, called S32, is used to color the subset P_{12} of P_1 that contains the long paths on rings r_1 and r_2 (see Fig. 18.11b). Let A and B be the sets of long paths on links $\{u, u^-\}$ and $\{u, u^+\}$, respectively. Then $Q_1 = A \cup B$. Define $A_0 \subseteq A$ (resp. $B_0 \subseteq B$) to be the set of paths on link $\{u, u^-\}$ (resp. on $\{u, u^+\}$), each of which has a color in $W_A \setminus W_B$ (see Fig. 18.11b). We construct a graph G_0 with

$$V(G_1) = P_{12} \cup A_1 \text{ and } E(G_1) = \{\{p, q\} \mid p \text{ and } q \text{ are edge disjoint}\}.$$

We find a maximum matching M_1 of G_1. For each pair $\{p, q\} \in M_1$, either $p \in P_{12}$ and $q \in A_1$ or $p, q \in P_{12}$. For each pair $\{p, q\}$ with $q \in A_1$, assign the color of q to p. For each pair $\{p, q\}$ with $p, q \in P_{12}$, assign the pair a same color.

Algorithm A2 follows the framework in Figure 18.9. Step 2 of A2 is the same as that in Algorithm A1. Step 3.1 uses Scheme S31. In Step 3.2, we first use Scheme S32 to color the long paths in P_{12}. Then we color the short paths on r_1 and those on r_2. Let Q' be the set of all long paths on u and r_1. We assign the short paths on r_1 the colors of $W \setminus W_{Q'}$ by the first-fit coloring such that the set of short paths is assigned distinct colors. Let Q'' be the set of all long paths on u and r_2. We assign the short paths on r_2 the colors of $W \setminus W_{Q''}$ by the first-fit coloring such that the set of short paths is assigned distinct colors.

Theorem 9 (Bian et al. [6]) *Algorithm A2 solves the WA problem on TR with n nodes and degree at most 6 using at most* $2w_{\mathrm{opt}}$ *colors.*

Algorithm A2 is a 2-approximation algorithm for the WA problem on *TR* of degree at most 6.

18.4.5.3 RWA Problem The RWA problem on trees of rings is NP-hard [21]. The edge avoidance routing approach has been used to solve this problem [29,37]. In this approach, one link (or a pair of links) is removed from each ring to get a tree network and the solution of the WA problem on the tree is used as the solution of the RWA problem on the tree of rings. By this approach and the upper bound of $3L/2$ colors for the WA problem on T_n, a 3-approximation algorithm is known for the RWA problem on *TR* [37]. For the RWA problem on $\vec{TR}$, by the upper bound of $5L/3$ colors on $\vec{T}_n$ [30] and the edge avoidance routing, a (10/3)-approximation algorithm can be obtained.

Algorithm A1 for the WA problem on *TR* can be used to obtain a 3-approximation algorithm for the RWA problem on *TR* as follows. First, for a given set of connection requests, a path for each request can be found efficiently such that L is minimized [18]. Then, the set of found paths is colored by Algorithm A1 using at most $3L$ colors. Since the load L is optimal, it is also a lower bound on the number of colors for the original RWA problem. In this way, the 3-approximation ratio is achieved without using the edge avoidance routing approach. Algorithm A1 can also be used for the RWA problem on $\vec{TR}$ but only guarantees an approximation ratio of 6.

18.5 TRAFFIC GROOMING PROBLEM

Traffic grooming is to multiplex/demultiplex low-rate traffic demands by SADMs to share a wavelength channel in SONET/WDM networks. A general goal in the traffic grooming problem is to realize a connectivity as a function of the network size and the functionality of network nodes. Major optimization goals are to minimize the number of SADMs and the number of colors (wavelength channels). It is known that the two optimization goals cannot be achieved simultaneously for many cases [5,25,35]. It has received much attention to minimize SADMs subject to using the minimum number of colors [5,28,35,45].

18.5.1 Traffic Grooming for Unidirectional Path-Switched Ring

A main network architecture for SONET/WDM networks is the Unidirectional Path-Switched Ring (UPSR) in which there are two optical fibers between each pair of adjacent nodes. These fibers constitute two unidirectional rings with one in the clockwise direction and the other in the counter-clockwise direction, where one ring (e.g., the clockwise ring) is used as the working ring and the other as the protecting ring.

A network traffic demand from node u to node v is routed on the unique path from u to v in the working ring.

A set of traffic demands is unitary if each demand requires one unit of bandwidth. A unitary demand from node u to node v is denoted by pair (u, v). We assume that every traffic demand is realized by one-hop of optical routing and use R to denote the set of traffic demands. A set R of traffic demands is *symmetric* if $(u, v) \in R$ implies $(u, v) \in R$. Symmetric traffic demands are very common in many applications, for example, TCP connections and IP telephony. We use $\{u, v\}$ to denote the unitary symmetric pair (u, v) and (v, u), and we say nodes u and v are *involved* in $\{u, v\}$. Given a set R of unitary symmetric traffic demands, the traffic grooming problem can be solved by partitioning R into subsets, each of which has at most k demand pairs, and multiplexing each subset into one wavelength channel. For each node involved in at least one symmetric pair of a subset carried by a color λ, we need one SADM for λ at the node, and minimizing the total number of used SADMs is equivalent to minimizing the sum of the number of distinct nodes involved in each subset. The traffic grooming problem for unitary symmetric traffic demands has been widely discussed [35,45]. For algorithms with guaranteed performance, a graph partition approach has been used [5,7,26]. In this approach, a simple undirected graph $G(V, E)$, called *traffic graph*, is constructed based on the set R, where node set $V(G)$ denotes the set of nodes in the UPSR and there is an edge $\{u, v\} \in E(G)$ between nodes u and v if and only if there is a unitary symmetric pair $\{u, v\} \in R$. The traffic grooming problem is then formulated as the following k-edge-partitioning problem on G: Given a positive integer k, partition the edge set $E(G)$ into a collection of subsets $\mathcal{E} = \{E_1, E_2, \ldots, E_{w_{\text{up}}}\}$ with $\bigcup_{i=1}^{w_{\text{up}}} E_i = E(G)$ and $E_i \cap E_j = \emptyset$ for $i \neq j$, such that $|E_i| \leq k$ for each $E_i \in \mathcal{E}$ and $\sum_{E_i \in \mathcal{E}} |V_i|$ is minimized, where V_i is the set of nodes in the subgraph induced by edge set $E_i \in \mathcal{E}$. It is observed that integer k corresponds to the grooming factor, w_{up} corresponds to the number of used colors and $\sum_{E_i \in \mathcal{E}} |V_i|$ corresponds to the total number of used SADMs.

A trivial lower bound on the number of used colors is $\lceil |E(G)|/k \rceil$ (i.e., $w_{\text{up}} \geq \lceil |E(G)|/k \rceil$). A lower bound on the number of used SADMs has been shown [5,26]. $\sum_{E_i \in \mathcal{E}} |V_i| \geq |E(G)|/g_{\max}(k)$, where $g_{\max}(k) = \max\{|E_i|/|V_i| \,|\, |E_i| \leq k\}$. The value of $|E_i|/|V_i|$ reaches the maximum when E_i forms a complete graph. For a complete graph of k edges, there are $(\sqrt{8k+1}+1)/2$ nodes and $|E_i|/|V_i| = 2k/(\sqrt{8k+1}+1)$.

It is NP-hard to find the minimum number of SADMs for arbitrary graph G [26]. The minimum number of SADMs and the minimum number of colors cannot be obtained simultaneously even for complete graph [5]. A number of heuristics have been known for partitioning $E(G)$ into subgraphs to minimize the number of SADMs. Those approaches including spanning tree based partitioning, Euler path based partitioning, skeleton based partitioning, and design theory based partitioning [5,7,26,43,44].

The spanning tree based partitioning algorithm [26] works as follows: First, a spanning tree T of the traffic graph G is found. Next, for every edge $\{u, v\}$ not included in T, a new node u_v and an edge $\{u, u_v\}$ are created to form a new tree T_G

Procedure `GraphPartition_SpanningTree(G,k)`
Input: An undirected graph G and grooming factor k.
Output: A partition $E_1, ..., E_{w_{up}}$ of E(G) s.t. $|E_i| \leq k$.
begin
  Find a spanning tree T of G.
  For each $\{u, v\} \in E(G) \setminus E(T)$, add node u_v and edge $\{u, u_v\}$ to
    T to get a tree T_G containing all edges of G.
  Partition T_G into subtrees $T_1, ..., T_{w_{up}}$ with $\lceil k/2 \rceil \leq |E(T_i)| \leq k$.
end.

FIGURE 18.12 Spanning tree based graph partitioning algorithm.

that contains all edges of G (viewing edge $\{u, u_v\}$ as edge $\{u, v\}$). Finally, tree T_G is partitioned into subtrees, each of which has at most k edges. The algorithm is shown in Figure 18.12. As shown in the work by Goldschmidt et al. [26], the number of edges in each subtree obtained from the partition is between $\lceil k/2 \rceil$ and k. This implies that $E(G)$ is partitioned into at most $\lceil 2|E(G)|/k \rceil$ subsets. Since each subtree is a connected graph, the subtree has at most $(k + 1)$ nodes. Thus, we have the following result.

Theorem 10 (Goldschmidt et al. [26]) *The traffic grooming problem on an arbitrary traffic graph G can be solved using at most $\lceil 2|E(G)|/k \rceil$ colors and at most $\lceil (1 + 2/k)|E(G)| \rceil$ SADMs.*

The number of colors used in the algorithm can be as twice as the minimum in the worst case (each subtree has $k/2$ edges).

The Euler path based partitioning algorithm [7] is given in Figure 18.13. In the algorithm, dummy edges are added to the traffic graph G to make every node of G having even degree and then an Euler path of G is found. The Eular path is partitioned into segments, each of which has exactly k real edges of G. This implies that $E(G)$ is partitioned into $\lceil |E(G)|/k \rceil$ subsets. If the subgraph reduced from the k edges of each subset is connected then there are at most $k + 1$ nodes in each subgraph. However, a subgraph may not be connected due to the removal of dummy edges. Removing one dummy edge increases the number of SADMs by 1 and there are $n_{odd}/2$ dummy edges, where n_{odd} is the number of odd degree nodes. Thus, we have the following result.

Theorem 11 (Brauner et al. [7]) *The traffic grooming problem on an arbitrary traffic graph G can be solved using at most $\lceil |E(G)|/k \rceil$ colors and at most $\lceil (1 + (1/k))|R| \rceil + n_{odd}/2$ SADMs, where n_{odd} is the number of odd-degree nodes in G.*

This algorithm uses the minimum number of colors (i.e., $w_{up} = w_{opt}$).

Intuitively, to achieve good solutions for the k-edge-partitioning problem, we need partition traffic graph G into subgraphs of at most k edges such that each subgraph contains as few nodes as possible. One key observation is that given a fixed number of

```
Procedure GraphPartition_EulerPath(G,k)
Input: An undirected graph G and grooming factor k.
Output: A partition E1,...,E_Wup of E(G) s.t. |Ei| ≤ k.
begin
  Adding dummy edges into G to make each node of G
    having even degree.
  Finding an Euler path of G.
  Partition the Euler path into subgraphs,
    each of which contains exactly k real edges of G.
end.
```

FIGURE 18.13 Euler path based graph partitioning algorithm.

edges of G, a subgraph induced by the edges more likely contains fewer nodes if there are fewer connected components in the subgraph. This is the basic idea behind the algorithms given in other studies [7,26]. The algorithm in the work by Goldschmidt et al. [26] guarantees that each subgraph is connected, while every subgraph might contain only $\lceil k/2 \rceil$ edges in the worst case. The algorithm in the work by Brauner et al. [7] does not guarantee that each subgraph is connected, instead it guarantees that the total number of connected components over all subgraphs is bounded above and each subgraph contains exactly k edges. Following a similar idea, an approach that partitions G into a special subgraphs called *skeletons* is proposed in the work by Wang and Gu [44].

A skeleton S of G is a connected subgraph of G that consists of a *backbone* and a set of *branches*, where the backbone is a path of G, and each branch is an edge of G such that the edge is incident to at least one node in the backbone. A *skeleton cover* $\mathcal{S}$ of graph G is a set of skeletons $\{S_1, \ldots, S_s\}$ that form an edge partition of G (i.e., $\bigcup_{i=1}^{s} E(S_i) = E(G)$ and $E(S_i) \cap E(S_j) = \emptyset$ for $i \neq j$). It is known that for any skeleton S and integer t with $0 < t < |E(S)|$, S can be partitioned into two skeletons S_1 and S_2 such that $|E(S_1)| = t$ and $|E(S_2)| = |E(S)| - t$. From this property, it is easy to transform a skeleton cover to a k-edge partition of G with each subgraph containing exactly k edges: we add $s - 1$ dummy edges to connect the s skeletons into one virtual skeleton and then partition the virtual skeleton into subgraphs, each of which contains exactly k real edges.

Based on the above approach, a skeleton based partitioning algorithm was proposed [44]. The algorithm is given in Figure 18.14.

Theorem 12 (Wang and Gu [44]) *The traffic grooming problem on an arbitrary traffic graph G of n nodes can be solved using at most $\lceil |E(G)|/k \rceil$ colors and at most $\lceil (1 + 1/k)|R| \rceil + (n/4)$ SADMs.*

The algorithm uses the minimum number of colors.

A special case of the traffic grooming problem is the all-to-all traffic pattern, in which there is a traffic demand pair $\{u, v\}$ for every two nodes u and v in the UPSR. For the all-to-all traffic pattern, the traffic graph is complete. Using the results

```
Procedure GraphPartition_TreeSkeleton(G, k)
Input: An undirected graph G and grooming factor k.
Output: A partition E_1, ..., E_{w_up} of E(G) s.t. |E_i| ≤ k.
begin
  Find a spanning tree T of G.
  Find a skeleton cover S with edges of E(T) as
    backbones and edges of E(G) \ E(T) as branches for
      S_i ∈ S.
  Add |S| − 1 dummy edges to connect the skeletons of
    S into one skeleton S.
  Partition S into subgraphs, each has k real edges.
end.
```

FIGURE 18.14 Skeleton based graph partitioning algorithm.

of design theory [11], the k-edge partitioning problem on complete graphs can be solved optimally if grooming factor k is a practical value or in the infinite congruence classes of values [5]. It was shown that for complete graph G, the minimum number of SADMs cannot be obtained using the minimum number of colors for some values of k and n [5]. For example, the minimum number of SADMs for $k = 6$ and $n = 13$ is 52 which is obtained with $w_{\mathrm{up}} = 13$. Any partition of the complete graph of 13 nodes into $w_{\mathrm{opt}} = 12$ subgraphs requires at least 54 SADMs. An open problem here is whether the minimum number of SADMs can be obtained using the minimum number of colors when $n(n-1)/2k$ is an integer for complete graph.

18.5.2 Traffic Grooming on Other Networks

The discussion on UPSR is based on the assumption that every traffic demand is realized by one hop of optical routing. If we relax this constraint and allow multihops of optical routing to minimize the number of SADMs, then finding the minimum number of SADMs in the traffic grooming problem becomes more difficult. It is shown that the problem is NP-hard even in the network topologies of path, star, and trees [15]. Ad hoc heuristics and integer linear programming have been main approaches for the traffic grooming problem on arbitrary networks but the performance of existing algorithms are not guaranteed.

18.6 SUMMARY

Routing is a critical issue for WDM networks. The routing problem on WDM networks is challenging due to the complex hierarchical structure for multiplexing communication channels. Algorithms with guaranteed performance are known only for simple and regular networks. This chapter introduced a number of such algorithms for rings, trees, and trees of rings. There are many open problems in the routing on the WDM networks. It is especially interesting to develop efficient algorithms with guaranteed performance for the RWA problem and traffic grooming problem on networks with more complex topologies than those discussed in this chapter. Such topologies may include those used in the backbone of the Internet and metropolitan area networks. The routing problem can be studied from a different point of view as well: to maximize the connectivity subject to the given resources in the networks.

REFERENCES

1. Aggarwal A, Bar-Noy A, Coppersmith D, Ramaswami R, Schieber B, Sudan M. Efficient routing and scheduling algorithms for optical networks. Proceedings of the ACM–SIAM Symposium on Discrete Algorithms (SODA93); 1993. p 412–423.
2. Beauquier B, Bermond JC, Gargano L, Hell P, Perennes S, Vaccaro U. Graph problems arising from wavelength-routing in all-optical networks. Proceedings of the 2nd Workshop on Optics and Computer Science (WOCS'97); 1997.
3. Berge C. Graphs. North-Holland; 1985.
4. Bermond JC, Gargano L, Perennes S, Rescigno A, Vaccaro U. Efficient collective communication in optical networks. Theor Comput Sci 2000;233:165–189.
5. Bermond JC, Coudert D. Traffic grooming in unidirectional WDM ring networks using design theory. Proceedings of the IEEE International Conference on Communications (ICC2003); 2003. p 11–15.
6. Bian Z, Gu Q, Zhou X. Tight bounds for wavelength assignment on trees of rings. Proceedings of the 19th International Parallel and Distributed Processing Symposium (IPDPS05). CD-ROM; 2005.
7. Brauner N, Crama Y, Finke G, Lemaire P, Wynants C. Approximation algorithms for SDH/SONET networks. RAIRO Oper Res 2003;37:235–247.
8. Caragiannis I, Kaklamanis C, Persiano P. Bounds on optical bandwidth allocation in directed tree topologies. Proceedings of the 2nd Workshop on Optics and Computer Science; 1997.
9. Caragiannis I, Kaklamanis C, Persiano P. Wavelength routing in all-optical networks: A survey. B Eur Assoc Theor Comput Sci 2002;76:104.
10. Carpenter T, Cosares S, Saniee I. Demand routing and slotting on ring networks. Technical Report No. TR-97-02. Bellcore; 1997.
11. Colbourn C, Dinitz J, editors. The CRC Handbooks of Combinatorial Design. Boca Raton: CRC Press; 1996.
12. Cheng C. A new approximation algorithm for the demand routing and slotting problem on rings unit demands. Lecture Notes in Computer Science. Volume 1671. New York: Springer-Verlag; 1999. p 209–220.
13. Choplin S, Jarry A, Perennes S. Virtual network embedding in the cycle. Discrete Appl Math 2005;145:368–375.
14. Deng X, Li G, Zang W, Zhou Y. A 2-approximation algorithm for path coloring on a restricted class of trees of rings. J Algor 2003;47(1):1–13.
15. Dutta R, Huang S, Rouskas GN. Traffic grooming in path, star, and tree networks: complexity, bounds, and algorithms. Proceedings of 2003 OPTICOMM; 2003.
16. Dutta R, Rouskas GN. Design of logical topologies for wavelength routed networks. In: Sivalingam KM, Subramaniam S, editors. Optical WDM Networks, Principle and Practice. Kluwer Academic Publishers; 2000. p 79–102.

17. Ellinas G, Bala K. Wavelength assignment algorithms for WDM protected rings. Proceedings of the 1998 International Conference on Communications (ICC98); 1998.
18. Erlebach T. Approximation algorithms and complexity results for path problems in trees of rings. Proceedings of the 26th International Symposium on Mathematical Foundations of Computer Science (MFCS01). Lecture Notes in Computer Science. Volume 2136. 2001. p 351–362.
19. Erlebach T, Jansen K. Call scheduling in trees, rings and meshes. Proceedings of the 30th Hawaii International Conference on System Science; 1997.
20. Erlebach T, Jansen K. Scheduling of virtual connections in fast networks. Proceedings of the 4th Workshop on Parallel Systems and Algorithms (PASA96); 1997. p 13–32.
21. Erlebach T, Jansen K. The complexity of path coloring and call scheduling. Theor Comput Sci 2001;255(1–2):33–50.
22. Garey M, Johnson D, Miller G, Papadimitriou C. The complexity of coloring circular arcs and chords. SIAM J Algebra Discr Method 1980;216–227.
23. Garey MR, Johnson DS. Computers and Intractability, a Guide to the Theory of NP-Completeness. New York: Freeman; 1979.
24. Gavril F. Algorithms on circular arc graphs. Networks 1974;4:357–369.
25. Gerstel O, Lin P, Sasaki G. Wavelength assignment in a WDM ring to minimize cost of embedded SONET rings. Proceedings of 1998 IEEE INFOCOM; 1998. p 94–101.
26. Goldschmidt M, Hochbaum DS, Levin A, Olinick EV. The SONET edge–partition problem. Networks 2003;41(1):13–23.
27. Gu Q, Peng S. Multihop all-to-all broadcast on WDM optical networks. IEEE Trans Parallel Distrib Syst 2003;5:477–486.
28. Hu JQ. Optimal traffic grooming for wavelength division multiplexing rings with all-to-all uniform traffic. J Opt Netw 2002;1(1):32–42.
29. Kaklamanis C, Mihail M, Rao S. Efficient access to optical bandwidth. Proceedings of the 36th Annual Symposium on Foundations of Computer Science (FOCS95); 1995; p 548–557.
30. Kaklamanis C, Persiano P, Erlebach T, Jansen K. Constrained bipartite edge coloring with applications to wavelength routing. Proceedings of the 24th International Colloquium on Automata, Language, and Programming (ICALP97); 1997. p 493–504.
31. Karapetian IA. On coloring of arc graphs. Dokladi Acad Sci Armenian Sov Socialist Repub 1980;70(5):306–311.
32. Khot S. Improved inapproximability results for maxclique, chromatic number, and approximate graph coloring. Proceedings of the 42nd IEEE Symposium on Foundations of Computer Science (FOCS01); 2001.
33. Kumar E, Schwabe E. Improved access to optical bandwidth in trees. Proceedings of the 8th ACM–SIAM Symposium on Discrete Algorithms (SODA97);1997. p 437–444.
34. Kumar V. Approximating circular arc coloring and bandwidth allocation in all-optical networks. Proceedings of International Workshop on Approximation Algorithms for Combinatorial Optimizations; 1998. p 147–158.
35. Modiano E, Chiu A. Traffic grooming algorithms for reducing electronic multiplexing costs in WDM ring networks. J Lightwave Technol 2000;18(1):2–12.
36. Opatrny J. Uniform multi-hop all-to-all optical routings in rings. Proceedings of LATIN00; 2000. Lecture Notes in Computer Science, Volume 1776 (LATIN00); 2000.

37. Ragavan P, Upfal E. Efficient routing in all-optical networks. Proceedings of the 26th Annual ACM Symposium on the Theory of Computing (STOC94); 1994. p 134–143.

38. Ramaswami R, Sivarajan KN. Optical Networks, A Practical Perspective. Morgan Kaufmann; 2002.

39. Shanoon CE. A theorem on coloring the lines of a network. J Math Phys 1949;28:148–151.

40. Design of logical topologies for wavelength routed networks. Sivalingam KM, Subramaniam S, editors. Optical WDM Networks: Principle and Practice. Kluwer Academic Publishers; 2000.

41. Stern T, Bala K. Multiwavelength Optical Networks. Addison Wesley; 1999.

42. Tucker A. Coloring a family of circular arcs. SIAM J Appl Math 1975;229(3):493–502.

43. Wang Y, Gu Q. Efficient algorithms for traffic grooming in SONET/WDM neworks. Proceedings of 2006 International Conference on Parallel Processing. CD ROM; 2006.

44. Wang Y, Gu Q. Grooming of symmetric traffic in unidirectional SONET/WDM rings. Proceedings of 2006 International Conference on Communication. CD ROM; 2006.

45. Zhang X, Qiao C. An effective and comprehensive approach for traffic grooming and wavelength assignment. IEEE/ACM Trans Network 2000;8(5):608–617.

INDEX

Handbook of Applied Algorithms: Solving Scientific, Engineering and Practical Problems
Edited by Amiya Nayak and Ivan Stojmenović